Introduction to Materials Management

Introduction to Materials Management

NINTH EDITION

Stephen N. Chapman, Ph.D., CPIM-F, CSCP
North Carolina State University

Ann K. Gatewood, CPIM-F, CIRM, CSCP-F, CLTD-F
Gatewood Associates, LLC

J. R. Tony Arnold, CPIM-F, CIRM

Lloyd M. Clive, CPIM-F

Content Management: Tara Warrens
Content Production: Deepali Malhotra
Product Management: Derril Trakalo
Rights and Permissions: Jenell Forschler

Please contact https://support.pearson.com/getsupport/s/ with any queries on this content

Cover Image by Aun Photographer/Shutterstock

Copyright © 2023, 2017, 2012 by Pearson Education, Inc. or its affiliates, 221 River Street, Hoboken, NJ 07030. All Rights Reserved. Manufactured in the United States of America. This publication is protected by copyright, and permission should be obtained from the publisher prior to any prohibited reproduction, storage in a retrieval system, or transmission in any form or by any means, electronic, mechanical, photocopying, recording, or otherwise. For information regarding permissions, request forms, and the appropriate contacts within the Pearson Education Global Rights and Permissions department, please visit www.pearsoned.com/permissions/.

Acknowledgments of third-party content appear on the appropriate page within the text.

PEARSON and ALWAYS LEARNING are exclusive trademarks owned by Pearson Education, Inc. or its affiliates in the U.S. and/or other countries.

Unless otherwise indicated herein, any third-party trademarks, logos, or icons that may appear in this work are the property of their respective owners, and any references to third-party trademarks, logos, icons, or other trade dress are for demonstrative or descriptive purposes only. Such references are not intended to imply any sponsorship, endorsement, authorization, or promotion of Pearson's products by the owners of such marks, or any relationship between the owner and Pearson Education, Inc., or its affiliates, authors, licensees, or distributors.

Library of Congress Cataloging-in-Publication Data
Names: Chapman, Stephen N., author. | Gatewood, Ann K., author. |
    Arnold, J. R. Tony, author. | Clive, Lloyd M., author.
Title: Introduction to Materials Management / Stephen N. Chapman, Ph.D., CPIM-F, CSCP,
    North Carolina State University, Ann K. Gatewood, CPIM-F, CIRM, CSCP-F, CLTD-F, Gatewood
    Associates, LLC, J. R. Tony Arnold, CPIM-F, CIRM, Lloyd M. Clive, CPIM-F.
Description: Ninth edition. | Hoboken : Pearson, 2023. | Includes
    bibliographical references and index.
Identifiers: LCCN 2021038394 | ISBN 9780137565504 (casebound)
Subjects: LCSH: Materials management.
Classification: LCC TS161 .A76 2022 | DDC 658.7—dc23
LC record available at https://lccn.loc.gov/2021038394

2 2022

ISBN-10:   0-13-756550-X
ISBN-13: 978-0-13-756550-4

Pearson's Commitment to Diversity, Equity, and Inclusion

Pearson is dedicated to creating bias-free content that reflects the diversity, depth, and breadth of all learners' lived experiences.

We embrace the many dimensions of diversity, including but not limited to race, ethnicity, gender, sex, sexual orientation, socioeconomic status, ability, age, and religious or political beliefs.

Education is a powerful force for equity and change in our world. It has the potential to deliver opportunities that improve lives and enable economic mobility. As we work with authors to create content for every product and service, we acknowledge our responsibility to demonstrate inclusivity and incorporate diverse scholarship so that everyone can achieve their potential through learning. As the world's leading learning company, we have a duty to help drive change and live up to our purpose to help more people create a better life for themselves and to create a better world.

Our ambition is to purposefully contribute to a world where:

- Everyone has an equitable and lifelong opportunity to succeed through learning.
- Our educational content accurately reflects the histories and lived experiences of the learners we serve.
- Our educational products and services are inclusive and represent the rich diversity of learners.
- Our educational content prompts deeper discussions with students and motivates them to expand their own learning (and worldview).

Accessibility

We are also committed to providing products that are fully accessible to all learners. As per Pearson's guidelines for accessible educational Web media, we test and retest the capabilities of our products against the highest standards for every release, following the WCAG guidelines in developing new products for copyright year 2022 and beyond.

 You can learn more about Pearson's commitment to accessibility at
https://www.pearson.com/us/accessibility.html

Contact Us

While we work hard to present unbiased, fully accessible content, we want to hear from you about any concerns or needs with this Pearson product so that we can investigate and address them.

 Please contact us with concerns about any potential bias at
https://www.pearson.com/report-bias.html

 For accessibility-related issues, such as using assistive technology with Pearson products, alternative text requests, or accessibility documentation, email the Pearson Disability Support team at **disability.support@pearson.com**

CONTENTS

Preface   xv

1 Introduction to Materials Management   1

Introduction   1
Operating Environment   1
The Supply Chain Concept   4
What Is Materials Management?   7
Summary   11
Key Terms   12
Questions   12
Problems   12
Case Study 1.1: Fran's Flowers   13

2 Production Planning System   16

Introduction   16
Manufacturing Planning and Control System   17
Sales and Operations Planning   22
Manufacturing Resource Planning   24
Enterprise Resource Planning   25
Making the Production Plan   27
Summary   36
Key Terms   36
Questions   37
Problems   38
Case Study 2.1: Medina Water Pumps   42
Case Study 2.2: Williams 3D Printers   43

3 Master Scheduling   45

Introduction   45
Relationship to Production Plan   46
Developing a Master Production Schedule   48
Production Planning, Master Scheduling, and Sales   53
Summary   59
Key Terms   60
Questions   60
Problems   60
Case Study 3.1: Acme Water Pumps   65
Case Study 3.2: The MasterChip Electronics Company   66
Case Study 3.3: Macarry's Bicycle Company   67

4 Material Requirements Planning 71

Introduction 71
Bills of Material 73
Material Requirements Planning Process 80
Using the Material Requirements Plan 91
Summary 95
Key Terms 95
Questions 96
Problems 97
Case Study 4.1: Apix Polybob Company 109
Case Study 4.2: Benzie Products Company 111

5 Capacity Management 113

Introduction 113
Definition of Capacity 113
Capacity Planning 114
Capacity Requirements Planning 115
Capacity Available 117
Capacity Required (Load) 120
Scheduling Orders 123
Making the Plan 125
Capacity Planning in the Lean Production Environment 126
Summary 127
Key Terms 127
Questions 128
Problems 129
Case Study 5.1: Wescott Products 131

6 Production Activity Control 134

Introduction 134
Data Requirements 137
Order Preparation 138
Scheduling 139
Load Leveling 144
Scheduling in a Nonmanufacturing Setting 145
Scheduling Bottlenecks 145
Theory of Constraints and Drum-Buffer-Rope 148
Implementation 151
Control 152
Production Reporting 157
Product Tracking 157
Measurement Systems 158
Summary 158
Key Terms 158

Questions   159
Problems   160
Case Study 6.1: Johnston Products   164
Case Study 6.2: Craft Printing Company   165
Case Study 6.3: Melrose Products   167

7 Fundamentals of Supply Chain Management   169

Introduction   169
Supply Chain Management   170
Supply Chain Strategies   172
Integrated Supply Chains   174
Supply Chain Risks   176
Sustainability   179
Life Cycle Assessment   182
Supply Chain Performance Measurements   183
Supply Chain Technology and Trends   184
Summary   186
Key Terms   187
Questions   187
Problems   188
Case Study 7.1   188
Case Study 7.2   189
Case Study 7.3   189

8 Purchasing   190

Introduction   190
Establishing Specifications   194
Functional Specification Description   195
Selecting Suppliers   197
Price Determination   200
Purchasing Trends   202
Environmentally Responsible Purchasing   205
Ethical Procurement and Sourcing   205
Some Organizational Implications of Supply Chain Management   206
Summary   207
Key Terms   208
Questions   208
Problems   209
Case Study 8.1: Let's Party!   209
Case Study 8.2: The Connery Company   210
Case Study 8.3: Keltox Fabrication   211

9 Forecasting and Demand Management   214

Introduction   214
Demand Management   214

Demand Forecasting 216
Characteristics of Demand 216
Principles of Forecasting 218
Collection and Preparation of Data 219
Forecasting Techniques 219
Some Important Intrinsic Techniques 221
Seasonality 224
Tracking the Forecast 227
Summary 234
Key Terms 234
Questions 234
Problems 235
Case Study 9.1: Northcutt Bicycles: The Forecasting Problem 241
Case Study 9.2: Hatcher Gear Company 243

10 Inventory Fundamentals 245

Introduction 245
Aggregate Inventory Management 245
Item Inventory Management 245
Inventory and the Flow of Material 246
Supply and Demand Patterns 247
Functions of Inventories 247
Objectives of Inventory Management 249
Inventory Costs 251
Financial Statements and Inventory 253
ABC Inventory Control 258
Summary 261
Key Terms 262
Questions 262
Problems 263
Case Study 10.1: Randy Smith, Inventory Control Manager 267

11 Order Quantities 269

Introduction 269
Economic Order Quantity 270
Variations of the EOQ Model 274
Quantity Discounts 275
Order Quantities for Families of Product When Costs Are Not Known 276
Period Order Quantity 277
Summary 280
Key Terms 280
Questions 280
Problems 281
Case Study 11.1: Jack's Hardware 284

12 Independent Demand Ordering Systems   288

Introduction   288
Order Point System   288
Determining Safety Stock   290
Determining Service Levels   297
Different Forecast and Lead Time Intervals   298
Determining When the Order Point Is Reached   299
Periodic Review System   300
Distribution Inventory   302
Summary   305
Key Terms   305
Questions   306
Problems   307
Case Study 12.1: Carl's Computers   313

13 Physical Inventory and Warehouse Management   316

Introduction   316
Warehousing Management   316
Physical Control and Security   322
Inventory Record Accuracy   323
Consignment Inventory and Vendor-Managed Inventory   329
Technology Applications   330
Summary   332
Key Terms   332
Questions   332
Problems   333
Case Study 13.1: Lesscost Warehouse   337

14 Physical Distribution   340

Introduction   340
Physical Distribution   343
Physical Distribution Interfaces   346
Transportation   346
Legal Types of Carriage   349
Transportation Cost Elements   350
Warehousing   355
Packaging   360
Material Handling   362
Multi-Warehouse Systems   362
Impact of Technology in Distribution   365
Summary   366
Key Terms   366
Questions   367
Problems   368
Case Study 14.1: Metal Specialties, Inc.   369
Case Study 14.2: Rictok Fabrication Products Delivery Problem   370

15 Products and Processes 373

Introduction 373
Need for New Products 373
Product Development Principles 374
Product Specification and Design 376
Process Design 379
Factors Influencing Process Design 380
Processing Equipment 381
Process Systems 382
Process Costing 384
Selecting the Process 384
Continuous Process Improvement 387
Summary 396
Key Terms 397
Questions 397
Problems 399
Case Study 15.1: Cheryl Franklin, Production Manager 402

16 Lean Production 404

Introduction 404
Lean Production 404
Waste 406
The Lean Production Environment 408
Manufacturing Planning and Control in a Lean Production Environment 415
Comparing ERP, Kanban, and Theory of Constraints 428
Expanding Lean Principles to Warehousing 432
Summary 432
Key Terms 433
Questions 433
Problems 434
Case Study 16.1: Maxnef Manufacturing 436
Case Study 16.2: Catskill Metal Products 438

17 Total Quality Management 440

Introduction 440
What Is Quality? 440
Total Quality Management 442
Quality Cost Concepts 446
Variation as a Way of Life 447
Process Capability 449
Process Control 453
Sample Inspection 456
ISO 9000 458
ISO 26000 459
ISO 14001 460
Benchmarking 460

Six Sigma   461
Quality Function Deployment   462
The Relationship of Lean Production, TQM, and ERP   464
Summary   465
Key Terms   465
Questions   466
Problems   467
Case Study 17.1: Accent Oak Furniture Company   469

Readings   473
Index   477

PREFACE

Introduction to Materials Management is an introductory text written for students in community colleges and universities as well as professionals involved in supply chain and other materials management positions. It is used in technical programs, such as industrial engineering and manufacturing engineering; in business, operations and supply chain management programs; and by those already in industry, whether or not they are working in materials management.

This text has been widely adopted by colleges and universities not only in North America but also in many other parts of the world. The Association for Supply Chain Management (ASCM) (formerly APICS) organization recommends this text as the reference for certification preparation for CPIM examinations. In addition, the text is used by production and inventory control societies around the world, including South Africa, Australia, New Zealand, Germany, France, and Brazil, and by consultants who present in-house courses to their customers.

Introduction to Materials Management covers all the basics of supply chain management, manufacturing planning and control systems, purchasing, physical distribution, lean and quality management. The material, examples, questions, and problems lead the student logically through the text. The writing style is simple and user-friendly—both instructors and students who have used the book attest to this.

NEW TO THIS EDITION

- All chapters have been updated to reflect new techniques and technology.
- Several additional case studies have been added.
- Additional special topic boxes have been added relating chapter topics to discussions of interest, including nonmanufacturing settings such as service industries, and recent responses of supply chains to risk events.
- Several end-of-chapter problems have been altered.
- A completely new chapter discussing the fundamentals of supply chain management has been added which gives an overall coverage of supply chain concepts as well as tying together several supply chain related issues that previously appeared in the chapters as individual topics.
- Additional information has been included on warehousing, including some impacts of technology and lean principles applied to warehouse management.
- Additional information has been added on the impact technology has on physical distribution.
- Management of special demand situations has been included as it applies to lean production.

In addition, we have retained several features from previous editions:

- Key terms listed at the end of each chapter
- Example problems within the chapters
- Chapter summaries
- Questions and problems at the end of each chapter

APPROACH AND ORGANIZATION

Materials management means different things to different people. In this textbook, materials management includes all activities in the flow of materials from the supplier to the consumer. Such activities include physical supply, operations planning and control, and physical distribution. Other terms sometimes used in this area are *business logistics* and *supply chain management*. Often, the emphasis in business logistics is on transportation and distribution systems with little concern for what occurs in the factory. Whereas some chapters in this text are devoted to transportation and distribution, emphasis is placed on operations planning and control.

Distribution and operations are managed by planning and controlling the flow of materials through them and by using the system's resources to achieve a desired customer service level. These activities are the responsibility of materials management and affect every department in a manufacturing business. If the materials management system is not well designed and managed, the distribution and manufacturing system will be less effective and more costly. Anyone working in manufacturing or distribution should have a good basic understanding of the factors influencing materials flow. This text aims to provide that understanding and also includes chapters on quality management and lean production.

ASCM (formerly APICS) defines the body of knowledge, concepts, and vocabulary used in production and inventory control. Establishing standard knowledge, concepts, and vocabulary is essential both for developing an understanding of production and inventory control and for making clear communication possible. Where applicable, the definitions and concepts in this text subscribe to ASCM vocabulary and concepts.

The first six chapters of *Introduction to Materials Management* cover the basics of production planning and control. Chapter 7 discusses important factors in the management of the supply chain. Chapter 8 examines specific topics in purchasing, and Chapter 9 discusses forecasting. Chapters 10, 11, and 12 look at the fundamentals of inventory management. Chapter 13 discusses physical inventory and warehouse management, and Chapter 14 examines the elements of distribution systems, including transportation, packaging, and material handling. Chapter 15 covers factors influencing product and process design. Chapter 16 looks at the philosophy and environment of just-in-time and lean production and explains how operations planning and control systems relate to just-in-time and lean production. Chapter 17 examines the elements of total quality management and six sigma quality approaches.

Alternate Versions

We are also excited about the ways readers can purchase this textbook to ensure that they have the most effective and affordable options:

- Print: Readers can choose between purchasing the textbook or using the affordable, rent-to-own print book option.
- EText: With the Pearson eText, students can search the text, use the study tools such as flashcards, make notes online, print out reading assignments that incorporate the notes they take during lectures, and bookmark important passages for later review. The mobile app lets students learn wherever life takes them, offline or online. For more information on Pearson eText, visit www.pearson.com/learner.

SUPPLEMENTS

To enhance the learning process, a full supplements package accompanies this text and is available to students and instructors using the text for a course.

Instructor Resources

Instructor Resources can be downloaded at www.pearsonhighered.com/irc. If you do not have a username and password for access, you can request access at www.pearsonhighered.com/irc.

ACKNOWLEDGMENTS

Although it has been several years since the passing of Tony Arnold, the original author, he developed and refined a clear vision for the concepts and goals that this book was intended to achieve. He also clearly expressed those points to the present authors, knowing that keeping those concepts and goals constantly in mind while new editions and new material was added would continue to let new editions be relevant. This ninth edition continues to reflect the original vision of providing a clear and understandable introductory look at the field of Materials Management.

Help and encouragement have come from a number of valued sources, among them friends, colleagues, and students. We thank the many users of the book who have provided comments and suggestions. We especially wish to thank members of the ASCM CPIM Committees who have provided specific guidance for the revision. Additionally, we would like to thank our reviewers, past and present, for their help in suggesting new content: Andrea Prud'homme (The Ohio State University), Steven Rudnicki (Westinghouse Electric Company), Jim Caruso (Covidien), Frank Montabon (Iowa State University), Mark Hardison (SIGA Technologies), and John Kanet (The University of Dayton).

Overall, this book is dedicated to those who have taught us the most—our students.

Stephen N. Chapman, Ph.D., CPIM-F, CSCP, Associate Professor Emeritus
Department of Business Management, Poole College of Management
North Carolina State University
Raleigh, North Carolina

Ann K. Gatewood, CPIM-F, CIRM, CSCP-F, CLTD-F
President, Gatewood Associates
Mooresville, North Carolina

CHAPTER ONE

INTRODUCTION TO MATERIALS MANAGEMENT

1.1 INTRODUCTION

What is the source of wealth? Wealth is measured by the amount of goods and services produced. For example, the wealth of a country is measured by its gross national product—the output of goods and services produced by the nation in a given time. Goods are physical objects, something one can touch, feel, or see. Services are the performance of some useful function such as banking, medical care, restaurants, clothing stores, or social services.

Although rich natural resources may exist in an economy, such as mineral deposits, farmland, and forests, these are only potential sources of wealth. A production function is needed to transform these resources into useful goods. The transformation process begins with extracting minerals from the earth, farming, lumbering, or fishing, and then using these resources to manufacture useful products.

There are many stages between the extraction of resource material and the final consumer product. At each stage in the development of the final product, value is added, thus creating more wealth. If ore is extracted from the earth and sold, wealth is gained from the efforts, but those who continue to transform the raw material will gain more and usually far greater wealth. Japan is a prime example of this. It has very few natural resources and imports most of the raw materials it needs. However, the Japanese have developed one of the wealthiest economies in the world by transforming the raw materials they purchase and adding value to them through manufacturing.

Manufacturing companies are in the business of converting raw materials to a form that is of far more value and use to the consumer than the original raw materials. Logs are converted into tables and chairs, iron ore into steel, and steel into cars and refrigerators. This conversion process, called *manufacturing* or *production*, makes a society wealthier and creates a better standard of living.

To get the most value out of resources, production processes must be designed so that they make products most efficiently. Once the processes exist, operations are managed so that goods are produced most economically. Managing the operation means planning for and controlling the resources used in the process: labor, capital, and material. All are important, but the foremost method in which management plans and controls operations is through the flow of materials. The flow of materials in turn controls the performance of the process. If the right materials in the right quantities are not available at the right time, the process cannot produce what it should. Labor and machinery will be poorly utilized. The profitability, and even the existence, of the company will be threatened.

1.2 OPERATING ENVIRONMENT

Operations management works in a complex environment affected by many factors. Among the most important are government regulation, the economy, competition, customer expectations, and quality.

> **Government.** Regulation of business by the various levels of government is extensive. Regulation applies to such areas as the environment, safety, product liability, and taxation. Government, or the lack of it, affects the way business is conducted.

Economy. General economic conditions influence the demand for a company's products or services and the availability of inputs. During economic recession, the demand for some products may decrease while demand for others may increase. Materials and labor shortages or surpluses influence the decisions management makes. Shifts in the age of the population, needs of ethnic groups, low population growth, increased free trade between countries, and increased global competition all contribute to changes in the marketplace.

Competition. Competition is more significant today than ever before.

- Manufacturing companies face competition from foreign competitors throughout the world due to the global economy.
- Transportation and the movement of materials are more efficient and less costly than they used to be.
- Worldwide communications are fast, effective, and cheap. Information and data can be moved instantly around the globe. The internet allows buyers to search out new sources of supply from anywhere in the world as easily as they can from local sources.

Customers. Both consumers and industrial customers have become much more demanding, and suppliers have responded by improving the range of characteristics they offer. Some of the characteristics and selection customers expect in the products and services they buy are:

- A fair price.
- Higher (perfect) quality products and services.
- Fast and accurate delivery lead time.
- Better presale and after-sale service.
- Product and volume flexibility.

Quality. Since competition is international and aggressive, successful companies provide quality that not only meets customers' high expectations but also exceeds them.

Order Qualifiers and Order Winners

Generally, a supplier must meet set minimum requirements to be considered a viable competitor in the marketplace. Customer requirements may be based on price, quality, delivery, and so forth and are called **order qualifiers**. For example, the price for a certain type of product must fall within a certain range for the supplier to be considered by potential customers. But being considered does not mean winning the order. To win orders, a supplier must have characteristics that encourage customers to choose its products and services over competitors'. Those competitive characteristics, or combination of characteristics, that persuade a company's customers to choose its products or services are called **order winners**. They provide a competitive advantage for the firm. Order winners change over time and may well be different for different markets. For example, fast delivery may be vital in one market but not in another. Characteristics that are order winners today probably will not remain so, because competition will try to copy winning characteristics, and the needs of customers will change.

It is very important that a firm understands the order winners and order qualifiers for each of its products or services and in each of its markets in order to drive the manufacturing and corporate strategy. Since it is virtually impossible to be the best in every dimension of competition, firms should in general strive to provide at least a minimal level of acceptance for each of the order qualifiers but should try to be the *best* in the market for the order winner(s).

One also should recognize that the order winners and qualifiers for any product/market combination are not static. Not only will customers change perspectives as competitors jockey for position, but the order winners and qualifiers will also often change based on the concepts of the product life cycle. Most products go through a life cycle, including introduction, growth, maturity, and decline. For example, in the introduction phase, design and availability are often much more important than price. Quality and delivery tend to have increased importance during growth, while price and delivery are often the order winners for mature products. This life cycle approach is complicated in that the duration of the life cycle will be very different for different products. Although some products have life cycles of many years, the life cycle of other products (certain toys or electronics, for example) can be measured in months or even weeks.

Manufacturing Strategy

A highly market-oriented company will focus on meeting or exceeding customer expectations and on order winners. In such a company, all functions must contribute toward a winning strategy. Thus, operations must have a strategy that allows it to supply the needs of the marketplace and provide fast on-time delivery.

Delivery lead time From the supplier's perspective, delivery lead time is the time from receipt of an order to the delivery of the product. From the customer's perspective, it may also include time for order preparation and transmittal. Most customers want delivery lead time to be as short as possible, and manufacturing must determine a process strategy to achieve this. There are five basic process strategy choices: engineer-to-order, make-to-order, configure-to-order, assemble-to-order, and make-to-stock. Customer involvement in the product design, delivery lead time, and inventory state are influenced by each strategy. Based on the type of products a company makes, and their customer base, a company may determine that more than one process strategy is required. Figure 1.1 shows the effect of each process strategy on lead time.

Engineer-to-order means that the customer's specifications require unique engineering design or significant customization. Usually the customer is highly involved in the product design. Inventory will not normally be purchased until needed by manufacturing. Delivery lead time is long because it includes not only purchase lead time but also design lead time.

Make-to-order means that the manufacturer does not start to make the product until a customer's order is received. The final product is usually made from standard items but may include custom-designed components as well. Delivery lead time is reduced because there is little design time required and inventory is held as raw material.

Configure-to-order means that the customer is allowed to configure a product based on various features and options. Each customer, and order, may be an entirely unique configuration that has never been done before, and the configuration often occurs at

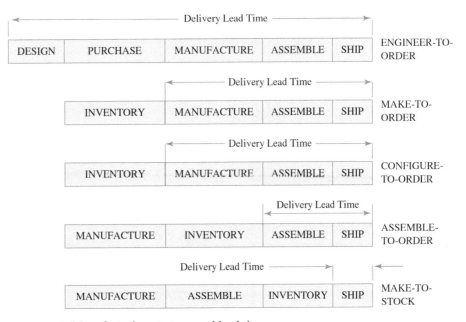

FIGURE 1.1 Manufacturing strategy and lead time.

the beginning of the process. Delivery lead time is reduced because there is no design time required and the different features and options may already be available. Customer involvement includes selecting the features and options desired.

Assemble-to-order means that the product is made from standard components or options that the manufacturer can inventory and assemble according to a customer order. This is usually done at a later stage in the process than configure-to-order. Delivery lead time is reduced further because there is no design time needed and inventory is held ready for assembly. Customer involvement in the design of the product is limited to selecting the assembly options needed.

Make-to-stock means that the supplier manufactures the goods and sells from a finished goods inventory. Delivery lead time is shortest as manufacturing and assembly have already been completed. The customer has little direct involvement in the product design.

Postponement is another application of assemble-to-order, described in the *APICS Dictionary,* 16th edition, as "a product design or supply chain strategy that deliberately delays final differentiation (assembly, production, packaging, tagging, etc.) until the latest possible time in the process. This shifts product differentiation closer to the consumer to reduce the anticipatory risk of producing the wrong product. The practice eliminates excess finished goods in the supply chain."

An example of postponement would be computer printers for a global market that use universal power supplies that can be switched to different voltages. Upon receipt of a customer's order, they are packaged with the appropriate cords, instructions, and labeling. This avoids filling an entire supply chain with expensive printers destined for many different countries. Some basic postponement can be done in a distribution center and often by third party logistics (3PL) providers. Foreign suppliers of appliances, such as vacuum cleaners destined for multiple customers, postpone the packaging of their products, applying customer-specific labels, bar codes, boxes, instructions, and so forth until after receipt of the customer order.

1.3 THE SUPPLY CHAIN CONCEPT

There are three phases in the flow of materials. Raw materials flow into a manufacturing company from a physical supply system, they are processed by manufacturing, and finally, finished goods are distributed to end consumers through a physical distribution system. Figure 1.2 shows this system graphically. Although this figure shows only one supplier and one customer, usually the **supply chain** consists of several companies linked in a supply–demand relationship. For example, the customer of one supplier buys a product, adds value to it, and supplies it to yet another customer. Similarly, one customer may have several suppliers and may in turn supply several customers. As long as there is a chain of supplier–customer relationships, they are all members of the same supply chain.

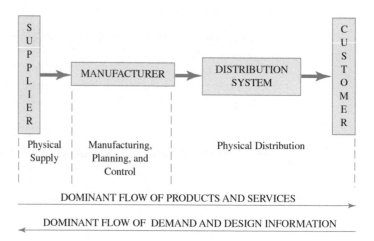

FIGURE 1.2 Supply–production–distribution system.

There are a number of important factors in supply chains:

- The supply chain includes all activities and processes to supply a product or service to a final customer.
- Any number of companies can be linked in the supply chain.
- A customer can be a supplier to another customer, so the total chain can have a number of supplier-customer relationships.
- Although the distribution system can be direct from supplier to customer, depending on the products and markets, it can contain a number of intermediaries (distributors) such as wholesalers, warehouses, and retailers.
- Product or services usually flow from supplier to customer; design, demand information, and cash usually flow from customer to supplier.

Although these systems vary from industry to industry and company to company, the basic elements are the same: supply, production, and distribution. The relative importance of each depends on the costs of the three elements.

Fundamentals of the management of the supply chain will be discussed in Chapter 7.

Performance Metrics

A **metric** is a verifiable measure stated in either quantitative or qualitative terms defined with a reference point. Without metrics, no firm can expect to function effectively or efficiently on a daily basis. Benefits of metrics include:

1. Reporting of data to superiors and external groups for better control.
2. Communicating priorities by the organization.
3. Drawing attention to opportunities for improvement.
4. Keeping processes under control.

Determining the right metrics to use is vital to a company, as metrics communicate expectations, identify problems, direct a course of action, and motivate people. Problems must be anticipated and corrective action taken before they become severe and costly. Companies cannot risk waiting to react until the order cycle is completed and feedback from customers is received.

Today, production control works in a demanding environment shaped by five major challenges:

1. Customers that are increasingly demanding.
2. A supply chain that is large and must be managed.
3. A product life cycle that is getting shorter and shorter.
4. A vast amount of data.
5. An increasing number of alternatives.

A firm typically has a corporate strategy that states how it will treat its customers and what services it will supply. This identifies how a firm will compete in the marketplace. It is the customer who assesses the firm's offering by its decision to buy or not to buy. How metrics link strategy to operations is shown in Figure 1.3. Focus describes the particular activity that is to be measured. Standards are the yardstick that is the basis of comparison on which performance is judged.

There is a difference between performance measurements and performance standards. A **performance measure** must be both quantified and objective and contain at least two

FIGURE 1.3 Metrics context.

parameters. For example, the number of orders per day consists of both a quantity and a time measurement.

Transforming company policies into objectives and specific goals creates **performance standards**. Each goal should have target values. An example of this would be to improve order fill rate to 98% measured by number of lines. Performance standards set the goal, while performance measures reveal how close to the goal the organization reached.

Many companies do not realize the potential benefits of performance measurement, nor do they know how to measure performance, and often try to use them without performance standards. This might occur when the concept of performance measurement and standards is new. Only when standards are put into use can management begin to monitor the company. The old saying "What you do not measure, you cannot control" is as valid today as it was when first stated.

The necessary steps in implementing such a program are as follows:

1. Establish company goals and objectives.
2. Define performance to be measured.
3. State the measurement to be used.
4. Set performance standards.
5. Educate the participants.
6. Make sure the program is consistently applied.
7. Review and adjust as necessary.

Although financial performance has traditionally been the measure of success in most companies, today the focus is on continuous improvement and, with this, an increase in standards. Emphasis should not be placed on a "one-shot" improvement but on such things as the rate of improvement in quality, cost, reliability, innovation, effectiveness, and productivity.

Conflicts in Traditional Systems

In the past, supply, production, and distribution systems were organized into separate functions that reported to different departments of a company. Often, policies and practices of the different departments maximized departmental objectives without considering the effect they would have on other parts of the system. Because the three systems are interrelated, conflicts often occurred. Although each system made decisions that were best for itself, overall company objectives suffered. For example, the transportation department would ship in the largest quantities possible so it could minimize per-unit shipping costs. However, this increased inventory and resulted in higher inventory-carrying costs.

To get the most profit, a company must have at least four main objectives. Examples of these include the following:

1. Provide best customer service.
2. Provide lowest production costs.
3. Provide lowest inventory investment.
4. Provide lowest distribution costs.

These objectives create conflict among the marketing, production, and finance departments because each has different responsibilities in these areas.

Marketing's objective is to maintain and increase revenue; therefore, it must provide the best customer service possible. There are several ways of doing this:

- Maintain high inventories so goods are always available for the customer.
- Interrupt production runs so that a non-inventoried item can be manufactured quickly if required by a customer.
- Create an extensive, and consequently costly, distribution system so goods can be shipped to the customer rapidly.

Introduction to Materials Management 7

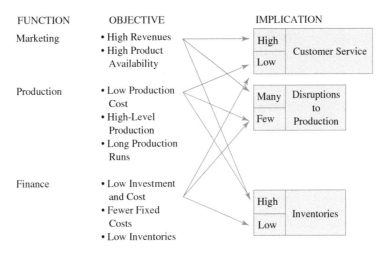

FIGURE 1.4 Conflicting objectives.

Finance must keep investment and costs low. This can be done in the following ways:

- Reduce inventory so inventory investment is at a minimum.
- Decrease the number of plants and warehouses.
- Produce large quantities using long production runs.
- Manufacture only to customer order.

Production must keep its operating costs as low as possible. This can be done in the following ways:

- Make long production runs of relatively few products. Fewer changeovers will be needed and specialized equipment can be used, thus reducing the cost of making the product.
- Maintain high inventories of raw materials and work-in-process so production is not disrupted by shortages.

These conflicts among marketing, finance, and production center on customer service, disruption of production flow, and inventory levels. Figure 1.4 shows this relationship.

Today, the concepts of lean production stress the need to supply customers with what they want, when they want it, and to keep inventories at a minimum. These objectives put further stress on the relationship among production, marketing, and finance. Chapter 16 will discuss the concepts of lean production and how it influences materials management.

One important means of resolving these conflicting objectives is to provide close coordination of the supply, production, and distribution functions. The aim is to balance conflicting objectives to minimize the total of all the costs involved and maximize customer service consistent with the goals of the organization. This requires some type of integrated materials management or logistics organization that is responsible for supply, production, and distribution. The functions of marketing, production, and distribution must coordinate the planning and control of these areas in order to operate under overall company objectives.

1.4 WHAT IS MATERIALS MANAGEMENT?

The concept of having one department responsible for the flow of materials, from supplier through production to consumer, thereby minimizing total costs and providing a better level of customer service, is known as **materials management**. Other names include distribution planning and control, supply chain management, and logistics management, but the one used in this text is materials management. As will be discussed in Chapter 16, lean production not only requires efficient individual operations but also requires all operations to work together. A materials management department can improve this coordination by having overall responsibility for material. Its objectives are as follows:

- Maximize the use of the firm's resources.
- Provide the required level of customer service.

Materials management can do much to improve a company's profit. An income (profit and loss) statement for a manufacturing company might look something like the following:

	Dollars	Percent of Sales
Revenue (sales)	$1,000,000	100
Cost of Goods Sold		
Direct Material	$500,000	50
Direct Labor	$200,000	20
Factory Overhead	$200,000	20
Total Cost of Goods Sold	$900,000	90
Gross Profit	$100,000	10

Direct labor and direct material are costs that increase or decrease with the quantity sold. Overhead (all other costs) does not vary directly with sales. For simplicity, overhead is assumed to be constant, even though it is initially expressed as a percentage of sales.

If, through a well-organized materials management department, direct materials can be reduced by 12%, the improvement in profit would be

	Dollars	Percent of Sales
Revenue (sales)	$1,000,000	100
Cost of Goods Sold		
Direct Material	$440,000	44
Direct Labor	$200,000	20
Overhead	$200,000	20
Total Cost of Goods Sold	$840,000	84
Gross Profit	$160,000	16

Profit has been increased by 60%. In other words, managing inventory effectively goes right to the bottom line of a company's profit. To get the same increase in profit ($60,000) by increasing revenue, sales would have to increase to $1.2 million.

	Dollars	Percent of Sales
Revenue (sales)	$1,200,000	100
Cost of Goods Sold		
Direct Material	$600,000	50
Direct Labor	$240,000	20
Overhead	$200,000	17
Total Cost of Goods Sold	$1,040,000	87
Gross Profit	$160,000	13

Example Problem

a. If the cost of direct material is 60%, direct labor is 10%, and overhead is 25% of sales, what will be the improvement in profit if cost of direct material is reduced to 55%?

b. How much will sales have to increase to give the same increase in profit? (Remember, overhead cost is constant.)

Introduction to Materials Management

ANSWER

a.

	Before Improvement	After Improvement
Revenue (sales)	100%	100%
Cost of Goods Sold		
Direct Material	60%	55%
Direct Labor	10%	10%
Overhead	25%	25%
Total Cost of Goods Sold	95%	90%
Gross Profit	5%	10%

b. Profit = sales − (direct material + direct labor + 0.25)
         = sales − (0.6 sales + 0.1 sales + 0.25)
         = sales − 0.7 sales − 0.25
0.1 = 0.3 sales − 0.25
0.3 Sales = 0.35
$$\text{Sales} = \frac{0.35}{0.3} = 1.17$$

Sales must increase 17% to give the same increase in profit.

Work-in-Process

Inventory not only accounts for the raw materials and purchased components, but is also made up of the product as it is processed into finished goods. This type of inventory is called **work-in-process (WIP)**. WIP is a major investment for many companies, and reducing the amount of time that inventory spends in production is a good way to reduce the costs associated with this investment. Labor, materials, and overhead are applied to goods continuously through-out production, which increases the value of WIP. Further discussion on WIP and reducing it is covered in Chapters 10 and 16.

Example Problem

On average, a company has a 12-week production lead time and an annual cost of goods sold of $36 million. Assuming the company works 50 weeks per year:

a. What is the dollar value of the WIP?
b. If the lead time could be reduced to five weeks, and the annual cost of carrying inventory was 20% of the inventory value, what would be the annual savings?

ANSWER

Weekly cost of goods sold = $36,000,000 per year/50 weeks per year
                                   = $720,000/week
WIP value at 12 weeks LT = 12 weeks × $720,000/week = $8,640,000
WIP value at 5 weeks LT = 5 weeks × $720,000/week = $3,600,000
Reduction in WIP = $8,640,000 − 3,600,000 = 5,040,000
Annual savings = $5,040,000 × 20% = $1,008,000

Reducing cost contributes directly to profit. Increasing sales increases direct costs of labor and materials so profit does not increase in direct proportion. Materials management can reduce costs by being sure that the right materials are in the right place at the right time and the resources of the company are properly used.

There are several ways of classifying this flow of material. An often-used classification, and the one used in this text, is manufacturing planning and control and physical supply/distribution.

Manufacturing Planning and Control

Manufacturing planning and control are responsible for the planning and control of the flow of materials through the manufacturing process. The primary activities carried out are as follows:

1. **Production planning.** Production must be able to meet the demand of the marketplace. Finding the most productive way of doing so is the responsibility of production planning. It must establish correct priorities (what is needed and when) and make certain that capacity is available to meet those priorities. It involves the following activities:

    a. Forecasting.

    b. Master planning.

    c. Material requirements planning.

    d. Capacity planning.

2. **Implementation and control.** These functions are responsible for putting into action and executing the plans made by production planning. These responsibilities are accomplished through production activity control (often called *shop floor control*) and purchasing.

3. **Inventory management.** Inventories are materials and supplies carried on hand either for sale or to provide material or supplies to the production process. They are part of the planning process and provide a buffer against the differences in demand rates and production rates.

Production planning, implementation, control, and inventory management work hand-in-hand. Inventories in manufacturing are used to support production or are the result of production.

Inputs to the manufacturing planning and control system There are five basic inputs to the manufacturing planning and control system:

1. **Product description.** The product description shows how the product will appear at some stage of production. *Engineering drawings* and *specifications* are methods of describing the product. Another method, and the most important for manufacturing planning and control, is the **bill of material**. As used in materials management, the bill of material does two things:
    - Describes the components and quantities used to make the product.
    - Describes the subassemblies at various stages of manufacture.

2. **Process specifications.** Process specifications describe the steps necessary to make the assembly or end product. They are a step-by-step set of instructions describing how the product is made. This information is usually recorded on a route sheet or in a **routing**. These provide information such as the following on the manufacture of a product:
    - Operations required to make the product.
    - Sequence of operations.
    - Equipment and accessories required.
    - Standard time required to perform each operation.

3. **Time.** The time needed to perform operations is usually expressed in **standard time**, which is the time taken by an average operator, working at a normal pace, to perform a task. It is needed to schedule work through the plant, load the plant, make delivery

promises, and cost the product. Standard times for operations are usually obtained from the routing information.

4. **Available facilities.** Manufacturing planning and control must know what plant, equipment, and labor resources will be available to process work. This information is usually found in the work center information.

5. **Quantities required.** This information will come from forecasts, customer orders, orders to replace finished goods inventory, and the material requirements plan.

Physical Supply/Distribution

Physical supply/distribution includes all the activities involved in moving goods, from the supplier to the beginning of the production process, and from the end of the production process to the consumer.

The activities involved are as follows:

- Transportation.
- Distribution inventory.
- Warehousing.
- Packaging.
- Material handling.

Materials management is a balancing act. The objective is to be able to deliver what customers want, when and where they want it, and to do so at minimum cost. To achieve this objective, materials management must make tradeoffs between the level of customer service and the cost of providing that service. As a rule, costs rise as the service level increases, and materials management must find that combination of inputs to maximize service and minimize cost. For example, customer service can be improved by establishing warehouses in major markets. However, that causes extra cost in operating the warehouse and in the extra inventory carried. To some extent, these costs will be offset by potential savings in transportation costs if lower cost transportation can be used.

By grouping all those activities involved in the movement and storage of goods into one department, the firm has a better opportunity to provide maximum service at minimum cost and to increase profit. The overall concern of materials management is the balance between priority and capacity. The marketplace sets demand and materials management must plan the firm's priorities (what goods to make and when) to meet that demand. Capacity is the ability of the system to produce or deliver goods. Materials management is responsible for planning and controlling priority and capacity to meet customer demand at minimum cost.

SUMMARY

Manufacturing creates wealth by adding value to goods. To improve productivity and wealth, a company must first design efficient and effective systems for manufacturing. It must then manage these systems to make the best use of labor, capital, and material. One of the most effective ways of doing this is through the planning and control of the flow of materials into, through, and out of manufacturing. There are three elements to a material flow system: supply, manufacturing planning and control, and physical distribution. They are linked, and what happens in one system affects the others.

Traditionally, there are conflicts in the objectives of a company and in the objectives of marketing, finance, and production. The role of materials management is to balance these conflicting objectives by coordinating the flow of materials so customer service is maintained and the resources of the company are properly used.

In subsequent chapters this text will examine some of the theories and practices considered to be the fundamental body of knowledge of materials and supply chain management.

KEY TERMS

Assemble-to-order 4
Bill of material 10
Configure-to-order 3
Engineer-to-order 3
Inventory management 10
Make-to-order 3
Make-to-stock 4
Materials management 7
Metric 5
Order qualifiers 2
Order winners 2
Performance measure 5
Performance standard 6
Postponement 4
Production planning 10
Routing 10
Specification 10
Standard time 10
Supply chain 4
Work-in-process (WIP) 9

QUESTIONS

1. What is wealth, and how is it created?
2. What is value added, and how is it achieved?
3. Name and describe five major factors affecting operations management.
4. What are an order qualifier and an order winner? What is the major difference between the two?
5. Describe the five primary manufacturing strategies. How does each affect delivery lead time?
6. What must manufacturing management do to manage a process or operation? What is the major way in which management plans and controls?
7. What are the four objectives of a firm wishing to maximize profit?
8. What is the objective of marketing? What three ways will help it achieve this objective?
9. What are the objectives of finance? How can these objectives be met?
10. What are the objectives of production? How can these objectives be met?
11. Describe how the objectives of marketing, production, and finance are in conflict over customer service, disruption to production, and inventories.
12. What is the purpose of materials management?
13. Name and describe the three primary activities of manufacturing planning and control.
14. Name and describe the inputs to a manufacturing planning and control system.
15. What are the five activities involved in the physical supply/distribution system?
16. Why can materials management be considered a balancing act?
17. What are metrics? What are their uses?
18. A computer carrying case and a backpack are familiar items to a student of manufacturing planning and control. Discuss the manufacturing planning and control activities involved in producing a variety of these products. What information from other departments is necessary for manufacturing planning and control to perform its function?
19. Which manufacturing strategies are used in a fast-food business? How does this affect the lead time from the customers' point of view?
20. Give an example of a postponement activity.

PROBLEMS

1.1 If the cost of manufacturing (direct material and direct labor) is 60% of sales and profit is 10% of sales, what would be the improvement in profit if, through better planning and control, the cost of manufacturing was reduced from 60% of sales to 55% of sales?

Answer. Profits would improve by 500%.

1.2 In problem 1.1, how much would sales have to increase to provide the same increase in profits (keeping the manufacturing cost the same at 60%)?

Answer. Sales would have to increase 12.5%.

1.3 On the average, a firm has an eight-week lead time for work-in-process, and annual cost of goods sold is $16 million. Assuming that the company works 50 weeks a year:

a. What is the dollar value of the work-in-process?

b. If the lead time could be reduced to five weeks, what would be the reduction in WIP?

Answer.   a. $2,560,000

b. $960,000

1.4 On the average, a company has a work-in-process lead time of 10 weeks and annual cost of goods sold of $30 million. Assuming that the company works 50 weeks a year:

a. What is the dollar value of the work-in-process?

b. If the work-in-process could be reduced to five weeks and the annual cost of carrying inventory was 20% of the WIP inventory value, what would be the annual savings?

1.5 Amalgamated Fenderdenter's sales are $10 million. The company spends $6.4 million for purchase of direct materials and $6.6 million for direct labor; overhead is $6 million and profit is $1,000,000. Direct labor and direct material vary directly with sales, but overhead does not. The company wants to double its profit.

a. By how much should the firm increase annual sales?

b. By how much should the firm decrease material costs?

c. By how much should the firm decrease labor cost?

CASE STUDY 1.1

Fran's Flowers

After Fran Fuller graduated with an undergraduate art degree in 2018, she decided to combine her knowledge and love of art with a second love—plants and flowers—toward developing a business. Her intent was to focus on a specialty niche in the flower shop business. She decided to concentrate her efforts on make-to-order special flower arrangements, like are typically found at banquets and weddings. Due to her talent and dedication to doing a good job, she was highly successful, and her business grew to where she now has a shop located in a highly visible and successful strip mall. As with many successful businesses, her success has produced unanticipated problems, some of which are normal growth pains, but others are relatively unique to the type of business. At a recent meeting with her business advisor, she outlined some of the major issues she faces:

1. **Business Focus.** When she moved into her new shop in the mall, she continued to specialize in the make-to-order specialty arrangements, but customers frequently walked into her shop requesting "spot" purchases, including gifts for sick friends and last-minute flower purchases for occasions such as birthdays, anniversaries, Valentine's Day, and so forth. As this business represented an attractive addition to the store revenue, she accommodated it with three large climate-controlled display cases stocked with ready-to-sell arrangements of various sizes, types, and costs. Even though she did not aggressively pursue this market with advertising, the heavy mall traffic where her store is located and word of mouth caused the walk-in business to steadily grow to where it now represents almost half of her total revenue. This business has brought her numerous headaches, however, due to several characteristics:

    a. Even though some days have predictably high demand (e.g., just prior to Valentine's Day, Mother's Day), most of the time she has no idea how many customers will come in for spot buys an any given day, nor does she have any idea as to the price range they will look for. Even such variables as the weather and the schedule of local sports teams appear to affect her demand. She knows she needs to manage this demand better, because not having what a customer wants could mean the permanent loss of a good potential customer. On the other side, flowers have a limited shelf life, and having too much of the wrong price range could mean a high spoilage rate. It would not take many lost sales on a daily basis to represent the difference between profit and loss for that part of the business.

b. Some customers have become irate that her delivery system, a major part of the make-to-order business, will not accommodate the delivery of a $20 ready-to-sell arrangement to a hospital, for example. Angry customers have even asked her how much they need to spend on an arrangement before she will deliver. She has never really thought about an answer to that question and has not known how to reply. Generally, she just states that she does not deliver premade flower arrangements. She knows this lack of delivery has cost her some goodwill, some business, and perhaps even some potential return customers.

c. Related to the point above, several customers have expressed serious dissatisfaction that she is not a member of some national delivery service, so they can have flowers delivered out of town. She is afraid such a business will pull her even further from her core business of make-to-order, as those services typically focus on catalogs of set designs. As those services are also expensive to belong to, she knows she would have to spend a lot more time and effort in that area to make it financially feasible.

d. Another group of customers wants her to extend her open shop hours, as they say they occasionally drop by for flowers on their way home from work and often find her closed for the day or at least not available while she is setting up a flower order in some other location.

2. **Personnel Issues.** As her business grew, Fran hired another skilled arranger, Miguel, to work with her. The unpredictability of the walk-in demand has caused her to bring people issues up as a problem, however. As walk-in customers demand immediate attention, she and Miguel are frequently called to the front of the shop to sell arrangement from the cases. This pulls them away from working on their orders, and while they have been late only on a couple of special orders within the last few weeks, several others were delivered before she was satisfied with their appearance, merely to avoid their being late. This worries her a great deal, as she has worked very hard to obtain a reputation for the quality of her arrangements. She thought about hiring a delivery person, but decided it was important that either she or Miguel deliver the orders so that they may put last-minute touches on the arrangement in case of disturbance during the delivery process.

Instead she opted to hire some part-time unskilled help for the shop to handle the walk-in shop sales. This has proved less than satisfactory, because of two reasons:

a. The unpredictability of demand has her constantly wondering about what hours and how many hours to schedule the help. The extra help adds to overall cost, and paying someone to stand around while no customers come in the shop makes the difference between profit and loss even more sensitive.

b. Customers frequently have questions about the type of flowers in an arrangement, how long they last, and so forth. The unskilled workers she hires often don't know what to answer. They will then frequently interrupt either Fran or Miguel with the question, and even when they get the answer the customer often is left with a poor impression, as they often expect more knowledge from a salesperson. The impression is even worse if both Fran and Miguel are out servicing orders, as the only answer the customer gets is "I'm not sure." Since she pays only slightly above minimum wage, her worker turnover is high. This means she is constantly trying to hire and train people, further distracting her from her main business. She knows she could reduce the turnover and hire more knowledgeable people if she paid her help more per hour, but that issue again pushes her closer to the loss column for many of the days the shop is open, so she feels she really can't afford to pay more.

3. **Expansion.** Several of her regular customers are encouraging her to open another operation on the other side of the city, as well as considering expansion to other cities. They claim several of their friends like her arrangements a great deal, but consider her location too inconvenient from where they live or work. That is typically not a problem for large orders, as she or Miguel will typically offer to visit the customer to obtain details for the arrangement. That does take a lot of time, however, so she finds herself more frequently asking the long distance customer to come to the shop if possible. Many decline to do so, and the order is sometimes lost. While expansion is attractive to her, she worries about control—not only for order servicing, but also for delivery. How can she possibly maintain control of quality and design in two or more locations at the same time?

4. **Supply.** As her purchases of flowers from the wholesaler has grown, the wholesaler has recommended that Fran make a purchasing contract instead of making spot bulk buys as she now does. This contract will give her significant quantity price discounts, but her delivered quantity has to be above a certain amount of each type of flower so that the wholesaler can reduce costs due to economies of scale. The quantities she needs to order are reasonable given her average demand, but the fluctuation around that average is large enough to present significant spoilage during certain periods. She wonders if she would be better off in the long run with the purchasing contract.

Assignment

1. What are the key issues in this case? In other words, analyze the case to try to determine the true problems from the symptoms of those problems. How do these issues relate to the issue of strategy?
2. What type of data would you suggest collecting to both verify the problem identifications are correct and to provide solution approaches and support? How would you organize and use that data?
3. What would you suggest she do with her business and why? Provide a comprehensive and integrated plan of action and provide support for your suggestions.
4. Develop an implementation plan for whatever changes you suggest she make. Prioritize the key steps if appropriate.

CHAPTER TWO

PRODUCTION PLANNING SYSTEM

2.1 INTRODUCTION

This chapter introduces the manufacturing planning and control (MPC) system. First, it discusses the overall system and then goes into detail on production planning. Subsequent chapters in this text discuss master scheduling, material requirements planning, capacity management, production activity control, purchasing, and forecasting.

The manufacturing of goods is complex. Some firms make a few different products, whereas others make many products. However, each uses a variety of processes, machinery, equipment, labor skills, and material. To be profitable, a firm must organize all these factors to make the right goods at the right time at top quality and do so as economically as possible. It is a complex problem, and it is essential to have a good planning and control system.

A good planning system must answer the following four questions:

1. What are we going to make or provide to customers?
2. What does it take to make it?
3. What do we already have?
4. What do we need to get?

These are questions of priority and capacity.

Priority relates to what products are needed, how many are needed, and when they are needed. The marketplace establishes the priorities. Manufacturing is responsible for devising plans to satisfy the market demand if possible. An important aspect of this is the fact that even though the market (potential customers) may demand specific quantities of specific products, the company may, and in fact should, make their plans as to what to produce based on the market demand but often significantly modified by several considerations, including:

- Existing resources, including the following:
  - People, both how many available and skill levels.
  - Equipment, both quantity and capability.
  - Funds, possibly needed to procure or change other resources.
  - Raw material availability and costs.
- Company strategy, involving several conditions, including other markets, customers, competition, and product mix.
- Possible changes in market base, including new customers, old customers leaving the market, additional competition, and possible changes in customer preferences.

Capacity is the capability of manufacturing to produce goods and services. Essentially it depends on the resources of the company—the machinery, labor, and financial resources, and the availability of material from suppliers. In the short term, capacity is the quantity of work that labor and equipment can perform in a given period. The relationship that should exist between priority and capacity is shown graphically in Figure 2.1.

In the long and short term, manufacturing must devise plans to balance the demands of the marketplace with its resources and capacity. For long-range decisions, such as the building of new plants or the purchase of new equipment, the plans must extend over a

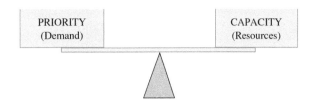

FIGURE 2.1 Priority–capacity relationship.

timeline of several years. For planning production over the next few weeks, the time span will be days or weeks. This hierarchy of planning, from long range to short range, is covered in the next section.

2.2 MANUFACTURING PLANNING AND CONTROL SYSTEM

There are five major levels in the manufacturing planning and control system:

- Strategic plan (strategy) and strategic business plan (a business plan based on the organizational strategy).
- Production plan (sales and operations plan).
- Master production schedule.
- Material requirements plan.
- Purchasing and production activity control.

Each level varies in purpose, time span, and level of detail. As the process moves from strategic planning to production activity control, the purpose changes from general direction to specific detailed planning, the time span decreases from years to days, and the level of detail increases from general categories to individual components and work centers.

Since each level is for a different time span and for different purposes, each differs in the following:

- Purpose of the plan.
- Planning horizon—the time span from now to some time in the future for which the plan is created.
- Level of detail—the detail about products required for the plan.
- Planning cycle—the frequency with which the plan is reviewed.

At each level, three questions must be answered:

1. What are the priorities—how much of what is to be produced and when?
2. What is the available capacity—what resources do we have?
3. How can differences between priorities and capacity be resolved?

Figure 2.2 shows the planning hierarchy. The first four levels are planning levels. The result of the plans is authorization to purchase or manufacture what is required. The final level is when the plans are put into action through purchasing and production activity control. Notice the arrows are pointing in two directions. This is an important issue in that it indicates that the information and planning activity goes both upward and downward. Specifically, the plan and execution levels are continually providing feedback to the planning level above it for consideration and possible modification. For example, once the production plan is developed issues of capacity, resources or demand may occur at the master schedule level that imply some evaluation and change is needed in the production plan. It is often not appropriate for a plan to be made and then not reviewed as conditions change significantly.

The following sections will examine each of the planning levels by purpose, horizon, level of detail, and planning cycle.

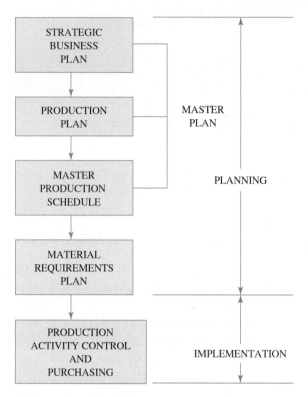

FIGURE 2.2 Manufacturing planning and control system.

The Strategic Plan

The **strategic plan** of a firm is a statement of the major goals and objectives the company expects to achieve over the next 2 to 10 years or more. It is a statement of the broad direction of the firm and shows the kind of business—product lines, markets, and so on—the firm wants to do in the future. The plan gives general direction about how the company hopes to achieve these objectives and represents a commitment to take various actions designed toward growth, defining and attracting customers, defining markets, and improving competitive and financial performance. It is based on long-range forecasts and includes participation from marketing, finance, production, engineering and all other major functions in the firm. In turn, the plan provides direction and coordination among the marketing, production, financial, engineering, and other functional plans.

Marketing and *Sales* are responsible for analyzing the marketplace and deciding the firm's response: the markets to be served, the products supplied, desired levels of customer service, pricing, promotion strategies, and so on.

Finance is responsible for deciding the sources and uses of funds available to the firm, cash flows, profits, return on investment, and budgets.

Production must satisfy the demands of the marketplace, by using plants, machinery, equipment, labor, and materials as efficiently as possible.

Engineering is responsible for research, development, and design of new products or modifications to existing ones. Engineering must work with marketing and production to produce designs for products that will sell in the marketplace and can be made most economically.

The development of the strategic plan is the responsibility of senior management. Using information from marketing, finance, production, and other functions, the strategic plan provides a framework that sets the goals and objectives for further planning by the marketing, finance, engineering, and production departments.

Some companies have adopted an approach of establishing vision statements and goals as part of the planning process. Tactical work plans are established for the visions to allow all parts of the organization to move systematically toward the overall goal.

Developed by Japanese companies, the approach is often referred to as **hoshin planning**. This approach includes the following steps:

1. Making the plan for what you wish to improve or accomplish.
2. Establishing subgoals.
3. Communicating the plan in the organization.
4. Measuring your results.
5. Analyzing data from the measures and taking corrective action as needed.
6. Repeating as necessary.

There are recent trends in business that sometimes impact the development and management of strategic plans. One of these is the issue of **sustainability**, meaning the capability to continue (sustain) operations into the long term. Much of the interest in sustainability evolved from the perspective of pollution control, saving the environment and **social responsibility** (establishing company policies that establish a positive relationship with society and strike some balance between the economy and the environment). Corporate social responsibility has become an important issue worldwide, as indicated by the **United Nations Global Compact**. This compact recognizes that business is a primary source of globalization, and the compact lays out ten principles for business strategy and operations to maintain appropriate human rights, treatment of labor, environmental issues, and anticorruption principles.

Sustainability is also based on the concepts of reduction of waste and inefficiency in production, leading not only to using fewer resources and producing less waste, but also less costs. Examples of waste reduction might include less need for packaging (often thrown away) and using resources to produce reusable outputs. Recycling and reusing material is also a major part of sustainability. Sometimes this activity is described as **remanufacturing** or **reverse logistics**. In some cases, companies will establish a formal supply chain used to retrieve a used product in order to dispose of it, reclaim materials from it, or reuse it in some fashion. This is sometimes referred to as a **reverse supply chain**. These concepts are explained in more detail in Chapters 7 (Fundamentals of Supply Chain Management) and 16 (Lean Production).

A second recent development impacting strategic planning is **risk management**. Often people consider risk to be a negative issue, and of course that is partially correct. Risks reflecting problems from some sort of failure of systems, people, or external events can result in loss of money, productivity, legal problems, and a reduction in the probability of successful implementation of the strategic plan. Risks can also be positive, and in this context are often called opportunities. Risk management is focused on establishing systems and measurements to try to quickly recognize risks and establish strategic mechanisms to minimize impacts from negative risks and take advantage of positive risks (opportunities). Risks related to supply chains will be discussed further in Chapter 7.

Effective strategic planning depends on obtaining appropriate measurements and feedback on how well the tactics related to the plan are working toward the overall set of strategic goals for the organization. These measures (both financial and nonfinancial) are sometimes referred to as **key performance indicators (KPI)**. It is important that one measure or subset of measures do not contradict with others—that they provide an overall balance in indicating the progress of the company toward the overall strategic plan and sustainability efforts including financial, societal, and environmental goals. The set of KPIs that are balanced are often referred to as a **balanced scorecard**, and detailed approaches have been developed to both establish and manage balanced scorecards. The scorecard tends to balance measures dealing with business processes, financial measures, customer focused measures, and learning and growth. All these perspectives and measures are, of course, developed as part of the overall strategic planning process.

FIGURE 2.3 Business plan.

The Strategic Business Plan (Business Plan)

Once the strategic plan has been established, the plan is often restated in financial terms, including projected revenues, a projected balance sheet, and a projected income statement. This financially based plan is often called the **business plan** or sometimes the strategic business plan. Each department produces its own plans to achieve the objectives set by the strategic business plan. These plans will be coordinated with one another and with the strategic business plan. Figure 2.3 shows this relationship.

The level of detail in the strategic business plan is low. It is concerned with general market and production requirements—total market for major product groups or product families, perhaps—and not sales of individual items. It is usually stated in dollars rather than units.

Strategic business plans are usually reviewed every six months to a year.

The Production Plan

Given the objectives set by the strategic business plan, production management is concerned with the following:

- The quantities of each product group that must be produced in each period.
- The projected and desired inventory levels.
- The resources of equipment, labor, and material needed in each period.
- The availability of the resources needed.

The level of detail of the **production plan** is low. For example, if a company manufactures children's bicycles, tricycles, and scooters in various models, each with many options, the production plan will show major product groups, or families, such as bicycles, tricycles, and scooters. Because the production plan tends to combine product groups or product families rather than individual products, it is sometimes referred to as the **aggregate production plan**.

Production planners must devise a plan to satisfy market demand within the resources available to the company. This will involve determining the resources needed to meet market demand, comparing the results to the resources available, and devising a plan to balance requirements and availability.

Production Planning System 21

This process of determining the resources required and comparing them with the available resources takes place at each of the planning levels and is the purpose of capacity management. For effective planning, there must be a balance between priority and capacity.

Along with the market and financial plans, the production plan is concerned with implementing the strategic business plan. The planning horizon is usually 6 to 18 months and is usually reviewed each month or quarter.

The Master Production Schedule

The **master production schedule (MPS)** is a plan for the production of individual end items. It breaks down the production plan to show, for each period, the quantity of each end item to be made. For example, it might show that 200 Model A23 scooters are to be built each week. Inputs to the MPS are the production plan, the forecast for individual end items, sales orders, inventories, and existing capacity.

The level of detail for the MPS is greater than for the production plan. Whereas the production plan was based upon families of products (e.g., tricycles), the master production schedule is developed for individual end items (each model of tricycle). The planning horizon usually extends from 3 to 18 months but primarily depends on the purchasing and manufacturing lead times. This is discussed in Chapter 3 in the section on master scheduling. The term **master scheduling** describes the process of developing a master production schedule. The term master production schedule is the end result of this process. Usually, the plans are reviewed and changed weekly or monthly, but this depends on the length of the MPS.

The Material Requirements Plan

The output of **material requirements planning (MRP)** is the **material requirements plan**, which is a plan for the production and purchase of the components used in making the items in the master production schedule. It shows the quantities needed and when manufacturing intends to make or use them. Purchasing and production activity control use MRP to execute the purchase or manufacture of specific items.

The level of detail of MRP is high. The material requirements plan establishes when the components and parts are needed to make each end item.

The planning horizon is at least as long as the combined purchase and manufacturing lead times. As with the master production schedule, it usually extends from 3 to 18 months.

Purchasing and Production Activity Control

Purchasing and **production activity control (PAC)** represent the implementation and control phase of the production planning and control system. Purchasing is responsible for establishing and controlling the flow of raw materials into the factory. PAC is responsible for planning and controlling the flow of work through the factory.

The planning horizon is very short, perhaps from a day to a month. The level of detail is high since it is concerned with individual components, work centers, and orders. Plans are reviewed and revised daily.

Figure 2.4 shows the relationship among the various planning tools, planning horizons, and levels of detail.

This chapter focuses on production planning. Later chapters deal with master scheduling, material requirements planning, purchasing and production activity control.

Capacity Management

At each level in the manufacturing planning and control system, the priority plan must be tested against the available resources and capacity of the manufacturing system. Chapter 5 describes some of the details of **capacity management**. For now, it is sufficient to understand that the basic process in capacity management is one of calculating the capacity needed to manufacture per the requirements of the priority plan and of finding methods

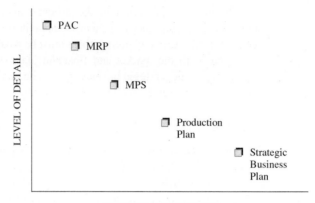

FIGURE 2.4 Level of detail versus planning horizon.

to make that capacity available. There can be no valid, workable production plan unless this is done. If the capacity cannot be made available when needed, then the plans must be changed, as inadequate resources equal missed production schedules. In addition, if resources significantly exceed what is required, they will sit idle and result in extra costs.

Determining the capacity required, comparing it to available capacity, and making adjustments (or changing plans) must occur at all levels of the manufacturing planning and control system.

Over several years, machinery, equipment, and plants can be added to or taken away from manufacturing. Some changes, such as changing the number of shifts, working overtime, subcontracting the work, and so on, can be accomplished in these time spans. However, in the time spans involved from production planning to production activity control, these kinds of large changes cannot be made.

2.3 SALES AND OPERATIONS PLANNING

The strategic business plan integrates the plans of all the departments in an organization and is normally updated annually. However, these plans should be updated as time progresses so that the latest forecasts and market and economic conditions are taken into account. **Sales and operations planning (S&OP)** is a process for continually revising the strategic business plan and coordinating plans of the various departments. S&OP is a cross-functional business plan that involves sales and marketing, product development, operations, and senior management. Operations represent supply, whereas marketing represents demand. The S&OP is the forum in which the production plan is developed.

Although the strategic plan and strategic business plan are usually updated annually, sales and operations planning is a dynamic process in which the company plans are updated on a regular basis, usually at least monthly. The process starts with the sales and marketing departments, which compare actual demand with the sales plan, assess market potential, and forecast future demand. The updated marketing plan is then reviewed with manufacturing, engineering, and finance, which adjust their plans to support the revised sales plan. If these departments find they cannot accommodate the new sales plan, then the sales plan must be adjusted. In this way the strategic business plan is continually revised throughout the year and the activities of the various departments are coordinated. Figure 2.5 shows the relationship between the strategic business plan and the sales and operations plan.

Sales and operations planning is medium range and includes the marketing, production, engineering, and finance plans. Sales and operations planning has several benefits:

- It provides a means of updating the strategic plan and the strategic business plan as conditions change.
- It provides a means of managing change. Rather than reacting to changes in market conditions or the economy after they happen, the S&OP forces management to look at the economy at least monthly and places it in a better position to plan changes.

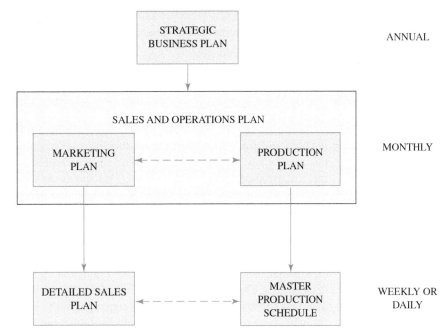

FIGURE 2.5 Sales and operations planning.

- It ensures that the various department plans are realistic and coordinated and support the business plan.
- It provides a realistic plan that can achieve the company objectives.
- It permits better management of production, inventory, and backlog (i.e., unfilled customer orders).

It is important to realize that effective sales and operations planning is an executive-level planning process that involves the potential to make significant trade-offs between the various functions/departments in the company. In this way executives involved in planning can ensure that the best approach to volume and mix are made and that supply and demand are balanced in the best approach possible.

It should be noted that the sales and operations plan does not actually schedule production, but instead focuses on producing a high-level plan for the use of company resources. These resources include not only production resources but also human resources, sales resources, financial resources, and virtually all other functions in the company. The sales and operations plan should reflect the vision and strategies developed in the strategic plan, and should serve as a focused, single plan across functional areas that provides the approach to running the entire business.

In his book *Sales and Operations Planning*, author Tom Wallace describes a five-step process in developing the S&OP, summarized here:

1. **Data gathering.** Actual past month sales, existing inventory levels, marketing/sales data, financial data, and so on.
2. **Demand planning.** Using data and other inputs from all appropriate sources to establish management forecasts. Statistical forecasts can be run, but should be evaluated in the context of other inputs. Those inputs may include new product introductions, price change impacts, competitive movements, and economic conditions. Demand planning and demand management is more fully described in Chapter 9 of this book.
3. **Supply planning.** Comparison of demand forecasts with capacity constraints.
4. **A pre-S&OP meeting.** This meeting should be used for balancing supply and demand, resolving differences (if possible), and developing action recommendations and an agenda for the top manager S&OP meeting.
5. **The executive meeting.** The final decisions are made as to how the company should proceed given all the data, recommendations, and how well the plan fits into the context of the strategic plan and the strategic business plan.

In many companies the resource discussions include some aspects of what is often called **green manufacturing**. Basically, these discussions will focus on issues such as

- Environmental impacts.
- Energy conservation.
- Material usages, such as waste reduction, reuse and recyclability.
- Scarcity of various resources.

The conclusions from these green resource discussions can help in both the design of products and process and in the most effective uses of production resources. As an example, a metal casting manufacturer in Canada, which uses a lot of electricity, schedules its melts to avoid periods of peak demand. This avoids the local energy supplier from having to increase their capacity and in turn the manufacturer is given a slightly preferred rate. It should be obvious that the concepts under the category of green manufacturing are a major part of sustainability discussed earlier in this chapter, and later in this text.

Given the fact that each level of the planning system obtains input from the level above it and also feedback from the systems below it in the hierarchy, one may be concerned that there needs to be continual evaluation and change based on something as detailed as a slight change in volume demanded from a specific customer, for example. That is not really the case, however. Referring again to Figure 2.4, notice that planning at the high levels tend to look quite far into the future, but with very little detail—and therefore need very little detail in terms of feedback from lower levels.

A brief example of a global automobile company may provide perspective. Just below the top level (overall strategic company plan) the next level planning may include general plans for an overall approach for each global region (one for Asia, another for Europe, another for North America, etc.) but do not need to project the sales of specific models, as the planning detail for the S&OP is done in product families. In the automobile company example, product families may aggregate the overall demand for SUVs as one family, overall demand for trucks as another family, overall demand for sedans as another family, and so forth. It is also possible that within each of those high-level families there can be further breakdown. Within the sedan family, for example, there could be a planning level for full-size sedans, another for compact models, and yet another for sub-compact models. At this point all the data is essentially forecast data only, since actual customer orders specify much more detail (engine size, transmission, color, options, etc.). Those specific customer order details are only captured at the master schedule level. The S&OP and other levels above the master schedule are used primarily for planning resources (type, quantity, timing, etc.). The S&OP planning level does not need to know differences in the specific details of customer orders in sedan purchases, for example, as long as the overall demand for sedans has not changed so much that it has a significant impact on resource needs to produce the sedans.

There are very practical reasons for much of this hierarchy. Typically, the development and procurement of resources (machinery, buildings, hiring and training people, for example) may take a fairly long time. The master schedule is focused on planning and delivering specific customer orders. If the type or quantity of the orders change significantly, there is usually insufficient time for the company to respond to those changes in resources needed for the order in time to meet specific customer orders. The planning for those resources should be done at the S&OP level (which needs less detail and is focused further into the future). If the S&OP is done correctly, the right resources should be available for the master schedule. If, however, a trend is observed at the master schedule level then that information needs to be communicated to the upper planning level to determine the nature of the trend and if the company needs to respond with resource allotment.

2.4 MANUFACTURING RESOURCE PLANNING

Because of the large amount of data and the number of calculations needed, the manufacturing planning and control system is usually computer-based. If a computer is not used, the time and labor required to make calculations manually is extensive and forces a company into compromises. Instead of scheduling requirements through the planning system, the company may have to extend lead times and build inventory to compensate for the inability to schedule quickly what is needed and when.

Initially, the integrated systems were referred to as **manufacturing resource planning (MRP II)** systems. The term *MRP II* was used to distinguish the manufacturing resource planning (MRP II) from material requirements planning (MRP).

MRP II provided coordination between marketing, finance and production, with everyone agreeing on a total workable plan expressed in the production plan. Marketing and production worked together on a weekly and daily basis to adjust the plan as changes occurred. Marketing managers and production managers changed master production schedules to meet changes in forecast demand. Senior management adjusted the production plan to reflect overall changes in demand or resources, all working through MRP II. It provided the mechanism for coordinating the efforts of marketing, finance, production, and other departments in the company, as well as the effective planning of all resources of a manufacturing company.

2.5 ENTERPRISE RESOURCE PLANNING

As MRP II systems evolved, they tended to take advantage of two changing conditions:

1. Computers and information technologies (IT) becoming significantly faster, more reliable, and more powerful. People in most companies had become at least comfortable, but often very familiar, with the advantages in speed, accuracy, and capability of integrated computer-based management systems.
2. Movement toward integration of knowledge and decision making in all aspects of direct and indirect functions and areas that impact materials flow and materials management. This integration not only included internal functions such as marketing, engineering, human resources, accounting, and finance but also the upstream activities in supplier information and the downstream activities of distribution and delivery. That movement of integration is what is now recognized as supply chain management.

As the needs of the organization grew in the direction of a truly integrated approach toward materials management, the development of IT systems matched that need. As these systems became both larger in scope and integration when compared to the existing MRP and MRP II systems, they were given a new name—**enterprise resource planning (ERP)**.

ERP is similar to the MRP II system except that the whole enterprise is taken into account, not just manufacturing. The *APICS Dictionary,* 16th edition, defines ERP as a "framework for organizing, defining, and standardizing the business processes necessary to effectively plan and control an organization so the organization can use its internal knowledge to seek external advantage." To fully operate, there must be applications for planning, scheduling, costing, and so forth, for all layers in an organization: work centers, sites, divisions, and corporate. The larger scope of ERP systems allows the tracking of orders and other important planning and control information throughout the entire company, from procurement to ultimate customer delivery. In addition, many ERP systems are capable of allowing managers to share data between firms, meaning that these managers can potentially have visibility across the complete span of the supply chain.

Due to the power and capability of these highly integrated ERP systems being extremely high, there are some large costs involved. The large data requirements (for both quantity and accuracy) tend to make the systems expensive, and time consuming to implement. Added to the complexity is that all major areas of the company (marketing, engineering, finance, human resources, etc.) are involved and impacted by ERP, and therefore significant coordination is needed to ensure an effective implementation.

ERP is intended to be a fully integrated planning and control system that works from the top down and has feedback from the bottom up. Strategic business planning integrates the plans and activities of marketing, finance, and production to create plans intended to achieve the overall goals of the company. In turn, master production scheduling, material requirements planning, production activity control, and purchasing are directed toward achieving the goals of the production and strategic business plans and, ultimately, the company. If priority plans have to be adjusted at any of the planning levels because of capacity problems, those changes should be reflected in the levels above. Thus, there must be feedback throughout the system.

The strategic business plan incorporates the plans of marketing, finance, and production. Marketing must agree that its plans are realistic and attainable. Finance must agree that the plans are desirable from a financial point of view, and production must agree that it can meet the required demand. The ERP system, as described here, is a master game plan for all departments in the company.

The feedback loops are quite important and as a result the overall approach is often called a "closed loop system." This implies that the higher-level planning activities are always generating plans and transmitting those plans to the lower levels for providing more detail to the plan, but simultaneously the lower levels are providing feedback to the upper-level planning systems to potentially impact both current and possible future plans.

Note the feedback (closed cycle) loops in the ERP system shown in Figure 2.6.

As the concept in supply chain grew, another planning approach was developed. Called **advanced planning and scheduling (APS)**, the approach has often included the

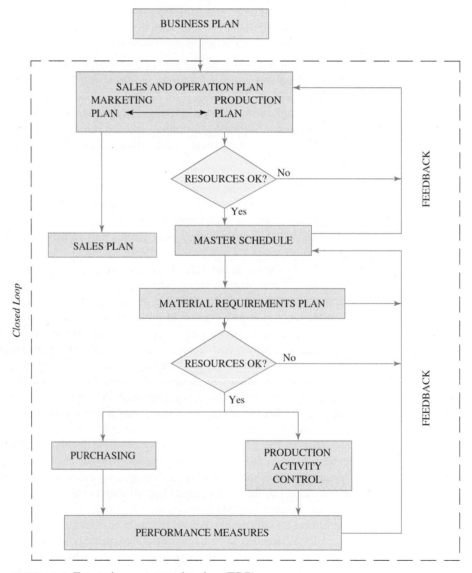

FIGURE 2.6 Enterprise resource planning (ERP).

suppliers and customers in the planning, thereby attempting to optimize the performance of the entire supply chain. The *APICS Dictionary*, 16th edition, defines APS as "any computer program that uses advanced mathematical algorithms or logic to perform optimization or simulation on finite capacity scheduling, sourcing, capital planning, resource planning, forecasting, demand management, and others." Extracting information from the entire supply chain, the system attempts to create a rapid and feasible schedule for satisfying customer demand. It includes mathematical optimization and analytic tools and the principle of finite scheduling (see Chapter 6). These same concepts can also be used internally in an operation of a single company in order to try to optimize or create a more feasible solution for the customers of that operation.

2.6 MAKING THE PRODUCTION PLAN

Thus far the purpose, planning horizon, and level of detail found in a production plan have been discussed. This section includes some details involved in making production plans.

Based on the market plan and available resources, the production plan sets the limits or levels of manufacturing activity for some time in the future. It integrates the capabilities and capacity of the factory with the market and financial plans to achieve the overall business goals of the company.

The production plan is developed as part of the S&OP process and sets the general levels of production and inventories over the planning horizon. Its prime purpose is to establish production rates that will accomplish the objectives of the strategic plan and the strategic business plan. These include inventory levels, backlogs, market demand, customer service, low-cost plant operation, labor relations, and so on. The plan must extend far enough in the future to plan for the labor, equipment, facilities, and material needed to accomplish it. Typically, this is a period of 6 to 18 months and is done in quarterly, monthly, or sometimes weekly periods depending on conditions.

The planning process at this level ignores such details as individual products, colors, styles, or options. With the time spans involved and the uncertainty of demand over long periods, the detail would not be accurate or useful, and the plan would be expensive to create. For planning purposes, a common unit or small number of product groups is what is needed.

Establishing Product Groups

Firms that make a single product, or products that are similar, can measure their output directly by the number of units they produce. A brewery, for instance, might use barrels of beer as a common denominator. Many companies, however, make several different products, and a common denominator for measuring total output may be difficult or impossible to find. Product groups or families need to be established. Marketing naturally looks at products from the customers' point of view of functionality and application, whereas manufacturing looks at products in terms of processes.

Manufacturing must provide the capacity to produce the goods needed. It is concerned more with the demand for the specific kinds of capacity needed to make the products than with the demand for the product. Capacity is the ability to produce goods and services. It means having the resources available to satisfy demand. For the time span of a production plan, it can be expressed as the time available or, sometimes, as the number of units or dollars that can be produced in a given period. The demand for goods must be translated into the demand for capacity. At the production planning level, where little detail is needed, this requires identifying product groups, or families, of individual products based on the similarity of manufacturing process and resources used. For example, several computer tablets might share the same processes and need the same kind of capacity, regardless of the variations in the models. They would be considered as a family group of products.

Over the time span of the production plan, large changes in capacity are usually not possible. Additions or subtractions in plant and equipment are impossible or very difficult to accomplish in this period. However, some things can be altered, and it is the responsibility of manufacturing management to identify and assess them. Usually the following can be varied:

- People can be hired and laid off, overtime and part time can be worked, and shifts can be added or removed.
- Inventory can be built up in slack periods and sold or used in periods of high demand.
- Work can be subcontracted or extra equipment leased.

Each alternative has its associated benefits and costs. Manufacturing management is responsible for finding the least-cost alternative consistent with the goals and objectives of the business.

Basic Strategies

In summary, the production planning problem typically has the following characteristics:

- A time horizon of 12 months or more is used, with periodic updating perhaps every month or quarter.
- Production demand consists of one or a few product families or common units.
- Demand is fluctuating or seasonal.
- Plant and equipment are fixed within the time horizon.
- A variety of management objectives are set, such as low inventories, efficient plant operation, good customer service, and good labor relations.

Figure 2.7 shows a hypothetical demand forecast for a product group. Note that the demand is seasonal.

There are three basic strategies that can be used in developing a production plan:

1. Chase strategy.
2. Production leveling.
3. Hybrid strategy.

Chase (demand matching) strategy   **Chase strategy** means producing the amounts demanded at any given time. Inventory levels remain stable while production varies to meet demand. Figure 2.8 shows this strategy. The firm manufactures just enough at any one time to meet demand exactly. In some industries, this is the only strategy that can be followed. The post office must process mail over a holiday rush and in slack seasons. Restaurants have to serve meals when the customers want them. These industries cannot stockpile or inventory their products or services and must be capable of meeting demand as it occurs.

In these cases, the company must have enough capacity to be able to meet the peak demand. Farmers must have sufficient machinery and equipment to harvest in the growing

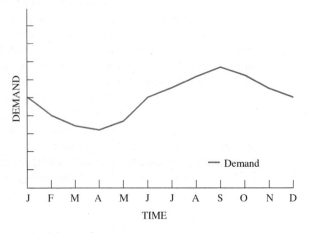

FIGURE 2.7 Hypothetical demand curve.

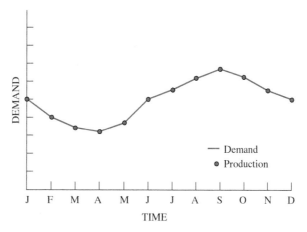

FIGURE 2.8 Chase (demand matching) strategy.

season, although the equipment will lie idle in the winter. Companies with seasonal or cyclical demand often have to hire and train people for the peak periods and lay them off when the peak is past. Sometimes they have to add extra shifts and overtime. All these changes add cost.

The advantage to the chase strategy is that inventories can be kept to a minimum. Goods are made when demand occurs and are not stockpiled. Thus, the costs associated with carrying inventories are avoided. Such costs can be quite high, as is shown in Chapter 10, on inventory fundamentals.

Production leveling **Production leveling** is continually producing an amount equal to the average demand. This relationship is shown in Figure 2.9. Companies calculate their total demand over the time span of the plan and, on the average, produce enough to meet it. Sometimes demand is less than the amount produced and an inventory builds up. At other times demand is greater and inventory is used up.

The advantage of a production leveling strategy is that it results in a smooth level of operation that avoids the costs of changing production levels. Firms do not need to have excess capacity to meet peak demand. They do not need to hire and train workers and lay them off in slack periods. They can build a stable workforce. The disadvantage is that inventory will build up in low-demand periods, which will cost money to carry. In addition, an understated forecast could result in not enough inventory being produced for the peak season.

Production leveling means the company will use its resources at a uniform rate and produce the same amount each day it is operating. The exact amount produced each month (or sometimes each week), however, will not be constant because the number of working days varies from month to month.

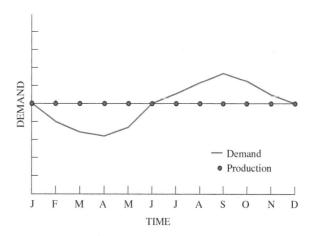

FIGURE 2.9 Production leveling strategy.

Example Problem

A company wants to produce 10,000 units of an item over the next three months at a uniform rate. Because of an annual shutdown in the third month, the first month has 20 working days; the second, 21 working days; and the third, 12 working days. On the average, how much should the company produce each day to level production?

Answer

$$\text{Total production} = 10{,}000 \text{ units}$$
$$\text{Total working days} = 20 + 21 + 12 = 53 \text{ days}$$
$$\text{Average daily production} = \frac{10{,}000}{53} = 188.7 \text{ units}$$

For some products for which demand is very seasonal, such as for holiday decorations, some form of production leveling strategy is necessary. The costs of idle capacity, and of hiring, training, and laying off, would be severe if a company employed a chase strategy.

As a pure strategy, production leveling would mean always producing at the defined level. Any demand above that level, assuming inventory is not available, would imply the demand would not be met or meeting the extra demand by some other method. One method already described in the discussion on chase strategy is to add people, work shifts, equipment or some combination.

Sometimes, however, conditions (e.g., no additional skilled workers available or a timing problem to add them quickly enough) imply that some other approach must be used to meet the demand. A common production choice in those cases is meeting any additional demand through **subcontracting**. Subcontracting can mean buying the extra amounts demanded from external sources. This situation is shown in Figure 2.10. The major advantage of subcontracting is internal production cost. Costs associated with excess production capacity are avoided, and because production is leveled, there are no costs associated with changing those production levels. The main disadvantage of subcontracting is that the cost of purchasing (item cost, purchasing, transportation, and inspection costs) may be greater than if the item were made internally. Subcontracting may also be used as part of a chase strategy.

An alternate choice is to purposely turn away extra demand. The latter can be done by increasing prices when demand is high or by extending lead times.

Few companies make everything or buy everything they need (this concept of make versus buy will be discussed further in Chapter 15). Subcontracting is one example of this type of decision. The decision about which items to subcontract and which to manufacture internally depends mainly on cost, but there are several other factors that may be considered.

As an example, some firms may wish to avoid subcontracting to keep confidential processes within the company, to maintain quality levels, and to maintain a workforce.

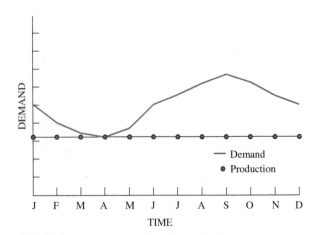

FIGURE 2.10 Subcontracting.

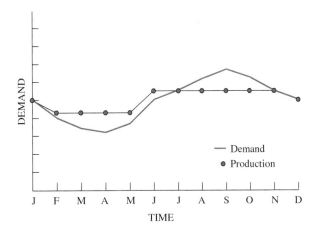

FIGURE 2.11 Hybrid strategy.

They may prefer to subcontract from a supplier who has special expertise in design and manufacture of a component, to allow the firm to concentrate on its own area of expertise, or to provide known and competitive prices. This is discussed further in Chapter 15.

For many items, such as nuts and bolts or components that the firm does not normally make, the decision is clear. For other items that are within the specialty of the firm, a decision must be made whether to subcontract or not.

Hybrid strategy The two strategies discussed so far are pure strategies. Each has its own set of costs: equipment, hiring/layoff, overtime, inventory, and subcontracting. In reality, there are many possible hybrid or combined strategies a company may use. Each will have its own set of cost characteristics. Production management is responsible for finding the combination of strategies that minimizes the sum of all costs involved, providing the level of service required, and meeting the objectives of the financial and marketing plans.

Figure 2.11 shows a possible **hybrid strategy** plan. Demand is matched to some extent, production is partially smoothed, and in the peak period some subcontracting takes place. The plan is only one of many that could be developed.

Developing a Make-to-Stock Production Plan

In a **make-to-stock** environment, products are made and put into inventory before an order is received from a customer. Sale and delivery of the goods are made from inventory. Off-the-rack clothing, frozen foods, and bicycles are examples of this kind of manufacturing.

Generally firms make to stock when

- Demand is fairly constant and predictable.
- There are few product options.
- Delivery times demanded by the marketplace are much shorter than the time needed to make the product.
- Product has a long shelf life.

The information needed to make a production plan is as follows:

- Forecast by period for the planning horizon.
- Opening inventory.
- Desired ending inventory.
- Any past-due customer orders. These are orders that are late for delivery and are sometimes called **backorders**. Note the difference between the terms backorders and backlogs (mentioned earlier in the chapter). The backlog is customer orders received but not yet shipped, while the backorders are customer orders that may be past due or due for immediate shipment but cannot because the inventory levels are too low to fill the order.

The objective in developing a production plan is to minimize the costs of carrying inventory, changing production levels, and stocking out (not having the products the customer wants when they are wanted).

The following sections develop a plan for leveling production and one for chase strategy.

Level production plan Following is the general procedure for developing a plan for level production.

1. Total the forecast demand for the planning horizon.
2. Determine the opening inventory and the desired ending inventory.
3. Calculate the total production required as follows:

   Total production = Total forecast + back orders
   + ending inventory − opening inventory

4. Calculate the production required each period by dividing the total production by the number of periods.
5. Calculate the ending inventory for each period.

Example Problem

Amalgamated Fish Sinkers makes a product group of fresh fish sinkers and wants to develop a production plan for them. The expected opening inventory is 100 cases, and the company wants to reduce that to 80 cases by the end of the planning period. The number of working days is the same for each period. There are no backorders. The expected demand for the fish sinkers is as follows:

Period	1	2	3	4	5	Total
Forecast (cases)	110	120	130	120	120	600

a. How much should be produced each period?
b. What is the ending inventory for each period?
c. If the cost of carrying inventory is $5 per case per period based on ending inventory, what is the total cost of carrying inventory?
d. What will be the total cost of the plan?

ANSWER

a. Total production required = 600 + 80 − 100 = 580 cases

   Production each period = $\frac{580}{5}$ = 116 cases

b. Ending inventory = opening inventory + production − demand

   Ending inventory after the first period = 100 + 116 − 110 = 106 cases

Similarly, the ending inventories for each period are calculated as shown in Figure 2.12. The ending inventory for period 1 becomes the opening inventory for period 2:

Ending inventory (period 2) = 106 + 116 − 120 = 102 cases

Period		1	2	3	4	5	Total
Forecast (cases)		110	120	130	120	120	600
Production		116	116	116	116	116	580
Ending Inventory	100	106	102	88	84	80	

FIGURE 2.12 Level production plan: Make-to-stock.

Period	0	1	2	3	4	5	Total
Demand (cases)		110	120	130	120	120	600
Production	100	90	120	130	120	120	580
Change in Production		10	30	10	10	0	60
Ending Inventory	100	80	80	80	80	80	

FIGURE 2.13 Chase strategy: Make-to-stock.

   c. The total cost of carrying inventory would be

$$(106 + 102 + 88 + 84 + 80)\$5 = \$2300$$

   d. Since there were no stockouts and no changes in the level of production, this would be the total cost of the plan.

Chase strategy Amalgamated Fish Sinkers makes another line of product called fish stinkers. Unfortunately, they are perishable, and the company cannot build inventory for sale later. They must use a chase strategy and make only enough to satisfy demand in each period. Inventory costs will be a minimum, and there should be no stockout costs. However, there will be costs associated with changing production levels.

Suppose in the preceding example that changing the production level by one case costs $20. For example, a change from 50 to 60 would cost

$$(60 - 50) \times \$20 = \$200$$

The opening inventory is 100 cases, and the company wishes to bring this down to 80 cases in the first period. The required production in the first period would then be

$$110 - (100 - 80) = 90 \text{ cases}$$

Assuming that production in the period before period 1 was 100 cases, Figure 2.13 shows the changes in production levels and in ending inventory.

The cost of the plan would be as follows:

$$\text{Cost of changing production level} = (60)\$20 = \$1200$$
$$\text{Cost of carrying inventory} = (80 \text{ cases})(5 \text{ periods})\$5 = \$2000$$
$$\text{Total cost of the plan} = \$1200 + \$2000 = \$3200$$

It should be noted that the previous examples provide a basic understanding of the dollar cost of possible production plans. But before a final selection of a plan is made, there are other considerations that should be evaluated, and some of those are difficult to estimate financially. Some of those issues might include the following:

- Impact on customers if cyclical demand in a level schedule causes shortages.
- Impact on production workers as they are moved into and out of production in a chase strategy. Such an impact will often reduce efficiency of the operation.
- Potential loss of profit if customers change buying preferences and seek a competitor's products.

Developing a Make-to-Order Production Plan

In a **make-to-order** environment, manufacturers wait until an order is received from a customer before starting to make the goods. Examples of this kind of manufacture are custom-tailored clothing, custom furniture, or any product made to customer specification. Very expensive items are usually made to order. Generally, firms make to order when

- Goods are produced to customer specification.
- The customer is willing to wait while the order is being made.
- Generally, the product is expensive to make and to store.
- Several product options are offered.

Assemble-to-order Where several product options exist, such as in automobiles, and where the customer is not willing to wait until the product is made, manufacturers produce and stock standard component parts. When manufacturers receive an order from a customer, they assemble the component parts from inventory according to the order. Since the components are stocked, the firm only needs time to assemble before delivering to the customer. Examples of assemble-to-order products include automobiles and computers. **Assemble-to-order** is a subset of make-to-order.

The following information is needed to make a production plan for make-to-order products:

- Forecast by period for the planning horizon.
- Opening backlog of customer orders.
- Desired ending backlog.

Backlog In a make-to-order environment, a company does not build an inventory of finished goods. Instead, it has a **backlog** of unfilled customer orders. The backlog normally will be for delivery in the future and does not represent orders that are late or past due. A custom woodwork shop might have orders from customers that will keep it busy for several weeks. This will be its backlog. If individuals want some work done, the order will join the queue or backlog. Manufacturers like to control the backlog so that they can provide a good level of customer service.

Level production plan Following is a general procedure for developing a make-to-order level production plan:

1. Total the forecast demand for the planning horizon.
2. Determine the opening backlog and the desired ending backlog.
3. Calculate the total production required as follows:

   Total production = total forecast + opening backlog − ending backlog

4. Calculate the production required each period by dividing the total production by the number of periods.
5. Spread the existing backlog over the planning horizon according to due date per period.

Example Problem

A local printing company provides a custom printing service. Since each job is different, demand is forecast in hours per week. Over the next five weeks, the company expects that demand will be 100 hours per week. There is an existing backlog of 100 hours, and at the end of five weeks, the company wants to reduce that to 80 hours. How many hours of work will be needed each week to reduce the backlog? What will be the backlog at the end of each week?

ANSWER

$$\text{Total production} = 500 + 100 - 80 = 520 \text{ hours}$$

$$\text{Weekly production} = \frac{520}{5} = 104 \text{ hours}$$

The backlog for each week can be calculated as

Projected backlog = old backlog + forecast − production

For week 1: Projected backlog = 100 + 100 − 104 = 96 hours

For week 2: Projected backlog = 96 + 100 − 104 = 92 hours

Figure 2.14 shows the resulting production plan.

Period	1	2	3	4	5	Total
Sales Forecast	100	100	100	100	100	500
Planned Production	104	104	104	104	104	520
Projected Backlog	100	96	92	88	84	80	

FIGURE 2.14 Level production plan: Make-to-order.

Product	Wood (board feet)	Labor (standard hours)
Tables	20	1.31
Chairs	10	0.85
Stools	5	0.55

FIGURE 2.15 Resource bill.

Resource Planning

Once the preliminary production plan is established, it must be compared to the existing resources of the company. This step is called resource requirements planning or **resource planning**. Two questions must be answered:

1. Are the resources available to meet the production plan?
2. If not, how will the difference be reconciled?

If enough capacity to meet the production plan cannot be made available, the plan must be changed.

A tool often used is the **resource bill**. This shows the quantity of *critical* resources (materials, labor, and bottleneck operations) needed to make one average unit of the product group. Figure 2.15 shows an example of a resource bill for a company that makes tables, chairs, and stools as a three-product family.

If the firm planned to make 500 tables, 300 chairs, and 1500 stools in a particular period, they could calculate the quantity of wood and labor that will be needed. For example, the amount of wood needed is

$$\text{Tables}: \quad 500 \times 20 = 10{,}000 \text{ board feet}$$
$$\text{Chairs}: \quad 300 \times 10 = 3000 \text{ board feet}$$
$$\text{Stools}: \quad 1500 \times 5 = 7500 \text{ board feet}$$
$$\text{Total wood required} = 20{,}500 \text{ board feet}$$

The amount of labor needed is

$$\text{Tables}: \quad 500 \times 1.31 = 655 \text{ standard hours}$$
$$\text{Chairs}: \quad 300 \times 0.85 = 255 \text{ standard hours}$$
$$\text{Stools}: \quad 1500 \times 0.55 = 825 \text{ standard hours}$$
$$\text{Total labor required} = 1735 \text{ standard hours}$$

The company must now compare the requirements for wood and labor with the availability of these resources. For instance, suppose the labor normally available in this period is 1600 hours. The priority plan requires 1735 hours, a difference of 135 hours, or about 8.4%. Extra capacity must be found, or the priority plan must be adjusted. In this example, it might be possible to work overtime to provide the extra capacity required. If overtime is not possible, the plan must be adjusted to reduce the labor needed. This might involve shifting some production to an earlier period or delaying shipments.

SUMMARY

Production planning is the first step in a manufacturing planning and control system and is often accomplished through the use of sales and operations planning, which is an executive-level planning process involving trade-offs across departments or functions in the company. The planning horizon usually extends for at least a year. The minimum horizon depends on the lead times to purchase materials and make the product. The level of detail is low. Usually, the plan is made for families of products based on the similarity of manufacturing processes or on some common unit. Most hierarchical planning also is "closed loop," in that it provides not only upper-level plans to drive the lower level detailed planning, but also incorporates feedback information from the lower-level planning activities in order to make all the plans more coordinated and effective.

Three basic strategies can be used to develop a production plan: chase, leveling production, or hybrid. Each has its operational and cost advantages and disadvantages. It is the responsibility of manufacturing management to select the best combination of these basic plans so total costs are minimized and customer service levels are maintained.

A make-to-stock production plan determines how much to produce in each period to meet the following objectives:

- Achieve the forecast.
- Maintain the required inventory levels.

Although demand must be satisfied, the plan must balance the costs of maintaining inventory with the cost of changing production levels.

A make-to-order production plan determines how much to produce in each period to meet the following objectives:

- Achieve the forecast.
- Maintain the planned backlog.

The cost of a backlog that is too large equals the cost of turning away business. If customers have to wait too long for delivery, they might take their business elsewhere. As with a make-to-stock production plan, demand must be satisfied, and the plan must balance the costs of changing production levels with the cost of a backlog that is larger than desired.

KEY TERMS

Advanced planning and scheduling (APS) 26
Aggregate production plan 20
Assemble-to-order 34
Backlog 34
Backorder 31

Balanced scorecard 19
Business plan 20
Capacity 16
Capacity management 21
Chase strategy 28
Enterprise resource planning (ERP) 25

Green manufacturing 24
Hoshin planning 19
Hybrid strategy 31
Key performance indicator (KPI) 19
Make-to-order 33
Make-to-stock 31
Manufacturing resource planning (MRP II) 25
Master production schedule (MPS) 21
Master scheduling 21
Material requirements plan 21
Material requirements planning (MRP) 21
Priority 16
Production activity control (PAC) 21
Production leveling 29

Production plan 20
Purchasing 21
Remanufacturing 19
Resource bill 35
Resource planning 35
Reverse logistics 19
Reverse supply chain 19
Risk management 19
Sales and operations planning (S&OP) 22
Social responsibility 19
Strategic plan 18
Subcontracting 30
Sustainability 19
United Nations Global Compact 19

QUESTIONS

1. What are the four questions a good planning system must answer?
2. Define capacity and priority. Why are they important in production planning?
3. Describe each of the following plans in terms of their purpose, planning horizon, level of detail, and planning cycle:
    a. Strategic plan.
    b. Business plan.
    c. Production plan.
    d. Master production schedule.
    e. Material requirements plan.
    f. Production activity control.
    g. Sales and operations planning
4. Describe the responsibilities and inputs of the marketing, production, finance, and engineering departments to the strategic business plan.
5. Describe the relationship among the production plan, the master production schedule, and the material requirements plan.
6. What is the difference between strategic business planning and sales and operations planning (S&OP)? What are the major benefits of S&OP?
7. What is MRP II?
8. What is ERP?
9. What two changing conditions led to the development of ERP systems?
10. In the short run, how can capacity be changed?
11. When making a production plan, why is it necessary to select a common unit or to establish product families?
12. On what basis should product groups (families) be established?
13. What are five typical characteristics of the production planning problem?
14. Describe each of the four basic strategies used in developing a production plan. What are the advantages and disadvantages of each?
15. What is a hybrid strategy? Why is it used?
16. Describe four conditions under which a firm would make-to-stock or make-to-order.
17. What information is needed to develop a make-to-stock production plan?
18. What are the steps in developing a make-to-stock production plan?
19. What is the difference between make-to-order and assemble-to-order? Give an example of each.

20. What information is needed to develop a make-to-order production plan? How does this differ from that needed for a make-to-stock plan?
21. What is the general procedure for developing a level production plan in a make-to-order environment?
22. What is a resource bill? At what level in the planning hierarchy is it used?
23. What kind of production environment would you expect to see if a company uses a chase strategy? What if it uses a level strategy?
24. What does the concept of green manufacturing mean? How will it potentially impact production planning?
25. Describe sustainability for production. What are some of the ways that a company can practice sustainability?
26. What should a company approach be to risks? What are some of the methods they can use to minimize the impact of negative risks?
27. Describe some of the possible advantages of a company developing a reverse supply chain.
28. What is hoshin planning?
29. What are product "families"? What is the potential value from using product families for sales and operations planning?
30. What is the primary output from sales and operations planning? What specifically is being planned?

PROBLEMS

2.1. If the opening inventory is 500 units, demand is 800 units, and production is 700 units, what will be the ending inventory?

   Answer.   400 units

2.2. A company wants to produce 420 units over the next three months at a uniform rate. The months have 19, 20, and 21 working days, respectively. On the average, how much should the company produce each day to level production?

   Answer.   Average daily production = 7 units

2.3. A company plans to produce 25,200 units in a 3-month period. The months have 22, 21, and 20 working days, respectively. What should the average daily production be?

2.4. In problem 2.2, how much will be produced in each of the three months?

   Answer.   Month 1: 133 Month 2: 140 Month 3: 147

2.5. In problem 2.3, how much will be produced in each of the three months?

2.6. A production line is to run at 1000 units per month. Sales are forecast as shown in the following. Calculate the expected period-end inventory. The opening inventory is 600 units. All periods have the same number of working days.

   Answer.   For period 1, the ending inventory is 770 units.

Period		1	2	3	4	5	6
Forecast		830	800	1050	1600	1000	850
Planned Production		1000	1000	1000	1000	1000	1000
Planned Inventory	600						

2.7. A company wants to develop a level production plan for a family of products. The opening inventory is 100 units, and an increase to 180 units is expected by the end of the plan. The demand for each period is given in what follows. How much should the company produce each period? What will be the ending inventories in each period? All periods have the same number of working days.

Period	1	2	3	4	5	6	Total
Forecast Demand	100	120	125	130	115	110	
Planned Production							
Planned Inventory	100						

Answer. Total Production = 780 units

Period production = 130 units

The ending inventory for period 1 is 130; for period 5, 160.

2.8. A company wants to develop a level production plan for a family of products. The opening inventory is 550 units, and a decrease to 200 units is expected by the end of the plan. The demand for each of the periods is given in what follows. All periods have the same number of working days. How much should the company produce each period? What will be the ending inventories in each period? Do you see any problems with the plan?

Period	1	2	3	4	5	6	Total
Forecast Demand	1300	800	600	1200	800	900	
Planned Production							
Planned Inventory	550						

2.9. A company wants to develop a level production plan. The beginning inventory is zero. Demand for the next four periods is given in what follows.

a. What production rate per period will give a zero inventory at the end of period 4?

b. When and in what quantities will backorders occur?

c. What level production rate per period will avoid backorders? What will be the ending inventory in period 4?

Period	1	2	3	4	Total
Forecast Demand	19	12	14	19	
Planned Production					
Planned Inventory	0				

Answer. a. 16 units

b. Period 1, minus 3

c. 19 units; 12 units

2.10. If the cost of carrying inventory is $60 per unit per period and stockout cost $500 per unit, what will be the cost of the plan developed in problem 2.9a? What will be the cost of the plan developed in 2.9c?

Answer. Total cost for plan in question 2.9a = $1740

Total cost for plan in question 2.9c = $1860

2.11. A company wants to develop a level production plan for a family of products. The opening inventory is 100 units, and an increase to 130 units is expected by the end of the plan. The demand for each month is given in what follows. Calculate the total production, daily production, and production and ending inventory for each month.

Month	May	Jun	Jul	Aug	Total
Working Days	21	19	20	10	
Forecast Demand	115	125	140	150	
Planned Production					
Planned Inventory	100				

Answer. Total monthly production for May = 168 units

The ending inventory for May = 153 units

2.12. A company wants to develop a level production plan for a family of products. The opening inventory is 500 units, and a decrease to 200 units is expected by the end of the plan. The demand for each of the months is given in what follows. How much should the company produce each month? What will be the ending inventory in each month? Do you see any problems with the plan?

Month	Jan	Feb	Mar	Apr	May	Jun	Total
Working Days	20	22	20	20	19	19	
Forecast Demand	1200	1200	800	800	900	800	
Planned Production							
Planned Inventory	500						

2.13. Because of its labor contract, a company must hire enough labor for 100 units of production per week on one shift or 200 units per week on two shifts. It cannot hire, lay off, or assign overtime. During the fourth week, workers will be available from another department to work part or all of an extra shift (up to 100 units). There is a planned shutdown for maintenance in the second week, which will cut production to half. Develop a production plan. The opening inventory is 200 units, and the desired ending inventory is 300 units.

Week	1	2	3	4	5	6	Total
Forecast Demand	120	160	240	240	160	160	
Planned Production							
Planned Inventory							

2.14. If the opening backlog is 350 units, forecast demand is 700 units, and production is 900 units, what will be the ending backlog?

Answer. 150 units

2.15. The opening backlog is 800 units. Forecast demand is as shown here. Calculate the weekly production for level production if the backlog is to be reduced to 100 units.

Week	1	2	3	4	5	6	Total
Forecast Demand	750	700	550	700	600	500	
Planned Production							
Projected Backlog	800						

Answer. Total production = 4500 units

Weekly production = 750 units

Backlog at end of week 1 = 800 units

2.16. The opening backlog is 1300 units. Forecast demand is as shown here. Calculate the weekly production for level production if the backlog is to be decreased to 900 units.

Week	1	2	3	4	5	6	Total
Forecast Demand	1200	1000	1200	1200	1100	1100	
Planned Production							
Projected Backlog	1300						

2.17. For the following data, calculate the number of workers required for level production and the resulting month-end inventories. Each worker can produce 15 units per day, and the desired ending inventory is 9000 units.

Month	1	2	3	4	Total
Working Days	20	24	12	19	
Forecast Demand	28,000	27,500	28,500	28,500	
Planned Production					
Planned Inventory	11,250				

Answer. Workers needed = 98 workers

First month's ending inventory = 12,650 units

2.18. For the following data, calculate the number of workers required for level production and the resulting month-end inventories. Each worker can produce 8 units per day, and the desired ending inventory is 500 units. Why is it not possible to reach the exact ending inventory target?

Month		1	2	3	4	5	6	Total
Working Days		20	24	12	22	20	19	
Forecast Demand		2800	3000	2800	3300	3000	3200	
Planned Production								
Planned Inventory	1000							

CASE STUDY 2.1

Medina Water Pumps

Olivia Lopez, president of Medina Water Pumps (a small water pump producer), was holding a meeting with her department managers. They were in the process of planning production of medium-sized pumps for the next six months. Olivia tolerated some of the arguments before she felt it necessary to stop the discussion as it was going so that she could direct it toward a solution. A summary of some of the arguments follows:

Marty Welch, marketing and sales manager: "My sales people are very good, but get very frustrated at times. Several times last year the sales people spent lots of their time trying to calm down frustrated customers. As they are supposed to do, the sales people sold as many of these pumps as they could, yet at times the production could not keep up with the orders. Production knows that we have some cyclicality in the demand, but we have plenty of machine capacity. They should be able to hire people so that we can meet the demand that we sell. Why can't they get their area to work correctly?"

Naveen Agarwal, production manager: "Come on, Marty, we know the sales are cyclical, but we never know exactly when the cycles happen. Even if we did, the Human Resource (HR) people always take too long to get us the people we need. By the time we get the new people hired and trained, the sales seem to drop again. What am I supposed to do? If we keep them and allow them to keep producing pumps, our inventory climbs and the finance people start yelling. I can't just let those new people sit around doing nothing. The only other alternative is to lay them off, but then the HR people get really angry."

Frank Conrad, human resource manager: "You bet we get angry. The production people will occasionally start pushing us to hurry and hire more people, yet get very impatient. It takes time to go through the interview process and get people hired and oriented to our business. Then we no sooner get them on site and working when production asks us to lay them off. That is a real problem for two reasons. First, there are costs involved. It takes an average of $100 to get a person hired, and another $100 to lay them off. Second, those people that we hire and then quickly lay off tend to not return. I can't blame them, since from their perspective it looks like we have no idea how to run our business. In addition, as those people complain to other people about our treatment of them, our reputation is getting to look bad, and that makes it increasingly difficult to find good people to hire."

Maylin Lin, finance manager: "Naveen is correct that I get upset when the inventory climbs. It costs us about $5 to keep one of these pumps in our inventory for a month. That cost comes right out of our profit. Since my job is to maximize profitability, I can't sit by and let those inventory dollars shrink that profitability. The same goes for all that hiring and layoffs. That money also hurts profitability. Can't we do better?"

At this point Olivia stopped the discussion and said "Enough of trying to blame each other. It is our job as managers to manage this process more effectively. Marketing has just completed a 6-month forecast of anticipated demand for this family of pumps, and we know from past history that their forecasts are pretty good. We should be able to come up with an approach that we all can live with and focus our efforts to meet. Let's get to work on it."

Assignment

Assume you have been given the job to develop an effective approach to the problem. First, here is the forecast developed by marketing:

Month	1	2	3	4	5	6
Forecasted Demand	600	750	1000	850	750	700

The production manager said there were currently 50 units in inventory, and they would like to end the six months with only 25 in inventory. He also said that currently each worker produces an average of 25 pumps in any given month. There are currently 20 workers in the medium-size pump area.

1. Using the data, develop a level production plan. How much extra cost (inventory and HR costs) are involved in this plan? What additional costs (both financial and nonfinancial) might be involved with such a plan?

2. Using the data, develop a chase production plan. How much extra cost (inventory and HR costs) are involved in this plan? What additional costs (both financial and nonfinancial) might be involved with such a plan?

3. Try to develop a possible hybrid plan that would accomplish the task with smaller total costs than either level or chase.

4. Based on your work, what would you recommend and why? What are some of the pros and cons of the solution you recommend?

CASE STUDY 2.2

Williams 3D Printers

The Williams 3D printer company was experiencing growing pains early in 2015. Jasper Williams had developed his own unique design for making a 3D printer with this relatively new technology that was growing fairly rapidly in interest and in competitors. He had started by using his engineering skills as an individual inventor, but with some borrowed money he was able to set up a small production facility. His sales the first year were modest, as he made and sold only five printers. Now that he had been in the business for three years, he noticed that near the end of his fiscal year, he was likely to sell more than 20 units. He only had three other people on his management staff: John Johnson, the financial officer, Paula Chavez, the production manager, and Yolanda Andrews, the marketing and sales manager. The following conversation took place during their most recent monthly planning meeting, where the key item on the agenda was to look at plans for the next fiscal year:

JASPER: "Yolanda, I think you mentioned that we are gaining a good reputation in a market that is growing rapidly, given that it is in the early stages of the life cycle. What do you think that means for sales this coming year?"

YOLANDA: "I think our good reputation is going to be a real plus. Not only do several of our original customers plan on buying another printer from us, but they have also told other potential customers that they like our design, and some of those potential customers are likely to buy as well. I think it is very likely that we could double our sales next year to possibly 40 or more printers."

PAULA: "We need to talk about that—perhaps you should hold back on making sales like that. We are already finding it tough to deliver on promises for this year. We have had a couple of late orders this year, and the only reason we didn't have more late orders was that our workers agreed to work on some weekends. Problem is, I don't know how agreeable they will be to that next year. While they like the extra money, they all have families and don't want to spend that much time away from home."

YOLANDA: "Look, Paula, we have worked hard to get the good reputation to increase our sales. What good does it do if we can't meet the needs of people who want to buy from us? Last I knew, we were in business to make sales and therefore make money. We have a good profit margin on the printers that should make the company very profitable."

PAULA: "Well the only way we can really expand to sell and make 40 units next year is to hire a lot more people. We possibly could hire a second shift, but our workers are skilled people who are making most of these printers in a somewhat unique design based on the specific needs of the customer. Skilled people like that often have several options as to where to work, and I am pretty sure most would not like to work a second shift where they could not spend evenings with their families. We could try to double the number of workers on the regular day shift, but that would mean duplicating all the current equipment as well. Also, while today we probably have the space in our facility to fit duplicate equipment, if we grow more the following year we will also have to expand our space requirements."

At this point John Johnson, the financial manager, had to break into the conversation:

JOHN: "We need to think long and hard about all this. After three years we are finally looking a little better financially, but adding a whole lot more people and equipment is going to cost us a lot. Also, you have to keep in mind that the printers have a pretty long lead time to produce, given that each customer specifies at least some unique aspect of design based on their individual needs, and also it takes a fair amount of time to build them. That means we get to see the money from the sale only after many weeks after the order is placed, but in the meantime we have to obtain materials and pay workers for today's new orders—which are larger in number in this growing market we have. This implies that even though the profit per unit is good, we have a struggle with cash flow. Paula, what about adding just one or two people as the sales grow, and then add some more later in the year as the sales continue to grow?"

PAULA: "I don't see how that can help. Adding a duplicate person without additional equipment for them to work with makes no sense, and even if they could do the work, the other people in areas we don't add would still have a lot more work to do without the time to do it."

JOHN: "Well we might be able to get another loan to help, but do we really want to do that just at the point we are starting to show some bottom-line profit?"

JASPER: "Okay, I understand each of your perspectives, but arguing back and forth doesn't help. We need to figure out something to do that we can all agree on. Let's get to work."

Assignment

Try to help the management team out. After listing the key issues and characteristics of the environment and the problem, list all the possible alternative approaches they could take to deal with the issues. For each alternative, try to list the pros and cons likely involved. Try to make a list of what data (information) they should attempt to collect to support analysis for each alternative and how that data would help them. Then select an approach you would recommend and attempt to justify it.

CHAPTER THREE

MASTER SCHEDULING

3.1 INTRODUCTION

After production planning, the next step in the manufacturing planning and control process is to prepare a **master schedule**. This chapter examines some basic considerations in making and managing a master schedule. It is an extremely important planning tool and forms the basis for communication between sales and manufacturing. The master schedule is a vital link in the production planning system. The master schedule uses demand data (both actual and forecasted) and existing inventories to help develop the primary output for the master schedule, called the **master production schedule (MPS)**. The MPS generally is shown as the final line on the master schedule and represents what is actually being planned to build and when it is scheduled.

- The master schedule forms the link between production planning and what manufacturing will actually build. From this perspective it forms the major link between customer demand and the production facility.
- The master schedule forms the basis for calculating the capacity and resources needed.
- The master schedule (specifically the MPS) drives the material requirements plan. As a schedule of items to be built, the MPS and bills of material determine what components are needed from manufacturing and purchasing.
- The master schedule keeps priorities valid. The MPS is a priority plan for manufacturing.

Whereas the production plan deals in families of products, the master schedule works with end items. It breaks down the production plan into the requirements for individual end items, in each family, by date and quantity. The production plan limits the master schedule, that is, the total of the items in the master schedule should not be different from the total shown on the production plan. For example, if the production plan shows a planned production of 1000 tricycles in a particular week, the total of the individual models planned for by the master schedule should be 1000. Within this limit, its objective is to balance the demand (priorities) set by the marketplace with the availability of materials, labor, and equipment (capacity) of manufacturing.

The end items made by the company are assembled from component and subcomponent parts. These must be available in the right quantities at the right time to support the MPS. The material requirements planning system plans the schedule for these components based on the needs of the MPS. Thus, the MPS drives the material requirements plan.

The master schedule is a plan for manufacturing. It reflects the needs of the marketplace and the capacity of manufacturing and forms a priority plan for manufacturing to follow. The master schedule forms a vital link between sales and production:

- It makes possible valid order promises. The MPS is a plan of what is to be produced and when. As such, it tells sales and manufacturing when goods will be available for delivery.
- It is an agreed-upon plan, and a contract between marketing and manufacturing.

The master schedule forms a basis to determine what is to be manufactured. It is not meant to be rigid. It is a device for communication and a basis to make changes that are consistent with the demands of the marketplace and the capacity of manufacturing.

The information needed to develop an MPS is provided by the following sources:

- The production plan—the aggregated production plan developed during the S&OP process.
- Forecasts for individual end items.
- Actual orders received from customers and for stock replenishment.

- Inventory levels for individual end items.
- Capacity restraints.

3.2 RELATIONSHIP TO PRODUCTION PLAN

Suppose the following production plan is developed for a family of three items:

Week	1	2	3	4	5	6
Aggregate Forecast (units)	160	160	160	160	215	250
Production Plan	205	205	205	205	205	205
Aggregate Inventory (units)	545	590	635	680	670	625

Opening inventories (units) are

Product A	350
Product B	100
Product C	50
Total	500

The next step is to forecast demand for each item in the product family. Keep in mind that the forecast data for the production plan was based on aggregated data (product families), while for the master schedule, the more specific forecast is needed for each item in the item family. Forecasts for the sales and operations planning (S&OP) process used product families since forecasts are often more accurate for product families than for individual products. Product family forecasts were all that was needed for the S&OP since it was being used to plan *resources* needed to produce the product family forecast and *not* individual item production. It should also be noted that while individual item forecasts might not add up to the exact product family forecast, that production plan for the product family represents a constraint on the total that can be produced for items in the family.

Week	1	2	3	4	5	6
Product A	70	70	70	70	70	80
Product B	40	40	40	40	95	120
Product C	50	50	50	50	50	50
Total	160	160	160	160	215	250

With this data, the master scheduler must now devise a plan to fit the constraints. The following illustrates a possible solution.

Master Schedule

Week	1	2	3	4	5	6
Product A						205
Product B	205	205	205			
Product C				205	205	
Total Planned	205	205	205	205	205	205

Inventory

Week	1	2	3	4	5	6
Product A	280	210	140	70	0	125
Product B	265	430	595	555	460	340
Product C	0	−50	−100	55	210	160
Total Planned	545	640	735	680	670	625

This schedule is satisfactory for the following reasons:

- It tells the plant when to start and stop production of individual items.
- Capacity is consistent with the production plan.

It is unsatisfactory for the following reasons:

- It has a poor inventory balance compared to total inventory.
- It results in a stockout for product C in periods 2 and 3.

The term master production schedule (MPS) refers to the last line of the matrix. The term **master scheduling** refers to the *process* of arriving at that line. Thus, the total matrix is called a master schedule.

Example Problem

The Hotshot Lightning Rod Company makes a family of two lightning rods, Models H and I. It bases its production planning on weeks. For the present month, production is leveled at 1000 units. Opening inventory is 500 units, and the plan is to reduce that to 300 units by the end of the month. The master schedule is made using weekly periods. There are four weeks in this month, and production is to be leveled at 250 units per week. The forecast and projected available for the two lightning rods follows. Calculate an MPS for each item.

ANSWER

Production Plan

Week		1	2	3	4	Total
Forecast		300	350	300	250	1200
Projected Available	500	450	350	300	300	
Production Plan		250	250	250	250	1000

Master Schedule: Model H

Week		1	2	3	4	Total
Forecast		200	300	100	100	700
Projected Available	200	250	200	100	100	
MPS		250	250		100	

Master Schedule: Model I

Week		1	2	3	4	Total
Forecast		100	50	200	150	500
Projected Available	300	200	150	200	200	
MPS				250	150	

3.3 DEVELOPING A MASTER PRODUCTION SCHEDULE

The objectives in developing a master schedule and an MPS are as follows:

- To maintain the desired level of customer service by maintaining finished-goods inventory levels or by scheduling to meet customer delivery requirements.
- To make the best use of material, labor, and equipment.
- To maintain inventory investment at the required levels.

To reach these objectives, the plan must satisfy customer demand, be within the capacity of manufacturing, and be within the guidelines of the production plan.

There are three steps in preparing a master schedule:

1. Develop a preliminary master schedule.
2. Check the preliminary master schedule against available capacity.
3. Resolve differences between the preliminary master schedule and capacity availability.

Preliminary Master Schedule

To show the process of developing a master schedule, an example is used that assumes that the product is made to stock, an inventory is kept, and the product is made in lots.

A particular item is made in lots of 100, and the expected opening inventory is 80 units. Figure 3.1 shows the forecast of demand, the projected available on hand, and the preliminary master schedule.

Period 1 begins with an inventory of 80 units. After the forecast demand for 60 units is satisfied, the projected available is 20 units. A further forecast demand of 60 in period 2 is not satisfied, and it is necessary to schedule an MPS receipt of 100 for week 2. This produces a projected available of 60 units ($20 + 100 - 60 = 60$) at the end of period 2. In period 3, the forecast demand for 60 is satisfied by the projected 60 on hand, leaving a projected available of 0. In period 4, a further 100 must be received, and when the forecast demand of 60 units is satisfied, 40 units remain in inventory.

This process of building a master schedule occurs for each item in the family. If the total planned production of all the items in the family and the total ending inventory do not

On hand = 80 units
Lot size = 100 units

Period		1	2	3	4	5	6
Forecast			60	60	60	60	60	60
Projected Available	80	20	60	0	40	80	20
MPS			100		100	100	

FIGURE 3.1 MPS example.

agree with the production plan, some adjustment to the individual plans must be made so the total production is the same.

Once the preliminary master schedules are made, they must be checked against the available capacity. This process is called **rough-cut capacity planning**.

Example Problem

Amalgamated Nut Crackers, Inc., makes a family of nut crackers. The most popular model is the walnut, and the sales department has prepared a six-week forecast. The opening inventory is 50 dozen (dozen is the unit used for planning). As master planner, you must prepare a master schedule. The nutcrackers are made in lots of 100 dozen.

ANSWER

Week		1	2	3	4	5	6
Forecast Sales		75	50	30	40	70	20
Projected Available	50	75	25	95	55	85	65
MPS		100		100		100	

Rough-Cut Capacity Planning

Rough-cut capacity planning checks whether critical resources are available to support the preliminary master schedules. Critical resources include bottleneck operations, labor, and critical materials (perhaps material that is scarce or has a long lead time).

The process is similar to resource requirements planning used in the production planning process. The difference is that now we are working with an individual product and not a family of products. The resource bill, used in resource requirements planning, assumes a typical product in the family. Here the resource bill is for a single product. As before, the only interest is in bottleneck work centers and critical resources.

One reason this method is described as "rough" is not only because it focuses primarily on critical resources but also because there are several other variables that can impact specific details of capacity usage. These include existing inventory, existing work orders that are partially complete, and lead times. Inclusion of those and other capacity variables are discussed in more detail in Chapter 5.

Suppose a firm manufactures four models of desktop computers assembled in a work center that is a bottleneck operation. The company wants to schedule to the capacity of this work center and not beyond. Figure 3.2 shows a resource bill for that work center showing the time required to assemble one computer.

Resource Bill	
Desktop Computer Assembly	
Computer	Assembly Time (standard hours)
Model D24	0.203
Model D25	0.300
Model D26	0.350
Model D27	0.425

FIGURE 3.2 Resource bill.

Suppose that in a particular week the MPSs show that the following computers are to be built:

$$\text{Model D24} \quad 200 \text{ units}$$
$$\text{Model D25} \quad 250 \text{ units}$$
$$\text{Model D26} \quad 400 \text{ units}$$
$$\text{Model D27} \quad 100 \text{ units}$$

The capacity required on this critical resource is

$$\begin{aligned}
\text{Model D24} \quad 200 \times 0.203 &= 40.6 \text{ standard hours} \\
\text{Model D25} \quad 250 \times 0.300 &= 75.0 \text{ standard hours} \\
\text{Model D26} \quad 400 \times 0.350 &= 140.0 \text{ standard hours} \\
\text{Model D27} \quad 100 \times 0.425 &= 42.5 \text{ standard hours} \\
\text{Total time required} &= 298.1 \text{ standard hours}
\end{aligned}$$

Example Problem

The Acme Tweezers Company makes tweezers in two models, medium and fine. The bottleneck operation is in work center 20. Following is the resource bill (in hours per dozen) for work center 20.

Work Center	Hours per Dozen	
	Medium	Fine
20	0.5	1.2

The MPS for the next four weeks is

Week	1	2	3	4	Total
Medium	40	25	40	15	120
Fine	20	10	30	20	80

Using the resource bill and the MPS, calculate the number of hours required in work center 20 for each of the four weeks. Use the following table to record the required capacity on the work center.

Answer

Week	1	2	3	4	Total
Medium	20	12.5	20	7.5	60
Fine	24	12	36	24	96
Total Hours	44	24.5	56	31.5	156

Resolution of Differences

The next step is to compare the total time required to the available capacity of the work center. If available capacity is greater than the required capacity, the MPS is workable. If not, methods of increasing capacity have to be investigated. Is it possible to adjust the available capacity with overtime, extra workers, routing through other work centers, or subcontracting? If not, it will be necessary to revise the MPS.

Finally, the MPS must be judged by three criteria:

1. **Resource use.** Is the MPS within capacity restraints in each period of the plan? Does it make the best use of resources?
2. **Customer service.** Will due dates be met and will delivery performance be acceptable?
3. **Cost.** Is the plan economical, or will excess costs be incurred for overtime, subcontracting, expediting, or transportation?

Master Schedule Decisions

The MPS should represent as efficiently as possible what manufacturing will make. If too many items are included, it will lead to difficulties in forecasting and managing the MPS. In each of the manufacturing environments—make-to-stock, make-to-order, assemble-to-order, configure-to-order, and engineer-to-order—master scheduling should take place where the smallest number of product options exists. Figure 3.3 shows the level at which items should be master scheduled.

Make-to-stock products In the **make-to-stock (MTS)** environment, a limited number of standard items are assembled from many components. Televisions and other consumer products are examples. The MPS is usually a schedule of finished goods items.

Make-to-order products In **the make-to-order (MTO)** environment, many different end items are made from a small number of components. Custom-tailored clothes are an example. The MPS is usually a schedule of raw material requirements, used not only for replenishment of raw materials but also for planning capacity needs. Technology advances are making this approach more attractive in many markets. For example, the growth of 3D (three dimensional) printing for products is growing in interest, and many of the 3D printed products are of the make-to-order category. As this market grows, it will become increasingly important to use a master schedule to plan for equipment capacity as well as the raw material used for production.

Assemble-to-order and configure-to-order products In the **assemble-to-order (ATO)** and **configure-to-order** environments, many end items can be made from combinations of basic components, subassemblies, features, and options. For example, suppose a company manufactures paint from a base color and adds tints to arrive at the final color. Suppose there are 10 tints and a final color is made by mixing any three of them with the base. There are 720 possible colors $(10 \times 9 \times 8 = 720)$. Forecasting and planning

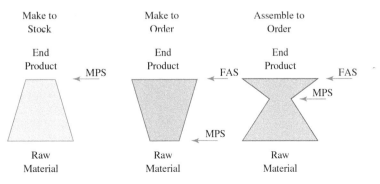

FIGURE 3.3 Different MPS environments.

production for 720 items is a difficult task. It is much easier if production is planned at the level of the base color and the 10 tints. There are then only 10 items with which to deal: the base color and each of the 10 tints. Once a customer's order is received, the base color and the required tints can be combined (assembled) according to the order.

Final assembly schedule (FAS) This step, assembly to customer order, is generally planned using a **final assembly schedule (FAS)**. This is a schedule of what will be assembled. It is used when there are many options and it is difficult to forecast which combination the customers will want. Master production scheduling is done at the component level, for example, the base color and tint level. The final assembly takes place only when a customer order is received.

The FAS schedules customer orders as they are received and is based on the components planned in the MPS. It is responsible for scheduling from the MPS through final assembly and shipment to the customer. The FAS is typically used in both the ATO and MTO environments.

Engineer-to-order The **engineer-to-order (ETO)** environment is a form of make-to-order (MTO) products. In this environment, the product is designed before manufacturing, based on the customer's very special needs. A bridge is an example.

Figure 3.4 shows the relationship of the MPS, the FAS, and other planning activities.

It should be noted that the master schedule is typically designed to plan production as "close" as possible to the customer. Make-to-stock products, for example, are master scheduled as the final product design to provide rapid response to customer demand. An additional advantage is that the production of the product itself is "insulated" from customer influence, implying more internal stability. Make-to-order products, on the other hand, have significant customer input on the final design. The master schedule for make-to-order products is often at the raw material stage, meaning customer influence and possible associated production disruptions are high. The assemble-to-order products are somewhere in-between. The master schedule is often for modules or options (and common parts for all products), while the exact customer-specified design is reflected in the final assembly schedule. The customer tends to have little influence on the specific design of the option, thereby insulating the actual production of that option. For example, you as a customer may buy a particular model of automobile. While you may have influence on whether the engine is a 6- or 8-cylinder engine, you have

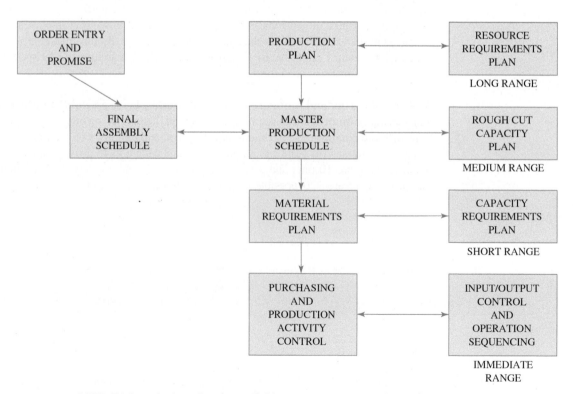

FIGURE 3.4 MPS, FAS, and other planning activities.

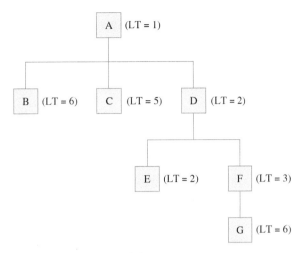

FIGURE 3.5 Product structure: critical lead time.

no influence on how either of those engines is produced. Some companies actively design products to shift customer influence on the final design as close as possible to the furthest point in the supply chain. This design strategy is called **postponement**.

Planning Horizon

The planning horizon is the time span for which plans are made. It must cover a period at least equal to the time required to accomplish the plan. For master production scheduling, the minimum planning horizon is the longest cumulative or end-to-end lead time (LT). For example, in Figure 3.5, the longest cumulative LT path is A to D to F to G. The cumulative LT is $1 + 2 + 3 + 6 = 12$ weeks. The minimum planning horizon must be 12 weeks; otherwise, raw material G would not be ordered in time to meet delivery.

The planning horizon is usually longer for several reasons. The longer the horizon, the greater the visibility and the better management's ability to avoid future problems or to take advantage of special circumstances. For example, firms might take advantage of economical purchase plans, avoid future capacity problems, or manufacture in more economical lot sizes.

As a minimum, the planning horizon for an FAS must include time to assemble a customer's order. It does not need to include the time necessary to manufacture the components. That time will be included in the planning horizon of the MPS.

3.4 PRODUCTION PLANNING, MASTER SCHEDULING, AND SALES

The production plan reconciles total forecast demand with available resources. It takes information from the strategic plan, the strategic business plan, and market forecasts to produce an overall plan of what production intends to make to meet forecast. It is a major output of the sales and operations planning process discussed in Chapter 2. It is dependent on the forecast and, within capacity limits, must plan to satisfy the forecast demand. It is not concerned with the detail of what will actually be made. It is intended to provide a framework in which detailed plans can be made in the MPS.

The MPS is built from forecasts and actual demands for individual end items. It reconciles demand with the production plan and with available resources to produce a plan that manufacturing can fulfill. The MPS is concerned with what items will actually be built, in what quantities, and when, to meet expected demand.

The production plan and the MPS uncouple the sales forecast from manufacturing by establishing a manufacturing plan. Together, they attempt to balance available resources of plant, equipment, labor, and material with forecast demand. However, they are not a sales forecast, nor are they necessarily what is desired. The MPS is a plan for what production *can and will do*.

Figure 3.6 shows the relationship among the sales forecast, production plan, and MPS.

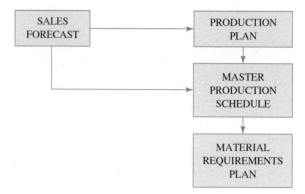

FIGURE 3.6 Sales forecast, production plan, and master production schedule.

The MPS must be realistic about what manufacturing can and will do. If it is not, it will result in overloaded capacity plans, past-due schedules, unreliable delivery promises, surges in shipments, and lack of accountability.

The MPS is a plan for specific end items or "buildable" components that manufacturing expects to make over some time in the future. It is the point at which manufacturing and marketing must agree what end items are going to be produced. Manufacturing is committed to making the goods, and marketing to selling the goods. However, the MPS is not meant to be rigid. Demand changes, problems occur in production, and, sometimes, components are scarce. These events may make it necessary to alter the MPS. Changes must be made with the full understanding and agreement of sales and production. The MPS provides the basis for making changes and a plan on which all can agree.

Some markets are changing due to competitive and technology evolutionary developments. One area of expected change is the increase in linking of some services with product sales. As that occurs, master plans linking product demand with service demand will need to be developed. As an example, some products that are sold offer periodic service calls for inspection, adjustments, or maintenance. Those services require capacity planning as well as some additional material planning (maintenance replacement parts, for example).

The MPS and Delivery Promises

In a make-to-stock environment, customer orders are satisfied from inventory. However, in make-to-order or assemble-to-order environments, demand is satisfied from production capacity. In either case, sales and distribution need to know what is available to satisfy customer demand. Since demand can be satisfied either from inventory or from scheduled receipts, the MPS provides a plan for doing either. Figure 3.7 illustrates the concept. As orders are received, they "consume" the available inventory or capacity. Any part of the plan that is not consumed by actual customer orders is available to promise to customers. In this way, the MPS provides a realistic basis for making delivery promises. It should be noted that for each period, the MPS quantity is assumed to be the quantity available at the beginning of that period.

Using the MPS, sales and distribution can determine the **available-to-promise (ATP)**. ATP is that portion of a firm's inventory and planned production that is not already committed and is available to the customer. This allows delivery promises to be made and customer orders and deliveries to be scheduled accurately.

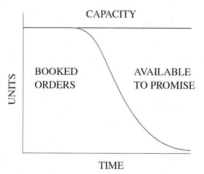

FIGURE 3.7 The MPS and delivery time.

Inventory on hand: 100 units

Period	1	2	3	4	5
Customer Orders	80	10	10		30
MPS		100		100	
ATP	20	80		70	

FIGURE 3.8 Available-to-promise calculation.

The ATP is calculated by adding **scheduled receipts** to the beginning inventory and then subtracting actual orders scheduled before the next scheduled receipt. A scheduled receipt is an order that has been issued either to manufacturing or to a supplier. Figure 3.8 illustrates a calculation of an ATP:

ATP for period 1 = on hand − customer orders due before next MPS
 = 100 − 80
 = 20 units
ATP for period 2 = MPS scheduled receipt − customer orders due before next MPS
 = 100 − (10 + 10)
 = 80 units
ATP for period 4 = 100 − 30 = 70 units

This method assumes that the ATP will be sold before the next scheduled receipt arrives. It is there to be sold, and the assumption is that it will be sold. If it is not sold, whatever is left forms an on-hand balance available for the next period.

Continuing with the example problem on page 49, Amalgamated Nut Crackers, Inc., has now received customer orders. Following is the schedule of orders received and the resulting ATP calculation (recall that there are 50 units in inventory):

Inventory on hand = 50 units

Week	1	2	3	4	5	6
Customer Orders	80	45	40	50	50	5
MPS	100		100		100	
ATP	25		10		45	

Sometimes, customer orders are greater than the scheduled receipts. In this case, the previous ATP is reduced by the amount needed. In this example, can the master planner accept an order for another 20 for delivery in week 3? Ten of the units are available from week 3, and 10 can be taken from the ATP in week 1, so the order can be accepted as shown in the following.

Inventory on hand = 50 units

Week	1	2	3	4	5	6
Customer Orders	80	45	60	50	50	5
MPS	100		100		100	
ATP	15		0		45	

Example Problem

Calculate the ATP for the following example. Can an order for 30 more be accepted for delivery in week 5? What will be the ATP if the order is accepted?

Week	1	2	3	4	5
Customer Orders	50	20	30	30	15
MPS	100		100		
ATP	30		25		

Answer

Week	1	2	3	4	5
Customer Orders	50	20	30	30	45
MPS	100		100		
ATP	25		0		

Projected Available Balance

Our calculations so far have based the projected available balance on the forecast demand. Now there are also customer orders to consider. Customer orders will sometimes be greater than forecast and sometimes less. Projected available balance is now calculated based on whichever is greater. For example, if the beginning projected available balance is 100 units, the forecast is 40 units, and customer orders are 50 units, the ending projected available balance is 50 units, not 60. The **projected available balance (PAB)** is calculated in one of two ways, depending on whether the period is before or after the **demand time fence**. The demand time fence is the number of periods, beginning with period 1, in which changes are not accepted due to excessive cost caused by schedule disruption.

For periods before the demand time fence, PAB is calculated as

PAB = prior period PAB or on-hand balance + MPS − customer orders

This process ignores the forecast and assumes that the only effect will be from the customer orders. Any new orders will have to be approved by senior management. For periods after the demand time fence, forecast will influence the PAB so it is calculated using either the forecast or customer orders, whichever is greater. Thus, the PAB becomes

PAB = prior period PAB or on-hand balance + MPS
  − greater of customer orders or forecast

Example Problem

Given the following data, calculate the PAB. The demand time fence is the end of week 3, the order quantity is 100, and 40 are available at the beginning of the period.

Week	1	2	3	4	5
Forecast	40	40	40	40	40
Customer Orders	39	42	39	33	23

ANSWER

Week		1	2	3	4	5
PAB	40	1	59	20	80	40
MPS			100		100	

So far we have considered how to calculate PAB and ATP. Using the Amalgamated Nut Cracker, Inc., example, we now combine the two calculations into one record. The demand time fence is at the end of three weeks.

Week		1	2	3	4	5	6
Forecast Demand		75	50	30	40	70	20
Customer Orders		80	45	40	50	50	5
PAB	50	70	25	85	35	65	45
ATP		25		10		45	
MPS		100		100		100	

Time Fences

Consider the product structure shown in Figure 3.9. Item A is a master-scheduled item and is assembled from B, C, and D. Item D, in turn, is made from raw material E. The lead times to make or to buy the parts are shown in parentheses. The lead time to assemble A is two weeks. To purchase B and C, the respective lead times are six and five weeks. To make D takes eight weeks, and the purchase lead time for raw material E is 16 weeks. The longest cumulative lead time is thus 26 weeks (A + D + E = 2 + 8 + 16 = 26 weeks).

Since the cumulative lead time is 26 weeks, the MPS must have a planning horizon of at least 26 weeks. A planning time fence of less than 26 weeks (say, for example, at 24 weeks) means that if a new MPS order is placed for week 24 (at the end of the planning horizon), any order for component E would already be two weeks late to make the MPS in week 24.

Suppose that E is a long-lead-time electronic component and is used in the assembly of other boards as well as D. When E is received 16 weeks after ordering, a decision must be made to commit E to be made into a D or to use it in another board. In eight weeks, a decision must be made to commit D to the final assembly of A. The company would not have to commit the E to making D until 10 weeks before delivery of the A. At each of

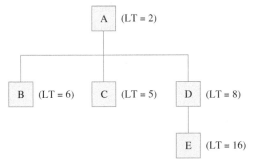

FIGURE 3.9 Product structure.

these stages, the company commits itself to more cost and fewer alternatives. Therefore, the cost of making a change increases and the company's flexibility decreases as production gets closer to the delivery time.

The establishment or use of a demand time fence is essentially a management decision. Depending on the nature of the product and the possible flexibility of the manufacturing process, management may set a time zone to essentially "freeze" the MPS. If they set that time zone (demand time fence), for example, at three weeks, then no additional customer orders or changes to the MPS should be made without managerial analysis and intervention. The advantage of using a demand time fence provides stability to the production process and will tend to minimize costs that might be associated with changing some aspect of an order so close to completion. The obvious disadvantage is a loss of flexibility and response during that time period. The costs and benefits should be evaluated prior to making the decision to use a demand time fence.

Changes to the master schedule will occur. For example,

- Customers cancel or change orders.
- Machines break down or new machines are added, changing capacity.
- Suppliers have problems and miss delivery dates.
- Processes create more scrap than expected.

A company wants to minimize the cost of manufacture and also be flexible enough to adapt to changing needs. Changes to production schedules can result in the following:

- Increased costs. Cost increases due to rerouting, rescheduling, extra setups, expediting, and buildup of work-in-process inventory.
- Decreased customer service. A change in quantity of delivery can disrupt the schedule of other orders.
- Reduced credibility. Loss of credibility for the MPS and the planning process.

Changes that are far off on the planning horizon can be made with little or no cost or disruption to manufacturing, but the nearer to delivery date, the more disruptive and costly changes will be. To help in the decision-making process, companies establish zones divided by time fences. Figure 3.10 shows how this concept might be applied to product A. The zones and time fences are as follows:

- **Frozen zone**. Capacity and materials are committed to specific orders. Since changes would result in excessive costs, reduced manufacturing efficiency, and poor customer service, senior management's approval is usually required to make changes. The extent of the frozen zone is defined by the demand time fence. Within the demand time fence, demand is usually based on customer orders, not forecast.
- **Slushy zone**. Capacity and material are committed to a less extent. This is an area for trade-offs that must be negotiated between marketing and manufacturing. In this zone, materials have been ordered and capacity established; these are difficult to change. However, changes in priorities are easier to change. The extent of the slushy zone is defined by the **planning time fence**. Within this time fence, the computer will generally not allow the automatic rescheduling of MPS orders. Any recommendation to change an MPS order in this zone will need to be evaluated before changes are made. Often, the minimum length of the planning time fence is determined by the cumulative lead time of the product.
- **Liquid zone**. Any change can be made to the MPS as long as it is within the limits set by the production plan. Changes are routine and are often made by the computer program without the need for input from the planner.

FIGURE 3.10 MPS and time fences.

Changes to the MPS will occur. They must be managed and decisions made with full knowledge of the costs involved.

Notice a fundamental difference between the development (and use) of these two common time fences. The planning time fence is really an essential number in that it provides a minimum time into the future that master planning needs to take place if major material shortages are to be prevented. The cumulative lead time is the lowest value for the planning time fence, but management may decide to extend it somewhat for increased visibility. The demand time fence, on the other hand, is not even required to be used and its development is somewhat subjective. Its use and time duration are established primarily on managerial preference based on the desire and ability to stabilize imminent production schedules. The trade-off for that stability is generally being less capable to respond to last-minute customer changes in the final design and/or quantity. Management must, therefore, decide on accepting somewhat decreased imminent responsiveness in order to stabilize production schedules, increase delivery reliabilities, and reduce production costs.

Error Management

Errors in customer orders occur all the time and require constant attention. Three general types of errors occur:

1. Wrong product or specification.
2. Wrong amount (too little or too much).
3. Wrong shipping date (too early or too late).

They require different responses, reengineering, alteration, negotiation of partial shipment, or expediting of shipment.

SUMMARY

The master schedule is a plan for the production of individual end items. It must match demand for the product in total, but it is not a forecast of demand. The master schedule must be realistic. It must be achievable and reflect a balance between required and available capacity. It is developed from the sales and operation plan (S&OP) and must essentially match the S&OP if the correct amount of the correct resources are available to meet the master schedule

The master schedule is the meeting ground for sales and production. It provides a plan from which realistic delivery promises can be made to customers. If adjustments have to be made in deliveries or the booking of orders, they are done through the master schedule. Since the S&OP plans primarily resources, the master schedule is the first and primary planning tool to address actual customer orders.

Master scheduling's major functions are to form a link between the production plan and the facility, as an input to plan capacity, as the major input to MRP, to help make order promises, keep priorities valid, and be a major link between production and sales.

The master schedule must be realistic and based on what production can and will do. If it is not, the results will be as follows:

- Overload or underload of plant resources.
- Unreliable schedules resulting in poor delivery performance.
- High levels of work-in-process (WIP) inventory.
- Poor customer service.
- Loss of credibility in the planning system.

KEY TERMS

Assemble-to-order (ATO) 51
Available-to-promise (ATP) 54
Configure-to-order 51
Demand time fence 56
Engineer-to-order (ETO) 52
Final assembly schedule (FAS) 52
Frozen zone 58
Liquid zone 58
Make-to-order (MTO) 51
Make-to-stock (MTS) 51

Master production schedule (MPS) 45
Master schedule 45
Master scheduling 47
Planning time fence 58
Postponement 53
Projected available balance (PAB) 56
Rough-cut capacity planning 49
Scheduled receipt 55
Slushy zone 58

QUESTIONS

1. What four functions does the master production schedule (MPS) perform in the production planning system?
2. What functions does the MPS perform between sales and production?
3. Does the MPS work with families of products or with individual items?
4. Where does the information come from to develop an MPS?
5. What are the three steps in making an MPS?
6. What is the purpose of a rough-cut capacity plan?
7. Where is the resource bill used?
8. At what level should master production scheduling take place?
   a. In a make-to-stock environment?
   b. In a make-to-order environment?
   c. In an assemble-to-order environment?
9. What is a final assembly schedule (FAS)? What is its purpose?
10. What is a planning horizon? What decides its minimum time? Why would it be longer?
11. How do the production plan and the MPS relate to sales and to the sales forecast?
12. What is the ATP (available-to-promise)? How is it calculated?
13. What is the purpose of time fences? Name and describe the three main divisions.
14. What would happen if the planning horizon for the master schedule were too short? Why?
15. What potential problem might arise if time fences are not used? Why?
16. What types of production environments might use both the FAS and the MPS? Why?
17. What is the primary purpose of the demand time fence? How does the demand time fence impact the way that forecast and customer demand numbers are treated?

PROBLEMS

3.1. The Wicked Witch Company manufactures a line of broomsticks. The most popular is the 36-inch model, and the sales department has prepared a forecast for six weeks. The opening inventory is 30. As master scheduler, you must prepare an MPS. The brooms are manufactured in lots of 100.

Week		1	2	3	4	5	6
Forecast Sales		10	50	25	50	10	15
PAB	30						
MPS							

Answer. There should be scheduled receipts in weeks 2 and 4.

3.2. The Shades Sunglass Company assembles sunglasses from frames, which it makes, and lenses, which it purchases from an outside supplier. The sales department has prepared the following six-week forecast for Ebony, a popular model. The sunglasses are assembled in lots of 200, and the opening inventory is 250 pairs. Complete the PAB and the MPS.

Week		1	2	3	4	5	6
Forecast Sales		200	300	300	200	150	150
PAB	250						
MPS							

3.3. The Amalgamated Mailbox Company manufactures a family of two mailboxes. The production plan and the MPS are developed on a quarterly basis. The forecast for the product group follows. The opening inventory is 270 units, and the company wants to reduce this to 150 units at the end of the year. Develop a level production plan.

Production Plan

Quarter		1	2	3	4	Total
Forecast Sales		220	300	200	200	
PAB	270					
Production Plan						

Answer. Quarterly production = 200 units.

The forecast sales for each of the mailboxes in the family also follow. Develop an MPS for each item, bearing in mind that production is to be leveled as in the production plan. For each mailbox, the lot size is 200.

Mailbox A. Lot size: 200

Quarter		1	2	3	4	Total
Forecast Sales		120	180	100	120	
PAB	120					
MPS						

Answer. Scheduled receipts in quarters 2 and 3.

Mailbox B. Lot size: 200

Quarter		1	2	3	4	Total
Forecast Sales		100	120	100	80	
PAB	150					
MPS						

Answer. Scheduled receipts in quarters 1 and 4.

3.4. Worldwide Can-Openers, Inc., makes a family of two hand-operated can openers. The production plan is based on months. There are four weeks in this month. Opening inventory is 2000 dozen, and it is planned to increase that to 4000 dozen by the end of the month. The MPS is made using weekly periods. The forecast and PAB for the two models follow. The lot size for both models is 1000 dozen. Calculate the production plan and the MPS for each item.

Production Plan

Week		1	2	3	4	Total
Forecast		3000	3500	3500	4000	
PAB	2000					
Production Plan						

Model A

Week		1	2	3	4	Total
Forecast		2000	2500	2000	2000	
PAB	1500					
MPS						

Model B

Week		1	2	3	4	Total
Forecast		1000	1000	1500	2000	
PAB	500					
MPS						

3.5. In the example given on page 47 earlier in the chapter, the MPS was unsatisfactory because there were poor inventory balances compared to the production plan. There was also a stockout for product C in periods 2 and 3. Revise the production plans for the three products to cut out or reduce these problems.

3.6. The Acme Widget Company makes widgets in two models, and the bottleneck operation is in work center 10. Following is the resource bill (in hours per part).

	Hours per Part	
Work Center	Model A	Model B
10	2.7	3.2

The MPS for the next five weeks is

Week	1	2	3	4	5
Model A	60	50	40	70	9
Model B	18	35	52	30	50

a. Using the resource bill and the MPS, calculate the number of hours required in work center 10 for each of the five weeks. Use the following table to record the required capacity on the work center.

Week	1	2	3	4	5
Model A					
Model B					
Total Hours					

Answer. The total hours required are as follows: week 1, 219.6; week 2, 247; week 3, 274.4; week 4, 285; and week 5, 189.3.

b. If the available capacity at work station 10 is 250 hours per week, suggest possible ways of meeting the demand in week 3.

3.7. Calculate the ATP using the following data. There are 90 units on hand.

Week	1	2	3	4	5	6
Customer Orders	70	70	20	40	10	
MPS		100		100		100
ATP						

Answer. ATP in week 1, 30; week 2, 10; week 4, 50; and week 6, 100.

3.8. Given the following data, calculate how many units are available-to-promise. There are 5 units on hand.

Week	1	2	3	4	5	6
Customer Orders	15	3	17	11		3
MPS	30		30	30		
ATP						

3.9. Using the following data, calculate the ATP. There are zero units on hand.

Week	1	2	3	4	5	6	7	8	9	10
Customer Orders	10		10		60	14			15	
MPS	50				50				50	
ATP										

3.10. Using the following data, calculate the ATP. There are 45 units on hand.

Week	1	2	3	4	5	6	7	8
Customer Orders	45	50	30	40	25	40	40	18
MPS		100		100		100		100
ATP								

3.11. Calculate the ATP using the following data. There are 40 units on hand.

Week	1	2	3	4	5	6
Customer Orders	20	50	35	30	50	30
MPS		100			100	
ATP						

3.12. Given the following data, can an order for 20 for delivery in week 4 be accepted? Calculate the ATP using the following table. On hand = 50 units.

Week	1	2	3	4	5	6	7	8
Customer Orders	50	50	30	40	50	40	30	15
MPS		100		100		100		100
ATP								

Answer. Yes. Ten can come from the ATP for week 4 and 10 from the ATP for week 2.

3.13. Given the following data, can an order for 40 more units for delivery in week 5 be accepted? If not, what do you suggest can be done? There are zero units on hand. Would it be possible to increase the MPS values or add new MPS to accommodate new orders? What information would you need to make that decision?

Week	1	2	3	4	5	6	7	8
Customer Orders	70	20	40	50	10	15	20	20
MPS	100		100			100		
ATP								

3.14. Given the following data, calculate the PAB and the planned MPS receipts. The lot size is 200. The demand time fence is two weeks.

Week	1	2	3	4
Forecast	80	80	80	75
Customer Orders	100	85	50	45
PAB	140			
MPS				

Answer. There is a planned MPS receipt in week 2.

3.15. Given the following data, calculate the PAB and the planned MPS receipts. The lot size is 100. The demand time fence is two weeks.

Week	1	2	3	4
Forecast	50	50	50	50
Customer Orders	60	25	70	15
PAB	60			
MPS				

3.16. Complete the following problem. The lead time is one week and the demand time fence is the end of week 3. There are 20 on hand. The lot size is 60.

Period	1	2	3	4	5	6
Forecast	20	21	22	20	28	25
Customer Orders	17	16	23	19	31	22
PAB	20					
MPS						
ATP						

3.17. Product A is an assemble-to-order product. It has a lot size of 150, and currently has an on-hand inventory of 110 units. There is a two-week demand time fence and a 12-week planning time fence. The following table gives the original forecast and the actual customer orders for the next 12 weeks:

Week	1	2	3	4	5	6	7	8	9	10	11	12
Forecast	80	80	80	70	70	70	70	70	70	70	70	70
Demand	83	78	65	61	49	51	34	17	11	7	0	0

a. Given this information, develop a realistic master schedule, complete with ATP logic.

b. Tell how you would respond to each of the following customer order requests. Assume these are independent requests, and do not have cumulative effects.
- units in week 3
- units in week 5
- units in week 7

3.18. a. Given the following master schedule, fill in the projected available and ATP rows:

On hand: 35	Planning time fence: 10	Lot size: 200
	Demand time fence: 2	

Period	1	2	3	4	5	6	7	8	9	10	11	12
Forecast	30	40	40	50	40	40	50	40	40	30	40	40
Customer Orders	35	34	29	20	17	14	31	10	5	3	3	2
Projected Available												
ATP												
MPS		200			200							

b. A customer wants an order of 100 in period 4. What can you tell them?

c. The customer from part (b) cancels the request, but then says they want 120 in period 5. What do you tell them now?

d. Sales has requested that you add an MPS of 200 in period 9 to cover their needs for a sales promotion. What do you tell them and why?

e. What action (if any) should be taken in period 11? Why is it okay to take the action?

CASE STUDY 3.1

Acme Water Pumps

The Acme Water Pump company has a problem. The pumps are fairly expensive to make and store, so the company tends to keep the inventory low. At the same time, it is important to respond to demands quickly, since a customer who wants a water pump is very likely to get one from a competitor if Acme doesn't have one available immediately. Acme's current policy to produce pumps is to produce 100 per week, which is the average demand. Even this is a problem, as the production manager has pointed out, since the equipment is also used for other products and the lot size of 300 would be much more efficient. She said she is currently set up for water pump production for the next week and states that she has capacity available to produce 300 at a time next week.

The following lists the forecasts and actual customer orders for the next 12 weeks

Week	1	2	3	4	5	6	7	8	9	10	11	12
Forecast	90	120	110	80	85	95	100	110	90	90	100	110
Customer Orders	105	97	93	72	98	72	53	21	17	6	2	5

The president of Acme has said that he wants to consider using a formal MPS with ATP logic to try to meet demand more effectively without a large impact on inventory. Acme has decided to use a demand time fence at the end of week 3 and has also found out that its current inventory is 25 units. Assume Acme will use the MPS lot size of 300 and that it will produce the first of those lots in week 1.

Assignment

1. Develop a master schedule using the information above.
2. A customer has just requested a major order of 45 pumps for delivery in week 5. What would you tell the customer about having such an order? Why? What, if anything, would such an order do to the operation?

CASE STUDY 3.2

The MasterChip Electronics Company

Sally Jackson, production manager of the MasterChip Electronics Company, was having another frustrating day. The final assembly area was woefully behind schedule, and several large orders were several days, and some several weeks, behind the promised delivery date. Customers were not happy and were giving lots of angry messages to the sales force. At the same time, some of the work areas in the early portions of the production process apparently did not have enough work. Sally viewed this as an equally important issue, since she could think of only two possible solutions—either let the people stand around and do nothing or have them work ahead on some of the components even though no order existed for those components. Working ahead was risky because their products competed in a market where customers could demand a lot of options for a basic product, and some of those options had highly variable demand (one option, for example, could go for months with no demand and then all at once have a very large demand as one customer ordered a large number of a product with that option). That was not likely to change since most of their customers were large retail chain stores. Letting people stand around was also bad, since she was evaluated on labor efficiency and utilization, and a worker not working would make those numbers look very bad.

She would like to be able to send some of the workers home for a day or part of a day, but the local union agreement prohibited that. She also liked to think about the possibility of using some of those workers to help out in another area (final assembly, in this current situation), but the union agreement also had specific work classifications for each worker, and those could not be violated. Even if that were possible, she knew it could be a problem since most of the production workers in the area with little work knew almost nothing about how the final assembly area worked, and that could generate lots of quality problems.

Sally made a note to herself to develop some specific numbers for her weekly meeting with the human resources manager. Every week she looked at the demand for each area and put together a set of recommendations for laying off some workers in one area and calling back some workers for another area. She knew that was allowed, on a week by week basis, under the union contract, but she still hated that task. Even though she could usually come up with some good numbers, she could not neglect the following impacts:

- These workers often were the sole source of income for their families, and even a week of layoff would likely imply hardships on their families.
- The longer a worker was not working, their skills were not allowed to remain at a high level of effectiveness. When they returned, they typically would not be able to work as efficiently as before, and also represented the potential for a larger number of quality problems.
- Even if they remained effective (if, for example, they had only been gone for a week), it was highly likely they would be resentful of the layoff, and why should they feel loyalty to the company when the company had not been loyal to them? The feelings of resentment might make them less efficient on purpose.
- Many of their best workers had skills that were in demand by several other companies. Why should a highly skilled worker with those skills in demand put up with those occasional layoffs when they had other choices? Just in the last few months, she had lost more than 10 of her best workers by having them go to work for one of the competitors of MasterChip.

Just as she was starting to work on the numbers for her meeting with the human resources manager, Keshawn Morgan (the sales manager) came into her office. The conversation went like this:

> KESHAWN: "Sally, I've got some good news and some bad news for you. First, the good news: I just got off the phone with the buyer for Ajax Department Stores. They want a very large order of over 1000 of the A77 product. They have some sort of promotion in the works and that product is to be featured."

SALLY: "When did you promise them that we would have the order done?"

KESHAWN: "I gave them our standard lead time for the product, six weeks."

SALLY: "That's going to be a problem for us. The A77 uses a power supply that is somewhat expensive, so we have only about 200 in stock. It generally takes us 8–10 weeks to get those in from our supplier. I suppose we could expedite a shipment, but that supplier would demand a much higher price since it disrupts their own operation so much to do an expedite. It might cost us enough extra to almost eliminate any profit on the order for us."

KESHAWN: "Why don't you people keep enough inventory—you know ours is a competitive business and we have to be responsive to our customers? If we can't make this order in six weeks, we are messing with a planned promotion from a major retail chain, and they won't be at all pleased. I wouldn't be a bit surprised if they started buying from one of our competitors. That point brings me to the bad news: I'm getting lots of angry phone calls about those orders you have behind schedule in final assembly. Remember, the customers of our customers tend to walk out of a store that doesn't have a product they want and go to a different store. Our customers are very sensitive to having their orders shipped on time. Can't your production people get your act together?"

SALLY: "You should know that we can't keep a lot of inventory sitting around. It is expensive to hold, since electronics are easily subject to being damaged in storage, and as the technology changes so fast it also may become obsolete before we can even use it. Top management would not like it too well if our inventory expense kills all our possible profit. Also, you taking an order like this without checking first if we can do it, is pretty dangerous. It's that kind of thing that causes the problems we have."

KESHAWN: "Sally, that's just silly. I have a customer on the phone that wants to spend a lot of money with us for a big order. How do you think it would sound if I told them to wait while I get permission from someone else to take the order? We can't mess around like that in sales; we need to work hard to get orders, and we did quote the standard lead time we give all our customers for that A77 product. You people have to work better. We can do our job to sell it, why can't you do your job to make it?"

All Sally could do after that conversation was to search for a pain killer for her newly developed headache, knowing she had to deal with that before she started to think on how she should deal with the problems she had in addition to the new one that was just handed to her by Keshawn.

Assignment

What are the key issues in this case? Be sure to classify them as much as possible as symptoms versus core causes. Be sure to keep in mind the constraints as defined by the type of customer and the internal conditions. Once you have analyzed and classified the issues, develop a comprehensive solution for MasterChip that can deal more effectively with their situation.

CASE STUDY 3.3

Macarry's Bicycle Company

Macarry's Bicycle Company makes and sells high-quality bicycles, primarily to larger North American bicycle retail outlets and to some wholesalers for smaller retail shops. They have several models, and most of those models have a fairly large number of options that can be mixed for a very large number of possible designs. The bicycles, for example, can be made in a number of colors, type of seats, number of speed settings (gears), type of tires and wheels, type of brakes, and handlebar styles. In addition, there are several options that can be included or not, including headlights and taillights, water bottle carriers, baskets, or kickstands.

In such an environment, it is clearly difficult to know how much inventory to carry or produce for each of the options. Several years ago the company decided they could not establish a master schedule for each combination of all the options. That would literally imply creating thousands of master schedules (one for each type of bicycle that is possible to make), and some of those combinations might, in fact, never be ordered. Instead they decided to make a master schedule for each of the options for a bicycle model and another one for the common parts for the model (e.g., a particular model has only one frame, and most of the connectors are common, such as nuts and bolts). The common part forecast was based on the total number of bicycles of a model type they planned to sell in a given period, and that allowed them to calculate a forecast for each of the options based on the historical percentage of the model sales that requested that option. Using this approach a final bicycle would never be produced except to a specific order from a specific customer.

The cumulative lead time to obtain or make all the parts for the bicycle was 20 weeks, so that is what the company used as their planning time fence. This is important to know since in many parts of the country the sales of bicycles were very seasonal. Bicycle shops in the North sold very few in winter, but in spring the demand was very high. The bicycle shops did not like to keep a lot of inventory of finished bicycles because of the cost and the fact that they did not know from year to year which type of options might be popular. They tended to wait as long as possible to place an order, but then were very sensitive that the order would be delivered in a timely manner. When a customer wanted to buy a bicycle they did not want to wait, especially since the season was short in some parts of the country.

The following charts show the forecast for one bicycle model (a hybrid heavy-duty bicycle), existing confirmed orders from customers, existing inventory, and master production schedule quantities of the common parts and a few of the options for the next 12 weeks. The forecasts for options are computed as follows: The 18-speed gear option was historically selected for this model bicycle 70% of the time. Since the forecast for this model of bicycle for the first week was 50, the forecast for the gear options could be calculated as 35 (70% of 50). The historical percentage of demand for the straight handlebars was 30%, and historically 20% of the orders included the head and tail light set. To understand the orders, for example, the first week there were orders for 56 of this bicycle model—37 of those orders wanted the 18-speed option, 16 of those orders wanted the straight handlebars, and 2 of those order wanted the light set. This data was taken from the late winter/early spring time frame, when the demand for the bicycles was starting to grow as bicycle shops started to prepare for their heavy sales period.

Common parts (Frame, etc.) Existing inventory 40

Week	1	2	3	4	5	6	7	8	9	10	11	12
Forecast	50	55	60	62	65	65	68	70	75	75	80	85
Cust. Orders	56	52	45	33	70	50	35	60	20	20	0	0
MPS	200				200			200		200		

18-speed gear option Existing inventory 25

Week	1	2	3	4	5	6	7	8	9	10	11	12
Forecast	35	39	42	44	46	46	48	49	53	53	56	60
Cust. Orders	37	38	40	33	50	20	25	40	5	5	0	0
MPS	150				150			150			150	

Straight handlebars Existing inventory 20

Week	1	2	3	4	5	6	7	8	9	10	11	12
Forecast	15	17	18	19	20	20	21	21	23	23	24	26
Cust. Orders	16	18	20	5	15	22	15	20	5	8	0	0
MPS		60			60			60		60		

Head and tail light set Existing inventory 5

Week	1	2	3	4	5	6	7	8	9	10	11	12
Forecast	10	11	12	13	13	13	14	14	15	15	16	17
Cust. Orders	2	12	10	8	15	9	7	11	2	1	0	0
MPS	30			30		30		30		30		30

Assignment

1. Fill in the master schedules listed below, taking the data from the above tables and adding row values for projected inventory and available-to-promise (ATP). Assume there is no demand time fence for this data.
2. Once you have completed the tables, examine the list of orders for this model bicycle and determine specifically what information should be given to the perspective customer. For example, if the order request was for 40 bicycles (with defined options) in week 4 and it appears that only 32 could be delivered, then you should be telling the customer that they can have only 32 in week 4 and then the rest could be delivered at a later week (you should be specific as to WHICH week). Assume the orders listed for evaluation are NOT cumulative—in other words, evaluate each requested order independently ignoring the existence of the other requested orders for the evaluation of this one order.
3. Suppose Macarry's Bicycle managers discover that a major competitor has had to shut down their production for the next three months due to a major fire. The Macarry's managers fully expect that many of the competitor's customers will turn to Macarry's Bicycles to fill their orders during this critical time for them. In fact, one of the competitor's customers has already asked about an order of 250 of the models for delivery in week 5. What actions should Macarry's take in this case? Be as specific as possible.

Common parts (Frame, etc.) Existing inventory 40

Week	1	2	3	4	5	6	7	8	9	10	11	12
Forecast	50	55	60	62	65	65	68	70	75	75	80	85
Cust. Orders	56	52	45	33	70	50	35	60	20	20	0	0
Projected Inven.												
MPS	200			200			200		200			
ATP												

18-speed gear option Existing inventory 25

Week	1	2	3	4	5	6	7	8	9	10	11	12
Forecast	35	39	42	44	46	46	48	49	53	53	56	60
Cust. Orders	37	38	40	33	50	20	25	40	5	5	0	0
Projected Inven.												
MPS	150				150			150			150	
ATP												

Straight handlebars Existing inventory 20

Week	1	2	3	4	5	6	7	8	9	10	11	12
Forecast	15	17	18	19	20	20	21	21	23	23	24	26
Cust. Orders	16	18	20	5	15	22	15	20	5	8	0	0
Projected Inven.												
MPS		60			60			60		60		
ATP												

Head and tail light set Existing inventory 5

Week	1	2	3	4	5	6	7	8	9	10	11	12
Forecast	10	11	12	13	13	13	14	14	15	15	16	17
Cust. Orders	2	12	10	8	15	9	7	11	2	1	0	0
Projected Inven.												
MPS	30		30		30		30		30		30	
ATP												

Here are the orders to evaluate. Again you are reminded to treat these independently. For example, when you evaluate order number 2, ignore the other order requests (1, 3, and 4), and so forth.

  a. A customer is asking about an order of 32 of the bicycles for week 3. All 32 are to be 18 speed, 12 are to have straight handlebars, and 14 are to have the light set.

  b. A customer is asking about an order of 60 of the bicycles for week 6. Fifty of them are to be 18 speed, 12 are to have straight handlebars, and 5 to have the light set.

  c. A customer is asking about an order of 20 of the bicycles for week 2. All are to be 18 speed, all are to have straight handlebars, and all are to have the light set.

  d. A customer is asking about an order of 110 of the bicycles in week 7. Sixty are to be 18 speed, 22 are to be straight handlebars, and 15 are to have the light set.

CHAPTER FOUR

MATERIAL REQUIREMENTS PLANNING

4.1 INTRODUCTION

Chapter 3 described the role of the master production schedule (MPS) in showing the end items, or major components, that manufacturing intends to build. These items are made or assembled from components that must be available in the right quantities and at the right time to meet the MPS requirements. If any component is missing, the product cannot be built and shipped on time. **Material requirements planning (MRP)** is the system used to avoid missing parts. It establishes a schedule (priority plan) showing the components required at each level of the assembly and, based on lead times, calculates the time when these components will be needed.

This chapter will describe bills of material (the major building block of MRP), detail the MRP process, and explain how the material requirements plan is used. But first, some details about the environment in which MRP operates will be discussed.

Nature of Demand

There are two types of demand: independent and dependent. **Independent demand** is usually not related to the demand for any other product. For example, if a company makes wooden tables, the demand for the tables is independent—it is essentially independent of any actions taken internally in the company. Instead, it is dependent only on the external demand for the table. MPS items (tables) are independent demand items. The demand for the sides, ends, legs, and tops to make the table *depends* on the demand for the finished tables, and these are **dependent demand** items.

Figure 4.1 depicts a product tree that shows the relationship between independent and dependent demand items. The figures in parentheses show the required quantities of each component.

Since independent demand is not related to the demand for any other assemblies or products, it must be forecast, unless actual external demand in known. However, since dependent demand is directly related to the demand for higher-level assemblies or products, it can be calculated. MRP is designed to do this calculation.

An item can have both a dependent and an independent demand. A service or replacement part has both. A manufacturer of vacuum cleaners uses flexible hose in the assembly of the units. In the assembly of the vacuums, the hose is a dependent demand item. However, the hose has a nasty habit of breaking, and the manufacturer must have

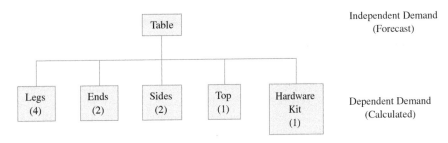

FIGURE 4.1 Product tree.

replacement hoses available. Demand for replacement hoses is independent since demand for them does not depend directly upon the number of vacuums manufactured but instead on external demand from a customer needing a replacement hose.

Dependency can be horizontal or vertical. The dependency of a component on its parent is vertical. However, components also depend on each other. If one component is going to be a week late, then the final assembly is a week late. The other components are not needed until later. This is also a dependency and is called horizontal dependency. Planners are concerned with horizontal dependency when a part is delayed or there is a shortage, for then other parts will have to be rescheduled.

Objectives of MRP

Material requirements planning has two major objectives: determine requirements and keep priorities current.

Determine requirements The main objective of any manufacturing planning and control system is to have the right materials in the right quantities available at the right time to meet the demand for the firm's products. The material requirements plan's objective is to determine what components are needed to meet the MPS and, based on lead time, to calculate the periods when the components must be available. It must determine the following:

- What to order.
- How much to order.
- When to order.
- When to schedule delivery.

Keep priorities current The demand for, and supply of, components change daily. Customers enter or change orders. Components get used up, suppliers are late with delivery, scrap occurs, orders are completed, and machines break down. In this ever-changing world, a material requirements plan must be able to reorganize priorities to keep plans current. It must be able to add and delete, expedite, delay, and change orders.

Linkages to Other Manufacturing Planning and Control Functions

The MPS drives the material requirements plan. The material requirements plan is a priority plan for the components needed to make the products in the MPS. The plan is valid only if the proper type and quantity of capacity is available when needed to make the components, and the plan must be checked against available capacity. The process of doing so is called capacity requirements planning and is discussed in Chapter 5.

MRP drives, or is an input to, production activity control (PAC) and purchasing. MRP plans the release and receipt dates for orders. PAC and purchasing must plan and control the performance of the orders to meet the due dates.

Figure 4.2 shows a diagram of the production planning and control system with its inputs and outputs.

MRP Software

If a company makes a few simple products, it might be possible to perform MRP manually. However, most companies need to keep track of thousands of components in a world of changing demand, supply, and capacity.

In the days before computers, it was necessary to maintain extensive manual systems and to have large inventories and long lead times. These were needed as a cushion due to the lack of accurate, up-to-date information and the inability to perform the necessary calculations quickly. Somehow, someone in the organization figured out what was required sooner or, very often, later than needed. "Get it early and get lots of it" was a good rule then.

Computers are incredibly fast, accurate, and ideally suited for the job at hand. With their ability to store and manipulate data and produce information rapidly, manufacturing now has a tool to use modern manufacturing planning and control systems properly.

Material Requirements Planning 73

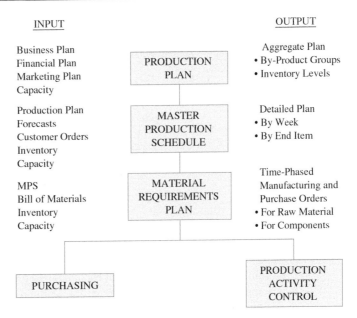

FIGURE 4.2 Production planning and control system.

There are many MRP software programs available, and while they may have some different looks, the processing logic is well-defined and tends to be the same for each of the different available programs.

Inputs to the MRP System

There are three primary inputs to MRP systems:

1. Master production schedule.
2. Inventory records.
3. Bills of material.

Master production schedule. The master production schedule is a statement of which end items are to be produced, the quantity of each, and the dates they are to be completed. It drives the MRP system by providing the initial input for the items needed.

Inventory records. A major input to the MRP system is inventory. When a calculation is made to find out how many are needed, the quantities available must be considered.

There are two kinds of information needed. The first is called **planning factors** and includes information such as order quantities, lead times, safety stock, and scrap. This information does not change often; however, it is needed to plan what quantities to order and when to order for timely deliveries.

The second kind of information necessary is the status of each item. The MRP system needs to know how much is available, how much is allocated, and how much is available for future demand. This type of information is dynamic and changes with every transaction that takes place.

This data is maintained in an **inventory record**, also called a part master or item master. Each item has a record, and all the records together form a file or table.

Bills of material. The bill of material is one of the most important documents in a manufacturing company. It is discussed next.

4.2 BILLS OF MATERIAL

Before something can be made, the components needed to make it must be known. To bake a cake, a recipe is needed. To mix chemicals together, a formula is needed. To assemble a wheelbarrow, a parts list is needed. Even though the names are different, recipes, formulas, and parts lists tell what is needed to make the end product. All of these are bills of material.

CHAPTER FOUR

Description: TABLE		
Part Number: 100		
Part Number	Description	Quantity Required
203	Wooden Leg	4
411	Wooden Ends	2
622	Wooden Sides	2
023	Table Top	1
722	Hardware Kit	1

FIGURE 4.3 Simplified bill of material.

The *APICS Dictionary*, 16th edition, defines a **bill of material** as "a listing of all the subassemblies, intermediates, parts, and raw materials that go into making the parent assembly showing the quantities of each required to make an assembly." Figure 4.3 shows a simplified bill of material. There are three important points around bills of material and part numbers:

1. The bill of material shows all the parts required to make *one* of the item.
2. Each part or item has one and only one part number. A specific number is unique to one part and is not assigned to any other part. Thus, if a particular number appears on two different bills of material, the part so identified is the same.
3. A part is defined by its form, fit, or function. If any of these change, then it is not the same part and it must have a different part number. For example, a part when painted becomes a different part and must have a different number. If the part could be painted in three different colors, then each must be identified with its unique number.

The bill of material shows the components that go into making the parent. It does not show the steps or process used to make the parent or the components. That information is recorded in a routing. This is discussed in Chapters 5 and 6.

Bills of Material Structure

Bills of material structure refer to the overall design for the arrangement of bills of material files. Different departments in a company use bills of material for a variety of purposes. Although each user has individual preferences for the way the bill should be structured, there must be only one structure, and it should be designed to satisfy most needs. However, there can be several formats, or ways, to present the bill. Following are some possible formats for bills.

Product tree Figure 4.4 shows a product tree for the bill of material shown in Figure 4.3. The **product tree** is a convenient way to think about bills of material, but it is seldom used except for teaching and testing. In this text, it is used for that purpose.

Parent–component relationship The product tree and the bill of material shown in Figures 4.1 and 4.3 are called single-level structures. An assembly is considered a **parent item**, and the items that comprise it are called its **components**. Figure 4.4 shows

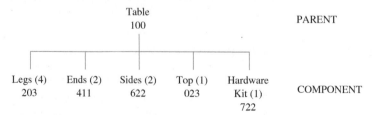

FIGURE 4.4 Product tree.

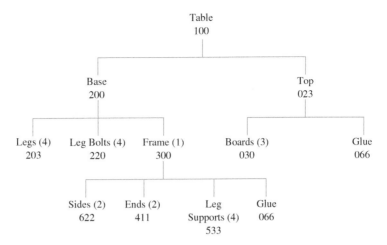

FIGURE 4.5 Multilevel bill.

the parent–component relationship of the table (P/N 100). Unique part numbers have also been assigned to each part. This makes identification of the part absolute.

Multilevel bill Figure 4.5 shows the same product as the single-level bill shown in Figures 4.3 and 4.4. However, the single-level components have been expanded into their components.

Multilevel bills of material are formed as logical groupings of parts into subassemblies based on the way the product is assembled. For example, a frame, chassis, doors, windows, and engine are required to construct an automobile. Each of these forms a logical group of components and parts and, in turn, has its own bill of material.

It is the responsibility of manufacturing engineering to decide how the product is to be made: the operations to be performed, their sequence, and their grouping. The subassemblies created are the result of this. Manufacturing has decided to assemble the sides, ends, and leg supports (part of the hardware kit) of the table (P/N 100) in Figure 4.4 into a frame (P/N 300). The legs, leg bolts, and frame subassembly are to be assembled into the base (P/N 200). The top (P/N 023) is to be made from three boards glued together. Note that the original parts are all there, but they have been grouped into subassemblies and each subassembly has its own part number.

One convention used with multilevel bills of material is that the last items on the tree (legs, leg bolts, ends, sides, glue, and boards) are all purchased items. Generally, a bill of material is not complete until all branches of the product structure tree end in a purchased part or a raw material.

Each level in the bill of material is assigned a number starting from the top and working down. The top level, or end product level, is level zero, and its components are at level one.

Multiple bill A **multiple bill** is used when companies usually make more than one product, and the same components are often used in several products. This is particularly true with families of products. Using our example of a table, the company makes two models. They are similar except the tops are different. Figure 4.6 shows the two bills of material. Because the boards used in the top are different, each top has a different part number. The balance of the components are common to both tables.

Single-level bill A **single-level bill of material** contains only the parent and its immediate components, which is why it is called a single-level bill. The tables shown in Figure 4.6 have six single-level bills, and these are shown in Figure 4.7. Note that many components are common to both tables.

The computer stores information describing the product structure as a single-level bill. A series of single-level bills is needed to completely define a product. For example, the table needs four single-level bills, one each for the table, base, top, and frame. These

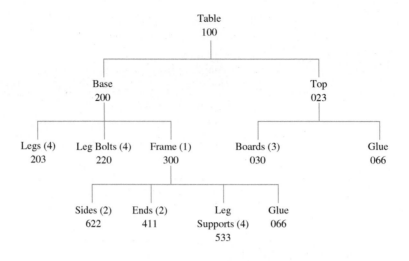
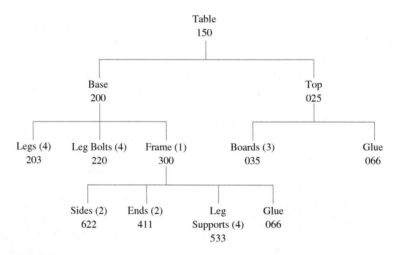

FIGURE 4.6 Multiple bills.

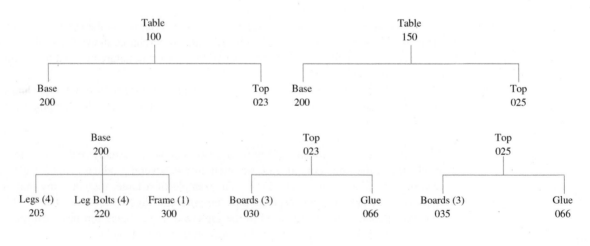
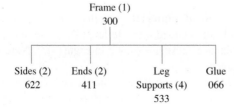

FIGURE 4.7 Single-level bills.

can be chained together to form a multilevel, or indented, bill. Using this method, the information has to be stored only once. For example, the frame (P/N 300) might be used on other tables with different legs or tops.

There are several advantages to using single-level bills, including the following:

- Duplication of records is avoided. For instance, base 200 is used in both table 100 and table 150. Rather than have two records of base 200, one in the bill for table 100 and one in the bill for table 150, only one record need be kept.
- The number of records and the file size are reduced by avoiding duplication of records.
- Maintaining bills of material is simplified. For example, if there is a change in base 200, the change needs be made in only one place.

Example Problem

Using the following product tree, construct the appropriate single-level trees. How many Ks are needed to make 100 Xs and 50 Ys?

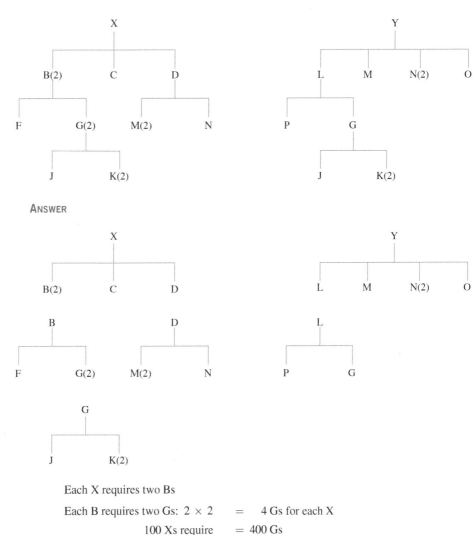

ANSWER

Each X requires two Bs

Each B requires two Gs: 2×2 = 4 Gs for each X

100 Xs require = 400 Gs

Each Y requires one L

Each L requires one G: 1×1 = 1 G

50 Ys require 50 Gs

Total Gs required = 450

Each G requires two Ks

Total Ks required $2 \times 450 = 900$

MANUFACTURING BILL OF MATERIAL
TABLE P/N 100

Part Number	Description	Quantity Required
200	Base	1
203	Legs	4
220	Leg Bolts	4
300	Frame	1
622	Sides	2
411	Ends	2
533	Leg Supports	4
066	Glue	
023	Top	1
030	Boards	3
066	Glue	

FIGURE 4.8 Indented bill of material.

Indented bill A multilevel bill of material can also be shown as an **indented bill of material**. This bill uses indentations as a way of identifying parents from components. Figure 4.8 shows an indented bill for the table in Figure 4.5.

The components of the parent table are listed flush left, and their components are indented. The components of the base (legs, leg bolts, and frame) are indented immediately below their parent. The components of the frame are further indented immediately below their parent. The components of the top are further indented immediately below their parent. Thus, the components are linked to their parents by indenting them as subentries and by listing them immediately below the parents.

Summarized bill of material The bill of material shown in Figure 4.3 is called a **summarized bill of material**. It lists all the parts needed to make one complete assembly. The parts list is produced by the product design engineer and does not contain any information about the way the product is made or assembled.

Planning bill A major use of bills of material is to plan production. **Planning bills** are an artificial grouping of components for planning purposes. They are used to simplify forecasting, master production scheduling, and MRP. They do not represent *buildable* products but an *average* product. Using the table example, suppose the company manufactured tables with three different leg styles, three different sides and ends, and three different tops. In total, they are making 27 ($3 \times 3 \times 3$) different tables, each with its own bill of material. For planning purposes, the 27 bills can be simplified by showing the percentage split for each type of component on one bill. Figure 4.9 shows how the product structure would look. The percentage usage of components is obtained from a forecast or past usage. Note that the percentage for each category of component adds up to 100%.

Since the percentages identified in the planning bill represent averages, in any given period the actual demand for the different options may differ somewhat from the average. Sometimes planners then artificially inflate the percentage values slightly from the actual average as a protection (hedge) against a period of higher-than-average demand. In those cases, the total for all the percentages will usually add up to more than 100%. This option overplanning is sometimes called a mix hedge, and often the actual demand information will reach the manufacturing facility before they have to commit to making the planned amount. Using a demand time fence will help prevent actual production of any option that is over planned based on forecast.

Where-Used and Pegging Reports

Where-used list A listing of all the parents in which a component is used is called a **where-used list**. Where-used lists give the same information as bills of material, but designate the parents for a component whereas the bill gives the components for a parent.

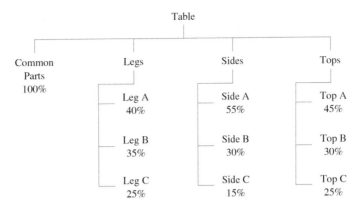

FIGURE 4.9 Planning bill.

A component may be used in making several parents. Wheels on an automobile, for example, might be used on several models of cars. This has several uses, such as in implementing an engineering change, or when materials are scarce, or in costing a product.

Pegging report A **pegging** report is similar to a where-used list. However, the pegging report shows only those parents for which there is an existing demand requirement, whereas the where-used report shows all parents for a component. The pegging report shows the parents creating the demand for the components, the quantities needed, and when they are needed. Pegging keeps track of the origin of the demand. Figure 4.10 shows an example of a product tree in which part C is used twice and a pegging report.

Uses for Bills of Material

The bill of material is one of the most widely used documents in a manufacturing company. Some major uses are as follows:

- **Product definition.** The bill specifies the components needed to make the product.
- **Engineering change control.** Product design engineers sometimes change the design of a product and the components used. These changes must be recorded and controlled. The bill provides the method for doing so.
- **Service parts.** Replacement parts needed to repair a broken component are determined from the bill of material.
- **Planning.** Bills of material define what materials have to be scheduled to make the end product. They define what components have to be purchased or made to satisfy the MPS.
- **Order entry.** When a product has a very large number of options (e.g., cars), the order-entry system very often configures the end product bill of materials. The bill can also be used to price the product.

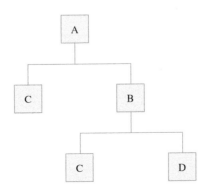

Pegged Requirements

Item Number	Week				
	1	2	3	4	5
C	50	125	25	50	150

Source of Requirements

A	50	25	25	50	50
B		100			100

FIGURE 4.10 Pegged requirements.

- **Manufacturing.** The bill provides a list of the parts needed to make or assemble a product.
- **Costing.** Product cost is usually broken down into direct material, direct labor, and overhead. The bill provides not only a method of determining direct material but also a structure for recording direct labor and distributing overhead.

This list is not complete, but it shows the extensive use made of the bill of material in manufacturing. There is scarcely a department of a company that will not use the bill at some time. Maintaining bills of material and their accuracy is extremely important. Again, the computer is an excellent tool for centrally maintaining bills and for updating them.

4.3 MATERIAL REQUIREMENTS PLANNING PROCESS

Each component shown on the bill of material is planned for by the MRP system. For convenience, it is assumed that each component will go into inventory and be accounted for. Whether the components actually go into a physical inventory or not is unimportant. However, it is important to realize that planning and control take place for each component on the bill. Raw material may go through several operations before it is processed and ready for assembly, or there may be several assembly operations between components and parent. These operations are planned and controlled by production activity control (PAC), not MRP.

The purpose of MRP is to determine the components needed, quantities, and due dates so items in the MPS are made on time. This section presents the basic MRP techniques for doing so. These techniques are discussed under the following headings:

- Exploding and offsetting.
- Gross and net requirements.
- Releasing orders.
- Capacity requirements planning.
- Low-level coding and netting.
- Multiple bills of material.

Exploding and Offsetting

Consider the product tree shown in Figure 4.11. It is similar to the ones used before but contains another necessary piece of information: lead time.

Lead time   **Lead time** is the span of time needed to perform a process. In manufacturing it includes time for order preparation, queuing, processing, moving, receiving and inspecting, and any expected delays. From the product tree shown in Figure 4.11, if B and C are available, it will take 1 week to assemble A. Thus, the lead time for A is 1 week.

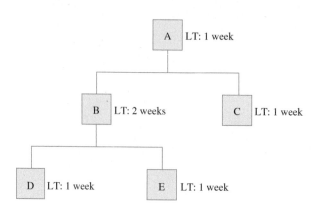

FIGURE 4.11 Product tree with lead time.

Similarly, if D and E are available, the time required to manufacture B is 2 weeks. The purchase lead times for D, E, and C are all 1 week.

In this particular product tree, the usage quantities—the quantity of components needed to make one of a parent—are all one. To make an A requires one B and one C, and to make a B requires one D and one E.

Exploding the requirements **Exploding** is the process of multiplying the requirements by the usage quantity and recording the appropriate requirements throughout the product tree.

Offsetting **Offsetting** is the process of placing the exploded requirements in their proper periods based on lead time. For example, if 50 units of A are required in week 5, the order to assemble the As must be released in week 4, and 50 Bs and 50 Cs must be available in week 4.

Planned orders If it is planned to receive 50 of part A in week 5 and the lead time to assemble an A is 1 week, the order will have to be released and production started no later than week 4.

Thus, there should be a **planned order receipt** for 50 in week 5 and a **planned order release** for that number in week 4. If an order for 50 As is to be released in week 4, 50 Bs and 50 Cs must be available in that week. Thus, there must be planned order receipts for those components in week 4. Since the lead time to assemble a B is 2 weeks, there must be a planned order release for the Bs in week 2. Since the lead time to make a C is 1 week, there must be a planned order release for 50 in week 3. The planned order receipts and planned order releases for the Ds and Es are determined in the same manner. Figure 4.12 shows when orders must be released and received so the delivery date can be met. Note that planned order releases and planned order receipts are paired with each other. Every time you have a planned order receipt, it should generate a planned order release offset by the lead time. Planned order releases and receipts are assumed to be orders for the item at the beginning of each period.

Part Number		Week				
		1	2	3	4	5
A	Planned Order Receipt Planned Order Release				50	50
B	Planned Order Receipt Planned Order Release		50		50	
C	Planned Order Receipt Planned Order Release			50	50	
D	Planned Order Receipt Planned Order Release	50	50			
E	Planned Order Receipt Planned Order Release	50	50			

FIGURE 4.12 Exploding and offsetting.

Example Problem

Using the product tree and lead times shown in Figure 4.11, complete the following table to determine the planned order receipts and releases. There are 50 As required in week 5 and 100 in week 6.

Part Number		Week					
		1	2	3	4	5	6
A	Planned Order Receipt Planned Order Release					50	100
B	Planned Order Receipt Planned Order Release						
C	Planned Order Receipt Planned Order Release						
D	Planned Order Receipt Planned Order Release						
E	Planned Order Receipt Planned Order Release						

ANSWER

Part Number		Week					
		1	2	3	4	5	6
A	Planned Order Receipt Planned Order Release				50	50 100	100
B	Planned Order Receipt Planned Order Release			50	100	50	100
C	Planned Order Receipt Planned Order Release				50	50 100	100
D	Planned Order Receipt Planned Order Release		50	50 100	100		
E	Planned Order Receipt Planned Order Release		50	50 100	100		

Gross and Net Requirements

The previous section assumed that no inventory was available for the As or any of the components. Often inventory is available and must be included when calculating quantities to be produced. If for example, in the preceding problem, there are 20 As in stock, at the beginning of period 5, only 30 need to be made. The requirements for component parts would be reduced accordingly. The calculation is as follows:

$$\text{Gross requirement} = 50$$
$$\text{Inventory available} = 20$$
$$\text{Net requirements} = \text{gross requirements} - \text{available inventory}$$
$$\text{Net requirements} = 50 - 20 = 30 \text{ units}$$

Since only 30 As need to be made, the gross requirement for Bs and Cs is only 30.
The planned order release of the parent becomes the gross requirement of the component.

The time-phased inventory record shown in Figure 4.12 can now be modified to consider any inventory available. For example, suppose there are 10 Bs available as well as the 20 As. The requirements for the components D and E would change. Figure 4.13 shows the change in the MRP record. Note that the projected available inventory shows the quantity projected on hand for the end of the period.

Example Problem

Complete the following table. Lead time for the part is 2 weeks. The order quantity (lot size) is 100 units.

Week		1	2	3	4
Gross Requirements			50	45	20
Projected Available	75				
Net Requirements					
Planned Order Receipt					
Planned Order Release					

ANSWER

Week		1	2	3	4
Gross Requirements			50	45	20
Projected Available	75	75	25	80	60
Net Requirements				20	
Planned Order Receipt				100	
Planned Order Release		100			

Releasing Orders

So far, we have looked at the process of planning when orders should be released so work is done in time to meet **gross requirements**. In many cases, requirements change daily. A computer-based MRP system automatically recalculates the requirements for subassemblies and components and recreates planned order releases to meet the shifts in demand.

Planned order releases are just planned; they have not been released. It is the responsibility of the material planner to release planned orders, not the computer.

Since the objective of MRP is to have material available when it is needed and not before, orders for material should not be released until the planned order release date arrives. Thus, an order is not normally released until the planned order is in the current week (week 1).

Releasing an order means that authorization is given to purchasing to buy the necessary material or to manufacturing to make the component.

Before a manufacturing order is released, component availability must be checked. The computer program checks the component inventory records to be sure that enough material is available and, if so, to allocate the necessary quantity to that work order. If the material is not available, the computer program will advise the planner of the shortage.

When the authorization to purchase or manufacture is released, the planned order receipt is canceled, and a scheduled receipt is created in its place. For the example shown in Figure 4.13, parts D and E have planned order releases of 20 scheduled for week 1. These orders will be released by the planner, and then the MRP records for parts D and E will appear as shown in Figure 4.14. Notice that scheduled receipts have been created, replacing the planned order releases.

When a manufacturing order is released the computer will **allocate** the required quantities of a parent's components to that order. This does not mean the components are withdrawn from inventory but that the projected *available* quantity is reduced. The allocated quantity of components is still in inventory, but they are not available for other orders. They will stay in inventory until withdrawn for use.

Part Number		Week				
		1	2	3	4	5
A	Gross Requirements Projected Available 20 Net Requirements Planned Order Receipt Planned Order Release	20	20	20	20 30	50 0 30 30
B	Gross Requirements Projected Available 10 Net Requirements Planned Order Receipt Planned Order Release	10	10 20	10	30 0 20 20	
C	Gross Requirements Projected Available Net Requirements Planned Order Receipt Planned Order Release		0	 30	30 0 30 30	
D	Gross Requirements Projected Available Net Requirements Planned Order Receipt Planned Order Release	0 20	20 0 20 20			
E	Gross Requirements Projected Available Net Requirements Planned Order Receipt Planned Order Release	0 20	20 0 20 20			

FIGURE 4.13 Gross and net requirements.

Part Number		Week				
		1	2	3	4	5
D	Gross Requirements Scheduled Receipts Projected Available Net Requirements Planned Order Receipt Planned Order Release	 0	20 20 0 0			
E	Gross Requirements Scheduled Receipts Projected Available Net Requirements Planned Order Receipt Planned Order Release	 0	20 20 0 0			

FIGURE 4.14 Scheduled receipts.

Material Requirements Planning

Scheduled receipts **Scheduled receipts** are orders placed on manufacturing or on a vendor and represent a commitment to make or buy. For an order in a factory, necessary materials are committed, and work-center capacity is allocated to that order. For purchased parts, similar commitments are made to the vendor. The scheduled receipts row shows the quantities ordered and when they are expected to be completed and available. They are generally expected to be due at the start of the period for which they are scheduled.

Open orders Scheduled receipts on the MRP record are **open orders** on the factory or a vendor and are the responsibility of purchasing and of PAC. These orders represent committed resources and are in process but not yet received. When the goods are received into inventory and available for use, the order is closed out, and the scheduled receipt disappears to become part of the on-hand inventory.

Net requirements The calculation for **net requirements** can now be modified to include scheduled receipts.

Net requirements = gross requirements − scheduled receipts − available inventory

Example Problem

Complete the following table. Lead time for the item is two weeks, and the order quantity is 200. What action should be taken?

Week	1	2	3	4
Gross Requirements	50	250	100	50
Scheduled Receipts		200		
Projected Available 150				
Net Requirements				
Planned Order Receipt				
Planned Order Release				

ANSWER

Week	1	2	3	4
Gross Requirements	50	250	100	50
Scheduled Receipts		200		
Projected Available 150	100	50	150	100
Net Requirements			50	
Planned Order Receipt			200	
Planned Order Release	200			

The order for 200 units should be released.

Basic MRP Record

Figure 4.15 shows a basic MRP record. There are several points that are important:

1. The current time is the beginning of the first period.
2. The top row shows periods, called **time buckets**. These are often a week but can be any length of time convenient to the company. Today's MRP applications typically use daily time buckets.

Part Number		Week				
		1	2	3	4	5
	Gross Requirements					35
	Scheduled Receipts				20	
	Projected Available 10	10	10	10	30	
	Net Requirements					5
	Planned Order Receipt					5
	Planned Order Release				5	

FIGURE 4.15 Basic MRP record.

3. The number of periods in the record is called the **planning horizon**, which shows the number of future periods for which plans are being made. It should be at least as long as the cumulative product lead time. Otherwise, the MRP system would not be able to release planned orders of items at the lower level at the correct time.

4. An item is considered available at the *beginning* of the time bucket in which it is required.

5. The quantity shown in the projected available row is the projected on-hand balance at the *end* of the period.

6. The immediate or most current period is called the **action bucket**. A quantity in the action bucket means that some action is needed now to avoid a future problem.

7. A **bucketless system** shows only the time buckets that have MRP activity, omitting the time periods with no activity.

Capacity Requirements Planning

As in the previous planning levels discussed, the MRP priority plan must be checked against available capacity. At the MRP planning level, the process is called capacity requirements planning (CRP). Chapter 5 examines this activity in some detail. If the capacity is available, the plan can proceed. If not, either capacity has to be made available or the priority plans must be changed.

Low-Level Coding and Netting

A component may reside on more than one level in a bill of material. If this is the case, it is necessary to make sure that all gross requirements for that component have been recorded before netting takes place. Consider the product shown in Figure 4.16. Component C occurs twice in the product tree and at different levels. It would be a mistake to net the requirements for the Cs before calculating the gross requirements for those required for parent B.

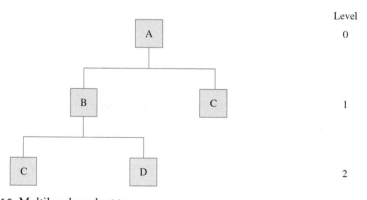

FIGURE 4.16 Multilevel product tree.

The process of collecting the gross requirements and netting can be simplified by using low-level codes. The **low-level code** is the lowest level on which a part resides in all bills of material. Every part has only one low-level code. The low-level codes for the parts in the product tree shown in Figure 4.6 are

Part	Low-Level Code
A	0
B	1
C	2
D	2

Low-level codes are determined by starting at the lowest level of a bill of material and, working up, recording the level against the part. If a part occurs again on a higher level, it is not assigned a code because its existence on the lower level has already been recorded.

Once the low-level codes are obtained, the net requirements for each part can be calculated using the following procedure. For the purpose of this exercise, there is a gross requirement for part A of 50 in week 5, all lead times are one week, and the following amounts are in inventory: A, 20 units; B, 10 units; and C, 10 units.

Procedure

1. Starting at level zero of the tree, determine if any of the parts on that level have a low-level code of zero. If so, those parts occur at no lower level, and all the gross requirements have been recorded. These parts can, therefore, be netted and exploded down to the next level, that is, into their components. If the low-level code is greater than zero, there are more gross requirements, and the part is not netted. In this example, A has a low-level code of zero so there is no further requirement for As; it can be netted and exploded into its components. Figure 4.17 shows the results.

2. The next step is to move down to level 1 on the product tree and to repeat the routine followed in step 1. Since B has a low-level code of 1, all requirements for B are

Low-Level Code	Part Number		Week				
			1	2	3	4	5
0	A	Gross Requirements Scheduled Receipts Projected Available 20 Net Requirements Planned Order Receipt Planned Order Release	20	20	20	20	50 0 30 30
						30	
1	B	Gross Requirements Scheduled Receipts Projected Available 10 Net Requirements Planned Order Receipt Planned Order Release				30	
2	C	Gross Requirements Scheduled Receipts Projected Available 10 Net Requirements Planned Order Receipt Planned Order Release				30	

FIGURE 4.17 Netting and exploding zero-level parts.

Low-Level Code	Part Number		\multicolumn{5}{c}{Week}
			1	2	3	4	5
1	B	Gross Requirements Scheduled Receipts Projected Available 10 Net Requirements Planned Order Receipt Planned Order Release	10	10	10 20	30 0 20 20	
2	C	Gross Requirements Scheduled Receipts Projected Available 10 Net Requirements Planned Order Receipt Planned Order Release			20	30	
2	D	Gross Requirements Scheduled Receipts Projected Available Net Requirements Planned Order Receipt Planned Order Release			20		

FIGURE 4.18 Netting and exploding first-level parts.

recorded, and it can be netted and exploded. The bill of material for B shows that it is made from a C and a D. Figure 4.18 shows the result of netting and exploding the Bs. Part C has a low-level code of 2, which signifies there are further requirements for Cs and at this stage they are not netted.

3. Moving down to level 2 on the product tree, part C has a low-level code of 2. This signifies that all gross requirements for Cs are accounted for and that the process can proceed and determine its net requirements. Notice there is a requirement for 30 Cs in week 4 to be used on the As and a requirement of 20 Cs in week 3 to be used on the Bs. Looking at its bill of material, it shows it is a purchased part and no explosion is needed.

Figure 4.19 shows the completed material requirements plan. The process of level-by-level netting is now completed using the low-level codes of each part. The low-level codes are used to determine when a part is eligible for netting and exploding. In this way, each part is netted and exploded only once. There is no time-consuming re-netting and re-exploding each time a new requirement is met.

Multiple Bills of Material

Most companies make more than one product and often use the same components in many of their products. The MRP system gathers the planned order releases from all the parents and creates a schedule of gross requirements for the components. Figure 4.20 illustrates what happens. Part F is a component of both C and B.

The same procedure used for a single bill of material can be used when multiple products are being manufactured. All bills must be netted and exploded level by level as was done for a single bill.

Figure 4.21 shows the product trees for two products. Both are made from several components, but, for simplicity, only those components containing an F are shown in the

Material Requirements Planning

Low-Level Code	Part Number		Week				
			1	2	3	4	5
0	A	Gross Requirements					50
		Scheduled Receipts					
		Projected Available 20	20	20	20	20	0
		Net Requirements					30
		Planned Order Receipt					30
		Planned Order Release				30	
1	B	Gross Requirements				30	
		Scheduled Receipts					
		Projected Available 10	10	10	10	0	
		Net Requirements				20	
		Planned Order Receipt				20	
		Planned Order Release			20		
2	C	Gross Requirements				20	30
		Scheduled Receipts					
		Projected Available 10	10	10	0	0	
		Net Requirements				10	30
		Planned Order Receipt				10	30
		Planned Order Release			10	30	
2	D	Gross Requirements				20	
		Scheduled Receipts					
		Projected Available	0	0	0		
		Net Requirements				20	
		Planned Order Receipt				20	
		Planned Order Release			20		

FIGURE 4.19 Completed material requirements plan.

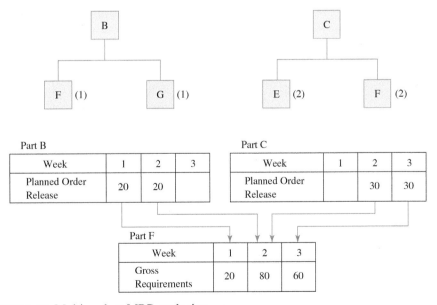

FIGURE 4.20 Multiproduct MRP explosion.

FIGURE 4.21 Multiproduct tree.

product tree. Note that both have F as a component but at different levels in their product tree. All lead times are one week. The quantities required are shown in parentheses; that is, two Cs are required to make an A, one F is required to make a B, and two Fs are needed to make a C. Figure 4.22 shows the completed material requirements plan that would result if 50 As were required in week 5 and 30 Bs in week 3.

Scrap is inherent in some processes due to errors or as a result of the process. As an example, the loss due to bones, juice, and evaporation when cooking a turkey is estimated at 50%. Industries that mix and pour liquids usually lose some product in the system. Due to scrap as shown in these examples, extra production must be scheduled to produce the desired end amount. To accommodate scrap MRP increases the planned order release from the amount needed for the planned order receipt. For example, a process may generate

Low-Level Code	Part Number			Week				
				1	2	3	4	5
0	A	Gross Requirements Scheduled Receipts Projected Available 20 Net Requirements Planned Order Receipt Planned Order Release		20	20	20	20 30	50 0 30 30
0	B	Gross Requirements Scheduled Receipts Projected Available 10 Net Requirements Planned Order Receipt Planned Order Release		10	10 20	30 0 20 20		
1	C	Gross Requirements Scheduled Receipts Projected Available 10 Net Requirements Planned Order Receipt Planned Order Release		10	10	10 50	60 0 50 50	
2	F	Gross Requirements Scheduled Receipts Projected Available Net Requirements Planned Order Receipt Planned Order Release		 40	40 0 40 40 50	50 0 50 50		

FIGURE 4.22 Partial material requirements plan.

15% scrap. If the planned order receipt for an item is for 400 units, then the planned order release for that item would be: $400 / (1 - 0.15) = 471 \text{ units}$.

$$\text{Planned order release} = \text{Planned order receipt} (1 - \text{scrap factor})$$

4.4 USING THE MATERIAL REQUIREMENTS PLAN

The people who manage the material requirement planning system are planners. They are responsible for making detailed decisions that keep the flow of material moving into, through, and out of the factory. In many companies where there are thousands of parts to manage, planners are usually organized into logical groupings based on the similarity of parts or supply.

The basic responsibilities of a planner are to:

1. Launch (release) orders to purchasing or manufacturing.
2. Reschedule due dates of open (existing) orders as required.
3. Reconcile errors and try to find their cause.
4. Solve critical material shortages by expediting or replanning.
5. Coordinate with other planners, master production schedulers, PAC, and purchasing to resolve problems.

The material planner works with three types of orders: planned, released, and firm.

Planned orders. Planned orders are automatically scheduled and controlled by the computer. As gross requirements, projected available inventory, and scheduled receipts change, the computer recalculates the timing and quantities of planned order releases. The MRP program recommends to the planner the release of an order when the order enters the action bucket but does not release the order.

MRP IN SERVICES

MRP systems have essentially been designed and implemented for use in manufacturing environments, where the number of dependent demand components in the bills of materials tends to be large and so do the number of calculations. Few service companies face such an environment, but the fundamental concepts inherent with MRP will still usually apply.

A cafeteria, such as might be found in many schools, can serve as an example. Most will use the principle of MRP, although it is not typically called that. The process starts with the development of a menu, perhaps on a monthly or weekly basis. That menu presents the "finished goods," in the form of specific completed food offers, and really is essentially a master schedule. As time draws nearer to when a food item is to be prepared (the cumulative lead time for making the food and the lead times to obtain the various ingredients), the cooks need to know the quantity of food to cook (how many people are expected to want that food item). Making too little can draw complaints, while making too much is wasteful as much of it ends up being thrown out or given away. There is also a lead time to consider—how long will it take to combine the ingredients and then cook the item (final assembly). Again, starting too late will mean the food will not be ready and starting too early could result in a product that sits too long and is no longer appealing. That information can be found on the "bill of material," which in this case is the recipe for making the food. That bill of material (recipe) contains quantities, lead times, and precedent relationships, just as a bill of material in manufacturing does. The recipe also usually calls for the steps used to combine the materials, which is essentially the same as what in manufacturing is represented by a separate document called the routing.

Once the quantity of each ingredient to make the right amount of the final food item is known, it is compared to the existing inventory of the ingredient. The difference, of course, needs to be obtained. There are lead times for each of those ingredients, and offsetting the lead times, just as in MRP, will tell when and how much of each ingredient needs to be ordered in order that everything needed to make the final item is present in its freshest possible form in time to start to actually combine and make the food item.

The important issue to remember here is that all companies, from the smallest service company to the largest manufacturing company, essentially have the same planning and control issues of long-range planning, master scheduling, inventory management, capacity planning and management, controlling production, and quality management. The primary difference lies in the vocabulary (what they call the activity) and how formally it is done. The similarities are that effective companies do these things well, and companies that do not do them well tend to be much less effective.

Released orders. Releasing, or launching, a planned order is the responsibility of the planner. When released, the order becomes an open order to the factory or to purchasing and appears on the MRP record as a scheduled receipt. It is then under the control of the planner, who may expedite, delay, or even cancel the order.

Firm planned orders. The computer-based MRP system automatically recalculates planned orders as the gross requirements change. At times, the planner may prefer to hold a planned order firm against changes in quantity and time despite what the computer calculates. This might be necessary because of future availability of material or capacity or special demands on the system. The planner can tell the computer that the order is not to be changed unless the planner advises the computer to do so. The order is "firmed" or frozen against the logic of the computer.

The MRP software nets, offsets, and explodes requirements and creates planned order releases. It keeps priorities current for all planned orders according to changes in gross requirements for the part. But it does not issue purchase or manufacturing orders or reschedule open orders. However, it does generate action or exception messages, suggesting that the planner should act and what kind of action might be appropriate to keep the supply and demand in balance.

Exception messages If the manufacturing process is under control and the MRP system is working properly, the system will work according to plan. However, sometimes there are problems that need the attention of the planner. An MRP system generates **exception messages** to advise the planner when some event needs attention. Following are some examples of situations that will generate exception messages.

- Components for which planned orders are in the action bucket and which should be considered for release.
- Open orders for which the timing or quantity of scheduled receipts does not satisfy the plan. Perhaps a scheduled receipt is timed to arrive too early or late, and its due date should be revised.
- Situations in which the standard lead times will result in late delivery of a zero-level part. This situation might call for expediting to reduce the standard lead times.

Transactions Transactions reflect that some event has occurred and are reflected in the software in order to ensure the MRP records are updated. For example, when the planner releases an order, or a scheduled receipt is received, or when any change to the data occurs, that action must be entered into the software. Otherwise, the records will be inaccurate, and the plan will become unworkable.

Material requirements planners must manage the parts for which they are responsible. This means not only releasing orders to purchasing and the factory, rescheduling due dates of open orders, and reconciling differences and inconsistencies but also finding ways to improve the system and removing the causes of potential error. If the right components are to be in the right place at the right time, the planner must manage the process.

Managing the Material Requirements Plan

The planner receives feedback from many sources such as

- Suppliers' actions through purchasing.
- Changes to open orders in the factory such as early or late completions or differing quantities.
- Management action such as changing the MPS.

The planner must evaluate this feedback and take corrective action if necessary. The planner must consider three important factors in managing the material requirements plan.

Priority Priority refers to maintaining the correct due dates by constantly evaluating the true due date need for released orders and, if necessary, expediting or de-expediting.

Consider the following MRP record. The order quantity is 300 units and the lead time is three weeks.

Week	1	2	3	4	5
Gross Requirements	100	50	100	150	200
Scheduled Receipts			300		
Projected Available 150	50	0	200	50	150
Net Requirements					150
Planned Order Receipt					300
Planned Order Release		300			

What will happen if the gross requirements in week 2 are changed from 50 to 150 units? The MRP record will look like the following.

Week	1	2	3	4	5
Gross Requirements	100	150	100	150	200
Scheduled Receipts			300		
Projected Available 150	50	−100	100	250	50
Net Requirements				50	
Planned Order Receipt				300	
Planned Order Release	300				

Note that there is a shortage of 100 units in week 2 and that the planned order release originally in week 2 is now in week 1. What can the planner do? One solution is to expedite the scheduled receipt of 300 units from week 3 to week 2. If this is not possible, the extra 100 units wanted in week 2 must be rescheduled into week 3. Also, there is now a planned order release in week 1, and this order should be released.

Bottom-up replanning Action to respond for changed conditions should occur as low in the product structure as possible. Suppose the part in the previous example is a component of another part. The first alternative is to expedite the scheduled receipt of 300 into week 2. If this can be done, there is no need to make any changes to the parent. If the 300 units cannot be expedited, the planned order release and net requirement of the parent must be changed. This is referred to as **bottom-up replanning**.

Reducing system nervousness Sometimes requirements change rapidly and by small amounts, causing the material requirements plan to change back and forth. The planner must judge whether the changes are important enough to react to and whether an order should be released. One method of reducing **system nervousness** is firm planned orders.

Example Problem

As the MRP planner, you arrive at work Monday morning and look at the MRP record for part 2876 as shown below.

Order quantity = 30 units
Lead time = 2 weeks

Week		1	2	3	4	5	6
Gross Requirements		35	10	15	30	15	20
Scheduled Receipts		30					
Projected Available	20	15	5	20	20	5	15
Net Requirements				10	10		15
Planned Order Receipt				30	30		30
Planned Order Release		30	30		30		

USING KNOWLEDGE OF THE SYSTEM TO EVALUATE PROBLEMS

ERP, and even basic MRP, systems are typically difficult to implement and, if not implemented correctly or not measured and corrected over time for problems, they can be the source of an ineffective system with frustrated users. They usually require timely and effective coordination between functions, accurate and timely recorded data, and knowledgeable and effective system management.

At times an organization may find that the systems generate so many exception messages that the people relying on the system information cannot possibly deal with all the messages during a normal work day. While this can happen on occasion even with a well-implemented and managed system, when it happens on a routine basis it is a strong symptom of a system that needs to be "repaired." Simply reacting to only "important" messages is a classic example of "solving the symptom" rather than attacking the core problem—but with a complex and highly integrated system, where should one start looking for the core set of problems?

The most logical place to start for most facilities is the master schedule. There are two reasons for this. First, in most systems there are significantly fewer master schedules to examine than there are MRP records. Second, for many products any small change in the master schedule has the potential to generate hundreds or even thousands of changes in the MRP records being driven by the master schedule (occasionally referred to as "system nervousness"). While one would expect to generate some changes in the master schedule based on changing customer and internal conditions, they should be considered very carefully for their potential impact before being accepted. If the master scheduling approach is not managed well, it should be the first candidate for process improvement. Sometimes that may require looking at the sales and operations planning (S&OP) approach. For example, did the S&OP provide for adequate resources of the right type at the right time?

Once it appears the master scheduling process is being managed well, the next major area to investigate for problems is data—is data being provided accurately and in a timely fashion? There are several data systems that need to be evaluated, including the purchasing area, capacity management, bill of material structures, item master data (lead time, for example), inventory data, and production activity control data.

In many cases those two issues—master scheduling and data management—represent the major cause of exception messages, and proper implementation and management of those issues should be able to make exception messages just that—exceptions—rather than a routine expectation.

Sometimes if the planner is faced with a large number of issues to resolve they may consider "short cuts" to free up some time. As an example, many modern software packages allow for the automatic releasing of orders, meaning the planner would avoid getting a message that an order needs to be released and then taking the time to release that order—it would automatically be released. While that would free up time and effort, it is not considered to necessarily be a wise move. Recall that planned orders in MRP are merely that—planned. They are nothing more than a number in the computer. Once released, however, they represent an actual financial commitment to either produce or purchase a component. If that is done in error because of some condition unknown to the computer software (e.g., a transportation tie-up or perhaps a supplier equipment problem), the erroneous financial obligation may end up costing a lot of money. If the planner were confronted about the error by the planner's manager, few managers would readily accept the excuse "the computer did it." As a wise general rule, any decision that commits actual resources of the company should at least be reviewed by a responsible person.

The computer draws attention to the need to release the planned order for 30 in week 1. Either you release this order, or there will be a shortage in week 3. During the first week, the following transactions take place:

a. Only 25 units of the scheduled receipt are received into inventory. The balance is scrapped.

b. The gross requirement for week 3 is changed to 10.

c. The gross requirement for week 4 is increased to 50.

d. The requirement for week 7 is 15.

e. An inventory count reveals there are 10 more in inventory than the record shows.

f. The 35 gross requirement for week 1 is issued from inventory.

g. The planned order release for 30 in week 1 is released and becomes a scheduled receipt in week 3.

As these transactions occur during the first week, you must enter these changes in the computer record. At the beginning of the next week, the MRP record appears as follows:

Order quantity = 30 units
Lead time = 2 weeks

Week		2	3	4	5	6	7
Gross Requirements		10	10	50	15	20	15
Scheduled Receipts			30				
Projected Available	20	10	30	10	25	5	20
Net Requirements				20	5		10
Planned Order Receipt				30	30		30
Planned Order Release		30	30		30		

The opening on-hand balance for week 2 is 20 (20 + 25 + 10 − 35 = 20). The planned order release originally set in week 4 has shifted to week 3. Another planned order has been created for release in week 5. More importantly, the scheduled receipt in week 3 will not be needed until week 4. You should reschedule this to week 4. The planned order in week 2 should be released and become a scheduled receipt in week 4.

SUMMARY

The job of material requirements planning is to produce the right components at the right time so that the MPS can be maintained. MRP depends on accurate bills of material and on accurate inventory records. Bills of material can be created in many ways, but one department (or individual) must be responsible for them. Inventory records are indispensable to MRP, and MRP is only as good as the inventory records.

The MRP exploding and offsetting processes outlined in this chapter are largely done by the computer. The logic used is repetitive and, while error prone when done by individuals, can be accomplished well by computer. Good MRP practice is achieved by planners being able to work with the system.

The MRP process uses the bill of materials that lists components used to make a product, the lead time to make or obtain those components, and the existing inventory of those components to calculate a series of planned order releases to obtain or make components to meet future product needs.

KEY TERMS

Action bucket 86
Allocation 83
Bill of material 74
Bottom-up replanning 93
Bucketless system 86
Component 74
Dependent demand 71
Exception messages 92
Explode 81
Firm planned order 92
Gross requirements 83
Independent demand 71
Indented bill of material 78
Inventory record 73
Lead time 80

Low-level code 87
Master production schedule 73
Material requirements planning (MRP) 71
Multilevel bill of material 75
Multiple bill 75
Net requirements 85
Offsetting 81
Open order 85
Parent item 74
Pegging 79
Planned order 91
Planned order receipt 81
Planned order release 81
Planning bills 78

Planning factors 73
Planning horizon 86
Product tree 74
Released order 92
Scheduled receipt 85

Single-level bill of material 75
Summarized bill of material 78
System nervousness 93
Time buckets 85
Where-used list 78

QUESTIONS

1. What is a material requirements plan?
2. What is the difference between dependent and independent demand?
3. Should MRP be used with dependent or independent demand items?
4. What are the objectives of MRP?
5. What is the relationship between the MPS and MRP?
6. Why is a computer necessary in an MRP system?
7. What are the major inputs to MRP?
8. What data is found in a part master file or an item master file?
9. What is a bill of material? What are two important points about bills of material?
10. To what does *bill of material structure* refer? Why is it important?
11. Describe the parent–component relationship.
12. Describe the following types of bills of material:
    a. Product tree.
    b. Multilevel bill.
    c. Single-level bill.
    d. Indented bill.
    e. Summarized parts bill of material.
    f. Planning bill.
13. Why do MRP computer programs store single-level bills?
14. Describe each of the seven uses of a bill of material described in the text.
15. What are where-used and pegging reports? Give some of their uses.
16. Describe the processes of offsetting and exploding.
17. What is a planned order? How is it created?
18. From where does the gross requirement of a component come?
19. Who is responsible for releasing an order? Describe what happens to the inventory records and to PAC and purchasing.
20. What is a scheduled receipt? From where does it originate?
21. What is an open order? How does it get closed?
22. What is the meaning of the term *low-level code?* How is the low-level code of an MPS part represented?
23. What are the responsibilities of a material requirements planner?
24. Give two examples of processes with inherent scrap. Hint, the use of natural products often involves some scrap.
25. What would make the planned order release for an item different from the planned order receipt?
26. Describe the differences among planned orders, released orders, and firm planned orders. Who controls each?
27. What are exception messages? What is their purpose?
28. What is a transaction message? Why is it important?
29. What are the three important factors in managing the material requirements plan? Why is each important?
30. Describe the problems that might come from using an incorrect bill of material in MRP.
31. Describe how MRP might be used to plan for a change in design for a product.

PROBLEMS

4.1. Using the following product tree, construct the appropriate single-level trees. How many Cs are needed to make 50 Xs and 100 Ys?

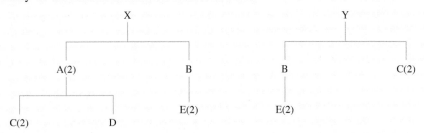

Answer. 400 Cs

4.2. Given the following parents and components, construct a product tree. Figures in parentheses show the quantities per item. How many Gs are needed to make one A?

Parent	A	B	C	E
Component	B(2)	E(2)	G(3)	G(4)
	C(4)	F(1)		F(3)
	D(4)			H(2)

4.3. Using the following product tree, determine the planned order receipts and planned order releases if 160 As are to be produced in week 5. All lead times are one week except for component D, which has a lead time of two weeks.

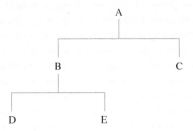

	Week	1	2	3	4	5
Part A	Planned Order Receipt					
Lead Time: 1 week	Planned Order Release					
Part B	Planned Order Receipt					
Lead Time: 1 week	Planned Order Release					
Part C	Planned Order Receipt					
Lead Time: 1 week	Planned Order Release					
Part D	Planned Order Receipt					
Lead Time: 2 weeks	Planned Order Release					
Part E	Planned Order Receipt					
Lead Time: 1 weeks	Planned Order Release					

4.4. Complete the following table. Lead time for the part is two weeks, and the order quantity is 50. What action should be taken?

Week		1	2	3	4
Gross Requirements		20	15	25	25
Projected Available	40				
Net Requirements					
Planned Order Receipt					
Planned Order Release					

Answer. An order for 50 should be released in week 1.

4.5. Given the following product tree, explode, offset, and determine the gross and net requirements. All lead times are one week, and the quantities required are shown in parentheses. The master production schedule calls for 100 As to be available in week 5. There are 50 Bs available. All other on-hand balances = 0.

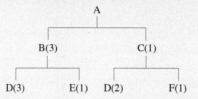

		Week	1	2	3	4	5
Part A Lead Time: 1 week		Gross Requirements					
		Scheduled Receipts					
		Projected Available					
		Net Requirements					
		Planned Order Receipt					
		Planned Order Release					
Part B Lead Time: 1 week		Gross Requirements					
		Scheduled Receipts					
		Projected Available					
		Net Requirements					
		Planned Order Receipt					
		Planned Order Release					
Part C Lead Time: 1 week		Gross Requirements					
		Scheduled Receipts					
		Projected Available					
		Net Requirements					
		Planned Order Receipt					
		Planned Order Release					
Part D Lead Time: 1 week		Gross Requirements					
		Scheduled Receipts					
		Projected Available					
		Net Requirements					
		Planned Order Receipt					
		Planned Order Release					
Part E Lead Time: 1 week		Gross Requirements					
		Scheduled Receipts					
		Projected Available					
		Net Requirements					
		Planned Order Receipt					
		Planned Order Release					
Part F Lead Time: 1 week		Gross Requirements					
		Scheduled Receipts					
		Projected Available					
		Net Requirements					
		Planned Order Receipt					
		Planned Order Release					

Answer. Planned order releases are

Part A: 100 in week 4
Part B: 250 in week 3
Part C: 100 in week 3
Part D: 950 in week 2
Part E: 250 in week 2
Part F: 100 in week 2

4.6. Complete the following table. Lead time for the part is two weeks. The lot size is 80. What is the projected available at the end of week 3? When is it planned to release an order?

Week	1	2	3	4
Gross Requirements	20	75	30	30
Scheduled Receipts		100		
Projected Available 30				
Net Requirements				
Planned Order Receipt				
Planned Order Release				

Answer. Projected available at the end of week 3 is 5.

An order release is planned for the beginning of week 2.

4.7. Complete the following table. Lead time for the part is two weeks. The lot size is 50. What is the projected available at the end of week 3? When is it planned to release an order?

Week	1	2	3	4
Gross Requirements	30	25	10	10
Scheduled Receipts	50			
Projected Available 10				
Net Requirements				
Planned Order Receipt				
Planned Order Release				

4.8. Given the following partial product tree, explode, offset, and determine the gross and net requirements for components H, I, J, and K. There are other components, but they are not connected to this problem. The quantities required are shown in parentheses. The master production schedule calls for the completion of 50 Hs in week 3 and 90 in week 5. There is a scheduled receipt of 120 Is in week 2. There are 400 Js and 300 Ks available. All lot sizes are lot-for-lot.

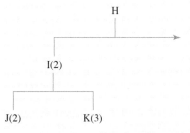

	Week	1	2	3	4	5
Part H	Gross Requirements					
	Scheduled Receipts					
Lead Time: 1 week	Projected Available					
	Net Requirements					
	Planned Order Receipt					
	Planned Order Release					

Part I						
Lead Time: 2 weeks	Gross Requirements					
	Scheduled Receipts					
	Projected Available					
	Net Requirements					
	Planned Order Receipt					
	Planned Order Release					
Part J						
Lead Time: 1 week	Gross Requirements					
	Scheduled Receipts					
	Projected Available	400				
	Net Requirements					
	Planned Order Receipt					
	Planned Order Release					
Part K						
Lead Time: 1 week	Gross Requirements					
	Scheduled Receipts					
	Projected Available	300				
	Net Requirements					
	Planned Order Receipt					
	Planned Order Release					

Answer. There is a planned order release for part K of 80 in week 1.

4.9. MPS parent X has planned order releases of 40 in weeks 2 and 4. Given the following product tree, complete the MRP records for parts Y and Z. Quantities required are shown in brackets.

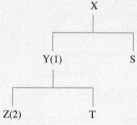

Part Y		Week			
Lead Time: 2 weeks Lot Size: 60		1	2	3	4
Gross Requirements					
Scheduled Receipts					
Projected Available	45				
Net Requirements					
Planned Order Receipt					
Planned Order Release					

Part Z		Week			
Lead Time: 1 Lot Size: 130		1	2	3	4
Gross Requirements					
Scheduled Receipts					
Projected Available	20				
Net Requirements					
Planned Order Receipt					
Planned Order Release					

4.10. Given the following product tree, explode, offset, and determine the gross and net requirements. The quantities required are shown in parentheses. The master production schedule calls for the completion of 100 As in week 5. There is a scheduled receipt of 100 Bs in week 1. There are 200 Fs available. All order quantities are lot-for-lot.

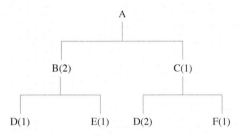

		Week	1	2	3	4	5
Part A Lead Time: 1 week	Gross Requirements						
	Scheduled Receipts						
	Projected Available						
	Net Requirements						
	Planned Order Receipt						
	Planned Order Release						
Part B Lead Time: 1 week	Gross Requirements						
	Scheduled Receipts						
	Projected Available						
	Net Requirements						
	Planned Order Receipt						
	Planned Order Release						
Part C Lead Time: 1 week	Gross Requirements						
	Scheduled Receipts						
	Projected Available						
	Net Requirements						
	Planned Order Receipt						
	Planned Order Release						
Part D Lead Time: 1 week	Gross Requirements						
	Scheduled Receipts						
	Projected Available						
	Net Requirements						
	Planned Order Receipt						
	Planned Order Release						
Part E Lead Time: 1 week	Gross Requirements						
	Scheduled Receipts						
	Projected Available						
	Net Requirements						
	Planned Order Receipt						
	Planned Order Release						
Part F Lead Time: 1 week	Gross Requirements						
	Scheduled Receipts						
	Projected Available						
	Net Requirements						
	Planned Order Receipt						
	Planned Order Release						

4.11. Given the following product tree, complete the MRP records for parts X, Y, W, and Z. Note that parts X and Y have specified order quantities.

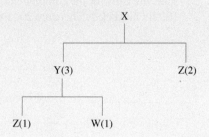

	Week	1	2	3	4	5
Part X	Gross Requirements	10	15	12	10	15
	Scheduled Receipts		20			
Lead Time: 1 week	Projected Available 10					
	Net Requirements					
Lot Size: 20	Planned Order Receipt					
	Planned Order					
Part Y	Gross Requirements					
	Scheduled Receipts		50			
Lead Time: 2 weeks	Projected Available 25					
	Net Requirements					
Lot Size: 50	Planned Order Receipt					
	Planned Order Release					
Part Z	Gross Requirements					
Lead Time: 1 week	Scheduled Receipts		90			
	Projected Available					
	Net Requirements					
Lot Size: lot-for-lot	Planned Order Receipt					
	Planned Order Release					
Part W	Gross Requirements					
	Scheduled Receipts					
Lead Time: 1 week	Projected Available					
	Net Requirements					
Lot Size: 400	Planned Order Receipt					
	Planned Order Release					

4.12. Given the following product tree, determine the low-level codes for all the components.

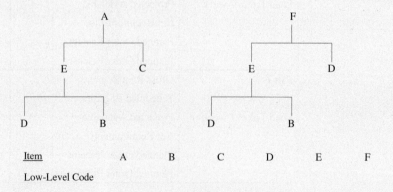

Item	A	B	C	D	E	F
Low-Level Code						

4.13. Given the following product tree, determine the low-level codes for all the components.

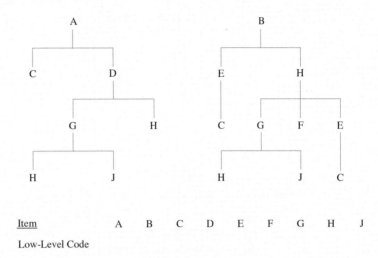

Item	A	B	C	D	E	F	G	H	J
Low-Level Code									

4.14. Given the following product tree, develop a material requirements plan for the components. Quantities per are shown in parentheses. The following worksheet shows the present active orders, the available balances, and the lead times.

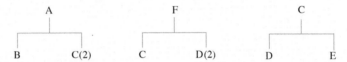

Low-Level Code		Week	1	2	3	4	5
0	Part A	Gross Requirements			50		80
		Scheduled Receipts					
	Lead Time: 1 week	Projected Available					
		Net Requirements					
	Lot-for-lot	Planned Order Receipt					
		Planned Order Release					
0	Part F	Gross Requirements				120	
		Scheduled Receipts					
	Lead Time: 1 week	Projected Available					
		Net Requirements					
	Lot-for-lot	Planned Order Receipt					
		Planned Order Release					
	Part B	Gross Requirements					
		Scheduled Receipts					
	Lead Time: 2 weeks	Projected Available 200					
		Net Requirements					
	Lot Size: 300	Planned Order Receipt					
		Planned Order Release					
	Part C	Gross Requirements					
		Scheduled Receipts	120				
	Lead Time: 2 weeks	Projected Available					
		Net Requirements					
	Lot Size: Lot-for-lot	Planned Order Receipt					
		Planned Order Release					

Low-Level Code		Week	1	2	3	4	5
	Part D Lead Time: 2 weeks Lot Size: 300	Gross Requirements Scheduled Receipts Projected Available Net Requirements Planned Order Receipt Planned Order Release	 300				
	Part E Lead Time: 3 weeks Lot Size: 500	Gross Requirements Scheduled Receipts Projected Available 400 Net Requirements Planned Order Receipt Planned Order Release					

Answer. The low-level code for part D is 2. There is a planned order release of 300 for part D in week 1. There are no planned order releases for part E. There is a planned order release of 100 for Part C in week 1 and 140 in week 2.

4.15. Given the following product tree, explode, offset, and determine the gross and net requirements. All lead times are one week, and the quantities required are shown in parentheses. The master production schedule calls for the completion of 90 As in week 4 and 80 in week 5. There are 300 Bs scheduled to be received in week 1 and 200 Ds in week 3. There are also 20 As available.

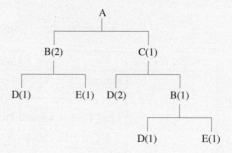

Low-Level Code		Week	1	2	3	4	5
	Part A Lead Time: 1 week Lot Size: lot-for-lot	Gross Requirements Scheduled Receipts Projected Available 20 Net Requirements Planned Order Receipt Planned Order Release					
	Part B Lead Time: 1 week Lot Size: lot-for-lot	Gross Requirements Scheduled Receipts Projected Available Net Requirements Planned Order Receipt Planned Order Release					
	Part C Lead Time: 1 week Lot Size: lot-for-lot	Gross Requirements Scheduled Receipts Projected Available Net Requirements Planned Order Receipt Planned Order Release					

Low-Level Code		Week	1	2	3	4	5
	Part D Lead Time: 1 week Lot Size: lot-for-lot	Gross Requirements Scheduled Receipts Projected Available Net Requirements Planned Order Receipt Planned Order Release					
	Part E Lead Time: 1 week Lot Size: lot-for-lot	Gross Requirements Scheduled Receipts Projected Available Net Requirements Planned Order Receipt Planned Order Release					

4.16. Given the following product tree, determine the low-level codes and the gross and net quantities for each part. There is a requirement for 100 As in week 4 and 50 Bs in week 5. There is a scheduled receipt of 100 Cs in week 2. Quantities required of each are also shown.

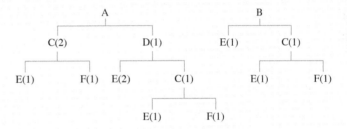

Low-Level Code		Week	1	2	3	4	5
	Part A Lead Time: 1 week Lot Size: lot-for-lot	Gross Requirements Scheduled Receipts Projected Available Net Requirements Planned Order Receipt Planned Order Release					
	Part B Lead Time: 1 week Lot Size: lot-for-lot	Gross Requirements Scheduled Receipts Projected Available Net Requirements Planned Order Receipt Planned Order Release					
	Part C Lead Time: 1 week Lot Size: lot-for-lot	Gross Requirements Scheduled Receipts Projected Available Net Requirements Planned Order Receipt Planned Order Release					
	Part D Lead Time: 1 week Lot Size: lot-for-lot	Gross Requirements Scheduled Receipts Projected Available Net Requirements Planned Order Receipt Planned Order Release					

	Part E	Gross Requirements					
	Lead Time: 1 week	Scheduled Receipts					
		Projected Available					
		Net Requirements					
	Lot Size: 500	Planned Order Receipt					
		Planned Order Release					
	Part F	Gross Requirements					
	Lead Time: 1 week	Scheduled Receipts					
		Projected Available					
		Net Requirements					
	Lot Size: lot-for-lot	Planned Order Receipt					
		Planned Order Release					

4.17. Complete the following MRP record. The lead time is four weeks, and the lot size is 200. What will happen if the gross requirements in week 3 are increased to 150 units? As a planner, what actions can you take?

Initial MRP

Week	1	2	3	4	5
Gross Requirements	50	125	100	60	40
Scheduled Receipts		200		200	
Projected Available 100					
Net Requirements					
Planned Order Receipt					
Planned Order Release					

Revised MRP

Week	1	2	3	4	5
Gross Requirements					
Scheduled Receipts					
Projected Available 100					
Net Requirements					
Planned Order Receipt					
Planned Order Release					

4.18. It is Monday morning, and you have just arrived at work. Complete the following MRP record as it would appear Monday morning. Lead time is two weeks, and the lot size is 100.

Initial MRP

Week	1	2	3	4	5
Gross Requirements	70	40	70	50	40
Scheduled Receipts	100				
Projected Available 50					
Net Requirements					
Planned Order Receipt					
Planned Order Release					

During the week, the following events occur. Enter them in the MRP record.

a. The planned order for 100 in week 1 is released.

b. Thirty of the scheduled receipts for week 1 are scrapped.

c. An order for 40 is received for delivery in week 3.

d. An order for 60 is received for delivery in week 6.

e. The gross requirements of 70 in week 1 are issued.

MRP record at the end of week 1

Week	2	3	4	5	6
Gross Requirements					
Scheduled Receipts					
Projected Available					
Net Requirements					
Planned Order Receipt					
Planned Order Release					

4.19. The following bill of materials represents the major components for a computer system.

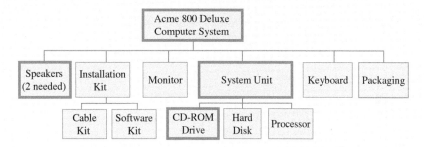

Complete the MRP records below. Note the following:

- Production plans (the MPS) for the 800 Deluxe computer system are as follows:
  Start assembling 2500 in week 2.
  Start assembling 3000 in weeks 3, 4, and 5.
  Start assembling 2000 in week 6.
- The gross requirements for the **system unit** have already been given to you. For the remaining items, you will need to figure out the gross requirements.
- All scheduled receipts, lead times, and beginning inventory levels are shown.

System Unit

Lead time = 1 week Minimum order quantity = 500						
Week	1	2	3	4	5	6
Gross Requirements		2500	3000	3000	3000	2000
Scheduled Receipts						
Projected Available	0					
Net Requirements						
Planned Receipts						
Planned Order Releases						

Speakers

Lead time = 1 week Minimum order quantity = 5000						
Week	1	2	3	4	5	6
Gross Requirements						
Scheduled Receipts	5000					
Projected Available	0						
Net Requirements						
Planned Receipts						
Planned Order Releases						

CD-ROM Drives

Lead time = 4 weeks						
Minimum order quantity = 5000						
Week	1	2	3	4	5	6
Gross Requirements						
Scheduled Receipts						
Projected Available 11,500						
Net Requirements						
Planned Receipts						
Planned Order Releases						

4.20. Acme Tool & Manufacturing Company makes a wide variety of lawn care products. Acme's products are sold under various brand names, and are available at retail outlets. One of Acme's products is the Model #540 Broadcast Spreader, shown below:

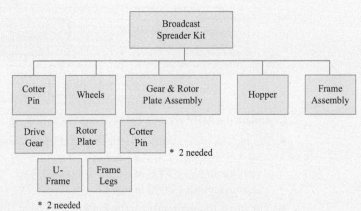

Complete the MRP records on the next page. Note the following:

- Acme intends to start assembling 2000 broadcast spreader kits in weeks 2, 4, and 6.
- The gross requirements for the gear & rotor plate assembly have already been given to you. For the remaining items, you will need to figure out the gross requirements.
- All scheduled receipts, lead times, and beginning inventory levels are shown.
- Note that cotter pins appear *twice* in the bill of material.

Gear and rotor plate assembly
Lead time = 1 week
Minimum order quantity = 2500

Week	1	2	3	4	5	6
Gross Requirements	0	2000	0	2000	0	2000
Scheduled Receipts	0	0	0	0	0	0
Projected Available 1000						
Net Requirements						
Planned Receipts						
Planned Order Releases						

Wheels

Lead time = 1 week

Minimum order quantity = 1

Week		1	2	3	4	5	6
Gross Requirements							
Scheduled Receipts		0	0	0	0	0	0
Proj. Available	0						
Net Requirements							
Planned Receipts							
Planned Order Releases							

Cotter Pins

Lead time = 3 weeks

Minimum order quantity = 15,000

Week		1	2	3	4	5	6
Gross Requirements							
Scheduled Receipts		0	0	0	0	0	0
Proj. Available	9000						
Net Requirements							
Planned Receipts							
Planned Order Releases							

CASE STUDY 4.1

Apix Polybob Company

Kimberly Mack, plant manager for the Apix Polybob Company, was having a heated discussion with Jack Gould, the production and inventory control manager. Kimberly was getting tired of frantic calls from Jim Uphouse, the marketing and sales manager, concerning late orders for their Polybob (polybobs are a fictitious product) customers and was once again after Jack to solve the problem. Some of the discussion points follow:

JACK: "Look, Kimberly, I'm not sure what more we can do. I've reexamined the EOQ (economic order quantity lot size) values and all the reorder points for all our inventory for all our Polybob models, including all component levels and purchased items. I've implemented strict inventory control procedures to ensure our accuracy levels to at least 80%, and I've worked with the production people to make sure we are maximizing both labor efficiency and utilization of our equipment. The real problem is with those salespeople. We no sooner have a production run nicely going, and they change the order or add a new one. If they'd only leave us alone for a while and let us catch up with our current late order bank, we'd be okay. As it is, everyone is getting tired of order changes, expediting, and making everything into a crisis. Even our suppliers are losing patience with us. They tend to disbelieve any order we give them until we call them up for a crisis shipment."

KIMBERLY: "I find it hard to believe that you really have the EOQ and reorder point values right. If they were, we shouldn't have all these part shortages all the time while our overall inventory is going up in value. I also don't see any way we can shut off the orders coming in. I can imagine the explosion of anger from Jim if I even suggested such a thing. He'll certainly remind me that our mission statement clearly points out that our number-one priority is customer service, and refusing orders and order changes certainly doesn't fit as good customer service."

JACK: "Then maybe the approach is to deal with Sofia Vasquez (the chief financial officer). She's the one who is always angerly pointing out that we have too much inventory, too much expediting cost, too much premium freight costs from suppliers, and poor efficiency. I've tried to have her authorize more overtime to relieve some of the late order conditions, but all she'll say is that we must be making the wrong models. She continually points to the fact that the production hours we are paying for currently are more than enough to make our orders shipped at standard, and that condition has held for over a year. She just won't budge on that point. Maybe you can convince her."

KIMBERLY: "I'm not sure that's the answer either. I think she has a point, and she certainly has the numbers to back her up. I'd have a real rough time explaining what we were doing to Jamar Marrison (the CEO). There's got to be a better answer. I've heard about a systems approach called material requirements planning or something like that. Why don't you look into that? Take a representative model and see if that approach could help us deal with what appears to be an impossible situation. I'm sure something would work. I know other factories have similar production conditions yet don't seem to have all our problems."

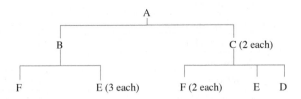

Following is the information about Polybob model A that Kimberly suggested as a representative model to use for the analysis:

Component	Lot Size	Inventory	Lead Time	Scheduled Receipts	Reorder Point
B	80	10	1	None	5
C	150	40	1	None	15
D	200	180	2	None	50
E	400	300	2	None	70
F	500	50	2	500, week 1	80

The following are the master schedule production lots for Model A:

Complete 50 units, week 3

Complete 50 units, week 5

Complete 60 units, week 7

Complete 60 units, week 9

Complete 50 units, week 11

Upon seeing this information, Jack stated, "Look at how regular our production schedule is for this model. The reorder points will more than cover requirements, and none have lead times that make it tough to respond. This analysis should show that all the work I did on EOQ and reorder points was right, and the real problem lies with those sales and finance people who don't understand our production needs."

Assignment

1. What are the key issues brought about in the conversation? What are the key symptoms, and what are the underlying problems? Be specific in your answers.

2. Use the product information to develop an MRP approach to the problems. Would MRP solve the problems? If so, show specifically how MRP would avoid the problems discussed by Ken and Jack.

3. Do any conditions bother you about the ability of MRP to deal with the problems? What specifically are those conditions?

4. Suppose it was discovered that only 250 of component E were in stock instead of the 300 listed on the inventory record. What problems would this cause (if any), and what are some of the ways that these problems could be addressed? How would (if at all) MRP help you when other methods might not?
5. Suppose that the design engineer advises that he has a new design for component F. It won't be ready until sometime after week 2, but he wants you to give a date for the first supplier shipment to come in, and you should be ready to tell the supplier how many to ship. Since the change is transparent to the customer, the design engineer advises you to go ahead and use up any existing material of the model. How will MRP help you to deal with this issue?
6. Can you think of any other "what if" questions that might be more easily addressed by a systematic approach such as MRP?

CASE STUDY 4.2

Benzie Products Company

Benzie Products Company produces several lines of products, but one (they call it "product X") uses unique parts to produce it and the demand is very seasonal. There are some possible variations in the design, so the company tends to use available-to-promise (ATP) logic to master schedule the product. Since the components to produce it are quite expensive, the company tries very hard to minimize any inventory of the product or its components during the seasons with very low sales. Product X is just now entering the low season, and the following chart represents the forecast data and actual customer orders for the next 10 weeks:

Week	1	2	3	4	5	6	7	8	9	10
Forecast	25	20	16	16	15	15	13	11	10	9
Customer Orders	27	21	16	13	11	9	7	4	3	2
MPS		50			50			50		

There are currently (at the start of week 1) 27 product X left in inventory. The following represents the product structure for product X:

```
        X
    ┌───┴───┐
    A      B(2)
    │       │
  ┌─┴─┐     │
  C   D    C(2)
```

The following table gives the relevant data for components A, B, C, and D at the start of week 1:

Component	A	B	C	D
Starting Inventory	0	2	212	63
Lead Time (weeks)	2	1	2	4
Lot Size	60	100	250	100
Safety Stock	0	0	0	0

In addition, component A has a scheduled receipt of 60 units for week 2.

Assignment

1. Complete the master schedule for product X, including the projected available inventory and the ATP numbers.
2. From the master schedule for product X and using the data given for components A, B, C, and D, create MRP grids for each of the components for the next 10 weeks.

3. Suppose that a customer for product X wants three additional units for their order scheduled for week 4. What would you tell them? Specifically, if you cannot promise them the three units for week 4, what is the best that you can do given the information you have. Assume that adequate capacity exists in all the production equipment.

4. The customer described in question 3 decides against placing the order for the extra units in week 4, but shortly thereafter you are informed that someone in the warehouse dropped a box of component C and broke 20 of them. They had to be scrapped. Describe the consequences and a plan of action to deal with the problem. Assume the lead time for component C is a firm two weeks.

5. It is clear that the company has set the safety stock (planning for extra material "just in case" something goes wrong) level for components at zero in order to minimize their inventory during the slow season. Discuss this policy, pointing out the pros and cons of such a policy. Develop what you might suggest as a policy for safety stock given the information available in the case.

CHAPTER FIVE

CAPACITY MANAGEMENT

5.1 INTRODUCTION

So far we have been concerned with planning priority, that is, determining what is to be produced and when. The system is hierarchical, moving from long planning horizons and few details (production plan) through medium time spans (master production schedule) to a high level of detail and short time spans (material requirements plan). At each level, manufacturing develops priority plans to satisfy demand. However, without the resources to achieve the priority plan, the plan will be unworkable. Capacity management is concerned with supplying the necessary resources. This chapter looks more closely at the question of capacity: what it is, how much is available, how much is required, and how to balance priority and capacity.

5.2 DEFINITION OF CAPACITY

Capacity is the amount of work that can be done in a specified time period. The *APICS Dictionary*, 16th edition defines capacity as "the capability of a worker, machine, work center, plant, or organization to produce output per time period." Capacity is a *rate* of doing work, not the *quantity* of work done. Two kinds of capacity are important: the capacity available and the capacity required. **Capacity available** is the capacity of a system or resource to produce a quantity of output in a given time period. **Capacity required** is the capacity of a system or resource needed to produce a desired output in a given time period. A term closely related to capacity required is **load**. This is the amount of released and planned work assigned to a facility for a particular time period. It is the sum of all the required capacities. These three terms—capacity required, load, and capacity available—are important in capacity management and will be discussed in subsequent sections of this chapter.

Capacity is often pictured as a funnel, as shown in Figure 5.1. Capacity available is the rate at which work can be withdrawn from the funnel. Load is the amount of work in the funnel.

Capacity management is responsible for determining the capacity needed to achieve the priority plans as well as providing, monitoring, and controlling that capacity so the

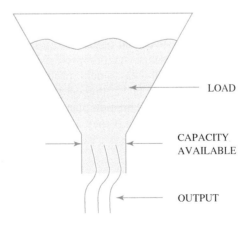

FIGURE 5.1 Capacity versus load.

priority plan can be met. The *APICS Dictionary*, 16th edition defines capacity management as "the function of establishing, measuring, monitoring, and adjusting limits or levels of capacity in order to execute all manufacturing schedules." As with all management processes, it consists of planning and control functions.

Capacity planning is the process of determining the resources required to meet the priority plan and the methods needed to make that capacity available. It takes place at each level of the priority planning process. Production planning, master production scheduling, and material requirements planning determine priorities, i.e., what is wanted and when. These priority plans cannot be implemented, however, unless the firm has sufficient capacity to fulfill the demand. Capacity planning links the various production priority schedules to manufacturing resources.

Capacity control is the process of monitoring production output, comparing it with capacity plans, and taking corrective action when needed. Capacity control will be examined further in Chapter 6.

5.3 CAPACITY PLANNING

Capacity planning involves calculating the capacity needed to achieve the priority plan and finding ways of making that capacity available. If the capacity requirement cannot be met, the priority plans are unachievable and have to be changed.

Priority plans are usually stated in units of product or some standard unit of output. Capacity can sometimes be stated in the same units, such as tons of steel or yards of cloth. If there is no common unit, capacity is stated as the hours available. The priority plan must then be translated into hours of work required and compared to the hours available. The process of capacity planning is as follows:

1. Determine the capacity available at each work center in each time period.

2. Determine the load at each work center in each time period.

   - Translate the priority plan into the hours of work required at each work center in each time period.

   - Sum up the capacities required for each item on each work center to determine the load on each work center in each time period.

3. Resolve differences between available capacity and required capacity. If possible, adjust available capacity to match the load. Otherwise, the priority plans must be changed to match the available capacity.

This process occurs at each level in the priority planning process, varying only in the level of detail and time spans involved.

Planning Levels

Resource planning involves long-range capacity resource requirements and is directly linked to production planning. Typically, it involves translating monthly, quarterly, or annual product priorities from the production plan into some total measure of capacity, such as gross labor hours. Resource planning involves changes in staffing, capital equipment, product design, or other facility changes that take a long time to acquire and eliminate. If a resource plan cannot be devised to meet the production plan, the production plan has to be changed. The two plans set the limits and levels for production. If they are realistic, the master production schedule (MPS) should be achievable. (See the Resource Planning section in Chapter 2, Production Planning System.)

Rough-cut capacity planning (RCCP) takes capacity planning to the next level of detail. The MPS is the primary information source for RCCP. The purpose of RCCP is to check the feasibility of the MPS, provide warnings of any bottlenecks, ensure utilization of work centers, and advise vendors of capacity requirements. (See the Rough-Cut Capacity Planning section of Chapter 3, Master Scheduling.)

Capacity Management

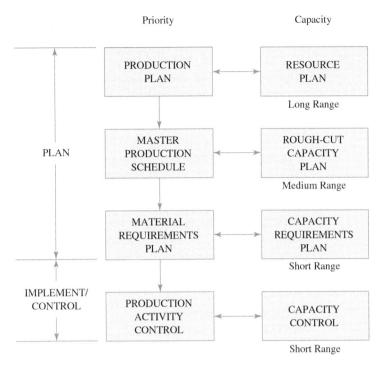

FIGURE 5.2 Planning levels.

Capacity requirements planning (CRP) is capacity planning at a more detailed level and is directly linked to the material requirements plan. Since this type of planning focuses on component parts, greater detail is involved than in rough-cut capacity planning. It is concerned with individual orders at individual work centers and calculates work center loads and labor requirements for each time period at each work center.

Figure 5.2 shows the relationship between the different levels of priority planning and capacity planning. Notice that, although the upper levels of priority planning are input to lower levels, the various capacity plans relate only to their level in the priority plan, not to subsequent capacity planning levels. Resource planning relates to production planning but is not an input to rough-cut capacity planning.

After the plans are completed, production activity control and purchasing must be authorized to process shop orders and purchase orders. Capacity must still be considered at this lowest level of detail. Work center capacity control will be covered in Chapter 6.

5.4 CAPACITY REQUIREMENTS PLANNING

Capacity requirements planning (CRP) occurs at the level of material requirements planning (MRP). It is the process of determining in detail the amount of labor and machine resources needed to achieve the required production. Planned orders from MRP and open **shop orders** (scheduled receipts) are converted into demand for time in each work center in each time period. This process takes into consideration the lead times for operations and offsets the operations at work centers accordingly. In considering open shop orders, it accounts for work already done on a shop order. CRP is the most detailed, complete, and accurate of the capacity planning techniques. This accuracy is most important in the immediate time periods. Because of the detail, a great amount of data and computation is required.

Inputs

The inputs needed for CRP are open shop orders, planned order releases, routings, time standards, lead times, and work center capacities. This information can be obtained from the following data:

- Open orders.
- Material requirements plan.

- Routings.
- Work centers.

Open orders An **open order** appears as a scheduled receipt on the MRP grid. It is a released order for a quantity of a part to be manufactured and completed on a specific date. It shows all relevant information, such as quantities, due dates, and operations. A record of all the active shop orders is maintained manually or as a computer file or table.

Planned order releases **Planned orders** are determined by the computer's MRP logic based upon the gross requirements for a particular part. They are inputs to the CRP process in assessing the total capacity required in future time periods.

Routings A **routing** is the path that work follows from work center to work center as it is completed. A routing is specified on a route sheet or in a computer routing file or table. Routing data should exist for every component manufactured and contain the following information:

- Operations to be performed.
- Sequence of operations.
- Work centers to be used.
- Possible alternate work centers.
- Tooling needed at each operation.
- Standard times: setup times and run times per piece.

Figure 5.3 shows an example of routing information.

Work centers A **work center** is comprised of a number of machines or workers capable of doing the same work. The equipment will normally be similar so there are no differences in the kind of work the machines can do or the capacity of each. Several sewing machines of similar capacity could be considered a work center. Work center data contains information on the capacity and move, wait, and queue times associated with the center.

The **move time** is the time normally taken to move material from one work center to another. The **wait time** is the time a job is at a work center after completion and before being moved. The **queue time** is the time a job waits at a work center before being processed. **Lead time** is the sum of queue, setup, run, wait, and move times.

Shop calendar Another piece of information needed is the number of working days available at a work center The Gregorian calendar, which is the one used in daily activities, has some serious drawbacks for manufacturing planning and control. The months do not have the same number of days, holidays are spread unevenly throughout the year, and the calendar does not work on a decimal base. Suppose that the lead time for an item is 35 working days and on December 13 an order is placed for delivery by January 22. This is about six weeks away, but with the Gregorian calendar, some calculations have to be made to decide if there is enough time to make the delivery. Holidays occur in that period, and the plant will be shut down for inventory the first week in January. How many working days really exist?

Part Name: Gear shaft Part Number: SG 123
Drawing Number: D123X

Operation No.	Work Center	S/U Time (standard hours)	Run Time/Piece (standard hours)	Operation
10	12	1.50	0.20	Turn shaft
20	14	0.50	0.25	Mill slot
30	17	0.30	0.05	Drill 2 holes
40	03	0.45	0.10	Grind
50	Stores			Inventory

FIGURE 5.3 Routing file.

MONTH	WEEK	MON.	TUES.	WED.	THURS.	FRI.	SAT.	SUN.
JULY	27	2 / 123	3 / 124	④	5 / 125	6 / 126	⑦	⑧
	28	9 / 127	10 / 128	11 / 129	12 / 130	13 / 131	⑭	⑮
	29	16 / 132	17 / 133	18 / 134	19 / 135	20 / 136	㉑	㉒
	30	23 / 137	24 / 138	25 / 139	26 / 140	27 / 141	㉘	㉙
	31	30 / 142	31 / 143	1 / 144	2 / 145	3 / 146	④	⑤

JULY 2 → [2 / 123] ← WORK DAY — 123   ◯ DEFINES NON-WORK DAYS

FIGURE 5.4 Planning calendar.
(*Source: The American Production and Inventory Control Society, Inc.*, Material Requirements Planning Training Aid, 5-21. Reprinted with permission.)

Because of these problems, it is desirable to develop a **shop calendar**. This can be set up in different ways, but the example shown in Figure 5.4 is typical.

5.5 CAPACITY AVAILABLE

We've already discussed that capacity available is the capacity of a system or resource to produce a quantity of output in a given time period. It is affected by the following:

- **Product specifications.** If the product specifications change, the process and time required to make the product may change, thus affecting the number of units that can be produced in a given time period.
- **Product mix.** Each product has its own set of processes and time requirements in what it takes to make the product. If the mix of products being produced changes, the total time, or capacity required, for the mix may change.
- **Plant and equipment.** This relates to the methods used to make the product. If the method is changed—for example, a faster machine is used—the rate of output may change. Similarly, if more machines are added to the work center, the capacity available will change.
- **Work effort.** This relates to the speed or pace at which the work is done. If the workforce changes pace, perhaps producing more in a given time, the capacity will be altered.

All of these elements have an impact on capacity. If these vary considerably, it is difficult to use units of product to measure capacity. So what units should be used to measure capacity?

Measuring Capacity

Units of output   If the variety of products produced at a work center or in a plant is not large, it is often possible to use a unit common to all products. Paper mills measure capacity in tons of paper, breweries in barrels of beer, and automobile manufacturers in numbers of cars. However, if a variety of products is made, a good common unit may not exist. In this case, the unit common to all products is time.

Standard time The work required to make a product is expressed as the time required to make the product using a given method of manufacture. Using time study techniques, the **standard time** (also known as **standard hours**) for a job can be determined, that is, the time it would take an average qualified operator working at a normal pace to do the job. It provides a yardstick for measuring work and a unit for stating capacity. It is also used in loading and scheduling work to be done.

Levels of Capacity

Capacity is usually measured on at least three levels:

- Machine or individual worker.
- Work center, production line or cell.
- Plant, which can be considered as a group of multiple work centers.

Determining Capacity Available

The factors of available time, utilization, and efficiency are used to determine the three types of capacity available: theoretical; calculated or rated; and demonstrated or measured.

Available time The **available time** is the number of hours a work center can be used. For example, a work center working one 8-hour shift for five days a week is available 40 hours a week. The available time depends on the number of machines, the number of workers, and the hours of operation.

Example Problem

A work center has three machines and is operated for 8 hours a day, five days a week. What is the available time?

Answer

$$\text{Available time} = 3 \times 8 \times 5 = 120 \text{ hours per week}$$

Utilization The available time is the maximum hours one can expect from the work center. However, it is unlikely this will be attained all the time. Downtime can occur due to machine breakdown, absenteeism, training, lack of material, and various problems that cause unavoidable delays or idle time. If the machine running is also dependent on an individual, time taken for lunch or breaks must also be considered. The percentage of time that the work center is active compared to the available time is called work center **utilization**:

$$\text{Utilization} = \frac{\text{hours actually worked}}{\text{available hours}} \times 100\%$$

Example Problem

A work center is available 120 hours but actually produces goods for only 100 hours. What is the utilization of the work center?

Answer

$$\text{Utilization} = \frac{100}{120} \times 100\% = 83.3\%$$

Utilization can be determined from historical records or by a work sampling study.

Efficiency It is possible for a work center to utilize 100 hours a week but not produce 100 standard hours of work. **Efficiency** measures the output as compared to the standard. The workers might be working at a faster or slower pace than the standard working pace, causing the efficiency of the work center to be more or less than 100%. For example, in a given shift, the expected output may be 100 hours, or the production of 50 units. But the actual output is 120 hours, which produced 60 units.

$$\text{Efficiency} = \frac{\text{standard hours produced}}{\text{hours actually worked}} \times 100\%$$

Example problem

A work center produces 120 standard hours of work in 100 hours. What is the efficiency?

$$\text{Efficiency} = \frac{120}{100} \times 100\% = 120\%$$

Productivity The *APICS Dictionary*, 16th edition defines **productivity** as "the overall measure of the ability to produce a good or a service." It compares the actual output of production to the actual input of all resources, incorporating the utilization of the time available and the efficiency during that time. There are many ways of measuring productivity. An example of a productivity calculation that can be applied to each work center uses the following formula:

$$\text{Productivity} = \text{utilization} \times \text{efficiency} \times 100\%$$

Example: A work center operates 32 hours of the 40 hours available. During that time, they produce output equivalent to 30 hours of work.

$$\text{Productivity} = \frac{32}{40} \times \frac{30}{32} \times 100\% = 75\%$$

Theoretical capacity is the maximum capacity available, with no regard for downtime, utilization, or efficiency. If a company uses two 8-hour shifts at a work center, the theoretical daily capacity would be 16 hours.

Rated capacity The available capacity at a work center for a period of time, and accounting for the average utilization and efficiency of that work center, is known as **calculated** or **rated capacity**.

$$\text{Rated capacity} = \text{available time} \times \text{utilization} \times \text{efficiency}$$

Example Problem

A work center consists of four machines and is operated 8 hours per day for five days a week. Historically, the utilization has been 85% and the efficiency 110%. What is the rated capacity?

ANSWER

Available time = 4 × 8 × 5 = 160 hours per week
Rated capacity = 160 × 0.85 × 1.10 = 149.6 standard hours

The expectation is to produce 149.6 standard hours of work from that work center in an average week.

Demonstrated Capacity

The historical output, or capacity, of a work center, is known as **demonstrated capacity**. This type of capacity examines the previous production records and uses that information as the available capacity of the work center. This is determined in part by the actual load input to the work center, and is not necessarily reflective of what the work center is capable of producing.

Example Problem

Over the previous four weeks, a work center produced 120, 130, 150, and 140 standard hours of work. What is the demonstrated capacity of the work center?

Answer

$$\text{Demonstrated capacity} = \frac{120 + 130 + 150 + 140}{4} = 135 \text{ standard hours}$$

Notice that demonstrated capacity is average, not maximum, output. It also depends on the utilization and efficiency of the work center, although these are not included in the calculation, as they would already have been taken into account in the production records used in the calculation.

Efficiency and utilization can be obtained from historical data if a record is maintained of the hours available, hours actually worked, and the standard hours produced by a work center.

Example Problem

Over a four-week period, a work center produced 540 standard hours of work, was available for work 640 hours, and actually worked 480 hours. Calculate the utilization and the efficiency of the work center.

Answer

$$\text{Utilization} = \frac{\text{hours actually worked}}{\text{available hours}} \times 100 = \frac{480}{640} \times 100\% = 75\%$$

$$\text{Efficiency} = \frac{\text{standard hours of work produced}}{\text{hours actually worked}} \times 100 = \frac{540}{480} \times 100\% = 112.5\%$$

Safety capacity One other type of capacity that is often used by companies is known as **safety capacity**. This capacity is available capacity that is planned to exceed capacity required. It is used to protect against unplanned activities, such as breakdowns, poor quality, preventive maintenance, and so forth. It is also referred to as a **capacity cushion**, and can be used as an alternative to safety stock. Care should be taken in using both safety capacity and safety stock, as this can cause additional costs for the firm.

5.6 CAPACITY REQUIRED (LOAD)

Capacity requirements are generated by the priority planning system and involve translating priorities, given in units of product or some common unit, into hours of work required at each work center in each time period. As mentioned previously, this translation takes place at each of the priority planning levels, from production planning to master production scheduling to material requirements planning. Figure 5.2 illustrates this relationship.

The level of detail, the planning horizon, and the techniques used vary with each planning level. In this chapter, the material requirements planning/capacity requirements planning level will be discussed.

Determining the capacity required is a two-step process. First, the time needed for each order at each work center is determined. Second, the capacity required for individual orders to obtain the load is totaled.

Time Needed for Each Order

The time needed for each order is the sum of the setup time and the **run time**. The run time is equal to the run time per piece multiplied by the number of pieces in the order.

Example Problem

A work center is to process 150 units of gear shaft SG 123 on work order 333. The setup time is 1.5 hours, and the run time is 0.2 hours per piece. What is the standard time needed to run the order?

Answer

$$\begin{aligned}
\text{Total standard time} &= \text{setup time} + \text{run time} \\
&= 1.5 + (150 \times 0.2) \\
&= 31.5 \text{ standard hours}
\end{aligned}$$

Example Problem

In the previous problem, how much actual time will be needed to run the order if the work center has an efficiency of 120% and a utilization of 80%?

$$\text{Capacity required} = (\text{actual time})(\text{efficiency})(\text{utilization})$$
$$\text{Actual time} = \frac{\text{capacity required}}{(\text{efficiency})(\text{utilization})}$$
$$= \frac{31.5}{(1.2)(0.8)}$$
$$= 32.8 \text{ hours}$$

Load

The load on a work center is the sum of the required times for all the planned and actual orders to be run on the work center in a specified period. The steps in calculating load are as follows:
1. Determine the standard hours of operation time for each planned and released order for each work center by time period.
2. Add all the standard hours together for each work center in each period. The result is the total required capacity (load) on that work center for each time period of the plan.

Example Problems

A work center has the following open orders and planned orders for week 20. Calculate the total standard time required (load) on this work center in week 20. Order 222 is already in progress, and there are 100 remaining to run.

	Order Quantity	Setup Time (hours)	Run Time (hours/piece)	Total Time (hours)
Released Orders				
222	100	0	0.2	
333	150	1.5	0.2	
Planned Orders				
444	200	3	0.25	
555	300	2.5	0.15	
Total Time				

Answer

Released orders 222 Total time = 0 + (100 × 0.2) = 20.0 standard hours
333 Total time = 1.5 + (150 × 0.2) = 31.5 standard hours
Planned orders 444 Total time = 3 + (200 × 0.25) = 53.0 standard hours
555 Total time = 2.5 + (300 × 0.15) = 47.5 standard hours
Total Time = 152.0 standard hours

In week 20, there is a load (requirement) for 152 standard hours.

The load must now be compared to the available capacity. One way of doing this is with a work center load profile report.

Work Center Load Profile Report

The work center **load profile** report shows future capacity requirements based on released and planned orders for each time period of the plan.

The load of 152 hours calculated in the previous example is for week 20. Similarly, loads for other weeks can be calculated and recorded on a load profile report such as the one shown in Figure 5.5. Figure 5.6 shows the same data in graphical form. Note that the

CAPACITY IN SERVICES

While few people in manufacturing tend to initially think of inventory as a luxury (many try to limit it because of the cost) from a capacity standpoint, it is a luxury. This is because manufacturing companies often can utilize their capacity to produce inventory ahead of the actual demand for that inventory. So, when the demand occurs, they can react very quickly by taking the inventory from stock. From that perspective, inventory can be thought of as "stored capacity."

Few service firms have that luxury of using inventory to store capacity. While many do carry inventory (retail shops, for example), the capacity they cannot "stock" is the capacity for the shop clerks to help customers. Since the demand for that service capacity is often erratic and difficult to predict, how do services respond while minimizing the time that capacity might be "wasted" (e.g., when there no customers to serve)?

There are at least three common methods used:

1. Employ multiskilled and flexible workers, e.g., a worker who is knowledgeable about restocking shelves, applying pricing tags, yet is also capable of serving customer needs. They can, for example, be restocking shelves when there is no demand for helping a customer, but immediately shift over to helping a customer when one needs help.

2. Utilize automation. Self-service checkout lines and ATMs in banks are examples.

3. In cases where the capacity of the service worker is often limited and very expensive (a physician or dentist, for example), rather than focus on automation or flexibility, the service will often control the demand itself so that the capacity can be highly utilized without pressure. Service appointments and reservations are the methods commonly used. This approach is also used in many automobile repair facilities and popular restaurants for the same reason.

Week	20	21	22	23	24	Total
Released Load	51.5	45	30	30	25	181.5
Planned Load	100.5	120	100	90	100	510.5
Total Load	152	165	130	120	125	692
Rated Capacity	140	140	140	140	140	700
(Over)/Under Capacity	(12)	(25)	10	20	15	8

FIGURE 5.5 Work center load profile report.

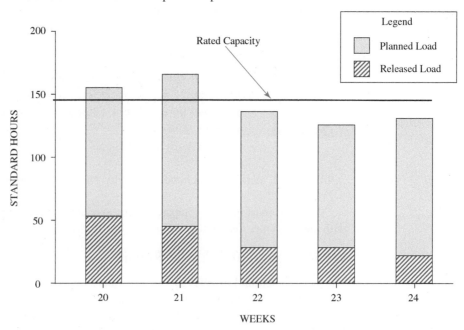

FIGURE 5.6 Graph of a load profile.

report shows released and planned load, total load, rated capacity, and over/undercapacity. The term *overcapacity* means that the work center is overloaded and the term *undercapacity* means the work center is underloaded. This type of display gives information used to adjust available capacity or to adjust the load by changing the priority plan. In this example, weeks 20 and 21 are overloaded, the rest are underloaded, and the cumulative load is less than the available. For the planner, this shows there is enough total capacity over the planning horizon, and available capacity or priority can be juggled to meet the plan.

5.7 SCHEDULING ORDERS

So far the assumption has been that CRP knows when an order should be run on one work center. Most orders are processed across a number of work centers, and it is necessary to calculate when orders must be started and completed on each work center so the final due date can be met. This process is called **scheduling**. The *APICS Dictionary*, 16th edition defines a schedule as "a timetable for planned occurrences."

Back scheduling The usual process is to start with the due date and, using the lead times, to work back to find the start date for each operation. This process is called **back scheduling** or **backward scheduling**. To back schedule, the following must be known for each order:

- Quantity and due date.
- Sequence of operations and work centers needed.
- Setup and run times for each operation.
- Queue, wait, and move times.
- Work center capacity available (rated or demonstrated).

The information needed is obtained from the following:

- Planned and open orders: Quantities and due dates.
- Routing: Sequence of operations, work centers needed, setup time, and run time.
- Work center data: Queue, move, and wait times and work center capacity.

The process is as follows:

1. For each work order, calculate the capacity required (time) at each work center.
2. Starting with the due date, schedule back to get the completion and start dates for each operation.

Example Problem

Suppose there is an order for 150 gear shaft SG 123. The due date is day 135. The route sheet, shown in Figure 5.3, gives information about the operations to be performed and the setup and run times. The work center file, shown in Figure 5.7, gives lead time data for each work center. Calculate the start and finish dates for each operation. Use the following scheduling rules.

Work Center	Queue Time (days)	Wait Time (days)	Move Time (days)
12	4	1	1
14	3	1	1
17	5	1	1
03	8	1	1

FIGURE 5.7 Lead time data from work center file.

124 CHAPTER FIVE

Operation Number	Work Center	Arrival Date (a.m.)	Queue (days)	Operation (days)	Wait (days)	Finish Date (p.m.)
10	12	95	4	4	1	103
20	14	105	3	5	1	113
30	17	115	5	1	1	121
40	03	123	8	2	1	133
50	Stores	135				

FIGURE 5.8 Work schedule.

a. Operation times are rounded up to the nearest 8 hours and expressed as days on a one-shift basis. That is, if an operation takes 6.5 standard hours, round it up to 8 hours, which represents one day.

b. Assume an order starts at the beginning of the day and finishes at the end of a day. For example, if an order starts on day 1 and is finished on day 5, it has taken five days to complete. If move time is one day, the order will be available to the next work center at the start of day 7.

ANSWER

The calculations for the operation time at each work center are as follows:

$$\text{Setup time} + \text{run time} = \text{total time (standard hours)}$$

Operation 10: Work center 12: $1.5 + 0.20 \times 150 = 31.5$ standard hours
$= 4$ days
Operation 20: Work center 14: $0.50 + 0.25 \times 150 = 38.0$ standard hours
$= 5$ days
Operation 30: Work center 17: $0.30 + 0.05 \times 150 = 7.8$ standard hours
$= 1$ day
Operation 40: Work center 03: $0.45 + 0.10 \times 150 = 15.45$ standard hours
$= 2$ days

The next step is to schedule back from the due date (day 135) to get the completion and start dates for each operation. To do so, not only must the operation times just calculated be known, but also the queue, wait, and move times. These are part of the work center data, such as those shown in Figure 5.7.

The back scheduling process starts with the last operation. The goods are to be in the stores at the beginning of day 135. It takes one day to move them, so the order must be completed on operation 40 at the end of day 133, leaving day 134 to move the product. Subtracting the wait, queue, and operation times (11 days), the order must be started at the beginning of day 123. With a move time of one day, it must be completed on operation 30 at the end of day 121. Using this process, the start and completion dates can be calculated for all operations. Figure 5.8 shows the resulting schedule and Figure 5.9 shows the same thing graphically.

An alternative approach is to use **forward scheduling**, which begins with a start date at the first operation, and moves forward through the operations calculating the start and completion dates for each operation to determine the completion date.

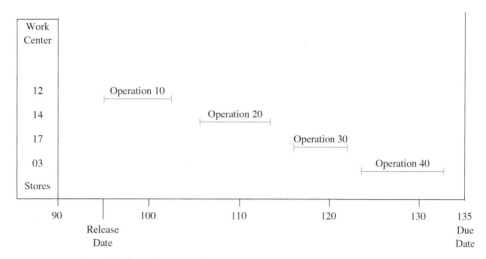

FIGURE 5.9 Graphical work schedule.

5.8 MAKING THE PLAN

So far the data needed for a capacity requirements plan, where the data comes from, and the scheduling and loading of shop orders through the various work centers has been determined. The next step is to compare the load with available capacity to see if there are imbalances and if so, to find possible solutions.

There are two ways of balancing capacity available and load: alter the load, or change the capacity available. Altering the load means shifting orders ahead or back so the load is leveled. If orders are processed on other work stations, the schedule and load on the other work stations have to be changed as well. It may also mean that other components should be rescheduled and the MPS changed.

Consider the bill of material shown in Figure 5.10. If component B is to be rescheduled to a later date, then the priority for component C is changed, as is the MPS for A. For these reasons, changing the load is usually not the preferred course of action. In the short timeframe, capacity can usually be adjusted easier than load. Some ways that this may be done are as follows:

- Schedule overtime or undertime. This will provide a temporary solution for cases where the load/capacity imbalance is not too large or long term.
- Adjust the level of the workforce by hiring or laying off workers. The ability to do so will depend on the availability of the skills required and the training needed. The higher the skill level and the longer the training needed, the more difficult it becomes to change the level of the workforce quickly.
- Shift workforce from underloaded to overloaded work centers. This may require a flexible cross-trained workforce, or adaptable equipment.
- Use alternate routings to shift some load to another work center. Often the other work center is not as efficient as the original. Nevertheless, the important thing is to meet the schedule, and this is a valid way of doing so.

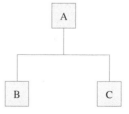

FIGURE 5.10 Simple bill of material.

- Subcontract work when more capacity is needed or bring in previously subcontracted work to increase load. It may be more costly to subcontract rather than make the item in-house, but again, it is important to maintain the schedule.

The result of CRP should be a detailed workable plan that meets the priority objectives and provides the capacity to do so. Ideally, it will satisfy the MRP and allow for adequate utilization of the workforce, machinery, and equipment.

5.9 CAPACITY PLANNING IN THE LEAN PRODUCTION ENVIRONMENT

Many manufacturing (and other) facilities have attempted to move their production environments to one typically identified as **lean production** (for a detailed description, see Chapter 16 on Lean Production). One of the general characteristics of successful lean production environments is a significant reduction in inventory at all stages (raw material, in-process material, and finished goods). The impact on capacity planning can be significant, especially considering that with far less inventory on hand any customer demand may need to be produced rather than being satisfied from finished goods. If the facility does not have the inventory to meet demand, they clearly should have the proper amount of capacity in the correct processes to produce the output for satisfying the customer demand.

An additional characteristic of a lean production environment is often a significant reduction in process throughput time. This, too, has an impact on capacity planning because scheduling work orders and developing detailed capacity plans is almost impossible to do. Specifically, the work almost needs to be completed and shipped to the customer before any detailed planning or scheduling can be completed.

These two characteristics imply a different focus on capacity planning. First, many would say that it is even more critical to plan for capacity in a lean production environment because if you don't have inventory to satisfy demand for your output you had better have the capacity to make that output in a very timely fashion. Secondly, since the processing time tends to happen much faster, it tends to negate the possibility (and often the need) to develop detailed capacity plans or schedules. Fortunately, an effective approach to rough-cut capacity planning paired with using the forward-looking power of ERP (MRP logic) organizations can still create capacity requirement plans from forecast driven master scheduling to identify some insight into capacity needs and timing.

Another recent response has been the development of **advanced planning and scheduling (APS)**. The *APICS Dictionary* (16th edition) describes APS as "techniques that deal with analysis and planning of logistics and manufacturing during short, intermediate, and long-term time periods. APS describes any computer program that uses advanced mathematical algorithms or logic to perform optimization or simulation on finite capacity scheduling, sourcing, capital planning, resource planning, forecasting, demand management, and others."

It should be noted that since capacity planning is often more critical and done differently in a lean environment, there is increased pressure on the data sources used to create these plans. For example, since a critical data source for capacity planning is standard processing times and setup times, these data sources should be monitored frequently. The very process of developing time standards is often based partially on estimates, and even if very accurate when developed, over time they will usually change due to processing conditions as well as learning curve impacts.

SUMMARY

Capacity management occurs at all levels of the planning process. It is directly related to the priority plan, and the level of detail and time spans will be similar to the related priority plan.

Capacity planning is concerned with translating the priority plan into the hours of capacity required in manufacturing to make the items in the priority plan and with methods of making that capacity available. Capacity available depends upon the number of workers and machines, their utilization, and efficiency.

Capacity requirements planning occurs at the material requirements planning level. It takes open shop orders and planned orders from the MRP and converts them to a load on each work center. It considers lead times and actual order quantities. It is the most detailed of the capacity planning techniques.

MRP and CRP should form part of a closed-loop system that not only includes planning and control functions but also provides feedback so planning can always be current. Figure 5.11 illustrates the concept.

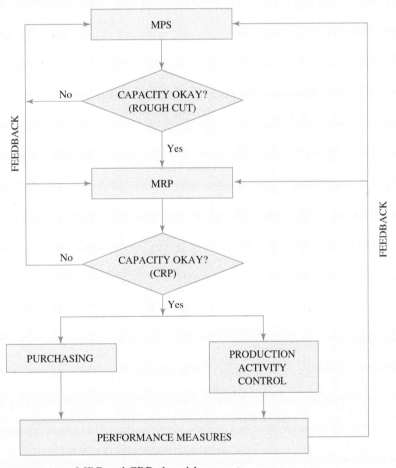

FIGURE 5.11 MRP and CRP closed-loop system.

KEY TERMS

Advanced planning and scheduling (APS) 126
Available time 118
Back scheduling 123
Backward scheduling 123

Calculated capacity 119
Capacity 113
Capacity available 113
Capacity control 114
Capacity cushion 120

Capacity management 113	Rated capacity 119
Capacity planning 114	Resource planning 114
Capacity required 113	Rough-cut capacity planning (RCCP) 114
Capacity requirements planning (CRP) 115	
Demonstrated capacity 119	Routing 116
Efficiency 118	Run time 120
Forward scheduling 124	Safety capacity 120
Lead time 116	Scheduling 123
Lean production 126	Shop calendar 117
Load 113	Shop order 115
Load profile 121	Standard hours 118
Move time 116	Standard time 118
Open order 116	Theoretical capacity 119
Planned order 116	Utilization 118
Productivity 119	Wait time 116
Queue time 116	Work center 116

QUESTIONS

1. What are the responsibilities of capacity management?
2. What is capacity planning?
3. Describe the three steps of capacity planning.
4. Relate the three levels of priority planning to capacity planning. Describe each level in terms of the detail and the time horizons used.
5. What is capacity requirements planning (CRP)? At what level of the priority planning process does it occur?
6. What are the inputs to the CRP process? Where is this information obtained?
7. Describe each of the following and the information they contain.
   a. Open order.
   b. Routing.
   c. Work center.
8. What is a shop calendar? Why is it needed?
9. Where would you find the following information?
   a. A scheduled receipt.
   b. A planned receipt.
   c. Efficiency and utilization.
   d. Sequence of operations on a part.
10. Define *capacity available*. What are the four factors that affect it?
11. Why is standard time usually used to measure capacity?
12. What are theoretical capacity, rated capacity, utilization, and efficiency? How are they related?
13. What is measured or demonstrated capacity? How is it different from rated capacity?
14. What is load?
15. What is a work center load profile report? What information does it contain?
16. What is a schedule?
17. Describe the process of backward scheduling.
18. What are the two ways of balancing capacity available and load? Which is preferred? Why?
19. What are some of the ways capacity available can be altered in the short run?
20. Why is feedback necessary in a control system?
21. What might be some of the problems in scheduling rated capacity too closely to the load?
22. How is safety capacity used?
23. How does the approach to capacity change in a "lean" production environment?

PROBLEMS

5.1. A work center consists of three machines each working a 16-hour day for five days a week. What is the weekly available time?

Answer. 240 hours per week

5.2. The work center in problem 5.1 is utilized 80% of the time. What are the hours per week actually worked?

5.3. If the efficiency of the work center in problem 5.1 is 120%, what is the rated capacity of the work center?

Answer. 230.4 standard hours per week

5.4. A work center consisting of seven machines is operated 16 hours a day for a five-day week. Utilization is 70%, and efficiency is 110%. What is the rated weekly capacity in standard hours?

Answer. 431.2 standard hours per week

5.5. A work center consists of four machines working 8 hours a day for a five-day week. If the utilization is 80% and the efficiency is 90%, what is the rated capacity of the work center?

5.6. Over a period of 4 weeks, a work center produced 50, 48, 44, and 52 standard hours of work. What is the demonstrated capacity of the work center?

Answer. 48.5 standard hours of work per week

5.7. In an 11-week period, a work center produces 1050 standard hours of work. What is the measured capacity of the work center?

5.8. In one week, a work center produces 75 standard hours of work. The hours scheduled are 80, and 72 hours are actually worked. Calculate the utilization and efficiency of the work center.

Answer. Utilization is 90%; efficiency is 104%

5.9. A work center consisting of three machines operates 40 hours a week. In a four-week period, it actually worked 360 hours and produced 470 standard hours of work. Calculate the utilization and efficiency of the work center. What is the demonstrated weekly capacity of the work center?

5.10. A firm wishes to determine the efficiency and utilization of a work center composed of five machines each working 16 hours per day for five days a week. A study undertaken by the materials management department found that over the past 50 weeks the work center was actually working for 16,000 hours, and work performed was 15,400 standard hours. Calculate the utilization, efficiency, and demonstrated weekly capacity.

5.11. How many standard hours are needed to run an order of 200 pieces if the setup time is 1.1 hours and the run time 0.3 hours per piece? How many actual hours are needed at the work center if the efficiency is 120% and the utilization is 70%?

Answer. 61.1 standard hours; 72.7 actual hours

5.12. How many standard hours are needed to run an order of 500 pieces if the setup time is 2.0 hours and the run time 0.3 hours per piece? How many actual hours are needed at the work center if the efficiency is 110% and the utilization is 85%?

5.13. A work center has the following open and planned orders for week 4. Calculate the total standard time required (load).

		Order Quantity	Setup Time (hours)	Run Time (hours/piece)	Total Time (hours)
Released Orders	120	300	1.00	0.10	
	340	200	2.50	0.30	
Planned Orders	560	300	3.00	0.25	
	780	500	2.00	0.15	
Total Time (standard hours)					

Answer. Total time = 248.5 standard hours

5.14. A work center has the following open and planned orders for week 4. Calculate the total standard time required (load).

		Order Quantity	Setup Time (hours)	Run Time (hours/piece)	Total Time (hours)
Released Orders	125	200	0.50	0.10	
	345	70	0.75	0.07	
Planned Orders	565	80	2.00	0.25	
	785	35	1.50	0.14	
Total Time (standard hours)					

5.15. Using the following route information, open order information, and MRP planned orders, calculate the load on the work center.

Routing: Part 123:   Setup time   = 2 standard hours
                              Run time per piece = 3 standard hours per piece

             Part 456:   Setup time   = 3 standard hours
                              Run time per piece = 1 standard hour per piece

Open Orders for parts

Week	1	2	3
123	13	11	4
456	16	6	7

Planned Orders for parts

1	2	3
0	5	10
0	10	12

Load profile report

Week		1	2	3
Released Load	123			
	456			
Planned Load	123			
	456			
Total Load				

5.16. Complete the following load profile report and suggest possible courses of action.

Week	18	19	20	21	Total
Released Load	160	155	100	70	485
Planned Load	0	0	70	80	150
Total Load					
Rated Capacity	150	150	150	150	600
(Over)/Under Capacity					

5.17. Back schedule the following shop order. All times are given in days. Move time between operations is one day, and wait time is one day. Due date is day 150. Assume orders start at the beginning of a day and finish at the end of a day.

Operation Number	Work Center	Operation Time (days)	Queue Time (days)	Arrival Date (a.m.)	Finish Date (p.m.)
10	111	2	4		
20	130	4	5		
30	155	1	2		
	Stores			150	

Answer. The order must arrive at work center 111 on day 126.

5.18. Back schedule the following shop order. All times are given in days. Move time between operations is one day, and wait time is one day. Due date is day 200. Assume orders start at the beginning of a day and finish at the end of a day.

Operation Number	Work Center	Operation Time (days)	Queue Time (days)	Arrival Date (a.m.)	Finish Date (p.m.)
10	110	3	2		
20	120	6	4		
30	130	3	2		
	Stores			200	

CASE STUDY 5.1

Wescott Products

Whenever Jason Roberts thought about going to work on Friday morning, he started to get a little knot in his stomach. Jason had recently accepted the job as operations manager for a small manufacturing company that specialized in a line of assemble-to-order products. When he accepted the job he was a recent graduate of a business program where he specialized in operations. He had done fairly well in his classes and had emerged as a confident, self-assured person who was sure he could handle such a job in a small company.

The company, Wescott Products, had recently experienced rapid growth from the original start in a two-car garage just five years earlier. In fact, Jason was the first person ever named as operations manager. Prior to that, the only production manager reporting to the owner, Judy Wescott, was Frank Adams, the production supervisor. While Frank was an experienced supervisor, he had been promoted to supervisor directly from his old job as a machine operator and had no formal training in planning and control. He soon found that planning was too complex and difficult for him to handle, especially since he also had full responsibility for all the Wescott workers and equipment. Jamarcus Stockard, the sales and marketing manager, had requested and finally applauded Judy Wescott's decision to hire Jason, since he felt production was having a much more difficult time in promising and delivering customer orders. Jamarcus was starting to spend more and more time on the phone with angry customers when they didn't get their orders at the time they expected them. The time away from developing new sales and the danger of losing established customers started to make him highly concerned about sustaining sales growth, to say nothing about his potential bonus check tied to new sales!

Once Jason was placed in the position, however, the "honeymoon" was short, and soon Jason started doubting how much he really did know. The company was still having trouble with promising customer orders and having the capacity to meet those orders. At first he thought it was the forecasting method he was using, but a recent analysis told him the total actual orders were generally within about 10% of what the forecast projected.

In addition, production never seemed to have any significant shortages in either subassemblies or components. In fact, many felt they had far too much material, and in the last couple of staff meetings Kali Chowdhury, the company controller, was grumbling that she thought the inventory turn ratio of just less than 3.5 was unreasonable and costing the company a lot of money. It must be something else, and Jason had to discover it quickly.

The first idea he thought about was to request the assembly areas to work overtime, but he soon found out that was a sensitive topic that could only be used as a last resort. The workers in that area were highly skilled and would be difficult, if not impossible, to replace in any reasonable time. Adding more employees would also be difficult for the same reason. A year earlier they were being worked a lot of overtime but had finally had enough. Even though Wescott had no union, the workers got together and demanded better overtime control or they would all quit to move to other jobs that were plentiful for skilled workers in this area. The agreement was that they were to be asked for no more than 4 hours of overtime per worker per week unless it was truly an emergency situation. They were well paid and all had families, and the time with their families was worth more to them than additional overtime pay. At least the high skill level had one advantage: Each of the workers in the assembly area could skillfully assemble any of the models, and the equipment was flexible enough to handle all the models.

Friday mornings were when Jason made his master schedule for the next week and no matter how hard he tried he never seemed to be able to get it right. Since the standard lead time for all assemblies was quoted as one week, the company had felt no need to schedule farther into the future when few orders were assumed to exist there. He was sure that he had to start the process by loading the jobs that were missed in the current week into the Monday and Tuesday time blocks and then hope that production could catch up with those in addition to the new jobs that were already promised. The promises came when Jamarcus would inform him of a customer request and ask for a promise date, which was often "as soon as possible." Jason would look at the order to see if the material to make it was in stock and if the equipment to make it was running. He would then typically promise to have it available when requested. Now that a lot of promises were not being met, however, Jamarcus was starting to demand that Jason get control of the operation. Jason tried to respond by scheduling a lot of each model to be run every week, but he often found he had to break into the run of a lot to respond to expediting from sales. He knew this made matters worse by using extra time to set up the equipment, but what else could he do? Even Judy Wescott was asking him what she could do to help him improve the performance. His normal high level of self-confidence was being shaken.

Jason started pouring over his old operations book looking for something he could use. He finally realized that what he needed was a more effective system to develop master schedules from which he could promise orders, order components, and plan capacity. Unfortunately, he also recalled that when that material was covered in his class he had taken off early for spring break. Even though he knew enough to recognize the nature of the problem, he didn't know enough to set up such a schedule. Humbly, he called his former instructor to ask for advice. Once she was briefed on the problem, she told him to gather some information that he could use to develop a sample master schedule and rough-cut capacity plan. Once he had the information, she would help show him how to use it.

The following describes what she asked him to collect:

1. Pick a work center or piece of equipment that has caused some capacity problem in the recent past. List all the product models that use that work center.
2. For each of the models, list the amount of run time they use the work center per item. Also list the setup time, if any. These times can be gathered from standards or, if the standard data is suspect or does not exist, use the actual average time from recent production.
3. For each of the models, list the usual lot size. This should be the same lot size used for the master schedule.
4. For each of the models, list the current inventory, the current forecast, and the current firm customer order quantities.

5. List the current capacity (hours) available for the equipment.

The following tables summarize the data Jason collected:

WORK CENTER 12

Model	Run Time (per item, in minutes)	Setup Time (per lot)	Lot Size (minimum qty.)	On-hand*
A	3.7	90 minutes	150	10
B	5.1	40 minutes	100	0
C	4.3	60 minutes	120	0
D	8.4	200 minutes	350	22
E	11.2	120 minutes	400	153

*Most of the on-hand was really forced when the lot size exceeded orders for the week for that model. They would then assemble the rest of the lot as "plain vanilla," such that they could easily add any subassembly options once the actual customer orders came in.

Two workers are currently assigned to the work center, and only to the first shift. Even though assembly workers are very flexible, Jason cannot take workers from another assembly area, as those work centers are also behind and therefore appear to be equally overloaded.

The following is the forecast and customer orders for each of the five models assembled in work center 12:

Model	Weeks	1	2	3	4	5	6	7	8	9	10
A	Forecast	45	45	45	45	45	45	45	45	45	45
	Customer Orders	53	41	22	15	4	7	2	0	0	0
B	Forecast	35	35	35	35	35	35	35	35	35	35
	Customer Orders	66	40	31	30	17	6	2	0	0	0
C	Forecast	50	50	50	50	50	50	50	50	50	50
	Customer Orders	52	43	33	21	14	4	7	1	0	0
D	Forecast	180	180	180	180	180	180	180	180	180	180
	Customer Orders	277	190	178	132	94	51	12	7	9	2
E	Forecast	200	200	200	200	200	200	200	200	200	200
	Customer Orders	223	174	185	109	74	36	12	2	0	0

Once Jason had gathered all the data, he immediately called his instructor, only to find out that by an ironic twist of fate she would be gone for more than a week on spring break.

Assignment

This leaves you to help Jason. Specifically, you need to do the following:

1. Discuss the nature and probable sources of the problem.
2. Examine the rough-cut capacity situation using the data Jason gathered. Discuss the results and how they are linked to the problems identified in question 1.
3. Use the information and your knowledge of the situation to develop a complete plan for Jason to use in the future. Part of this plan should be to build and demonstrate the approach to master scheduling for the data given in the case.

CHAPTER SIX

PRODUCTION ACTIVITY CONTROL

6.1 INTRODUCTION

After all the planning and scheduling has been completed, it is time for the plans to be put into action. Production activity control (PAC) is responsible for executing the master production schedule (MPS) and material requirements planning (MRP). At the same time, it must make good use of labor and machines, minimize work-in-process inventory, and maintain customer service.

The material requirements plan authorizes PAC to:

- Release work orders to manufacturing.
- Manage work orders and make sure they are completed on time.
- Be responsible for the immediate detailed planning of the flow of orders through manufacturing, carrying out the plan, and controlling the work as it progresses to completion.
- Manage day-to-day activity and provide the necessary support.
- Report on schedule completion, work-in-process progress, scrap/yield, and any other information necessary.

Figure 6.1 shows the relationship between the planning system and PAC.

The activities of the PAC system can be classified into planning, implementation, and control functions.

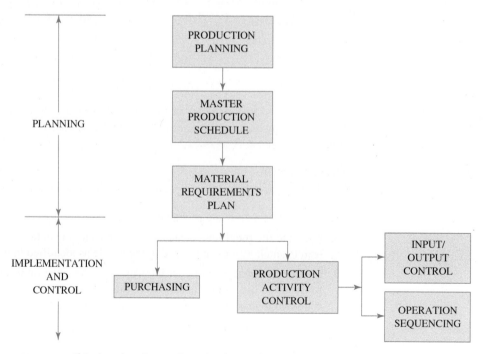

FIGURE 6.1 Priority planning and production activity control.

Planning

The flow of work through each of the **work centers** must be planned to meet delivery dates, which means production activity control must do the following:

- Ensure that the required materials, tooling, personnel, and information are available to manufacture the components when needed.
- Schedule start and completion dates for each production order at each work center so the scheduled completion date of the order can be met. This will include developing a load profile for the work centers.

Implementation

Once the plans are made, production activity control must execute the plan by advising manufacturing what must be done. Instructions can be given by issuing a production order with the relevant information, or by simply producing a schedule that shows product information, quantities, and dates. PAC will:

- Gather the information needed by manufacturing to make the product.
- Release orders to manufacturing as authorized by the material requirements plan. This is called dispatching.

Control

Once plans are made and production orders released, the process must be monitored to analyze what is actually happening. The results are compared to the plan to decide whether corrective action is necessary. PAC will do the following:

- Rank the production orders in desired priority sequence by work center and establish a dispatch list based on this information.
- Track the actual performance of work orders and compare it to planned schedules. Where necessary, PAC must take corrective action by replanning, rescheduling, or adjusting capacity to meet final delivery requirements.
- Monitor and control work-in-process, lead times, and work center queues.
- Report work center efficiency, operation times, order quantities, and scrap.

The functions of planning, implementing, and controlling are shown in Figure 6.2.

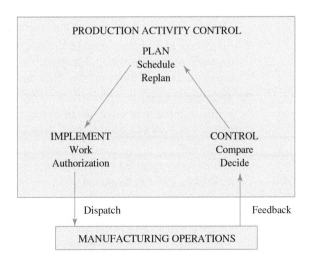

FIGURE 6.2 Diagram of a production activity control system.

Manufacturing Systems

The particular type of production control system used varies from company to company, but all should perform the functions mentioned. However, the relative importance of these functions will depend on the type of manufacturing process. Manufacturing processes can be conveniently organized into three categories:

1. Flow manufacturing.
2. Intermittent manufacturing.
3. Project manufacturing.

Flow manufacturing **Flow manufacturing** is concerned with the production of high-volume standard products. If the units are discrete (e.g., cars and appliances), the process is usually called **repetitive manufacturing**, and if the goods are made in a continuous flow (e.g., gasoline), the process is called **continuous manufacturing**. There are four major characteristics of flow manufacturing:

1. Routings are fixed, and work centers are arranged according to the routing. The time taken to perform work at one work center is almost the same as at any other work center in the line, enabling a constant flow and eliminating a queue in front of the next work center.
2. Work centers are dedicated to producing a limited range of similar products. Machinery and tooling are especially designed to make the specific products.
3. Material flows from one work center to another using some form of transfer. There is little buildup in work-in-process inventory, and throughput times are low.
4. Capacity is fixed by the line.

Production activity control concentrates on planning the flow of work and making sure that the right material is fed to the line as stated in the planned schedule. Since work flows from one work center to another automatically, implementation and control are relatively simple.

Intermittent manufacturing **Intermittent manufacturing** is characterized by many variations in product design, process requirements, and order quantities. This type of manufacturing is also referred to as **job shop**. This kind of manufacturing is characterized by the following:

1. Flow of work through the shop is varied and depends on the design of a particular product. As orders are processed, they may take more time at one work center than at another. Thus, the work flow is not balanced.
2. Machinery and workers must be flexible enough to do the variety of work involved in intermittent manufacturing. Machinery and work centers are usually grouped according to the function they perform, for example, all lathes in one department.
3. Throughput times are generally long. Scheduling work to arrive just when needed is difficult, the time taken by an order at each work center varies, and work queues before work centers, causing long delays in processing. Work-in-process inventory is often large.
4. The capacity required depends on the particular mix of products being built and is sometimes difficult to predict.

Production activity control in intermittent manufacturing is complex. Because of the number of products made, the variety of routings, and scheduling problems, PAC is a major activity in this type of manufacturing. Planning and control are typically exercised using production orders or detailed schedules for each batch being produced. Most of the discussion of PAC in this text assumes this kind of environment.

Project manufacturing **Project manufacturing** usually involves the creation of one unit or a small number of units. Complex shipbuilding is an example. Because the design of a product is often carried out or modified as the project develops, there is close coordination between manufacturing, marketing, purchasing, and engineering.

Project manufacturing or management uses many of the same techniques as PAC, but also has some unique characteristics. Activities typically included in project management are:

- Initiating the project, which includes identifying the project requirements.
- Planning the project, including the scope, schedule and tasks, budget, resources, and risks.
- Executing the project by carrying out the tasks.
- Monitoring and controlling the project tasks and resources, and communicating the status of the project to stakeholders.
- Closing the project, which includes documenting the results, as well as any variances in time and costs.

6.2 DATA REQUIREMENTS

To plan the processing of materials through manufacturing, PAC must have the following information:

- What and how much to produce.
- When parts are needed so the completion date can be met.
- What operations are required to make the product and how long the operations will take.
- What the available capacities of the various work centers are.

Planning Information

The types of information typically required by production activity control for planning are item master, product structure, routing, work center, and work order information.

Item master There is one record in the **item master** database for each item number, containing all of the pertinent data related to the part. For PAC, this includes the following:

- Item number. A unique number assigned to a component or finished goods.
- Item description.
- Manufacturing lead time. The normal time needed to make this part.
- Lot size quantity. The quantity normally ordered at one time.

Product structure (bill of material) The **product structure (bill of material)** contains a list of the single-level components and quantities needed to assemble a parent. It forms a basis for a **picking list** to be used by storeroom personnel to collect the parts required to make the assembly. There are a variety of ways of displaying a bill of material, including a single level bill of material, an indented bill of material, a multilevel bill of material, a phantom bill of material, a matrix bill of material, a modular bill of material, and a costed bill of material. In some industries, in particular the process industry, the bill of material is called a formula, recipe, or ingredients list.

Routing The **routing** contains a record for each part manufactured, consisting of a series of operations required to make the item. For each product, a step-by-step set of instructions is provided describing how the product is made. It gives details of the following:

- The operations required to make the product and the sequence in which those operations are performed.
- A brief description of each operation.

- Equipment, tools, and accessories needed for each operation.
- Setup times.
- Run times.
- Lead times for each operation.

Work center master The **work center master** collects all of the relevant data on a work center. For each work center, it gives details on the following:

- Work center number.
- Capacity.
- Number of shifts worked per week.
- Number of machine hours per shift.
- Number of labor hours per shift.
- Efficiency.
- Utilization.
- Queue time.
- Alternate work centers that may be used as alternatives if additional capacity is required.

Control Information

Control in intermittent manufacturing is exercised through **production orders** and the data contained on these orders. These orders may also be referred to as manufacturing orders, work orders or shop orders.

Production order master Each active manufacturing order has a record that provides summarized data on each production order, such as the following information:

- Order number—a unique number identifying the order.
- Order quantity.
- Quantity completed.
- Quantity scrapped.
- Quantity of material issued to the order.
- Due date—the date the order is expected to be finished.
- Balance due—the quantity not yet completed.

The details of each production order contain records for each operation needed to make the item. Each record contains the following information:

- Operation number.
- Setup hours, planned and actual.
- Run hours, planned and actual.
- Quantity reported complete at that operation.
- Quantity reported scrapped at that operation.
- Due date or lead time remaining.

6.3 ORDER PREPARATION

Once authorization to process an order has been received, PAC is responsible for planning and preparing its release to manufacturing. The order should be reviewed to be sure that the necessary tooling, material, and capacity are available. If they are not, the order cannot be completed and should not be released. Enterprise resource planning (ERP) software tools enable the checking of the availability of materials, and the pre-allocation of them to the order.

If capacity requirements planning (CRP) has been used, necessary capacity should be available. However, there may still be some differences between planned capacity and what is actually available, due to product that is behind schedule, daily changes in workforce, and so forth. The rechecking of capacity availability is done through the use of scheduling, as well as determining the current load on necessary work centers.

6.4 SCHEDULING

The objective of scheduling is to meet delivery dates and to make the best use of manufacturing resources. It involves establishing start and finish dates for each operation required to complete an item. To develop a reliable schedule, the planner must have information on the routing, required and available capacity, competing jobs, and manufacturing lead times at each work center involved.

Manufacturing Lead Time

Manufacturing lead time is the time normally required to produce an item in a typical lot quantity. As discussed previously, it typically consists of five elements:

1. **Queue time**—the amount of time the job is waiting at a work center before the operation begins.
2. **Setup time**—the time required to prepare the work center for the operation.
3. **Run time**—the time needed to run the complete order through the operation.
4. **Wait time**—the amount of time the job is at the work center before being moved to the next work center.
5. **Move time**—the transit time between work centers.

The total manufacturing lead time is the sum of order preparation and release plus the manufacturing lead time for each operation. Figure 6.3 shows the elements making up manufacturing lead time. Determining setup time and run time is the responsibility of the industrial engineering department. Queue, wait, and move times are under the control of manufacturing and PAC.

The largest of the five elements is generally queue time. Typically, in an intermittent manufacturing operation, it may account for 80–90% of the total lead time. Production activity control is responsible for managing the queue by regulating the flow of work into and out of work centers. If the number of orders waiting to be worked on (load) is reduced, so is the queue time, the lead time, and work-in-process. Increasing capacity also reduces queue. PAC must manage both the input of orders to the production process and the available capacity to control queue and work-in-process.

A term that is closely related to manufacturing lead time is cycle time. The *APICS Dictionary*, 16th edition, defines **cycle time** as "the time between the completion of two discrete units of production." A synonym of cycle time is **throughput time**.

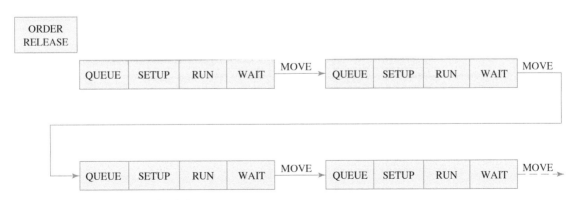

FIGURE 6.3 Manufacturing lead time.

Example Problem

An order for 100 of a product is processed on work centers A and B. The setup time on A is 30 minutes, and run time is 10 minutes per piece. The setup time on B is 50 minutes, and the run time is 5 minutes per piece. Wait time between the two operations is 4 hours. The move time between A and B is 10 minutes. Wait time after operation B is 4 hours, and the move time into stores is 15 minutes. There is no queue at either work center. Calculate the total manufacturing lead time for the order.

Answer

Work center A operation time	$= 30 + (100 \times 10) =$	1030 minutes
Wait time	$=$	240 minutes
Move time from A to B	$=$	10 minutes
Work center B operation time	$= 50 + (100 \times 5) =$	550 minutes
Wait time	$=$	240 minutes
Move time from B to stores	$=$	15 minutes
Total manufacturing lead time	$=$	2085 minutes
	$=$	34 hours, 45 minutes

Scheduling Techniques

There are many techniques used to schedule orders through a plant, but all of them require an understanding of forward and backward scheduling as well as finite and infinite loading.

Forward scheduling assumes that material procurement and operation scheduling for a component start when the order is received, whatever the due date, and that operations are scheduled forward from this date. The first line in Figure 6.4 illustrates this method. The result is completion before the due date, which may result in a buildup of finished goods inventory. This method is used to decide the earliest delivery date for a product.

Forward scheduling is used to calculate how long it will take to complete a task. The technique is used for purposes such as developing promise dates for customers or figuring out whether an order behind schedule can be caught up.

Backward scheduling is illustrated by the second line in Figure 6.4. The last operation on the routing is scheduled first and is scheduled for completion at the due date. Previous operations are scheduled back from the last operation. This schedules items to be available as needed and uses the same logic as the MRP system. Work-in-process inventory is reduced, but because there is little slack time in the system, customer service may suffer.

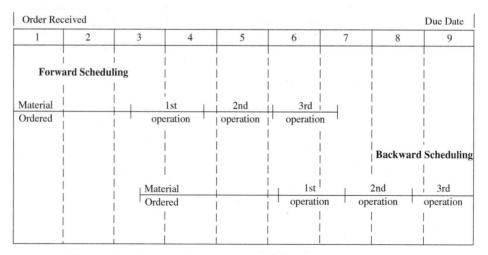

FIGURE 6.4 Forward and backward scheduling: infinite loading.

Production Activity Control

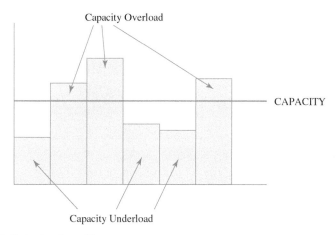

FIGURE 6.5 Infinite load profile.

Backward scheduling is used to determine when an order must be started, and is common in a make-to-stock environment because it schedules materials to be available only when needed.

Many companies use a hybrid approach, beginning with backward scheduling. If the necessary lead time is not available to complete the order on time, forward scheduling is then used to determine the earliest possible delivery date.

Infinite loading is also illustrated in Figure 6.4. The assumption is made that the work centers in which operations 1, 2, and 3 are processed have capacity available when required. It does not consider the existence of other production orders competing for capacity at these work centers. It assumes infinite capacity will be available. Figure 6.5 shows a load profile for infinite capacity. Notice the over- and underload.

Finite loading assumes there is a defined limit to available capacity at any work center. If there is not enough capacity available at a work center because of other shop orders, the order has to be scheduled in a different time period. Figure 6.6 illustrates the condition.

In the forward scheduling example shown in Figure 6.6, the first and second operations cannot be performed at their respective work centers when they should be because the required capacity is not available at the time required. These operations must be rescheduled to a later time period. Similarly, in the example of back scheduling, the second and first operations cannot be performed when they should be and must be rescheduled to an earlier time period. Figure 6.7 shows a load profile for finite loading. Notice the load is smoothed so there is no overload condition.

Chapter 5 gives an example of backward scheduling as it relates to capacity requirements planning. The same process is used in PAC.

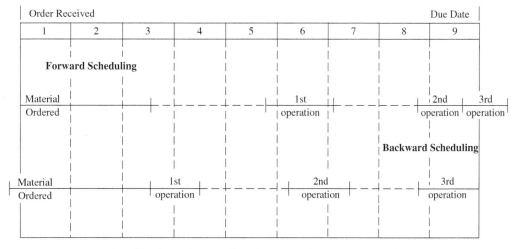

FIGURE 6.6 Forward and backward scheduling. Finite loading.

142 CHAPTER SIX

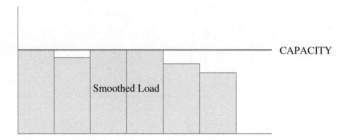

FIGURE 6.7 Finite load profile.

Example Problem

A company has an order for 50 brand X to be delivered on day 100. Draw a backward schedule based on the following:

a. Only one machine is assigned to each operation.
b. The factory works one 8-hour shift five days a week.
c. The parts move in one lot of 50.

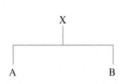

Part	Operation	Time (days)
A	10	5
A	20	3
B	10	10
X	Assembly	5

ANSWER

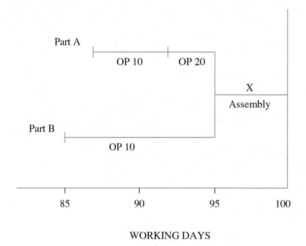

WORKING DAYS

Operation Overlapping

In **operation overlapping**, the next operation is allowed to begin before the entire lot is completed on the previous operation. This reduces the total manufacturing lead times because the second operation starts before the first operation finishes all the parts in the order. Figure 6.8 shows an example of how it works and the potential reduction in lead time.

A concept used in operation overlapping is the difference between a process batch and a transfer batch. A **process batch** is the total lot size that has been released to production. A **transfer batch** is that quantity that moves from work center to work center. A process batch may consist of one or more transfer batches.

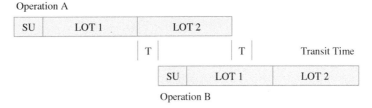

FIGURE 6.8 Operation overlapping.

To perform operation overlapping, an order is divided into at least two lots. When the first lot is completed on operation A, it is transferred to operation B. As seen in Figure 6.8, it is assumed that operation B cannot be set up until the first lot is received, but this is not always the case. While operation A continues with the second lot, operation B starts on the first lot. When operation A finishes the second lot, it is transferred to operation B. If the lots are sized properly, there will be no idle time at operation B. The manufacturing lead time is reduced by the overlap time and the elimination of queue time.

Operation overlapping is a method of expediting an order, but there are some costs involved. First, move costs are increased, especially if the overlapped operations are not close together. Second, it may increase the queue and wait times for other orders. Third, it does not increase capacity but potentially reduces it if the second operation is idle waiting for parts from the first operation.

The difficulty is deciding the size of the sublot. If the run time per piece on operation B is shorter than that on A, the first batch must be large enough to avoid idle time on operation B.

Example Problem

Refer to the data given in the example problem in the section on manufacturing lead time. It is decided to overlap operations A and B by splitting the lot of 100 into two lots of 70 and 30. Wait time between A and B and between B and stores is eliminated. The move times remain the same. Setup on operation B cannot start until the first batch arrives. Calculate the manufacturing lead time. How much time has been saved?

Answer

Operation time for A for lot of 70 = 30 + (70 × 10)	= 730	minutes
Move time between A and B	= 10	minutes
Operation time for B for lot of 100 = 50 + (100 × 5)	= 550	minutes
Move time from B to stores	= 15	minutes
Total manufacturing lead time	= 1305	minutes
	= 21	hours, 45 minutes
Time saved	= 13	hours

Operation Splitting

Operation splitting is a second method of reducing manufacturing lead time. The order is split into two or more lots or transfer batches and run on two or more machines simultaneously. If the lot is split in two, the run-time component of lead time is effectively cut in half, although an additional setup is incurred. Figure 6.9 shows a sample of operation splitting.

Operation splitting is practical when:

- Setup time is low compared to run time.
- A suitable work center is idle.
- It is possible for an operator to run more than one machine at a time.
- Duplicate tooling or equipment is available.

144 CHAPTER SIX

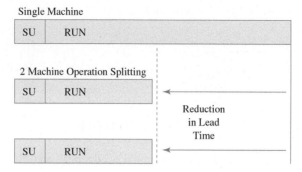

FIGURE 6.9 Operation splitting.

The last condition often exists when a machine cycles through its operation without human intervention, leaving the operator time to set up another machine. The time needed to unload and load must be shorter than the run time per piece. For example, if the unload/load time was 2 minutes and the run time was 3 minutes, the operator would have time to unload and load the first machine while the second was running.

Example Problem

A component made on a particular work center has a setup time of 100 minutes and a run time of 3 minutes per piece. An order for 500 is to be processed on two machines simultaneously. The machines can be set up at the same time. Calculate the elapsed operation time.

ANSWER

Elapsed operation time = $100 + 3 \times 250 = 850$ minutes
$\phantom{\text{Elapsed operation time}} = 14$ hours and 10 minutes

6.5 LOAD LEVELING

Load profiles were discussed in Chapter 5 in the section on capacity requirements planning. The **load profile** for a work center is constructed by calculating the standard hours of operation for each order in each time period and adding them together by time period. Figure 6.10 shows an example of a load report.

This report tells PAC what the load is on the work center. There is a capacity shortage in week 20 of 30 hours. This means there would be no point in releasing all of the planned orders that week. Perhaps some could be released in week 18 or 19, and perhaps some overtime could be worked to help reduce the capacity crunch. Rescheduling operations in order to more evenly distribute the load is referred to as **load leveling**.

Week	18	19	20	21	22	23	Total
Released Load	105	100	80	30	0	0	315
Planned Load			60	80	130	80	350
Total Load	105	100	140	110	130	80	665
Rated Capacity	110	110	110	110	110	110	660
(Over)/Under Capacity	5	10	(30)	0	(20)	30	(5)

FIGURE 6.10 Work center load profile report.

6.6 SCHEDULING IN A NONMANUFACTURING SETTING

All industries deal with the challenges of scheduling resources, balancing demand and supply, and available versus required capacity. For example, in the transportation industry, fleets of trucks must be scheduled and routed to minimize the total cost while ensuring timely deliveries, minimal downtime, and the nonproductive time equated with vehicles returning empty.

In the healthcare industry, organizations must balance the available capacity of doctors, surgeons, nurses, technicians, operating rooms, hospital rooms, and so forth, with a dynamic capacity required by patients, emergency vehicles, and major traumas. While some of these may be planned, such as the scheduling of office visits and annual physicals, much of the load comes from unplanned events such as illnesses and natural disasters, and is difficult to predict. Many hospitals have begun forecasting the load by looking at past history of patient days by month to determine trends or seasonality in order to better plan capacity. Time studies can also be done to determine standards for activities such as lab work and surgical prep to better determine available capacity for specific resources.

The planning of resources is also critical in service industries such as retail, food, airlines, and so forth. Scheduling of service personnel often occurs at a weekly, daily, and hourly level based on predictions of when customers are most likely to need the service. The component of capacity in this case is human resources, but can also include equipment, tools, and time. Some industries, such as airlines and transportation, have to also deal with limitations of working hours for personnel, for example, not working more than a certain number of hours in a day.

Being able to adapt the level of work force in a service industry by cross-training employees, utilizing part-time workers, or adopting automation tools such as self-help kiosks, a company can optimize resources and increase capacity. Nonurgent tasks, such as cleaning and maintenance, can be performed during periods of low or no demand to utilize personnel. One popular food chain developed its own method of scheduling that included breaks, projections of how much food to prepare, and when to cut back on the production of its baked goods and begin offering samples to customers.

6.7 SCHEDULING BOTTLENECKS

In intermittent manufacturing, it is almost impossible to balance the available capacities of the various work centers with the demand for their capacity. As a result, some work centers are overloaded and some underloaded. The overloaded work centers are called **bottlenecks** and, by definition, are those work centers where the required capacity is greater than the available capacity. The *APICS Dictionary*, 16th edition, defines a bottleneck as "a facility, function, department, or resource whose capacity is less than the demand placed upon it."

Throughput

Throughput is the total volume of production passing through a facility. Bottlenecks control the throughput of all products processed by them, as total throughput cannot be more than can be processed through the bottleneck. If work centers feeding bottlenecks produce more than the bottleneck can process, excess work-in-process inventory is built up. Therefore, work should be scheduled through the bottleneck at the rate it can process the work. Work centers fed by bottlenecks have their throughput controlled by the bottleneck, and their schedules should be determined by that of the bottleneck.

Example Problem

Suppose a manufacturer makes wagons composed of a box body, a handle assembly, and two wheel assemblies. Demand for the wagons is 500 a week. The wheel assembly capacity is 1200 sets a week, the handle assembly capacity is 450 a week, and final assembly can produce 550 wagons a week.

a. What is the capacity of the factory?
b. What limits the throughput of the factory?
c. How many wheel assemblies should be made each week?
d. What is the utilization of the wheel assembly operation?
e. What happens if the wheel assembly utilization is increased to 100%?

Answer

a. units a week.
b. Throughput is limited by the capacity of the handle assembly operation.
c. wheel assemblies should be made each week. This matches the capacity of the handle assembly operation.
d. Utilization of the wheel assembly operation is 900 ÷ 1200 = 75.
e. Excess inventory builds up.

The service sector also deals with throughput, such as the length of time a patient stays at a hospital, the number of times a restaurant turns tables during the dinner hour, or the amount of time a customer waits in line at a bank. One of the difficulties for service organizations is the variability in the time a service may take. The time it takes to wait on a customer at a bank, for example, may vary considerably, depending on the number and type of transactions.

Bottleneck Principles

Since bottlenecks control the throughput of a facility, some important principles should be noted:

1. **Utilization of a nonbottleneck resource is not determined by its potential but by another constraint in the process.** In the previous example problem, the utilization of the wheel assembly operation was determined by the handle assembly operation.

2. **Using a non-bottleneck resource 100% of the time does not produce 100% utilization.** If the wheel assembly operation was utilized 100% of the time, it would produce 1200 sets of wheels a week, 300 sets more than needed. Because of the buildup of inventory, this operation would eventually have to stop.

3. **The capacity of the facility depends on the capacity of the bottleneck.** If the handle assembly operation breaks down, the throughput of the factory is reduced.

4. **Time saved at a non-bottleneck does not save capacity elsewhere.** If the industrial engineering department increased the capacity of the wheel assembly operation to 1500 units a week, the extra capacity could not be utilized, and nothing would be gained.

5. **Capacity and priority must be considered together.** Suppose the wagon manufacturer made wagons with two styles of handles. During setup, nothing is produced, which reduces the capacity of the system. Since handle assembly is the bottleneck, every setup in this operation reduces the throughput of the system. Ideally, the company would run one style of handle for six months, then switch over to the second style. However, customers wanting the second style of handle might not be willing to wait six months. A compromise is needed whereby runs are as long as possible but priority (demand) is satisfied.

6. **Loads can, and should, be split.** Suppose the handle assembly operation (the bottleneck) produces one style of handle for two weeks, then switches to the second style. The batch size is 900 handles. Rather than waiting until the 900 are produced before moving them to the final assembly area, the manufacturer can move a day's production (90) at a time. The process batch size (900) and the transfer batch size (90) are different. Thus, delivery to the final assembly is matched to usage, and work-in-process inventory is reduced.

7. **Focus should be on balancing the flow through the facility.** The key is total throughput that ends up in saleable goods.

Managing Bottlenecks

Since bottlenecks are so important to the throughput of a system, scheduling and controlling them is extremely important. The following must be done:

1. **Establish a time buffer before each bottleneck.** A time buffer is an inventory (queue) before each bottleneck. Because it is of the utmost importance to keep the bottleneck working, it must never be starved for material by disrupting the flow from feeding work centers. The time buffer should be only as long as the time of any expected delay caused by feeding work centers. In this way, the time buffer ensures that the bottleneck will not be shut down for lack of work and this queue will be held at a predetermined minimum quantity.

2. **Control the rate of material feeding the bottleneck.** A bottleneck must be fed at a rate equal to its capacity so the time buffer remains constant. The first operation in the sequence of operations is called a **gateway operation**. This operation and any other operations prior to the bottleneck control the work feeding the bottleneck and must operate at a rate equal to the output of the bottleneck so the time buffer queue is maintained.

3. **Do everything possible to provide the needed bottleneck capacity.** Anything that increases the capacity of the bottleneck increases the capacity of the process. Better utilization, fewer setups, and improved methods to reduce setup and run time are some methods for increasing capacity.

4. **Adjust loads.** This is similar to point 3 but puts emphasis on reducing the load on a bottleneck by using such things as alternate work centers and subcontracting. These may be more costly than using the bottleneck, but utilization of non-bottlenecks and throughput of the total facility are increased, which will result in more efficient operations, and a potential for increased sales and profits.

5. **Change the schedule.** As discussed earlier, this should be done as a final resort, but may be necessary in order to provide accurate delivery promises.

Once the bottleneck is scheduled according to its available capacity and the market demand it must satisfy, the non-bottleneck resources can be scheduled. When a work order is completed at the bottleneck, it can be scheduled on subsequent operations.

Operations that feed the bottleneck have to protect the time buffer by scheduling backward from the bottleneck. If the time buffer is set at four days, the operation immediately preceding the bottleneck is scheduled to complete the required parts four days before they are scheduled to run on the bottleneck. Each preceding operation can be back scheduled in the same way so the parts are available as required for the next operation.

Any disturbances in the feeding operations are absorbed by the time buffer, and throughput is not affected. Also, work-in-process inventory is reduced. Since the queue is limited to the time buffer, lead times are reduced.

Bottlenecks occur in every process, including manufacturing, hospitals, transportation, and restaurants. They must be managed, if possible, to retain customer loyalty. If the bottleneck is a result of not enough personnel, extra hours or additional shifts may be possible. However, in some cases, increasing capacity, such as adding extra hotel rooms or airline seats, cannot be easily accomplished.

One large airline determined that a bottleneck existed in the turnaround of an aircraft, which was limiting the number of flights available, and causing too much downtime of their limited resource—flight crews. From the arrival at the airport terminal, the time it took to disembark passengers, clean out the airplane, resupply the catering items, add fuel to the aircraft, unload and reload baggage, and get ready to board the next set of passengers, took a minimum of 45 minutes. They determined that there was also a direct correlation between aircraft turnaround efficiency and schedule punctuality. Through the use of technology, the staging of luggage carts, and a more robust communication system between the flight crews and ramp personnel, they were able to cut the time down to an average of 20 minutes. This also had a positive impact on the on-time departure percentage, which improved by 30%.

6.8 THEORY OF CONSTRAINTS AND DRUM-BUFFER-ROPE

The section on managing bottlenecks was developed based on the work of Eliyahu M. Goldratt in his *Theory of Constraints*. It has allowed many people to rethink their approaches to improving and managing their production processes. The fundamental concept behind **theory of constraints (TOC)** is that every operation producing a product or service is a series of linked processes. Each process has a specific capacity to produce the given defined output for the operation, and in virtually every case, there is one process that limits or constrains the throughput from the entire operation. Refer to Figure 6.11 for an example of an operation producing product A.

The total operation is constrained by process 3 at a capacity of four per hour. No matter how much efficiency there is in the other processes and how many process improvements are made in processes 1, 2, and 4, it will never be possible to exceed the overall operational output of four per hour. Increased efficiency and utilization in processes 1 and 2 will only increase inventory, not sales.

Identifying the constraint in a process can actually be fairly simple. There is always a set of defined actions (processes) that are needed to create a finished product. When one process is discovered that is working to full capacity while inventory is growing behind the process waiting for the process, and processes downstream from that one process tend to have idle time with respect to their need to process the inventory, then the constraint has been identified. If all orders are scheduled and all raw material for those orders released, yet all processes in a production sequence have idle time for the required production, then sales is said to be the constraint.

Manage the Constraint

Several fundamental guidelines have been developed for understanding how to manage a constraining process or bottleneck, which were discussed in the section on bottleneck principles. These principles of balancing the overall flow, and maintaining steady work at the constraint, are critical to TOC.

Improve the Process

Once a constraint has been identified, there is a five-step process that is recommended to help improve the performance of the operation. The five steps are summarized as follows:

1. **Identify the constraint.** The entire process must be examined to determine which process limits the throughput. The concept does not limit this process examination to merely the operational processes. For example, in Figure 6.11, suppose the sales department was selling the output only at the rate of three per hour. In that case, sales would be the constraint and not process 3. It must be remembered that a constraint limits throughput, not inventory or production.

2. **Exploit the constraint.** Find methods to maximize the utilization of the constraint toward productive throughput. For example, in many operations all processes are shut down during lunchtime. If a process is a constraint, the operation should consider rotating lunch periods so that the constraint is never allowed to be idle.

3. **Subordinate everything to the constraint.** Effective utilization of the constraint is the most important issue. Everything else is secondary.

4. **Elevate the constraint.** This denotes finding ways to increase the available hours of the constraint until it is no longer the constraint.

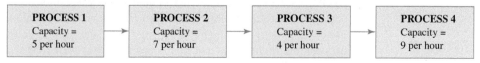

FIGURE 6.11 Product A.

5. **Once the constraint is no longer a constraint, find a new one and repeat the steps.** As the effective utilization of the constraint increases, it may cease to be a constraint as another process becomes one. In that case the emphasis shifts to the new process constraint.

Scheduling with the Theory of Constraints

The scheduling system developed for the theory of constraints has its own specific approach. It is often described as **drum-buffer-rope (DBR)**:

- **Drum**. The drum of the system refers to the "drumbeat" or pace of production. It represents the master schedule for the operation, which is focused around the pace of throughput as defined by the constraint. Note that once the pace of the constraint has been defined, it does no good to schedule more production than the constraint can handle. To do so will merely add in-process inventory and may actually decrease its effectiveness.

- **Buffer**. Since it is so important that the constraint never be starved for needed inventory, a time buffer is often established in front of the constraint. It is called a time buffer because it represents the amount of time that the inventory in the buffer protects the constraint from disruptions. In many systems there are actually three buffers: one for the constraint, one for assembly, and one for shipping. The constraint buffer often represents the processing time to protect the buffer from unexpected process variation. For example, a two-day time buffer would necessitate upstream operations to complete processing and have material in the buffer two days before actually needed. The initial time buffer used is often the total processing time from raw material release to reaching the constraint process. The assembly buffer often represents the time from raw material release to a process where components that do not go through the constraint process have to be assembled with components that do have to go through the constraint process. Finally, the buffer for shipping represents the processing time from the point the material leaves the constraint process to completion of the final product.

- **Rope**. The analogy is that the rope "pulls" production to the constraint for necessary processing. Although this may imply a reactive replenishment system, such as a reorder point, it can be done by a well-coordinated release of material into the system at the right time. The rope schedules release of raw material into production at a pace that maintains the buffer, ensures the constraint is not waiting for material, and that excessive inventory does not build up. It is basically defined by the processing capability of the constraint process.

Organizations using theory of constraints focus scheduling activities primarily on effective management of the organization's constraint to throughput and sales.

Four primary plant types are defined, and they are used to specify the flow of materials through a production process. They can therefore be helpful in understanding how to manage the operation using TOC. The four types are:

- I-plant, where one raw material is used to make one final product. Processing is usually done in a straight line.
- A-plant, where numerous subassemblies merge into a single final assembly.
- V-plant, where few raw materials can be made into several end products.
- T-plant, where multiple straight lines can split into several assemblies.

The theory of constraints also includes a process to help develop and implement change in an organization. The first step to this is to identify core conflicts, which are then validated by building what is called a *current reality tree*. After those undesirable effects are identified from the core conflicts, a *future reality tree* is developed, which will lay out a strategy to resolve the problems. The final major step in the process is to build a tactical objective map that will define a strategy to accomplish the future reality.

Example Problem

Parent X requires 1 each of component Y and Z. Both Y and Z are processed on work center 20, which has an available capacity of 40 hours. The setup time for component Y is 1 hour and the run time 0.3 hour per piece. For component Z, setup time is 2 hours and the run time is 0.20 hour per piece. Calculate the number of Ys and Zs that can be produced.

ANSWER

Let x = number of Ys and Zs to produce

$$\text{Time}_Y + \text{Time}_Z = 40 \text{ hours}$$
$$1 + 0.3x + 2 + 0.2x = 40 \text{ hours}$$
$$0.5x = 37 \text{ hours}$$
$$x = 74$$

Therefore, work center 20 can produce 74 Ys and 74 Zs.

Example Problem

In this problem, parent A is made of one B and 2 Cs. As and Bs are both made on work center 1, which has a capacity of 40 hours per week. The Cs are made on work center 2, which also has a capacity of 40 hours per week.

Product	Setup Time (hours)	Run Time (hours/unit)
A	2	0.1
B	2	0.2
C	1	0.3

Based on the above information, calculate the maximum number of As, Bs, and Cs that should be produced per week.

ANSWER

The number of Bs produced should equal the number of As produced to avoid over production. Therefore, the number of Bs can be expressed in a formula as the number of As.

Work center 1

$$\text{Time}_A + \text{Time}_B = 40 \text{ hours}$$
$$2 + 0.1A + 2 + 0.2A = 40$$
$$0.3A = 36$$
$$A = 120$$

Work center 1 has the capacity to produce enough As and Bs to make 120 As per week.

Work center 2

The number of Cs produced should be twice the number of As produced to avoid over production. Therefore, the number of Cs can be represented by 2 × A.

$$\text{Time}_C = 40 \text{ Hours}$$
$$1 + 2 \times 0.3A = 40$$
$$0.6A = 39$$
$$A = 65$$

Work center 2 has the capacity to make enough Cs to support production of only 65 As per week, and in this case, this is the constraint. To avoid over production, work center 1 should produce 65 As and 65 Bs per week. Work center 2 should produce 130 Cs per week (enough for 65 As).

In this example, work center 1 will have very low utilization. However, producing more than 65 Bs per week will only build inventory and work center 1 will be starved by work center 2, which has the capacity to produce only 130 Cs per week.

6.9 IMPLEMENTATION

Orders that have the necessary tooling, material, and capacity available have a good chance of being completed on time and can be released to production. Other orders that do not have all of the necessary elements should not be released because they only cause excess work-in-process inventory and may interrupt work on orders that can be completed. The process for releasing an order is shown in Figure 6.12.

Implementation is performed by issuing an order or schedule to manufacturing, which is the authorization for them to proceed with making the item. A shop packet that contains the information that is needed by manufacturing can be compiled. It may include any of the following:

- Order header showing the shop order number, item number, name, description, and quantity.
- Engineering drawings.
- Bills of material.
- Route sheets showing the operations to be performed, equipment and accessories needed, materials to use, and setup and run times.
- Material issue tickets that authorize manufacturing to get the required material from stores. These are also used for charging the material against the order.
- Tool requisitions authorizing manufacturing to withdraw necessary tooling from the tool crib.
- Job tickets for each operation to be performed. As well as authorizing the individual operations to be performed, they also can function as part of a reporting system. The

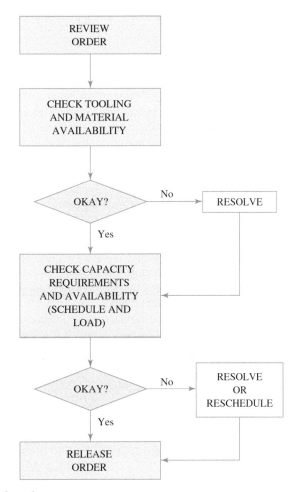

FIGURE 6.12 Order release process.

worker can log on and off the job using the job ticket, and it then becomes a record of that operation.
- Move tickets that authorize and direct the movement of work between operations.

Many manufacturing companies today use a paperless system, which authorizes production via a production schedule, rather than the release of a shop order. The shop packet is replaced by access to the same information electronically. Online reporting of material movement and labor reporting are used in exchange for tickets.

6.10 CONTROL

Once work orders have been issued to manufacturing, their progress has to be controlled. To control progress, performance has to be measured and compared to what is planned. If what is actually happening (what is measured) varies significantly from what was planned, either the plans have to be changed or corrective action must be taken to bring performance back to plan.

The objectives of PAC are to meet delivery dates and to make the best use of company resources. To meet delivery dates, a company must control the progress of orders on the production floor, which means controlling the lead time for orders. As discussed previously in this chapter, the largest component of lead time is queue. If queue can be controlled, delivery dates can usually be met. Intermittent operations have many different products and order quantities and many different routings, each requiring different capacities. In this environment, it is almost impossible to balance the load over all the work centers. Queue exists because of this erratic input and output.

To control queue and meet delivery commitments, PAC must:

- Control the work going into and coming out of a work center. This is generally called **input/output control (I/O)**.
- Set the correct priority of orders to run at each work center, which is referred to as dispatching.

Input/Output Control

Production activity control must balance the flow of work to and from different work centers. This is to ensure that queue, work-in-process, and lead times are controlled. The input/output control system is a method of managing queues and work-in-process lead times by monitoring and controlling the input to, and output from, a facility or work center. It is designed to balance the input rate in hours with the output rate.

The input rate is controlled by the release of orders to production. If the rate of input is increased, queue, work-in-process, and lead times increase. The output rate is controlled by increasing or decreasing the capacity of a work center. Capacity change can be attained by overtime or undertime, shifting workers, and so forth. Figure 6.13 shows the idea graphically.

Input/output report To control input and output, a plan must be devised, along with a method for comparing what actually occurs against what was planned. This information is shown on an input/output report. Figure 6.14 shows an example of such a report. The values are in standard hours.

The cumulative variance is the difference between the total planned for a given period and the actual total for that period. It is calculated as follows:

Cumulative variance = previous cumulative variance + actual − planned

Cumulative input variance week 2 = −4 + 32 − 32 = −4

Backlog is the same as queue and expresses the work to be done in hours. It is calculated as follows:

Planned backlog for period 1 = previous backlog + planned input − planned output
$$= 32 + 38 - 40$$
$$= 30 \text{ hours}$$

Production Activity Control 153

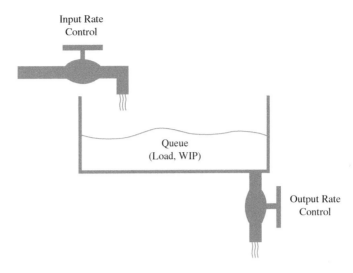

Figure 6.13 Input/output control.

Period	1	2	3	4	5	Total
Planned Input	38	32	36	40	44	190
Actual Input	34	32	32	42	40	180
Cumulative Variance	−4	−4	−8	−6	−10	−10

Planned Output	40	40	40	40	40	200
Actual Output	32	36	44	44	36	192
Cumulative Variance	−8	−12	−8	−4	−8	−8

Planned Backlog	32	30	22	18	18	22	
Actual Backlog	32	34	30	18	16	20	

FIGURE 6.14 Input/output report.

The report shows that the plan was to maintain a level output in each period and to reduce the queue and lead time by 10 hours, but input and output were lower than expected.

Planned and actual inputs monitor the flow of work coming to the work center. Planned and actual outputs monitor the performance of the work center. Planned and actual backlogs monitor the queue and lead time performance.

Example Problem

Complete the following input/output report for weeks 1 and 2.

Dispatching rules should be simple to use and easy to understand. As shown in Figure 6.16, each rule produces a different sequence and has its own advantages and disadvantages. Whichever rule is selected, it should be consistent with the objectives of the planning system.

6.11 PRODUCTION REPORTING

Production reporting provides feedback of what is actually happening on the plant floor. It allows PAC to maintain valid records of on-hand and on-order balances, job status, scrap, material shortages, and so on. PAC needs this information to establish proper priorities and to answer questions regarding deliveries, shortages, and the status of orders. Manufacturing management needs this information to make decisions about plant operations. Sales and customer service need this information to provide timely information to customers.

Data must be collected, sorted, and reported. The particular data collected depends upon the needs of the various departments. The methods of data collection vary. Sometimes the operator reports the start and completion of an operation, order, movement, and so on, using an online system directly reporting events as they occur. In other cases, the operator, or supervisor reports this information on an operation reporting form included in the shop packet. In addition, some manufacturing facilities have the capability for the information to be recorded automatically as production occurs using technology such as scanners, barcode readers, RFID, etc. Information about inventory withdrawals and receipts must be reported as well.

Once the data is collected and recorded, appropriate reports are produced. Types of information needed for the various reports include:

- Order status.
- Weekly input/output by department or work center.
- Exception reports on such things as scrap, rework, and late orders.
- Inventory status.
- Performance summaries on order status, work center and department efficiencies, and so on.

6.12 PRODUCT TRACKING

Production control is often responsible for product tracking or **lot traceability**. This is the process of tracking parts and materials back to their origins. It may have a very practical application, such as the matching of colors in fabrics and paint ensuring the consumer that different units of the product match in color at the time of manufacturing and during the product's lifetime. Traceability is also legislated in industries such as food, pharmaceutical, or aerospace to ensure the safety of the product. Should a product prove unsafe, it is possible for the manufacturer to trace back to find the source of all materials and recall all finished products that used that particular lot. Software and technology have allowed this to change from a very meticulous process to a more streamlined approach of collecting information along the product's supply chain.

Lot control **Lot control** is a primary method used for traceability of products, from finished goods back through raw materials. Each purchase or production run is identified with a unique batch number in order to maintain information and integrity as it moves from supplier to end customer. These unique numbers are often referred to as **serial numbers**, which are unique numbers assigned to a single piece or lot. For example, a production run of a product may be identified with a number that identifies the date and location of the production (042121NC39). The raw materials and components may also be serialized by the supplier, enabling the tracing back from the finished goods to a particular component or food substance that has a defect, or is being recalled.

6.13 MEASUREMENT SYSTEMS

As mentioned previously, to control progress, as well as adjust plans, performance has to be measured and compared to what was planned. In the area of production activity control, there are many types of performance measurement systems available. The primary purpose of these measurements is to provide an objective means of evaluating performance, and to take corrective action if necessary. It is also important to make sure whatever type of measurement is used, it is aligned with the overall performance measurement of the organization.

In addition to those already discussed in this text, such as utilization, efficiency, productivity, demonstrated capacity, and input/output control, some of the more common measurements used are as follows:

Actual versus planned lead time: A comparison of the actual throughput time to the stated lead time.

Percent orders completed on time: Percentage of orders completed on the due date, rather than early or late.

Performance to schedule: A measure of the quantity and date produced as compared to the master schedule.

Measurement systems will be discussed further in Chapters 16 and 17.

SUMMARY

Production activity control (PAC) is concerned with converting the material requirements plan into action, reporting the results achieved, and when required, revising the plans and actions to meet the required results. Order release, dispatching, and progress reporting are the three primary functions. To accomplish the plans, PAC must establish detailed schedules for each order, set priorities for work to be done at each work center, and keep them current. PAC is also responsible for managing the queue and lead times. Nonmanufacturing industries must also control capacity and inventory in order to monitor progress, manage resources, and derive appropriate schedules.

The theory of constraints (TOC) modifies the approach to PAC since it views a production facility, the suppliers, and the market as a series of interdependent functions. TOC attempts to optimize the constraints (bottlenecks) in a system as they affect the overall throughput. As a result, traditional lot sizing rules should be modified to increase the throughput of the entire process and not just the individual work centers. Drum-buffer-rope describes how the TOC works by setting an overall pace of material flow, ensuring bottlenecks never run out of material, and linking the output of one work center to another. Measurement systems are used by PAC to monitor and control progress, meet delivery dates, utilize labor and equipment efficiently, and keep inventory levels down.

KEY TERMS

Backward scheduling 140
Bill of material 137
Bottleneck 145
Buffer 149
Continuous manufacturing 136
Critical ratio (CR) 156
Cycle time 139
Dispatching 154
Drum 149
Drum-buffer-rope (DBR) 149

Earliest due date (EDD) 155
Earliest operation due date (ODD) 155
Finite loading 141
First-come-first-served (FCFS) 155
Flow manufacturing 136
Forward scheduling 140
Gateway operation 147
Infinite loading 141
Input/output control (I/O) 152
Intermittent manufacturing 136

Item master 137
Job shop 136
Load leveling 144
Load profile 144
Lot control 157
Lot traceability 157
Manufacturing lead time 139
Move time 139
Operation overlapping 142
Operation splitting 143
Operations sequencing 154
Picking list 137
Process batch 142
Production order 138
Product structure 137
Project manufacturing 137

Queue time 139
Repetitive manufacturing 136
Rope 149
Routing 137
Run time 139
Serial number 157
Setup time 139
Shortest processing time (SPT) 155
Slack time 156
Theory of constraints (TOC) 148
Throughput 145
Throughput time 139
Transfer batch 142
Wait time 139
Work center 135
Work center master 138

QUESTIONS

1. What is the responsibility of production activity control?
2. What are the major functions of planning, implementation, and control?
3. What are the major characteristics of flow, intermittent, and project manufacturing?
4. Why is production activity control more complex in intermittent manufacturing?
5. To plan the flow of materials through manufacturing, what four things must production activity control know? Where will information on each be obtained?
6. What are the four types of planning data used in production activity control? What information does each contain?
7. What information is used for controlling production?
8. What should production activity control check before releasing a production order?
9. What is manufacturing lead time? Name and describe each of its elements.
10. Describe forward and backward scheduling. Why is backward scheduling preferred?
11. Describe infinite and finite loading.
12. What is operation overlapping? What is its purpose?
13. What is operation splitting? What is its purpose?
14. What information does a load report contain? Why is it useful to production activity control?
15. What is a bottleneck operation?
16. What is the definition of throughput?
17. What are the seven bottleneck principles discussed in the text?
18. What are the five things that are important in managing bottlenecks?
19. What is a shop order? What kind of information does it usually contain?
20. What two things must be done to control queue and meet delivery commitments?
21. What is input/output control designed to do? How is input controlled? How is output controlled?
22. What is dispatching? What is a dispatch list?
23. Describe each of the following dispatching rules giving their advantages and disadvantages.
    a. First-come-first-served.
    b. Earliest due date.
    c. Earliest operation due date.
    d. Shortest processing time.
    e. Critical ratio.

24. If the time remaining to complete a job is 10 days and the lead time remaining is 12 days, what is the critical ratio? Is the order ahead of schedule, on schedule, or behind schedule?

25. Would critical ratio be better utilized as a static ratio or a dynamic ratio, and why?

26. What is the purpose of production reporting? Why is it needed?

27. A student of production inventory management has decided to apply critical ratio to their homework assignments. Describe what is happening if the critical ratio for various assignments is:
    a. negative.
    b. zero.
    c. between zero and 1.
    d. greater than 1.

28. Bottlenecks exist in many business processes that serve the public and are usually indicated by lineups. Choose a business that experiences lineups and identify the constraints in the system. Give specific examples of each of the seven bottleneck principles that apply to that business. Suggest a way to increase the throughput of the bottleneck and describe the benefits to the business and to the customers.

29. What is the purpose of lot traceability, and how is it done?

30. Choose a service industry and describe the scheduling and bottleneck issues that must be controlled in order to maintain customer service.

31. Provide an explanation of drum-buffer-rope and give an example of how it would be used.

PROBLEMS

6.1. Production order 7777 is for 600 of part 8900. From the routing file, it is found that operation 20 is done on work center 300. The setup time is 3.5 hours, and run time is 0.233 hours per piece. What is the required capacity on work center 300 for order 7777?

Answer. 143.3 standard hours

6.2. An order for 100 of a product is processed on work centers A and B. The setup time on A is 40 minutes, and run time is 4 minutes per piece. The setup time on B is 50 minutes, and the run time is 5 minutes per piece. Wait time between the two operations is 5 hours. The move time between A and B is 30 minutes. Wait time after operation B is 5 hours, and the move time into stores is 3 hours. Queue at work center A is 20 hours and at B is 30 hours. Calculate the total manufacturing lead time for the order.

Answer. 80 hours

6.3. In problem 6.2, what percentage of the time is the order actually running?

Answer. 18.75%

6.4. An order for 50 of a product is processed on work centers A and B. The setup time on A is 45 minutes, and run time is 4 minutes per piece. The setup time on B is 30 minutes, and the run time is 5 minutes per piece. Wait time between the two operations is 8 hours. The move time between A and B is 60 minutes. Wait time after operation B is 7 hours, and the move time into stores is 2 hours. Queue at work center A is 40 hours and at B is 30 hours. Calculate the total manufacturing lead time for the order.

6.5. In problem 6.4, what percentage of time is the order actually running?

6.6. Amalgamated Skyhooks, Inc., has an order for 200 Model SKY3 Skyhooks for delivery on day 200. The Skyhook consists of three parts. Components B and C form subassembly A. Subassembly A and component D form the final assembly. Following are the work centers and times for each operation. Using a piece of graph paper, draw a backward schedule based on the following. When must component C be started to meet the delivery date?

a. Only one machine is assigned to each operation.

b. The factory works one 8-hour shift, five days a week.

c. All parts move in one lot of 200.

Part	Operation	Standard Time (days)
D	10	5
	20	7
B	10	5
	20	7
C	10	12
	20	5
Subassembly A		7
Final Assembly SKY3		5

Answer. Day 171

6.7. International Door Slammers has an order to deliver 500 door slammers on day 130. Draw up a backward schedule under the following conditions:

a. Only one machine is assigned to each operation.

b. Schedule one 8-hour shift per day for five days per week.

c. All parts are to move in one lot of 500 pieces.

d. Allow 8 hours between operations for queue and move times.

A slammer consists of three parts. Purchased components C and D form subassembly A. Subassembly A and component B form the final assembly. Part B is machined in three operations. No special tooling is required except for part B, operation 20. It takes 24 hours to make the tooling. Material is available for all parts.

Standard times for the lot of 500 are as follows:

Part	Operation	Standard Time (days)
B	10	10
	20	8
	30	6
Subassembly A		18
Final assembly		5

6.8. An order for 100 of a product is processed on operation A and operation B. The setup time on A is 50 minutes, and the run time per piece is 9 minutes. The setup time on B is 30 minutes, and the run time is 6 minutes per piece. It takes 20 minutes to move a lot between A and B. Since this is a rush order, it is given top priority (president's edict) and is run as soon as it arrives at either work center.

It is decided to overlap the two operations and to split the lot of 100 into two lots of 60 and 40. When the first lot is finished on operation A, it is moved to operation B where it is set up and run. Meanwhile, operation A completes the balance of the 100 units (40) and sends the units over to operation B. These 40 units should arrive as operation B is completing the first batch of 60; thus, operation B can continue without interruption until all 100 are completed.

a. Calculate the total manufacturing lead time for operation A and for B without overlapping.

b. Calculate the manufacturing lead time if the operations are overlapped. How much time is saved?

Answer. a. Total manufacturing lead time = 1600 minutes

b. Total manufacturing lead time = 1240 minutes

Saving in lead time = 360 minutes

6.9. An order for 250 bell ringers is processed on work centers 10 and 20. The setup and run times are as follows. It is decided to overlap the lot on the two work centers and to split the lot into two lots of 100 and 150. Move time between operations is 30 minutes. Work center 20 cannot be set up until the first lot arrives. Calculate the saving in manufacturing lead time.

Setup on A = 50 minutes Run time on A = 5 minutes per piece

Planned backlog = 20 units Actual backlog = 18 units

6.10. An order for 100 of a product is processed on operation A. The setup time is 50 minutes, and the run time per piece is 9 minutes. Since this is a rush order, it is to be split into two lots of 50 each and run on two machines in the work center. The machines can be set up simultaneously.

a. Calculate the manufacturing lead time if the 100 units are run on one machine.

b. Calculate the manufacturing lead time when run on two machines simultaneously.

c. Calculate the reduction in lead time.

Answer. a. 950 minutes

b. 500 minutes

c. 450 minutes

6.11. What would be the reduction in manufacturing lead time if the second machine could not be set up until the setup was completed on the first machine?

6.12. An order for 100 of a product is run on work center 40. The setup time is 4 hours, and the run time is 4 minutes per piece. Since the order is a rush and there are two machines in the work center, it is decided to split the order and run it on both machines. Calculate the manufacturing lead time before and after splitting.

6.13. In problem 6.12, what would be the manufacturing lead time if the second machine could not be set up until the setup on the first machine was completed? Would there be any reduction in manufacturing lead time?

6.14. Complete the following input/output report. What are the planned and actual backlogs at the end of period?

Period	1	2	3	4	Total
Planned Input	35	38	36	39	
Actual Input	33	33	31	40	
Cumulative Variance					

Planned Output	40	40	40	40	
Actual Output	38	35	40	38	
Cumulative Variance					

Planned Backlog	32				
Actual Backlog	32				

Answer. Planned backlog = 20 units. Actual backlog = 18 units

6.15. Complete the following input/output report. What is the actual backlog at the end of period 5?

Period	1	2	3	4	5	Total
Planned Input	78	78	78	78	78	
Actual Input	86	80	78	84	80	
Cumulative Variance						

Planned Output	80	80	80	80	80	
Actual Output	85	84	77	83	84	
Cumulative Variance						

Planned Backlog	45					
Actual Backlog	45					

6.16. Complete the following table to determine the run sequence for each of the sequencing rules.

Job	Process Time (days)	Arrival Date	Due Date	Operation Due Date	Sequencing Rule			
					FCFS	EDD	ODD	SPT
A	5	123	142	132				
B	2	124	144	133				
C	3	131	140	129				
D	6	132	146	135				

6.17. Jobs A, B, and C are in queue at work center 10 before being completed on work center 20. The following information pertains to the jobs and the work centers. For this problem, there is no move time. Today is day 1. If the jobs are scheduled by the earliest due date, can they be completed on time?

Job	Process Time (days)		Due Date
	Work Center 10	Work Center 20	
A	7	3	12
B	5	2	24
C	9	4	18

Job	Work Center 10		Work Center 20	
	Start Day	Stop Day	Start Day	Stop Day
A				
C				
B				

6.18. Calculate the critical ratios for the following orders and establish in what order they should be run. Today's date is 74.

Order	Due Date	Lead Time Remaining (days)	Actual Time Remaining (days)	CR
A	89	11		
B	95	25		
C	100	22		

CASE STUDY 6.1

Johnston Products

No matter how many times Justin Wang, the master scheduler for Johnston Products, tried, he couldn't seem to get it through people's minds. They kept trying to "front load" the production schedule by attempting to catch up with production they failed to make the previous week, and the problem appeared to be getting worse. It seemed to happen every week, and the only way Justin could get things back to a realistic position was to completely reconstruct the entire master schedule—usually about every three weeks.

Last month could serve as an example. The first week of the month Justin had scheduled production equal to 320 standard hours in the assembly area. The assembly area managed to complete only 291 hours that week because of some equipment maintenance and a few unexpected part shortages. The assembly supervisor then had the workers complete the remaining 29 hours from week 1 at the start of week 2. Since week 2 already had 330 standard hours scheduled, the additional 29 hours really put them in a position of attempting to complete 359 hours. The workers actually completed 302 hours in week 2, leaving 57 hours to front load into week 3, and so forth. Usually by the time Justin came to his three-week review of the master schedule, it was not uncommon for the assembly area to be more than 100 standard hours behind schedule.

Clearly, something needed to be done. Justin decided to review some of the areas that could be causing the problem:

1. **Job standards.** Although it had been at least four years since any job standards had been reviewed or changed, Justin felt the standards could not be the problem—quite the opposite. His operations course had taught him about the concept of the learning curve, implying that if anything the standard times for the jobs should be too high, allowing the average worker to complete even more production per hour than that implied by the job standard.

2. **Utilization.** The general manager was very insistent on high utilization of the area. She felt that it would help control costs, and consequently used utilization as a major performance measure for the assembly area. The problem was that customer service was also extremely important. With the problems Justin was having with the master schedule, it was difficult to promise order delivery accurately, and equally difficult to deliver the product on time once the order promise was made.

3. **The workers.** In an effort to control costs, the hourly wage for the workers was not very high. This caused a turnover in the workforce of almost 70% per year. In spite of this, the facility was located in an area where replacement workers were fairly easy to hire. They were assigned to the production area after they had a minimum of one week's worth of training on the equipment. In the meantime, the company filled vacant positions with temporary workers brought in by a local temporary employment service.

4. **Engineering changes.** The design of virtually all the products was changing, with the average product changing with respect to some aspect of the design about every two months. Usually this resulted in an improvement to the products, however, so Justin quickly dismissed the changes as a problem. There were also some engineering changes on the equipment, but in general little in the way of process change had been made.

The setup time for a batch of a specific design had remained at about 15 minutes. That forced a batch size of about from 50 to 300 units, depending on the design. The equipment was getting rather old, however, forcing regular maintenance as well as causing an occasional breakdown. Each piece of equipment generally required about three hours of maintenance per week.

Since the computer had done most of his calculations in the past, Justin decided to check to see if the computer was the source of the problem. He gathered information to conduct a manual calculation on a week when there were eight people assigned to the assembly area (one person for each of eight machines) for one shift per day. With no overtime, that would allow 320 hours of production.

Product	Batch Size	Standard Assembly Time (minutes per item)
A174	50	17
G820	100	9
H221	50	19.5
B327	200	11.7
C803	100	21.2
P932	300	14.1
F732	200	15.8
J513	150	17.3
L683	150	12.8

Assignment

1. With this information, Justin calculated the total standard time required to be within the 320 hours available. Is he correct? Calculate the time required and check the accuracy of his calculation.
2. List the areas you think are causing trouble in this facility.
3. Develop a plan to deal with the situation and try to get the production schedule back under control under the constraints listed.

CASE STUDY 6.2

Craft Printing Company

John Burton was not a happy man. He was a supervisor for the Craft Printing Company, having been recently promoted from lead printer. While he felt very comfortable with his knowledge and success in the printing business, this managerial position was starting to wear on him. He was determined not to let it get him down, however, as he felt he surely had the knowledge, experience, and respect of the workers. He had been asking general manager Stella Torres for months for a chance at management, and he certainly wasn't about to let it get the better of him.

His current problem had to do with scheduling. Since he had become supervisor, the sales people always asked him about an order before they promised delivery to a customer. He thought that would be quite simple; after all, who knew more about the printing business than he did? Based on his knowledge of the processes and what was already in progress, he gave what he thought were reasonable, even conservative, estimates of promise dates. Unfortunately, his track record was not too good. There had been many late deliveries since his managerial appointment, and nobody in the organization was too happy about it.

At first, he thought it must be the other workers. "They're just jealous about my selection as supervisor and want to make me look bad" was his initial reaction. Jerel Hurley, another long-time machine operator, was John's best friend. One afternoon over a beer, John asked Jerel about the problem in a confidential discussion. Jerel said he was sure that John had been trying to get it right but somehow it didn't seem to be going well. Jerel assured John that the workers were trying their best. In fact, according to Jerel, the workers had been putting in extra effort. They viewed John's promotion as a positive sign that there was a possible future for them in management as well. John's failure would have been, in fact, greatly discouraging to most of the workers.

John then thought he might be the problem when it came to giving estimates. The sales people would almost always contact him about a possible job to ask him when it should be promised to the customer. His great knowledge of the printing business allowed him, he thought, to quickly come to a good estimate. Perhaps he was not as good at estimating as he thought. To check this out, he looked at most of the jobs done during the last couple of weeks. In almost every case, the work recorded against a job was almost exactly what he had estimated. What little error existed was certainly not large enough to cause the problem.

John trusted Jerel and believed his account of the situation, and his analysis of the estimates convinced him the problem wasn't there. If it wasn't the workers and wasn't the estimates then what could it be? He must do something. Stella Torres was a patient woman, but there was a limit. He was worried about alienating their best customers, and at the same time knew he must be concerned about efficiency as a way to control cost.

John decided there was a need to take drastic action to ease the situation, or at least to find out what the cause was. On a Friday he scheduled overtime for Saturday to finish all jobs in progress. On Monday, therefore, he could start with a clean slate. There were several jobs already promised, but not yet started. He figured that on Monday he could start with all new jobs and really figure out the source of his problem.

The jobs were all promised within four days, but he figured there should be no problem. He had three operations, and most of the jobs went through all three, but not all jobs needed all operations. He had one worker assigned to each operation. Over the next four days that represented 96 hours of available work time (3 operations × 4 days × 8 hours per day), he had eight jobs promised. The total estimated time for all eight jobs was only 88 hours, giving him a buffer of 8 hours over the next four days. To make sure there would be no problem, he decided there would be no new jobs even scheduled to start during those four days, with the only exception being if any operation completed all the necessary work for all eight jobs before the end of the four days, they could start another. In any case, he wanted to make sure that, if necessary, all 96 hours would be reserved for just the 88 hours of scheduled work.

John had learned that a good priority rule to use was the critical ratio rule, primarily because it took into account both the customer due date and the amount of processing time for a job. He therefore used that rule to prioritize the jobs. The following table shows the eight jobs, together with processing time estimates and due dates. All due dates are at the end of the day indicated. Processing times for all jobs at all three operations are in hours.

Job	Operation 1	Operation 2	Operation 3	Total Time	Day Due
A	5 hours	3 hours	4 hours	12 hours	Tuesday
B	0	6	2	8	Wednesday
C	4	2	5	11	Tuesday
D	7	0	3	10	Thursday
E	2	8	0	10	Thursday
F	0	6	3	9	Thursday
G	3	3	4	10	Thursday
H	6	5	6	17	Thursday

Assignment

1. Using the CR rule, establish the priority for the eight jobs.
2. Use a chart to load the operation according to the priority rule established. In other words, load the most important job in all three work centers, then the next most important, and so forth. This is the method that John used.
3. Analyze John's approach and try to determine if he has a problem, and if he does determine the source of the problem.
4. Try to provide a solution to John that will ease the problem, and perhaps eliminate it.

CASE STUDY 6.3

Melrose Products

Jim Hartough was not in a good mood. He worked his way through the ranks when supervisors did supervision and workers did what they were told. He was now faced with the fact that the new president of Melrose Products was investigating the possible use of self-directed work teams. As the manufacturing manager, Jim was ultimately responsible to not only meet production needs, but also to do so in the most efficient and cost-effective manner possible. To him, that meant specific allocation of work. In his experience it had always worked that way and he saw nothing new to tell him it shouldn't continue to do so.

Part of the problem, Jim realized, was that the business environment was changing. Changes in the product design were becoming more frequent and the customers were expecting more service. While they were still sensitive to price (the competition had not disappeared), they wanted quick delivery, high quality, and the product designed more specifically to their need. To Jim, that meant putting more pressure on the engineers (many of whom Jim believed to be not too effective and even perhaps pampered) to make better designs. He also believed that additional pressure needed to be put on production line workers, many of whom he considered to lack dedication to their jobs enough to meet production needs. With better designs, he could more easily allocate the work to his workforce to meet the customer demands. He felt he had truly kept up with the times by recognizing the customer was important. The fact that the customer expected more meant little more than how to get them what they wanted from production. Jim felt that making production work was merely a case of making sure everyone delivered on the job the way they were supposed to.

While the owner of Melrose had avoided the need to become a public company and had resisted unionization, he had still apparently gone "soft," at least according to Jim. He had recently appointed Cindy Lopez as the new president, passing over Jim. She not only had an MBA (Jim had always thought the real business learning was done "on the firing line"), but also had never even been a supervisor. She had come from, of all places, the human resources department! Jim's opinion of human resources was that they had never done anything for him other than send him lots of people who didn't appear (at least to Jim) ready to work hard with real dedication. Some of those people had, in his mind, no chance of ever becoming useful. As far as he was concerned, the only real value of a human resources department was to keep the government from interfering with production.

So, now Jim was in the position to try to "change with the times," as Cindy had said. She wanted to gradually move the company toward flexible self-directed work teams. Jim, of course, felt that most workers really wanted was to get their paycheck on Friday night, and cared very little about having any say in the product or the customer. His real opinion was how was he ever going to get anything done with someone in charge who really didn't understand production as well as he did?

The Current Situation

Cindy had suggested that Jim start the process of changing to teams by looking at the K-line. The K-line of product was a fairly standard product that had recently undergone heavy competitive pressure in the form of delivery speed and design enhancements. Melrose had been gradually losing market share in the K-line. Jim had responded, before the naming of Cindy as president, by putting additional pressure on workers to be more efficient and reducing their task times. As Jim said, "there's always some slack time we can squeeze out of any process if we really put our minds to it."

They are using carefully developed time standards, much as Jim learned in his Industrial Engineering home study courses. He feels they are quite good, including a liberal 10% allowance. Since the K-line is a fairly standard product, Jim not only uses the time standard to develop cost figures for labor, but also uses those cost figures to allocate overhead.

There are currently seven labor tasks to make one of the K-line products.

Task	Standard Time (Min.)	Estimated Labor Cost/Minute
1	7.5	$0.24
2	2.3	$0.22
3	4.7	$0.28
4	5.1	$0.29
5	17.8	$0.26
6	19.1	$0.18
7	8.4	$0.25

The overhead allocation is currently at 230% of direct labor. Material costs are $9.35 per unit. They currently have enough labor to produce 20 of the K-line per shift. Each shift has one supervisor costing about $24 per hour, accounted for in the overhead account.

From this information, Jim was being asked to develop teams, and without direct supervision. From his standpoint, the effort was doomed to failure. Jim, however, always considered himself a "company man" and would do what he could to make it happen.

Assignment

1. What is the standard cost of the K-line product?
2. What specific steps would you undertake to make the self-directed teams? How, specifically, would you deal with the cost and time standard issues?
3. Do you agree with Cindy? Do you agree with Jim? Is there some other alternative approach that might be better in this situation? Explain.
4. What do you do with the supervisor in this situation? Be specific in your approach.
5. How do you deal with Jim? Develop a specific plan to deal with a situation such as the one described.
6. Are self-directed work teams the answer? Where should or shouldn't they be used. Discuss the pros and cons of such teams and where, or where not, they should be used, and how they would be used in this situation, if appropriate.

Week	1	2
Planned Input	45	40
Actual Input	42	46
Cumulative Variance		
Planned Output	40	40
Actual Output	42	44
Cumulative Variance		
Planned Backlog	30	
Actual Backlog	30	

Answer

Cumulative input variance week 1 = 42 − 45 = −3
Cumulative input variance week 2 = −3 + 46 − 40 = 3
Cumulative output variance week 1 = 42 − 40 = 2
Cumulative output variance week 2 = 2 + 44 − 40 = 6
Planned backlog week 1 = 30 + 45 − 40 = 35
Planned backlog week 2 = 35 + 40 − 40 = 35
Actual backlog week 1 = 30 + 42 − 42 = 30
Actual backlog week 2 = 30 + 46 − 44 = 32

Operations Sequencing

The *APICS Dictionary*, 16th edition, defines **operations sequencing** as "a technique for short-term planning of actual jobs to be run in each work center based on capacity (i.e., existing workforce and machine availability) and priorities." Priority, in this case, is the sequence in which jobs at a work center should be worked on.

The enterprise resource plan establishes proper need dates and quantities for orders. Over time, these dates and quantities change for a variety of reasons. Customers may require different delivery quantities or dates. Deliveries of component parts, either from suppliers or internally, may not be met. Scrap, shortages, and overages may occur. In addition, multiple orders may have the same due date, or be scheduled to run on a particular work center the same day, but need to be sequenced. Control of priorities is exercised through dispatching.

Dispatching Dispatching is the function of selecting and sequencing available jobs to be run at individual work centers. The dispatch list is the instrument of priority control. It is a listing by operation of all the jobs available to be run at a work center with the jobs listed in priority sequence. It normally includes the following information and is updated at least daily:

- Plant, department, and work center.
- Item number, order number, operation number, and operation description of jobs at the work center.
- Standard hours.
- Priority information.
- Jobs coming to the work center.

Figure 6.15 shows an example of a daily dispatch list.

Production Activity Control

```
                              DISPATCH LIST
Work Center: 10
Rated Capacity: 16 standard hours per day
Shop Date: 250
```

Order Number	Part Number	Order Quantity	Setup Hours	Run Hours	Total Hours	Quantity Completed	Load Remaining	Operation Start	Dates Finish
123	6554	100	1.5	15	16.5	50	8	249	250
121	7345	50	0.5	30	30.5	10	24	249	251
142	2687	500	0.2	75	75.2	0	75	250	259
		Total Available Load in Standard Hours					107		
Jobs Coming									
145	7745	200	0.7	20	20.7	0	20.7	251	253
135	2832	20	1.2	1.0	2.7	0	2.7	253	254
		Total Future Load in Standard Hours					23.4		

FIGURE 6.15 Dispatch list (based on two machines working one 8-hour shift per day).

Dispatching rules The ranking of jobs for the dispatch list is created through the application of dispatching or priority rules. There are many rules, some attempting to reduce work-in-process inventory, others attempting to minimize the number of late orders or maximize the output of the work center. None is perfect or will satisfy all objectives. Some commonly used rules are:

- **First-come-first-served (FCFS).** Jobs are performed in the sequence in which they are received. This rule ignores due dates and processing time.
- **Earliest due date (EDD).** Jobs are performed according to their due dates. Due dates are considered, but processing time is not.
- **Earliest operation due date (ODD).** Jobs are performed according to their operation due dates. Due dates and processing time are taken into account. In addition, the operation due date is easily understood on the production floor.
- **Shortest processing time (SPT).** Jobs are sequenced according to their process time. This rule ignores due dates, but it maximizes the number of jobs processed. Orders with long process times tend to be delayed.

Figure 6.16 illustrates how these sequencing rules work. Notice that each rule usually produces a different sequence.

Job	Process Time (days)	Arrival Date	Due Date	Operation Due Date	Sequencing Rule			
					FCFS	EDD	ODD	SPT
A	4	223	245	233	2	4	1	3
B	1	224	242	239	3	2	2	1
C	5	231	240	240	4	1	3	4
D	2	219	243	242	1	3	4	2

FIGURE 6.16 Application of sequencing rules.

Critical ratio (CR). Critical ratio considers due dates and process time, and is an index of the relative priority of an order to other orders at a work center. It is based on the ratio of time remaining to work remaining and is expressed as:

$$CR = \frac{\text{due date} - \text{present date}}{\text{lead time remaining}} = \frac{\text{actual time remaining}}{\text{lead time remaining}}$$

Lead time remaining includes all elements of manufacturing lead time that have not yet been processed and expresses the amount of time the job normally takes to complete.

If the actual time remaining is less than the lead time remaining, it implies there is not sufficient time to complete the job and the job is behind schedule. Similarly, if lead time remaining and actual time remaining are the same, the job is on schedule. If the actual time remaining is greater than the lead time remaining, the job is ahead of schedule. If the actual time remaining is less than 1, the job is late already. The following summarizes these facts and relates them to the CR:

CR less than 1 (actual time less than lead time).   Order is behind schedule.
CR equal to 1 (actual time equal to lead time).   Order is on schedule.
CR greater than 1 (actual time greater than lead time). Order is ahead of schedule.
CR zero or less (today's date greater than due date). Order is already late.

Using the critical ratio dispatching rule, orders are listed in order of their CR with the lowest one first. The CR of an order may change as the actual time remaining and lead time remaining change.

Example Problem

Today's date is 175. Orders A, B, and C have the following due dates and lead time remaining. Calculate the actual time remaining and the CR for each.

Order	Due Date	Lead Time Remaining (days)
A	185	20
B	195	20
C	205	20

ANSWER

Order A has a due date of 185, and today is day 175. There are 10 actual days remaining. Since the lead time remaining is 20 days,

$$\text{Critical ratio} = \frac{10}{20} = 0.5$$

Similarly, the actual time remaining and the CRs are calculated for orders B and C. The following table gives the results:

Order	Due Date	Lead Time Remaining (days)	Actual Time Remaining (days)	CR
A	185	20	10	0.5
B	195	20	20	1.0
C	205	20	30	1.5

Order A has less actual time remaining than lead time remaining, so the CR is less than 1. It is, therefore, behind schedule. Order B has a CR of 1 and is exactly on schedule. Order C has a CR of 1.5—greater than 1—and is ahead of schedule.

An additional principle sometimes used for sequencing is **slack time**. Slack time is the result of adding up the remaining setup and run times for an order, and subtracting that from the time remaining. The job with the least slack would be scheduled first. Slack time can also be divided by the number of remaining operations, which is called slack per operation, where the job with the smallest value would be the priority.

CHAPTER SEVEN

FUNDAMENTALS OF SUPPLY CHAIN MANAGEMENT

7.1 INTRODUCTION

As mentioned in Chapter 1, the **supply chain** includes all activities and processes required to supply a product or service to a final customer. This begins with the first raw materials, and includes all companies involved in the production and distribution of that product or service until it reaches the end consumer. This concept of the supply chain has four major components that must be managed:

1. The flow of physical materials from suppliers, downstream through the company itself, and finally to distributors and/or customers.
2. The flow of money upstream from customers back to the companies and suppliers.
3. The flow of information up and down through the stream.
4. The flow of products back (upstream) from the customers, typically for recovery, repairs, recycling or reuse of material. This is known as **reverse logistics**, which is covered in Chapter 14. Another term often used is **reverse supply chain**, which, according to the *APICS Dictionary*, 16th edition, is "the planning and controlling of the processes of moving goods from the point of consumption back to the point of origin for repair, reclamation, recycling, or disposal." Reverse supply chain is also sometimes used to reflect the reverse flow of information, from the consumer back to the producer.

Figure 7.1 portrays a basic supply chain, showing the flow of products and services, information and funds.

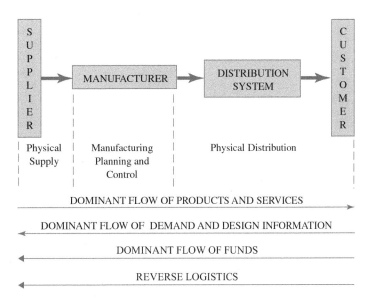

FIGURE 7.1 Basic supply chain flow.

7.2 SUPPLY CHAIN MANAGEMENT

In recent years there has been a great deal of attention given to **supply chain management**. It is important to understand fundamental concepts of supply chain management and its impact on materials management.

Historical Perspective

In the past, many company managers placed most of their attention on the issues that were internal to their companies. Of course, they were aware of the impact of suppliers, customers, and distributors, but those entities were often viewed as business entities only. Specialists in purchasing, sales, and logistics were assigned to deal with those outside entities, often through formal legal contracts that were negotiated regularly and represented short-term agreements. For example, suppliers were often viewed as business adversaries. A key responsibility of a purchasing agent was to negotiate the best financial and delivery conditions from a supplier, whose job was to maximize company profit.

The first major change in that perspective for most companies can be traced to the explosive growth in **just in time (JIT)** concepts, originally developed by Toyota and other Japanese companies in the 1970s. Supplier partnerships were felt to be a major aspect of successful JIT. With that concept, suppliers were viewed as partners as opposed to adversaries, meaning the supplier and the customer had mutually linked destinies, and each was linked to the success of the other. Great emphasis was put on trust between the partners, and many of the formal boundary mechanisms, such as the receiving/inspection activity of incoming parts, were changed or eliminated altogether. As the partnership concept grew, there were many other changes in the relationship, including:

- **Mutual analysis for cost reduction.** Both parties examined the process used to transmit information and deliver parts, with the idea that cost reductions would be shared between the two parties.
- **Mutual product design.** In the past, the customer often submitted complete designs to the supplier, who was obligated to produce according to design. With partnering, both companies worked together. Often the supplier would know more about how to make a specific product, whereas the customer would know more about the application for which the design was intended. Together, they could produce a superior design compared to what either could do alone.
- **Enhanced information flow.** JIT incorporated the concept of greatly reduced inventory in the process and the need for rapid delivery according to need; therefore, the speed of accurate information flow became critical. Formal paper-based systems gave way to electronic data interchange (EDI) and more informal communication methods between individuals at the supplier and customer.

The Growth of the Supply Chain Concept

As the world continued to change, additional factors impacted the supply chain:

- The explosive growth in computer capability and associated software applications. Highly effective and integrated systems such as **enterprise resource planning (ERP)** and the ability to link companies electronically (through the internet, integration of applications, etc.) have allowed companies to share large amounts of information quickly and easily. The ability to acquire information rapidly has become a competitive necessity for many companies.
- There has been a large growth in global competition. Very few companies can still say they have only local competition, and many of the global competitors are forcing existing companies to find new ways to be successful in the marketplace.
- There has been a growth in technological capabilities for products and processes. Product life cycles for many products are shrinking rapidly, forcing companies to not only become more flexible in design but also to communicate changes and needs to suppliers and distributors.

- The changes prompted by JIT in the 1980s have continued to mature and become more accurately defined as lean production. Now many companies have new approaches to interorganizational relationships as a normal form of business.
- Partially in response to the preceding conditions, more and more companies are subcontracting more of their work to suppliers, keeping only their most important **core competencies** as internal activities.

What is the current supply chain philosophy? Companies adopting the supply chain concept now view the entire set of activities from raw material production to final customer purchase, to final disposal, as a linked chain of activities. Each entity and level, or **echelon**, of the supply chain, has activities, inventory, and cycle time. To yield optimal performance for customer service and cost, it is necessary for the entire supply chain of activities to be managed as an extension of the partnership.

This new philosophy reaches beyond the walls of an organization's own facilities, and incorporates the following into the management strategy:

- **Procurement**—the activities of sourcing and purchasing materials and components needed for operations. These activities are discussed in detail in Chapter 8.
- **Warehousing**—the storage of inventory required in the supply chain, whether it is one or more warehouses owned by the organization itself, or a network of warehouse and distribution center partners that take over the warehousing function. This is discussed further in Chapter 13.
- **Physical distribution**—the channels and logistics processes by which product is moved along the supply chain, especially from production to the end consumer. Most companies rely on partners to perform this functionality, such as transportation carriers or global distributors. These are discussed in Chapter 14.

The primary supply chain management approach is a virtual one. All portions of the material production or distribution, from raw materials to final customer, are considered to be in a linked chain. The most efficient and effective way to manage the activities along the chain is to view each separate organization in the chain as an extension of one's own organization. There can be many organizations in a supply chain. Take as an example the chain of organizations that represents the flow from raw silicon used to make computer chips to the delivery and disposal of the computer itself in Figure 7.2.

What is illustrated here is but one chain of a set of different component chains that represent a network of suppliers and distributors for a product.

Most companies work with a network of supply chains, obtaining a variety of materials from multiple suppliers and sending products to multiple customers. Even a grocery store has to deal with suppliers of dry goods, magazines, frozen and fresh products, and small suppliers of local produce or specialty goods. Supply chain management requires moving from only managing first-tier suppliers, to second-tier and even more distant suppliers along the supply chain.

The many independent businesses that make up a supply chain have individual profit motives and do not naturally cooperate to gain savings. This requires someone to take the initiative. Any member of the supply chain can work with other members to show the

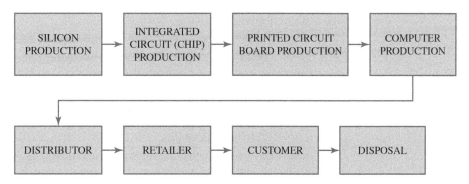

FIGURE 7.2 Supply chain organizations.

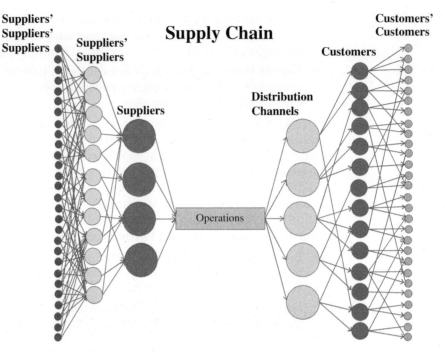

FIGURE 7.3 Extended supply chain.

benefits of sharing information on forecasts, sales information, or schedules. *Orchestrator* or *channel master* are two terms that describe the individual or company that takes the initiative to integrate both the upstream and downstream supply chain, getting members to work cooperatively to lower total costs and achieve greater efficiency. This is often the nucleus firm within the supply chain. The result is a network of companies that openly share information.

To manage a supply chain, one must not only understand the network of suppliers and customers along the chain but also try to efficiently plan material and information flows along each chain to maximize cost efficiency, effectiveness, delivery, and flexibility. This clearly implies not only taking a different approach to suppliers and customers but also a highly integrated information system and a different set of performance measures. Overall, the key to managing such a concept is with rapid flows of accurate information and increased organizational flexibility.

Depending on the complexity of the product, or the global nature of the end-to-end supply chain, there may be many customer–supplier relationships, such as displayed in Figure 7.3, all of which must be monitored to ensure that demand information, inventory, order status, payments, returned product, etc. are all managed to provide delivery of the products and services as efficiently and cost-effectively as possible.

7.3 SUPPLY CHAIN STRATEGIES

Because the supply chain is one way that companies compete, it is important that organizations spend time designing and developing the supply chain strategy that will best fulfill the goals of the overall business strategy.

Some of the areas where supply chains can be used strategically include:

- **Customer experience and value**—how and where the customer interacts along the supply chain. This includes an analysis between the cost to serve and profitability. For example, does the customer prefer to interact with a more regional warehouse, as opposed to remotely online? If so, what is the cost of securing more localized distribution partners? Supply chain objectives must be defined that determine speed of delivery, planning processes, sales channels, flexibility, ordering capabilities, and delivery methods.

- **Availability**—ensuring that the supply meets the demand. This may dictate where and how much inventory is kept throughout the supply chain network, and how much visibility is required on inventory quantities and location.

- **Innovation**—how does the supply chain support getting new products to market quickly. This may require a more integrative design approach with suppliers in order to shorten the time.
- **Quality and traceability**—many supply chains require that the materials and components used be traceable back to the point of origin, which will mandate an integrated network of information among supply chain partners.
- **Cost**—supply chain performance in this area includes the utilization of assets, inventory on hand, product costs, and total supply chain costs. What is the cost/benefit analysis for locating supply chain facilities in certain geographical areas? How much functional and technical integration will be required between supply chain partners?

Globalization

An additional decision that must be made regarding supply chains is whether the company intends to market, sell and/or produce its products and services globally as opposed to locally or regionally. This includes determining whether the presence will be strictly sales, with production remaining domestic, or whether production will be distributed into different regions or countries in order to be closer to the market. Today, countries and cultures are interconnected, trading and learning from each other, and contributing to the economic advancement of the world. Governments themselves can have an impact on the supply chain, both positively and negatively. In addition to **globalization**, **glocalization** is a concept being used by organizations as a "form of postponement where a product or service is developed for distribution globally but is modified to meet the needs of a local market. The modifications are made to conform with local laws, customs, cultures and preferences" (*APICS Dictionary*, 16th edition).

Several questions must be asked, such as:

- How will the business service customers not geographically located near the company?
- What are the tax advantages or disadvantages of having a business entity in a particular state, province, or country?
- What are the cultural differences that must be overcome in order to do business in another country?
- If the company decides to produce goods within foreign countries, is there a supplier base within that same country or region?
- Is it better to staff with local specialists to handle trade, customs, and political issues?
- What are the labor rules and costs?
- Are there regional requirements for products based on governance, culture, customs, etc.?
- What are the duties, tariffs, and other transportation costs for distributing to another region or country?
- How much flexibility is required to react to sudden changes in component availability, distribution or shipping channel issues, import duties, or currency exchange rates?

Supply Chain Design

Once the strategic questions have been addressed, other design decisions will follow. These include determining which supply chain functions will be performed internally, which will be performed by other intermediaries or partners, and what products will be sourced from which locations. The number of warehouses or distribution centers needed, where they will be located, and the types of transportation partners required must be determined, in order to execute the strategy for providing customer value. This will be discussed further in Chapter 14.

Other decisions include what services will be required from third-party entities, such as assembly, consignment, training, point-of-sale information, promotions and advertising, or forecasting. In some cases, the supply chain design will incorporate having suppliers **drop ship** product directly to the customer. Questions must be addressed on how much real-time information is necessary across the supply chain, and communication tools that are necessary between supply chain echelons must be put in place that fulfill the requirements.

> **SUPPLY CHAINS USE AGILITY TO RESPOND TO PANDEMIC**
>
> During the pandemic of 2020, many organizations and supply chains had to become agile in order to survive and not go out of business. Creativity and flexibility became the norm, not the exception. Countless manufacturers and distributors partnered together and used their expertise to provide much needed supplies to the public and healthcare industries worldwide. Companies worked around the clock to produce essential supplies, donate space for makeshift hospitals, and brainstorm ideas to boost production.
>
> Many manufacturers, including General Motors and their suppliers, altered production to begin making much-needed ventilators instead of their normal automotive components. Smaller companies, such as Iconex, repurposed their label operations to begin making sanitation supplies for hospitals, food tamper-resistant seals, floor decals, and secured door seals.
>
> Fashion and textile companies, such as Fruit of the Loom and Hanesbrands, built supply chains overnight to begin collectively producing medical grade face masks. A blue jeans manufacturer in France began producing sanitary masks within hours of discovering the need.
>
> Perfumeries such as Givenchy and Christian Dior began supplying hand sanitizer. Brewers and distillers like Anheuser-Busch produced and distributed sanitizer through the Red Cross. Pernod Ricard SA and the British Honey Company not only switched production to sanitizers, but also supplied alcohol to other manufacturing companies to do the same.
>
> The delivery of the coronavirus vaccine required an integrated supply chain consisting of scientists, manufacturers, distributors, transportation companies, and healthcare workers. Vaccine producer Pfizer assembled a cold supply chain that included building a staging area equipped with freezers to stage the vaccine for distribution, and developing GPS-trackable, temperature-controlled shipping containers. Truck companies, distribution hubs and airports contributed by delivering and storing the vaccine at ultra-cold temperatures. FedEx, UPS, and DHL then delivered billions of vaccines to healthcare and pharmaceutical companies for dispensing to the public.
>
> Cruise ship companies, some of the first to be hit by the pandemic, offered temporary housing for foreign workers who were not able to travel home due to immigration restrictions in their home countries. Others, such as the cruise ship Splendid, were converted to makeshift hospital rooms off the coast of Spain. Airlines like Deutsche Lufthansa used some of its suspended flights to deliver essential goods. Hotels, including the Dan Panorama and Dan Hotel in Israel, transformed their facilities into quarantine shelters, as did many universities with their dorms.
>
> These are just a few of the examples of how the world's supply chains worked together, demonstrated their agility and resiliency, and broke down technology and trust barriers in order to respond to a worldwide crisis.

Lastly, will one supply chain work for all products and services? Depending on the number of products and services offered by the organization, and the types of markets served, multiple supply chains may be necessary, or preferred.

Supply Chain Agility

Supply chain **agility**, meaning the ability to quickly adapt and respond to changes in planning, sourcing, making and delivering products and services, is crucial to supply chain management. It is dependent on: the quality of the relationships among supply partners and other third-parties, such as transportation carriers; a high level of shared information and connectivity between the organizations in the supply chain; and the sharing of knowledge and information about the marketplace. Moving toward a more integrated supply chain increases the responsiveness and improves the customer experience.

7.4 INTEGRATED SUPPLY CHAINS

Even though the objective of supply chain management is for all parties within a particular supply chain to work together toward a common goal—providing products or services to the end customer—it is often difficult to accomplish the objective without intentional integration between entities. Frequent barriers to a truly integrated supply chain include technology, the sharing of information among trading partners, cultural issues, an aversion to change, and trust and a willingness to collaborate.

As computers and software (ERP, for example) have become more powerful and effective, information flows have become easier, and the ability to handle large amounts of data has become more feasible. This condition has allowed companies to expand their planning and control perspectives to include upstream (suppliers) and downstream (distributors and customers) entities.

Two processes in particular have arisen to assist in the management of relationships across the supply chain. These can be accomplished internally through the use of manual methods, but are often put in place through software applications specifically designed with the functionality needed.

- **Customer relationship management (CRM)** includes several activities with the intent to build and maintain a strong customer base. Customer wants and needs are assessed and cross-functional teams from the company work to align company activities around those customer needs.
- **Supplier relationship management (SRM)** is similar to CRM, with the focus for these activities being the building and maintaining of close, long-term relationships with key suppliers.

Bullwhip Effect

One critical reason for developing formal links and relationships in the supply chain is to help control the **bullwhip effect**. This occurs when there is uncertainty in the supply chain based on the use of forecasts, and that uncertainty is then exaggerated as material moves through the supply chain.

The effect can produce large fluctuations in demand for raw materials based on relatively small changes in demand from the customer end of the supply chain. This is illustrated in Figure 7.4. A small fluctuation at the customer causes a ripple effect as the change passes through each node of the supply chain, exaggerated by lead times and differences in lot sizes. An example of this occurred during the recent pandemic, as panic buying created material shortages, which was magnified as it moved back through the supply chain. Managing the supply chain with visibility of data (information flow) and building flexibility and agility across the supply chain, can substantially reduce large fluctuations.

Integration Techniques

One of the simplest methods for integrating is the coordination and cooperation between supply chain entities. This allows for communication and infrastructure to make it possible to know what each supplier and customer needs, and how everyone can participate to achieve those needs, working together toward a common goal.

One method of coordinating information is quick response. Quick-response programs arose in the 1980's by textile companies who were looking for a way to reduce stock replenishment lead times. Per the *APICS Dictionary*, 16th edition, a **quick-response program (QRP)** is "a system of linking final retail sales with production and shipping schedules back through the chain of supply; employs a point-of-sale scanning and electronic data interchange, and may use direct shipment from a factory to a retailer." **Point of sale (POS)** and QRPs utilize UPC and QR codes to enable immediate relaying of sales information down through the supply chain to trigger replenishment.

Collaboration builds on the coordination and cooperation foundation by providing a greater level of connectivity. **Collaborative planning, forecasting and replenishment (CPFR)** is one program that enables collaboration between organizations. CPFR is discussed further in Chapter 9.

Most supply chain technology today has the capability of being integrated with other applications, especially through the use of cloud computing, which is discussed later in this chapter. However, people are the key to a successfully integrated supply chain. This

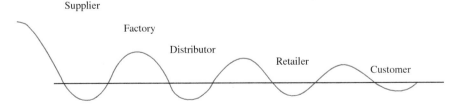

FIGURE 7.4 Bullwhip effect.

begins at the managerial level, where alliances must be developed between organizations in order to drive the willingness to share risks and rewards down through the enterprise, and across the supply chain. Once the personnel roadblocks have been overcome, firms can begin to move toward being truly customer focused, and begin to reap the benefits such as:

- Increased responsiveness.
- Consistent on-time deliveries.
- Customer satisfaction.
- Shorter order fulfillment lead times.
- Reduced purchasing and inventory costs.
- Better asset utilization.
- Ability to manage unexpected events.

7.5 SUPPLY CHAIN RISKS

Today's complex supply chains extend over increasingly longer distances and organizational boundaries. With the increased complexity comes increased risk, both internal and external. International trade, economic uncertainty, market volatility, shorter product life cycles, and complex networks of suppliers and third-party service providers all contribute to supply chain risks.

Demand risks Demand risks may stem from customer or market issues, such as consumer spending, forecasting, new products or a new customer base, competitor activities, or social media. Other demand risks are related to products or the company itself, such as current trends or brand reputation.

Supply risks Supply risks include those things that disrupt the flow of supply, whether it is quality, communication, supplier processes, transportation, etc. Supply chain disruption can be minor, moderate or major. Most minor disruptions occur as part of normal operations, and have limited impact on the supply chain. Often these risks can be mitigated by simply using safety stock to ensure an uninterrupted flow.

Moderate disruptions may come in the form of an unavailable raw material or component, which must then be obtained from a different source, such as a certain color of mined clay, or components from a supplier who is no longer in business. Other examples might include an outage in the internet infrastructure being used by many partners to house their computing networks.

On the other hand, major disruptions have a very high impact, are difficult to predict, and may have long-term effects, such as natural disasters, government or political crises, or pandemics. Some of the major disruptions in the recent past have included European volcanic eruptions which closed down air space, tsunamis in Southeast Asia, a breakout of mad cow disease in Canada, a ship blocking the Suez Canal, and the worldwide pandemic of COVID-19. All of these had a severe impact on the world's supply chains, and consequently, the delivery of goods and services to consumers.

Operational or process risks Operational risks occur internally, and are often related to quality problems, inventory shortages, poor forecasting, and other operational issues such as IT outages, equipment failure, or poor performance. Disruptions may occur due to labor issues or even security and cyber-security risks, such as the theft of intellectual property or ransomware.

Financial risks occur when participants within the supply chain have financial difficulties, and are not able to deliver product or fulfill financial obligations. An assessment of the financial viability of suppliers, as well as customers, should be part of the supply chain design sessions.

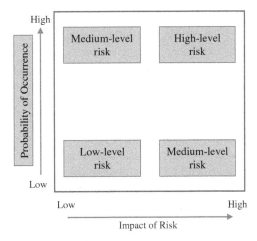

FIGURE 7.5 Risk management matrix.

Risk Management

Supply chain strategy should include how resilient the supply chain will be during atypical events, which is determined by how risk will be managed. **Risk management** in the supply chain includes four steps:

1. **Risk identification.** The type of risk, potential causes, and who owns the risk must be determined.
2. **Risk assessment.** This is the process of quantifying and assessing the potential risk. Risks are evaluated to determine not only the likelihood or probability that it will occur, but also the consequences (severity and impact) of the occurrence. This can be charted simply in a matrix to give a visual of how significant the risk is, as in Figure 7.5.

In addition, numerical values can be applied, in order to provide a means of prioritizing which risks need the most attention.

For example, a matrix such as in Figure 7.6 might use the following values:

If a risk is determined to be possible, but only have a minor impact, the risk score would be: $3 \times 2 = 6$. Another risk may be rare, but if it occurs would be major. That risk score would be $2 \times 4 = 8$.

3. **Risk response.** Once the assessment has been done, the next step is to determine what the response to the risk will be. Several types of responses are possible:

- **Risk mitigation**—how the organization will respond to a risk event by lessening the impact. For example, mitigation for a risk of a supplier's inability to deliver on-time might be to have an alternate supplier available, or buffer stock on hand.
- **Risk avoidance**—how the organization plans to avoid the risk, such as avoiding sourcing items from a part of the world where supply is uncertain.

		Impact				
		Insignificant	Minor	Moderate	Major	Catastrophic
Probability	Unlikely	1	2	3	4	5
	Rare	2	4	6	8	10
	Possible	3	6	9	12	15
	Likely	4	8	12	16	20
	Certain	5	10	15	20	25

Risk score = impact x probability

FIGURE 7.6 Numerical risk management matrix.

- **Risk prevention**—taking action to make sure that the risk does not occur, or will not have an impact. For example, certain safety measures may be put in place to ensure that accidents do not occur.
- **Risk acceptance**—the organization has determined that it will accept the risk should it occur. This may be due to the analysis that the cost of mitigating the risk outweighs the impact of the risk itself.

Risk sharing is one way that supply chain partners can distribute the costs of risk across multiple entities to limit the cost to one organization. **Risk pooling** is a type of sharing inventory risk, where "manufacturers and retailers that experience high variability in demand for their products can pool together common inventory components associated with a broad family of products to buffer the overall burden of having to deploy inventory for each discrete product" (*APICS Dictionary*, 16th edition).

A new methodology that has arisen in the risk management arena is **threatcasting**. Similar to military strategic planning, threatcasting uses inputs from history, social science, technical research, experts, economics, and trends to determine potential models of the future, with the goal of anticipating where and how future threats will occur, and to develop strategies to reduce their impact, mitigate the risk or recover from the event. Indicators or flags can also be identified to warn when the situation is progressing toward

RECENT SUPPLY CHAIN RISK EVENTS HAVE SEVERE IMPACT ON THE GLOBAL SUPPLY CHAIN

Several events in the past few years have highlighted the importance, and the fragility, of global supply chains. Many organizations faced employee, manufacturing, and financial crises due to unexpected events.

The pandemic of 2020 caused havoc on the global supply chain that had not been seen in modern history, and put it in the spotlight. Within days of each country shutting down due to the crisis, shortages of consumer products were felt across the globe, such as food, paper products, and cleaning supplies. In addition, the early shut down of China caused shortages in personal protective equipment and prescription drugs. Ventilators, virus tests, and hospital beds became critical supply chain needs overnight.

A surge of online purchasing caused Amazon to hire 100,000 people for its warehouses and distribution centers. Supermarkets and food delivery services also saw huge increases in demand.

The shutdown of factories and warehouses, and then the sudden turn-on of the supply chain, created a pronounced bullwhip effect. Shortages occurred on cars, appliances, fitness equipment, mattresses, furniture, boats, RVs, and power tools due to the increased demand of people staying home.

Service industries such as food service, travel, tourism, and special events industries also experienced supply chain issues. Layoffs of personnel, closing of venues, and the selling off of rental automobiles were suddenly reversed, causing a shortage of available workers, booking dates, and rental vehicles.

A freak snowstorm in Texas shut down an oil refinery that feeds 85% of the resin producers for automobile manufacturers in the United States. The unavailability of supply shut down production at several plants and resin had to be flown in from overseas suppliers. Ford, Toyota, Volkswagen, Nissan, and Tesla have begun to rethink the JIT concept of not stocking inventory along the supply chain, and have begun to stockpile smaller items. Some began vertical integration by building factories to produce some items in-house, such as batteries, to avoid potential supply issues in the future. Plastic prices in other industries increased due to supply shortfalls.

Roughly 60% of the world's semiconductors are manufactured in Asia-Pacific countries. Government sanctions, pandemic lockdowns, strikes, earthquakes, fires, and delays in global shipments all contributed to a semiconductor shortage which forced car firms to shut down plants around the world. In addition, the severe shortage caused a disruption for smartphone and chip-making companies such as Samsung, who also had to close plants.

The shortage of workers due to the pandemic, along with a sudden restocking drive after the global crisis, caused a backlog at California ports, with as many as 40 container ships waiting for dock space, delaying raw materials to American manufacturers.

The Suez Canal, one of the world's most important waterways, was closed for almost a week due to a 200,000-ton container ship going aground. Twelve percent of the world's global trade passes through the 120-mile canal. By the time the canal was cleared, more than 350 ships had been delayed, amounting to roughly $60 billion in global trade.

It is not only manufacturing that felt the pain. In March of 2021, 44% of small businesses reported supply chain disruptions, from gloves to food products. Some of this was due to the trickling effect of the coronavirus pandemic, which caused a shortage of employees and temporary business closures. Once restrictions eased and businesses reopened, the demand for their products and services, as well as hiring staff, increased faster than they could react.

With the imbalance of supply and demand comes price increases for those items that are difficult to procure. Steel tubing costs increased by 125%, and the cost of lumber to build crates and pallets for packaging supplies rose by 50%–100%. A ransomware attack on a fuel pipeline in the southeast United States caused a fuel shortage and price increases across several states.

All of these events have put a focus on supply chain resiliency, as organizations remap their supply chains to identify vulnerabilities, enable quick responses to disruptions, and strategize on long-term risk mitigation techniques.

one of the envisioned threats. In light of the recent pandemic, and its impact on the supply chain, current supply chain professionals are collaborating with threatcasting leaders to explore additional potential threats, the impacts, and steps that should be taken to mitigate the effects in order for their organizations to stay resilient.

4. **Risk monitoring**. Risk monitoring refers to the activities of monitoring risks as they occur, or as the probability of the risk increases or decreases. For example, over the life of a project, certain risks that were inherent at the beginning of a project may no longer be a threat.

7.6 SUSTAINABILITY

Introduced in Chapter 2, **sustainability** is an integral part of today's supply chain management. Supply chains are tasked with taking the necessary steps to be sustainable long-term, employing such methods as reduction of wastes, using fewer resources, producing output that is reusable, and reducing energy consumption. According to the *APICS Dictionary*, 16th edition, an **environmentally responsible business** is "a firm that operates in such a way as to minimize detrimental impacts on society." Consumers of goods and services are becoming more and more concerned and demanding that their suppliers focus on sustainable processes, materials, etc. In order to be considered a **green supply chain**, the entities must consider the environmental impacts each operation has, promote the development of sustainable practices, and take action throughout the supply chain to comply with all environmental and safety regulations.

Triple Bottom Line

In addition to a focus on the environment, sustainability also includes principles related to the economic and social impact of an organization. This is often referred to as the **triple bottom line (TBL)** and displayed as shown in Figure 7.7.

Other terms used for TBL include the three P's (people, plant, and profit) or three E's (economics, equity, and environment).

FIGURE 7.7 Triple bottom line.

Waste Hierarchy

Known as the **waste hierarchy**, reduce, reuse, recycle, and recover is a widely used phrase to guide people in lowering the impact they have on the environment. It is a tool that ranks the options for managing waste, according to what is most environmentally sound. Properly understood and implemented, the 4 Rs can also reduce expenses and increase profits. It is usually depicted as shown in Figure 7.8.

Reduce Reducing the use or generation of materials, whether hazardous or scrap, is the most environmentally friendly of the waste management options. Suppliers within the supply chain are often the first to learn of new environmentally friendly materials. Lead-free

FIGURE 7.8 Waste hierarchy.

solder and water-based solvents are just two examples of materials that have been developed by suppliers to help their customers reduce their environmental impact and reduce their costs. Lean principles (discussed further in Chapter 16), when applied to suppliers, involve reducing waste for all stages of the supply chain. The use of returnable racks or packaging is widely used in many industries, reducing both costs and environmental impact.

Reuse The next most effective step is to **reuse** materials wherever possible. Scrap from one process may be reused directly within the organization or can be slightly processed for reuse in another process. Corrugated cartons can be slit and crushed for use as packing material in the shipping department. Many manufactured products are cut from a continuous sheet resulting in waste material. However, smaller products can often be nested between the cut-outs to make other, smaller products, reducing the need for raw materials and also the amount of material to be sent for disposal. Another category of reuse is **by-products**, which are salable products made from what was previously considered waste. It was not too long ago that butchers had to find a use for a nonsalable product, chicken wings!

An example from the food packaging industry is the shipping of glass jars in boxes preprinted with finished product artwork. The empty jars are removed from their containers as they enter the cleaning and filling station. The empty boxes are sent to the end of the filling process, where the finished product is placed in the boxes. A final description and batch code are then printed on each box. The company never has to deal with the handling and disposal of the boxes used to ship the empty jars. The boxes have two uses: the shipment from glass manufacturer to the food processor and from the food processor as finished product.

Another example of reuse is the development of packaging that can be converted for reuse by the consumer for returning product as part of the reverse logistics cycle.

Recycle Suppliers are often the best source of information on the disposal of scrap materials and often will buy materials back to **recycle** or reprocess. This does require good management to keep materials in their most useful form. Some liquids, notably chlorinated solvents, are difficult to dispose of. They should be kept separate from other valuable liquids, such as used machine oils. Contamination of a waste stream can turn the resale of a valuable waste product into an expense and should be avoided at all costs. All materials destined for recycling should be kept separate from other materials where possible.

Recover Recovery is a reverse logistics strategy which involves collecting used and discarded components, products and materials in order to recover as much of the economic value as possible. In particular, it is used to refer to the recovery of energy from waste, in order for it to be reused, such as the generation of fuel from solid waste, creation of clean energy from old tires, or converting scrap plastic into oil.

Circular Economy

Another concept born out of sustainability is that of the **circular economy**. According to the *APICS Dictionary*, 16th edition, it is "an economic system intended to minimize waste and maximize the use of resources through a regenerative process achieved through long-lasting design, maintenance, repair, reuse, remanufacturing, refurbishing, recycling, and upcycling." Figure 7.9 shows an example of some of the actions included in a circular economy.

An example of the application of circular economy is the recycling of paper. Products made from paper are sourced from trees, which are a natural, renewable resource. The North American paper industry plants roughly twice as many trees as it harvests each year, resulting in 20% more trees than there were 50 years ago. When trees are harvested, every part of the tree is used, including the creation of energy for paper mills from the leaves and bark. Paper generated from trees is one of the most recycled materials in the world, and paper fibers can be recycled five to seven times. At the end of its life cycle, paper is also biodegradable and compostable, returning it to earth.

Carbon Footprint

The supply chain has a direct impact on the **carbon footprint** of the products it provides. The carbon footprint measures the amount of carbon emissions, such as greenhouse gases (carbon dioxide), generated by a person, organization, operations, etc. The supply chain is responsible for as much as 50% of a product's carbon footprint, impacted by decisions such as transportation mode, shipment size, and location.

Corporate Social Responsibility

Supply chain partners fully engaged in sustainability activities are committed to corporate **social responsibility**, requiring ethical behavior, and a focus on their impact on the social and cultural aspects of people. This includes human rights issues, community activities and citizenship. As mentioned in Chapter 2, these principles are addressed by the **United Nations Global Compact**. The 10 principles outlined by the compact are:

Human Rights

1) Business should support and respect the protection of internationally proclaimed human rights; and
2) make sure that they are not complicit in human rights abuses.

Labor

3) Businesses should uphold the freedom of association and the effective recognition of the right to collective bargaining;
4) the elimination of all forms of forced and compulsory labor;
5) the effective abolition of child labor; and
6) the elimination of discrimination in respect of employment and occupation.

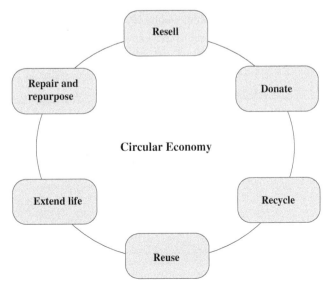

FIGURE 7.9 Circular economy.

Environment

7) Businesses should support a precautionary approach to environmental challenges;
8) undertake initiatives to promote greater environmental responsibility; and
9) encourage the development and diffusion of environmentally friendly technologies.

Anti-corruption

10) Businesses should work against corruption in all its forms, including extortion and bribery.

People in supply chain positions often have responsibility for managing the flow of a great deal of money, and as a result it becomes quite important to have honest and ethical conduct on the part of those individuals. Many companies, and countries, have developed strict codes and rules of conduct to ensure consistent and ethical treatment of the supply chain activities and reported measures. More will be discussed on ethics in Chapter 8.

7.7 LIFE CYCLE ASSESSMENT

The supply chain for a product or service, the risks associated with it, and the impacts on the environment, are all associated with the life cycle of that product or service. As the life cycle of a product progresses, sales increase or decrease based on the phase, such as shown in Figure 7.10.

Life cycle assessment (LCA) is "understanding the human and environmental impacts during the life of a produce, process or service, including energy, material, and environmental inputs and outputs" (*APICS Dictionary*, 16th edition). LCA is often referred to as a cradle-to-grave analysis, as it applies from product development through the disposal and recycling of components at the end of a product's life. It encourages supply chain partners to make sustainability decisions that do not simply shift problems from one part of the supply chain to another.

Examples of applying LCA within the supply chain include:

- **Introduction**—during the launch process, sourcing decisions are made as part of the supply chain design of that product or service. Mutual cost-reduction efforts between partners can be discussed, as well as alternate materials that might be more environmentally friendly.
- **Growth**—volume leveraging occurs during this phase, as well as negotiations and decisions on stocking levels, ordering policies, etc. Procedures for sharing forecasts of demand, or actual sales information, should be put in place in order to minimize the bullwhip effect and excess inventory as demand changes are frequent.
- **Maturity**—long-term contracts with supply chain partners can be put in place due to more certainty about demand and longevity.
- **Decline/end of life**—reverse supply chain activities may begin to be incorporated more frequently, in order to recycle, refurbish, or remanufacture products.

Product life cycles are covered in more detail in the section on product development in Chapter 15.

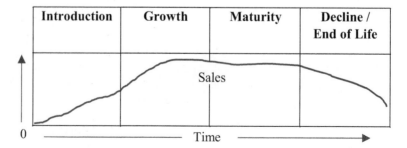

FIGURE 7.10 Life cycle of a product.

7.8 SUPPLY CHAIN PERFORMANCE MEASUREMENTS

In addition to the performance measures discussed in Chapter 1, there are several metrics that should be considered when measuring the performance of the supply chain. Examples of some of these can be found in the table below.

Supply Chain Area	Metric
Order fulfillment	On-time delivery
	Cycle time reduction
	Internal customer satisfaction
	Responsiveness
	Time to market for new products
	Order processing
Logistics	Logistics costs as % of sales revenue
	Inbound transportation costs
	Premium transportation % of overall transportation costs
	Actual transportation cost versus planned
	Actual warehouse cost versus planned
Quality	% and # of defects per supplier
	Failure rates in the field
	Value of rejected items versus total value of shipments
	Warranty costs from supplier defects
	% of quality-certified suppliers to total # of suppliers
Supplier management	Early supplier involvement in new product design
	Technical integration of key suppliers
	Reductions in costs of doing business with a supplier
	Supplier satisfaction measures
	Supplier relationship management (SRM) programs
	Joint cost-reduction programs
Technology/Innovation	% of suppliers using e-transactions
	Pull systems/shared schedules relationships
	% of standardization of components
	Vendor-managed inventory (VMI) programs
Environment	% of recycled and renewable materials used or purchased
	Measuring of carbon footprint
	Reductions in water contamination
	Decrease in energy consumption
	Management of hazardous and solid waste
	Reuse and recycle programs in place
	Use of recyclable or reusable packaging
	Minimization of packaging volume and weight
	Increase in use of alternate fuels
	Coordination of inbound and outbound shipments
Social responsibility	Tracking of supplier safety conditions
	Supplier adherence to human and labor rights and workplace standards
	Annual spend with minority, women and small business
	Number of volunteer hours by employees
	Employee satisfaction

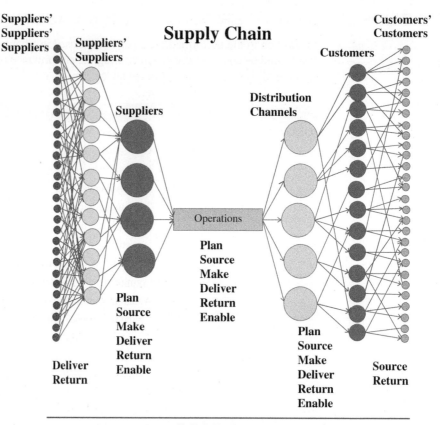

FIGURE 7.11 SCOR model.

Supply Chain Operations Reference (SCOR) Model

The **Supply Chain Operations Reference (SCOR) model** is used by many organizations as a means of evaluating their supply chains. The SCOR model provides standard performance metrics associated with supply chains, broken down into the primary activities of plan, source, make, deliver, return, and enable, as noted in Figure 7.11.

7.9 SUPPLY CHAIN TECHNOLOGY AND TRENDS

As supply chains evolve, new trends and technology are enabling functionality such as automation of communication, improved decision-making and end-to-end visibility, creating what some refer to as an *intelligent supply chain*. As new technology and methods for managing the increasing amounts of data continue to emerge, the value and efficiency of supply chains will continue to improve.

Cloud Computing

The ability for companies to store data and access applications over the internet, known as **cloud computing**, has changed the way organizations manage their operations and supply chains. Data is now universally accessible across an enterprise, as well as to other echelons within the supply chain, making it faster and easier to do business anywhere in the world. Many ERP suppliers now provide **software-as-a-service (SaaS)**, enabling organizations to house their systems online, providing lower implementation costs and upgrades. Other benefits of cloud computing include the speed of access to information, visibility to supply chain data analytics, unlimited data capacity, and an ability to easily collaborate with supply chain partners.

Big Data

Businesses are collecting and processing immense amounts of data, and require rapid information, analysis, and decision-making. Traditional methods of storing and organizing data no longer provide the tools necessary to accomplish the management, sharing and analyzing of information. **Big data**, as defined by the *APICS Dictionary,* 16th edition, is "collecting, storing and processing massive amounts of data for the purpose of converting it into useful information." This functionality is normally used for volumes of data greater than a petabyte of data. One petabyte of data equates to approximately 20 million four-drawer filing cabinets. Google estimates that over 20 petabytes of information are processed by its search engines daily.

Control Towers

In addition to accessing the data within an organization, supply chain management processes require a comprehensive view of data across the end-to-end supply chain. **Control towers** provide the functionality of collecting, monitoring, and directing information into a centralized hub from all the unrelated systems of supply chain entities. This allows for the distribution of information for monitoring activities, enhances the ability to get accurate information on supply location and availability, and supports cross-functional decision-making and measurement.

Artificial Intelligence

Artificial intelligence (AI), or the ability for computer programs to learn and reason similar to humans, is becoming more mainstream across the supply chain. Using the data collected, artificial intelligent agents can detect demand patterns, generate forecasts, analyze distance between customer and supply chain points, automate supplier ranking and selection, and optimize the balance of supply and demand within the supply chain. In addition, AI can be used to perform tasks that are too difficult or time consuming for humans, which improves productivity and minimizes errors.

Interorganizational Systems

The coordination of processes between supply chain echelons and parties utilizes the concept of **business-to-business (B2B) commerce**. Interorganizational systems facilitate the ability for business to be conducted over the internet, creating a **virtual organization** that acts as a single entity in order to improve quality and performance, while reducing lead time, and costs.

Some of these interorganizational systems that take advantage of cloud computing, web portals, or private electronic networks include trading exchanges, e-hubs and e-markets, **decision support systems (DSS)**, and **group decision support systems (GDSS)**. Threatcasting, discussed earlier in this chapter, utilizes decision support technology in order for management to make decisions and take actions should a modeled threat event occur.

Blockchain

At the forefront of supply chain trends is the recent use of blockchains. A **blockchain** is "a continuously growing list of records, called blocks, which are linked and secured using cryptography" (*APICS Dictionary,* 16th edition). It consists of a shared, unchangeable ledger that joins all transactions in a digital chain. The transactions, or blocks, are located in a private, secure network, with no single supply chain partner controlling the information. Once the transaction occurs, participants of that blockchain have access and visibility to the transaction history real-time. The transactions are secure, tamper-resistant, and irreversible, providing an accurate, permanent "chain" of events.

Benefits of blockchain technology include:

- Scalability, allowing unlimited transactions, members, and touchpoints.
- Immediate product and material traceability through the audit trail, from point of origin to final destination. This can also be used in an effort to ensure products are coming from socially responsible organizations.

- Tracking of root causes of quality issues back through the supply chain, such as food-borne illnesses being traced back to the particular batch of product affected.
- Risk mitigation through data integrity, prevention of data breaches or counterfeiting, and increased security. Blockchains are proving to be more secure for data backup since they are immune to hackers. Pharmaceutical companies use blockchains to manage and track the shipment of drugs.

Internet of Things (IoT)

The **Internet of Things (IoT)** assumes that there is information everywhere, in every object, that can be shared. In the supply chain world, this provides the ability for sensors and technology to provide information without human intervention, such as monitoring of carbon emissions, or tracking of transportation vehicles for real-time location information. IoT is discussed further in Chapter 13.

Supply Chain Event Management (SCEM)

Supply chain event management (SCEM) refers to software that monitors events throughout the supply chain and triggers alerts when unplanned and unexpected events occur. It notifies decision makers when these events occur so that action can be taken. This visibility allows supply chain personnel to allocate their time to exceptions only, rather than transactions that are occurring to plan. This can be used as a type of supply chain risk mitigation since it provides a notice when something unanticipated occurs early on, so that a response can be initiated. SCEM software can also be used to simulate an event to determine the impact downstream. When integrated with other applications, it can also be programmed to perform additional functions based on certain events.

Virtual Try-On

Due to the pandemic-fueled online shopping surge, many companies have witnessed as much as a 70% increase in returns from online purchases. Even with superb reverse supply chain procedures in place, retailers are experiencing huge costs in order to receive and restock inventory being returned. Larger companies that offer free shipping, such as Walmart, Inc. and Amazon, may have consumers keep unwanted items in order to cut down on reverse logistics costs. Others, including Ulta Beauty, Levi Strauss and CO, Tommy Hilfiger, Unspun and H&M, are virtually replicating the in-store experience. Consumers use technology to visualize how makeup, clothing, sunglasses, etc. will look before buying. Others use questionnaires to create a pattern for customized clothing, or utilize 3-D scanners to create a digital avatar for consumers to be able to see how the clothes fit. Some companies are seeing a drop in returns from 30% to 5%, which directly affects the bottom line.

SUMMARY

The management required to effectively manage the supply chain is heavily based on managing data and inventory, but there are other aspects as well. Strategic supply chain planning and design allow a supply chain to better service the end customer and increase the likelihood of agility and resiliency. Integration techniques among supply chain partners can break down barriers between supply chain entities and minimize the bullwhip effect. Supply chain management also focuses on anticipating and preparing for risks such as supply chain disruptions, and sharing those risks among other entities in the supply chain. Today's consumers are demanding that supply chains incorporate sustainability principles in order to be environmentally and socially responsible. Reducing waste, raw material consumption and carbon footprints are three ways companies can demonstrate corporate responsibility. Supply chain metrics can be used to measure supply chain performance across multiple areas. As supply chains continue to advance, technology increases the ability to integrate processes, share information, and provide an improved experience to the end customer.

KEY TERMS

Agility 174	Quick-response program (QRP) 175
Artificial intelligence (AI) 185	Recovery 180
Big data 185	Recycle 180
Blockchain 185	Reuse 180
Bullwhip effect 175	Reverse logistics 169
Business-to-business (B2B) commerce 185	Reverse supply chain 169
By-product 180	Risk acceptance 178
Carbon footprint 181	Risk avoidance 177
Circular economy 180	Risk management 177
Cloud computing 184	Risk mitigation 177
Collaborative planning, forecasting and replenishment (CPFR) 175	Risk pooling 178
	Risk prevention 178
Control tower 185	Risk response 177
Core competencies 171	Social responsibility 181
Customer relationship management (CRM) 175	Software-as-a-service (SaaS) 184
	Supplier relationship management (SRM) 175
Decision support system (DSS) 185	
Drop ship 173	Supply chain 169
Echelon 171	Supply chain event management (SCEM) 186
Enterprise resource planning (ERP) 170	
Environmentally responsible business 179	Supply chain management 170
Globalization 173	Supply Chain Operations Reference (SCOR) model 184
Glocalization 173	
Green supply chain 179	Sustainability 179
Group decision support system (GDSS) 185	Threatcasting 178
	Triple bottom line (TBL) 179
Internet of Things (IoT) 186	United Nations Global Compact 181
Just in time (JIT) 170	
Life cycle assessment (LCA) 182	Virtual organization 185
Point of sale (POS) 175	Waste hierarchy 179

QUESTIONS

1. What are the four flows in a supply chain?
2. What is meant by reverse logistics? Why might it be important in manufacturing today?
3. Describe the current philosophy of supply chain management?
4. What are five ways organizations can use supply chains strategically?
5. What are some of the questions an organization must ask when determining their global strategy?
6. How can CRM and SRM be used to manage supply chain relationships?
7. What is the bullwhip effect, and how could it have been prevented using supply chain management principles during the recent pandemic?
8. Select an organization you are familiar with, and describe an example of how they could demonstrate supply chain agility?
9. How are point of sale and quick response programs used to integrate supply chain information?
10. What are some examples of supply risk?
11. What are the four steps of risk management?
12. What is the difference between risk mitigation, risk avoidance, risk prevention, and risk acceptance?

13. You are a supply chain manager for a company that has a major disruption in the supply chain of one of your products. What steps should your organization take to respond? What are some options for redesigning your supply chain to avoid a similar disruption?
14. What are the three principles of the triple bottom line?
15. What are the four levels of waste hierarchy? Describe which level has the most beneficial impact on the environment and why.
16. What does a supply chain's carbon footprint measure?
17. What are the 10 principles of the United Nations Global Compact?
18. What are some examples of how life cycle assessment can be applied to the supply chain?
19. Give some examples of supply chain performance measures.
20. What are the primary activities of the SCOR model?
21. What improvements has cloud computing brought to the supply chain?
22. What is a blockchain? What are the benefits of using blockchain?

PROBLEMS

7.1. You are the supply chain manager for a global supply chain. Your organization uses the following scoring to prioritize risks that are being assessed.

Probability:	Unlikely – 1	Impact:	Insignificant – 1
	Rare – 2		Minor – 2
	Possible – 3		Moderate – 3
	Likely – 4		Major – 4
	Certain – 5		Catastrophic – 5

The following risks have been identified for your supply chain:
- A supplier of a B class item is having quality issues which impact your supply: Likely, minor.
- One of your A class items is sourced from a country in the middle of a political conflict: Possible, major.
- Forecasting demand has proved very difficult for a new item in its growth phase: Certain, insignificant.

1. Using risk assessment techniques, which risk should be the priority?
2. What are some risk responses that could be put in place to mitigate the risks?
3. Are there any risks that could be avoided or prevented, and if so, how?

CASE STUDY 7.1

In the late 1970's and early 1980's the technological and market development in the computer industry was growing very rapidly. Not only were more companies purchasing and using larger computers, but the people around the world were creating a large demand for personal computers. The result was a much accelerated growth in demand for the integrated circuits (chips) used to make the computers. The normal lead time quoted by most of the producers of chips was about six weeks. Once they realized the large increase in orders, however, they naturally thought "If we have a lot more orders then it may take longer for us to satisfy them all—we had better quote eight weeks instead of six." As soon as they did that then the customers naturally thought "If it takes us longer to get orders, we had better put in more orders to cover us for at least eight weeks and not just six." This generated lots of new orders, prompting the chip producers to once again be concerned and extend the quoted lead time yet again, which again generated more orders as the chip customers grew concerned about the supply. This back-and-forth activity continued until at one point the quoted lead time for some of the chips exceeded two years! The computer makers were now frustrated because they had no real way to determine what the actual demand would be a full two years away. The chip makers were also frustrated as they were subjected to a very large number of order changes as time passed and the customers were able to better understand their real needs.

Assignment

1. Discuss how something like this could be prevented using our present understanding of supply chains, and given the situation what might be a good solution?

CASE STUDY 7.2

In early 2020 many of the countries in the world declared a "lockdown" condition and required their citizens to stay home unless they were deemed an "essential" worker. This was in response to the declaration of a global pandemic caused by the COVID-19 virus. The response from many people was to purchase more than a usual quantity of supplies. The resulting shortage in many goods led to a hoarding response by many people, and soon there were extreme shortages. One of the most notable in some parts of the world was a shortage of bathroom tissue, and the shops that sell the tissue were unable to obtain replacement quantities for several months.

Assignment

1. Discuss the supply chain concepts that this condition illustrates. Then discuss alternatives of how this condition could be addressed more effectively if it were to occur again. For each alternative, try to discuss the benefits and costs involved. This should be done from the perspective of both the retailer and the producer. What method would you recommend and why?

CASE STUDY 7.3

In early 2020 a mid-sized hospital was faced with a major problem, and the supply chain manager was expected to solve it. Until that time the hospital used a few hundred protective face masks per day, mainly for surgical procedures and healthcare workers working with patients having contagious diseases. Now the COVID-19 pandemic was declared and suddenly all hospital employees and patients were required to wear protective masks. The usage went from less than a thousand a day up to as high as 4000 per day. In addition, the hospital typically used a lean replenishment approach where they kept only a two-week supply of most supplies. They had regular contracts with suppliers to replenish every week up to the two-week requirement level. They soon discovered that the suppliers suddenly had many customers (including many other hospitals) who were in a similar condition and all were pressuring the supplier to significantly increase their order size. The condition was made worse in this particular hospital because the administration decided there was too much risk in the lean approach (because of this supply problem) so they changed the policy and now wanted the supply chain manager to keep a minimum of three month's supply of all important supplies. The manager was faced with not only trying to keep needed supplies available from day to day but also now had to build an inventory, and he also realized that the hospital simply did not have the proper storage space to maintain three month's supplies of all key items.

Assignment

1. What steps would you suggest the supply chain manager take in such a condition? What are the costs and benefits potentially involved? What alternatives exist to your solution, if any? What would the order of implementation steps be and why would you proceed in that manner? Whatever solution you suggest, would you recommend it be permanent or should it later be modified. If modified, what criteria would you use?

CHAPTER EIGHT

PURCHASING

8.1 INTRODUCTION

Purchasing can simply be considered the process of buying. Many assume purchasing is solely the responsibility of the purchasing department. However, the function is much broader and, if carried out effectively, all departments in the company may be involved. Obtaining the right material, in the right quantities, with the right delivery (time and place), from the right source, and at the right price are all purchasing functions.

Choosing the right material requires input from the marketing, engineering, manufacturing, and purchasing departments. Quantities and delivery of finished goods are established by the needs of the marketplace. However, manufacturing planning and control must decide when to order which raw materials so that marketplace demands can be satisfied. Purchasing is then responsible for placing the orders and for ensuring that the goods arrive on time.

The purchasing department has the major responsibility for locating suitable sources of supply and for negotiating prices. Input from other departments may be required in finding and evaluating sources of supply and to help the purchasing department in price negotiation. Environmental responsibility is becoming a major consideration in business due to potential costs and consumer demand. Purchasing departments are in a position to take the lead role in reducing a company's environmental impact since they are familiar with all materials purchased and have excellent contacts with suppliers for product information. Purchasing, in its broad sense, is everyone's business.

Purchasing and Profit Leverage

On the average, manufacturing firms spend about 50% of their sales dollars in the purchase of raw materials, components, and supplies. This gives the purchasing function tremendous potential to reduce costs and increase profits. As a simple example, suppose a firm spends 50% of its revenue on purchased goods and shows a net profit before taxes of 10%. For every $100 of sales, it receives $10 of profit and spends $50 on purchases. Other expenses are $40. For the moment, assume that all costs vary with sales. These figures are shown in the following as a simplified income statement:

INCOME STATEMENT

Sales		$100
Cost of Goods Sold		
Purchases	$50	
Other Expenses	$40	$90
Profit Before Tax		$10

To increase profits by $1, a 10% increase in profits, sales must be increased to $110. Purchases and other expenses increase to $55 and $44. The following modified income statement shows these figures:

INCOME STATEMENT (SALES INCREASE)

Sales		$110
Cost of Goods Sold		
Purchases	$55	
Other Expenses	$44	$99
Profit Before Tax		$11

However, if the firm can reduce the cost of purchases from $50 to $49, a 2% reduction, it would gain the same 10% increase in profits. In this particular example, a 2% reduction in purchase cost has the same impact on profit as a 10% increase in sales.

INCOME STATEMENT (REDUCED PURCHASE COST)

Sales		$100
Cost of Goods Sold		
Purchases	$49	
Other Expenses	$40	$89
Profit Before Tax		$11

Purchasing Objectives

Purchasing is responsible for the flow of materials into the firm, following up with the supplier, and expediting delivery. Missed deliveries can create havoc for manufacturing and sales, and purchasing can reduce problems for both areas, further adding to the profit.

The objectives of purchasing can be divided into five categories:

- Obtaining goods and services of the required quantity and quality.
- Obtaining goods and services at the lowest total cost.
- Ensuring the best possible service and prompt delivery by the supplier.
- Developing and maintaining good supplier relations and developing potential suppliers.
- Selecting products and suppliers that minimize the impact on the environment.

To satisfy these objectives, some basic functions must be performed:

- Determining purchasing specifications: right quality, right quantity, and right delivery (time and place).
- Selecting supplier (right source).
- Negotiating terms and conditions of purchase (right price).
- Issuing and administration of purchase orders and agreements.

Outsourcing

The *APICS Dictionary*, 16th edition, defines **outsourcing** as "the process of having suppliers provide goods and services that were previously provided internally." One method of outsourcing is to use a company **offshore**, which is defined as "outsourcing a business function to another company in a different country than the original company's country" (*APICS Dictionary*, 16th edition). This is a growing trend with many companies as lower labor costs and an increasingly educated workforce are becoming available offshore. Internet communications and efficient multimodal supply chains can make outsourcing and offshore sourcing very attractive. Companies are under pressure to reduce costs and to focus on their core competencies, which can turn outsourcing into a competitive advantage.

The purchasing department is directly affected by the growth in outsourcing. Operations departments add value to goods through efficiently using people, machines, and materials. Many of these components are now purchased rather than produced internally, and there is a shift from internal management of staff to working with outside suppliers through the purchasing department. One of the keys to determining whether outsourcing should be used for a particular product, component or service is performing a **make-or-buy cost analysis** which compares the cost of making an item versus the cost of buying it. Other considerations may be the level of quality, reliability or expertise of suppliers, or the proprietary knowledge of the internal process. Make-or-buy decisions are discussed further in Chapter 15.

The outsourcing trend is not just occurring in manufacturing but is also affecting service departments as companies may outsource maintenance, information technology, logistics, finance, and customer service. Outside contractors can often do the same work better, faster, and cheaper.

Purchasing departments have an increasing responsibility with the management of outside operations and the development and administration of contracts. Contracts and legal terms are beyond the scope of this text.

Purchasing Cycle

The purchasing cycle consists of the following steps:

1. Receiving and analyzing purchase requisitions.
2. Selecting suppliers, including researching and finding potential suppliers, issuing requests for quotations, receiving and analyzing quotations, and selecting the right supplier.
3. Determining the right price.
4. Issuing purchase orders and agreements.
5. Following up to ensure delivery dates are met.
6. Receiving and accepting goods.
7. Approving supplier's invoice for payment.

Receiving and analyzing purchase requisitions **Purchase requisitions** start with the department or person who will be the ultimate user. In the **enterprise resource planning (ERP)** environment, the planner or buyer/planner releases a planned order authorizing the purchasing department to go ahead and process a purchase order. For items not used in the manufacturing process, such as **maintenance, repair, and operating (MRO)** items, office supplies, or capital equipment, a paper or electronic requisition is sent to the purchasing department. At a minimum, the purchase requisition contains the following information:

- Identity of requestor, signed approval, and account to which cost is assigned.
- Material specification.
- Quantity and unit of measure.
- Required delivery date and place.
- Any other supplemental information needed.

Electronic requisition systems are now widely used as part of ERP software. The minimum requisition information is still required, and the system can supply much of the details and control of the information based on predetermined settings. For example, the requisitioner can enter the desired part number and ERP will provide the appropriate description, specification, suggested suppliers, shipping instructions, and so on. The system will then forward the requisition for the appropriate approvals with controls in place for account number and spending limits. Once all the approvals have been completed, the requisition is sent to the purchasing department to produce the purchase order without reentering all the information. For items of small value (see C items covered in Chapter 10) that are ordered frequently, the system may send an electronic release of material directly to the approved supplier. The benefits of these tools to the company are ease of entry for the requisitioner, reduced paperwork, decreased turnaround time of requisitions, and improved accuracy of information.

Selecting suppliers Identifying and selecting suppliers are important responsibilities of the purchasing department. For routine items or those that have been purchased before, a list of approved suppliers is maintained. If the item has not been purchased before or there is no acceptable known supplier, a search for appropriate sources must be made. Other items may require input from the engineering and design departments for suggested suppliers. More on selecting suppliers will be discussed later in this chapter.

Requesting quotations For high-cost items, it is usually desirable to issue a **request for quote (RFQ)**. This is a written inquiry that is sent to enough suppliers to be sure competitive and reliable quotations are received. It is not a purchase order. After the suppliers have completed and returned the quotations to the buyer, the quotations are analyzed for price, total cost, compliance to specification, terms and conditions of sale, delivery, and payment terms.

Determining the right price This is the responsibility of the purchasing department and is closely tied to the selection of suppliers. The purchasing department is also responsible for price negotiation and will try to obtain the best price from the supplier. Price negotiation will be discussed in a later section of this chapter.

Issuing a purchase order A purchase order is a legal offer to purchase. Once accepted by the supplier, it becomes a legal contract for delivery of the goods according to the terms and conditions specified in the purchase agreement. The purchase order is prepared from the purchase requisition or the quotations and from any other additional information needed. A copy is sent to the supplier and is retained by purchasing and sent to other departments such as accounting, the originating department, and receiving. ERP purchase orders are submitted electronically to the supplier, and are retained internally as electronic files that are accessible by all departments, as a substitute for paper purchase orders.

Following up and delivery The purchasing department is responsible for ensuring that suppliers deliver the items ordered on time. If there is doubt that delivery dates can be met, purchasing must be notified in time to take corrective action. This might involve expediting transportation, alternate sources of supply, working with the supplier to solve its problems, or rescheduling production.

The purchasing department is also responsible for working with the supplier on any changes in delivery requirements. Demand for items changes with time, and it may be necessary to expedite certain items or push delivery back on some others. The buyer must keep the supplier informed of the true requirements so that the supplier is able to provide what is wanted when it is wanted. Outputs from MRP provide messages when items have been re-prioritized, so that the buyer can determine what action is necessary.

Receiving and accepting goods When the goods are received, the receiving department inspects the goods to be sure the correct items have been sent, are in the right quantity, and have not been damaged in transit. The receiving department then accepts the goods. Variances are noted manually on the packing slip, or automatically calculated using the receiving software. Provided the goods are in order and require no further inspection, they will be sent to the requisitioning department or to inventory.

If further inspection is required, such as by quality control, the goods are sent to quality control or held in receiving for inspection. If the goods are received damaged, the receiving department will advise the purchasing department and hold the goods for further action. Purchasing is notified once the goods have either been inspected or rejected. Purchasing is then responsible for notifying the supplier and determining the action necessary to return and/or replace the items.

A copy of the receiving report is provided to the purchasing department, noting any variance or discrepancy from the purchase order. If the receipt completes the purchase order, the PO moves to a closed status and is no longer available for receiving. If the purchase order has not been received complete, either due to quantities remaining or additional line items, the purchase order remains open.

Approving supplier's invoice for payment When the supplier's invoice is received, there are three pieces of information that are matched: the purchase order, the receiving information, and the invoice. The items and the quantities should be the same on all; the prices, and quantity extensions to prices, should be the same on the purchase order and the invoice. All discounts and terms of the original purchase order must be checked against the invoice. It is the job of the purchasing department to work with accounts payable to verify these and to resolve any differences. Once approved, the invoice is sent to accounts payable for payment.

8.2 ESTABLISHING SPECIFICATIONS

Determining what exactly to purchase is not necessarily a simple decision. For example, someone deciding to buy a car should consider how the car will be used, how often, how much one is willing to pay, and so on. Only then can an individual specify the type of car needed to make the appropriate purchase. This section looks at the problems that organizations face when developing specifications of products and the types of specifications that may be used.

When purchasing an item or a service from a supplier, several factors must be taken into consideration when specifications are being developed. These can be divided into three broad categories:

- Quantity requirements.
- Price requirements.
- Functional requirements.

Quantity Requirements

The balance of supply and demand determines the quantity needed. The quantity is important because it will be a factor in the way the product is designed, specified, and manufactured. For example, if the demand was for only one item, it would be designed to be made at the least cost, or a suitable standard item would be selected. However, if the demand were for several thousand, the item would be designed to take advantage of economies of scale, thus satisfying the functional needs at a better price.

Price Requirements

The price specification represents the economic value that the buyer puts on the item, and the amount the company is willing to pay. If the item is to be sold at a low price, the manufacturer will not want to pay a high price for a component part. The economic value placed on the item must relate to the use of the item and its anticipated selling price.

Functional Requirements

Functional specifications are concerned with the end use of the item and what the item is expected to do. By their very nature, functional specifications are the most important of all categories and govern the others. They are also the most difficult to define. To be successful, they must satisfy the real need or purpose of an item. In many cases, the real need has both practical and aesthetic elements to it. A coat is meant to keep one warm, but under what circumstances does it do so and what other functions is it expected to perform? How cold must it get before one needs a coat? On what occasions will it be worn? Is it for working or dress wear? What color and style should it be? In the same way, the question must be asked of what practical and aesthetic needs a component of a manufactured item is expected to satisfy.

Functional specifications and quality Functional specifications are closely tied to the quality of a product or service. There are many definitions of quality, but they all center on the idea of user satisfaction. On this basis, it can be said that an item has the required quality if it satisfies the needs of the user.

There are four phases to providing user satisfaction:

1. Quality and product planning.
2. Quality and product design.
3. Quality and manufacturing.
4. Quality and use.

Product planning is involved with decisions about which products and services a company is going to market. It must decide the market segment to be served, the product features and quality level expected by that market, the price, and the expected sales volume. The basic quality level is specified by senior managers according to their understanding of the needs and wants of the marketplace. The success of the product depends on how well they do this.

The result of the firm's market studies is a general specification of the product outlining the expected performance, appearance, price, and sales volume of the product. It is then the job of the product designer to build into the design of the product the quality level described in the general specification. If this is not properly done, the product may not be successful in the marketplace.

For manufactured products, it is the responsibility of manufacturing, at a minimum, to meet the specifications provided by the product designer. If the item is bought, it is purchasing's responsibility to make sure the supplier can provide the required quality level. For purchasing and manufacturing, quality means conforming to specifications or requirements.

To final users, quality is related to their expectations of how the product should perform. Customers do not care why a product or service is defective. They expect satisfaction. If the product is what the customer wants, well designed, well made, and well serviced, the quality is satisfactory.

Functional specifications should define the quality level needed. They should describe all those characteristics of a product determined by its final use.

Function, quantity, service, and price are interrelated. It is difficult to specify one without consideration of the others. The final specification is a compromise of all four, and the successful specification is the best combination. However, functional specifications ultimately are the ones that drive the others. If the product does not perform adequately for the price, it will not sell.

Value analysis **Value analysis** as defined by the *APICS Dictionary*, 16th edition, is "the systematic use of techniques that identify a required function, establish a value for that function, and finally provide that function at the lowest overall cost." Teams of engineers, users, production personnel, and suppliers analyze parts to challenge current specifications and identify redundant or unnecessary features. This form of supply chain collaboration during the design process can be instrumental in reducing the cost and, more importantly, improving the overall functionality of the part. A good example of value analysis is the evolution of the milk bottle as it went from a heavy glass bottle to a plastic jug. The result is a much cheaper package with improvements in sterility, transportation, and breakage. This will be discussed in further detail in Chapter 15.

8.3 FUNCTIONAL SPECIFICATION DESCRIPTION

Functional specification can be described in the following ways or by a combination of them:

1. By brand.
2. By specification of physical and chemical characteristics, material and method of manufacture, and performance.
3. By engineering drawings.
4. By miscellaneous attributes.

Description by Brand

Description by brand is most often used in wholesale or retail businesses but can also be used extensively in manufacturing. This is particularly true under the following circumstances:

- Items are patented, or the process is secret.
- The supplier has special expertise that the buyer does not have.
- The quantity bought is so small that it is not worth the buyer's effort to develop specifications.
- The supplier, through advertising or direct sales effort, has created a preference on the part of the buyer's customers or staff.

When buying by brand, the customer is relying on the reputation and integrity of the supplier. The assumption is that the supplier wishes to maintain the brand's reputation and will maintain and guarantee the quality of the product so repeat purchases will give the buyer the same satisfaction.

Most of the objections to purchasing by brand center on cost. Branded items, as a group, usually have price levels that are higher than nonbranded items. It may be less costly to develop specifications for generic products than to rely on brands. The other major disadvantage to specifying by brand is that it restricts the number of potential suppliers and reduces competition. Consequently, the usual practice, when specifying by brand, is to ask for the item by brand name or equivalent. In theory, this allows for competition.

Description by Specification

There are several ways of describing a product, but whatever method is used, description by **specification** depends on the buyer describing in detail exactly what is wanted. One or more of the following is typically used:

- **Physical and chemical characteristics.** The buyer must define the physical and chemical properties of the materials wanted. Petroleum products, pharmaceuticals, and paints are often specified in this way.
- **Material and method of manufacture.** Sometimes the method of manufacture determines the performance and use of a product. For example, hot- and cold-rolled steels are made differently and have different characteristics.
- **Performance.** This method is used when the buyer is primarily concerned with what the item is required to do and is prepared to have the supplier decide how performance is to be attained. For example, a water pump might be specified as having to deliver so many gallons per minute. Performance specifications are relatively easy to prepare and take advantage of the supplier's special knowledge.

Whatever the method of specification, there are several characteristics of description by specification:

- To be useful, specifications must be carefully designed. If they are too loosely drawn, they may not provide a satisfactory product. If they are too detailed and elaborate, they are costly to develop, are difficult to inspect, and may discourage possible suppliers.
- Specifications must allow for multiple sources and for competitive bidding.
- If performance specifications are used, the buyer is assured that if the product does not give the desired results, the seller is responsible. They provide a standard for measuring and checking the materials supplied.
- Not all items lend themselves to specification. For example, it may not be easy to specify color schemes or the appearance of an item.
- An item described by specification may be no more suitable, and a great deal more expensive, than a supplier's standard product.
- If the specifications are set by the buyer, they may be expensive to develop. They will be used only when there is sufficient volume of purchases to warrant the cost or where it is not possible to describe what is wanted in any other way.

Standard specifications *Standard specifications* have been developed as a result of much study and effort by governmental and nongovernmental agencies. They usually apply to raw or semifinished products, component parts, or the composition of material.

In many cases, they have become de facto standards used by consumers and by industry. When SAE 10W30 motor oil is purchased for a car, a standard grade of motor oil is being specified as established by the Society of Automotive Engineers. Most of the electrical products purchased in the United States are manufactured to Underwriters Laboratory (UL) standards.

There are several advantages to using standard specifications. First, they are widely known and accepted and, because of this, are readily available from most suppliers. Second, because they are widely accepted, manufactured, and sold, they are lower in price than nonstandard items. Finally, because they have been developed with input from a broad range of producers and users, they are usually adaptable to the needs of many purchasers.

Market grades are a type of standard specification usually set by the government and used for commodities and foodstuffs. Eggs, for example, are purchased by market grade—small, medium, or large.

Engineering Drawings

Engineering drawings describe in detail the exact configuration of the parts and the assembly. They also give information on such things as finishes, tolerances, and material to be used. These drawings are a major method of specifying what is wanted and are widely used because often there is no other way to describe the configuration of parts or the way they are to fit together. They are produced by the engineering design department and are expensive to produce, but they give an exact description of the part required. They can also be used a means for inspection to ensure compliance.

Miscellaneous Attributes

There are a variety of other methods of specification, including the famous phrase, "Give me one just like the last one." Sometimes samples are used, for example, when colors or patterns are to be specified. Often a variety of methods can be used, and the buyer must select the best one.

The method of description is determined by communication with the supplier. How well it is described will affect the success of the purchase and sometimes the price paid.

8.4 SELECTING SUPPLIERS

The objective of purchasing is to get all the right things together: quality, quantity, delivery, and price. Once the decision is made about what to buy, the selection of the right supplier is the next most important purchasing decision. A good supplier is one that has the technology to make the product to the required quality, has the capacity to make the quantities needed and deliver on time, and can run the business well enough to make a profit and still sell a product competitively. It is also beneficial if they can be involved and contribute in the improvement of the product.

Sourcing

There are three types of sourcing: sole, multiple, and single.

1. **Sole sourcing** implies that only one supplier is available because of patents, technical specifications, raw material, location, and so forth.
2. **Multiple sourcing** is the use of more than one supplier for an item. The potential advantages of multiple sourcing are that competition will result in lower price and better service and that there will be a continuity of supply.
3. **Single sourcing** is a planned decision by the organization to select one supplier for an item when several sources are available. It is intended to produce a long-term partnership. This is discussed at more length in Chapter 16, in the section on supplier partnerships.

Factors in Selecting Suppliers

The previous section discussed the importance of function, quantity, service, and price specifications. These are what the supplier is expected to provide and are the basis for selection and evaluation. Considering this, there are several factors in selecting a supplier.

Technical ability Does the supplier have the technical ability to make or supply the product wanted? Does the supplier have a program of product development and improvement? Can the supplier assist in improving the products? These questions are important since, often, the buyer will depend upon the supplier to provide product improvements that will enhance or reduce the cost of the buyer's products. Sometimes the supplier can suggest changes in product specification that will improve the product and reduce cost.

Manufacturing capability Manufacturing must be able to meet the specifications for the product consistently while producing as few defects as possible. This means that the supplier's manufacturing facilities must be able to supply the quality and quantity of the products wanted. The supplier must have a good quality assurance program, competent and capable manufacturing personnel, and good manufacturing planning and control systems to ensure timely delivery. These are important in ensuring that the supplier can supply the quality and quantity wanted.

Reliability In selecting a supplier, it is desirable to pick one that is reputable, stable, and financially strong. If the relationship is to continue, there must be an atmosphere of mutual trust and assurance that the supplier is financially strong enough to stay in business.

After-sales service If the product is of a technical nature or likely to need replacement parts or technical support, the supplier must have a good after-sales service. This should include a good service organization and inventory of service parts.

Supplier location Sometimes it is desirable that the supplier be located near the buyer, or at least maintain an inventory locally. A close location helps shorten delivery time and means emergency shortages can be delivered quickly.

Lean capabilities Companies competing in a lean environment depend on suppliers to quickly deliver small quantities of product, to keep inventories at a low level. Today's companies operate with very little inventory of raw materials and require accurate, on-time deliveries from their suppliers. Suppliers who simply keep extra inventory to meet these demands will soon have increased costs and pressures to increase their prices. Buyers in a lean environment need suppliers who value their new relationship, working in partnership to remove waste from the system. As a result, these suppliers need to have information and delivery systems in place that allow them to quickly ship exactly what the customer needs without increased cost or effort. Lean production is discussed further in Chapter 16.

Other considerations Sometimes other factors such as credit terms, reciprocal business, supplier health and safety record, and willingness of the supplier to hold inventory for the buyer should be considered. Supplier sustainability initiatives should be considered, which are discussed later in this chapter.

Price The supplier should be able to provide competitive prices. This does not necessarily mean the lowest price. It is one that considers the ability of the supplier to provide the necessary goods in the quantity and quality wanted, at the time wanted, as well as any other services needed.

The total **landed cost** of an item includes the price paid plus all the handling and delivery costs associated with getting the product to production. A buyer will often get a price and per unit transportation discount by ordering in larger quantities. However, the total cost may increase when the costs of storage and inventory are included.

A low landed cost still may not be a good decision when the **total cost of ownership (TCO)** to the company is considered. For example, a carpenter will pay a lower price for a lower grade of wood. However, the time spent on sorting out knots and defects and the decreased yield of good material will incur production-related costs, which will increase the total cost of the final product, perhaps canceling any savings made in price. The total cost concept looks at the total costs of an item, and not at just the price paid for materials.

In today's supply chain environment, the type of relationship between the supplier and the buyer is crucial to both. Ideally, the relationship will be ongoing with a mutual dependency. The supplier can rely on future business, and the buyer will be ensured of a supply of quality product, technical support, and product improvement. Communications between buyer and supplier must be open and complete so both parties understand the problems of the other and can work together to solve problems to their mutual advantage. Thus, supplier selection and supplier relations are of the utmost importance.

Identifying Suppliers

A major responsibility of the purchasing department is to continue to research all available sources of supply. Some aids for identifying sources of supply include the following:

- Salespersons of the supplier company.
- Internet.
- Catalogues.
- Trade magazines or directories.
- Expos or exhibitions.
- Information obtained by the salespeople of the buyer's firm.

Final Selection of Supplier

Some factors in evaluating potential suppliers are quantitative, and a dollar value can be put on them. Price and landed cost are obvious examples. Other factors are qualitative and require some judgment to determine them. These are usually specified in a descriptive fashion. The supplier's technical competence might be an example, as well as the ongoing relationship, mutual benefit of the parties, and problem-solving.

The challenge is finding some method of combining these two major factors that will enable a buyer to pick the best supplier. One method involves a supplier ranking method, as follows:

1. Select those factors that must be considered in evaluating potential suppliers.
2. Assign a weight to each factor. This weight determines the importance of the factor in relation to the other factors. Usually, a scale of 1 to 10 is used. If one factor is assigned a weight of 5 and another factor a weight of 10, the second factor is considered twice as important as the first. When developing the factors and their weights, the buyer can use input from the people who will be affected by the supplier selection. This will help the buyer in making a more informed decision and will improve the acceptance of the new supplier by the users.
3. Rate the suppliers for each factor. This rating is not associated with the weight. Rather, suppliers are rated on their ability to meet the requirements of each factor. Again, usually a scale of 1 to 10 is used.
4. Rank the suppliers. For each supplier, the weight of each factor is multiplied by the supplier rating for that factor. For example, if a factor had a weight of 8 and a supplier was rated 3 for that factor, the ranking value for that factor would be 24. The supplier rankings are then added to produce a total ranking. The suppliers can then be listed by total ranking and the supplier with the highest ranking chosen.

Figure 8.1 shows an example of this method of selecting suppliers. Supplier B has the highest total of 223; however, supplier D comes in a very close second with 222. The normal practice when using the ranking method is to eliminate the bottom ranking suppliers from consideration, allowing management to make a simpler decision.

Factor	Weight	Rating of Suppliers				Ranking of Suppliers			
Suppliers		A	B	C	D	A	B	C	D
Function	10	8	10	6	6	80	100	60	60
Cost	8	3	5	9	10	24	40	72	80
Service	8	9	4	5	7	72	32	40	56
Technical Assistance	5	7	9	4	2	35	45	20	10
Credit Terms	2	4	3	6	8	8	6	12	16
Total (rank of suppliers)						219	223	204	222

FIGURE 8.1 Supplier rating.

The supplier ranking method is an attempt to quantify those things that are not quantifiable by nature. It attempts to put figures on subjective judgment. It is not a perfect method, but it forces the buying company to consider the relative importance of the various factors. When the method includes the input of many people in determining the relative weights, agreement on the final selection will be improved.

8.5 PRICE DETERMINATION

As mentioned previously, price is not the only factor in making purchasing decisions, but can be the determining factor if all other things are equal. In the average manufacturing company, purchases account for about 50% of the cost of goods sold, and any savings made in purchase cost has a direct influence on profits. Best price would include the best mixture of function, quantity, service, and price characteristics.

Basis for Pricing

The term *fair price* is sometimes used to describe what should be paid for an item. But what is a fair price? One answer is that it is the lowest price at which the item can be bought. However, there are other considerations, especially for repeat purchases where the buyer and seller want to establish a good working relationship. One definition of a fair price is one that is competitive, gives the seller a profit, and allows the buyer ultimately to sell at a profit. Sellers who charge too little to cover their costs will not stay in business. To survive, they may attempt to cut costs by reducing quality and service. In the end, both the buyer and seller must be satisfied.

Since the objective is to pay a fair price and no more, it is good to develop some basis for establishing what a fair price is.

Prices have an upper and a lower limit. The market decides the upper limit, as what buyers are willing to pay is based on their perception of demand, supply, and their needs. The seller sets the lower limit, and it is determined by the costs of manufacturing and selling the product and profit expectation. If buyers are to arrive at a fair price, they must develop an understanding of market demand and supply, competitive prices, and the methods of arriving at a cost.

One widely used method of analyzing costs is to break them down into fixed and variable costs. **Fixed costs** are costs incurred no matter the volume of sales. Examples are equipment depreciation, taxes, insurance, and administrative overhead. **Variable costs** are those directly associated with the quantity produced or sold. Examples are direct labor, direct material, and commissions of the sales force.

$$\text{Total cost} = \text{fixed cost} + \text{variable cost per unit (number of units)}$$

$$\text{Unit cost} = \frac{\text{total cost}}{\text{number of units}}$$

$$= \frac{\text{fixed cost}}{\text{number of units}} + \text{variable cost per unit}$$

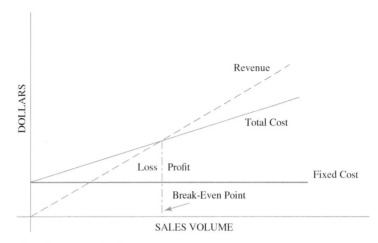

FIGURE 8.2 Break-even analysis.

The preceding formula shows that as the number of units produced increases, the unit cost decreases. This is an important factor when determining price. Buyers can lower the unit price paid by increasing the volume per order, using longer-term contracts or through the standardization of parts. Sellers will offer quantity discounts to encourage larger orders, also taking advantage of this reduction in unit cost. Quantity discounts are discussed later in Chapter 11, and standardization is discussed in Chapter 15.

Figure 8.2 shows the relationship of fixed and variable costs to sales volume and how revenue will behave. The sum of the fixed and variable costs is labeled Total Cost on the graph. The third line represents the sales revenue. Where this line intercepts the total cost line, revenue equals total cost, and profit is zero. This is called the *break-even point*. When the volume is less than the break-even point, a loss is incurred; when the volume is greater, a profit is realized. The break-even point occurs where the revenue equals the total cost.

$$\text{Revenue} = \text{Total cost}$$
$$(\text{Price per unit})(\text{number of units sold}) = \text{fixed cost} + (\text{variable cost per unit})$$
$$\times (\text{number of units})$$

Because a seller must have a sufficient volume to make a profit, knowing the seller's break-even point is useful in negotiations, as they may be willing to lower the price in order to receive a larger order.

Example Problem

To make a particular component requires an overhead (fixed) cost of $5000 and a variable unit cost of $6.50 per unit. What is the total cost and the average cost of producing a lot of 1000? If the selling price is $15 per unit, what is the break-even point?

ANSWER

$$\text{Total cost} = \$5000 + (\$6.50 \times \$1000) = \$11,500$$
$$\text{Average cost} = \$11,500 \div 1000 = \$11.50 \text{ per unit}$$
$$\text{Break-even point: Let } X = \text{number of units sold}$$
$$\$15X = \$5000 + \$6.5X$$
$$\$8.50X = \$5000$$

Break-even occurs when 588.2 units are made and sold.

Competitive Bidding

Competitive bidding occurs when a buyer compares the price of a product from various suppliers and simply chooses the lowest price. This can be the formal process of sending out quotations and analyzing the results or simply comparing catalogue or

> **HEALTHCARE PRICE NEGOTIATIONS**
>
> As in other industries, price is dependent on various factors. However, in healthcare, one factor that needs to be realized is the amount the institution will reimburse for that particular product. The cost to provide patient care includes the cost of goods and the cost to administer care. While most contracts within healthcare are negotiated toward a fixed cost, organizations need to make certain they cover those costs and have an understanding of what reimbursement they will receive from providers.
>
> The second piece is the ability to benchmark pricing. Some institutions have access to pricing data that is available from various sources. By using this data, it gives the institution the ability to see what the upper and lower limits are on particular products. This data can be used to begin price negotiation to ensure you are covering all the costs.
>
> The final portion is the availability of the product. Organizations need to partner with vendors/suppliers who sufficiently supply to meet their needs and deliver it in a timely, cost-effective manner.
>
> Quality products at the right price and delivered at the right time.
>
> Submitted by:
>
> Luis A. Richard
> AVP, Supply Chain Management Health-Quest Systems, Inc.
> Poughkeepsie, New York

advertised prices. The process does take time, and a number of sources must be available. At least three sources are desired to make a good comparison. Competitive bidding also requires that the product be well specified and widely available. Items such as nuts and bolts, gasoline, bread, and milk are usually sourced using competitive bidding.

Price Negotiation

Prices can be negotiated if the buyer has the knowledge and the clout to do so. A small retailer probably has little of the latter, but a large buyer may have much. Through negotiation, the buyer and seller try to resolve conditions of purchase to the mutual benefit of both parties. Skill and careful planning are required for the negotiation to be successful. It also takes a great deal of time and effort, so the potential profit must justify the expense.

One important factor in the approach to negotiation is the type of product. There are four categories of products:

1. **Commodities.** A **commodity** is a material such as copper, coal, wheat, meat, and metals. Price is set by market supply and demand and can fluctuate widely. Negotiation is concerned with contracts for future prices.
2. **Standard products.** These items are provided by many suppliers. Since the items are standard and the selection of suppliers to choose from is large, prices are determined on the basis of published prices. There is not much room for negotiation except for large purchases.
3. **Items of small value.** These are items such as maintenance or cleaning supplies and represent purchases of such small value that price negotiation is of little purpose. The prime objective should be to keep the cost of ordering low. Firms will negotiate a contract with a supplier that can supply many items and set up a simple ordering system that reduces the cost of ordering.
4. **Made-to-order items.** This category includes items made to specification or for which quotations from several sources are received. These can generally be negotiated.

8.6 PURCHASING TRENDS

Purchasing can be separated into two types of activities: (1) procurement and (2) supplier scheduling and follow-up. Much of what has been covered in this chapter is in the area of procurement. Procurement includes the functions of establishing specifications, selecting suppliers, price determination, and negotiation. Supplier scheduling and follow-up are concerned with the release of orders to suppliers, working with suppliers to schedule delivery, and follow-up. The goals of supplier scheduling are the same as those of production activity control: to execute the master production schedule (MPS) and the material requirements plan, ensure good use of resources, minimize work-in-process inventory, and maintain the desired level of customer service.

Planner/buyer concept In a traditional system, the material requirements planner releases an order either to production activity control or to purchasing. Purchasing issues purchase orders based on the material requirements plan. Production activity control prepares shop orders, schedules components into the work flow, and controls material progress through the plant. When plans change, as they invariably do, the production planner must advise the buyer of the change, and the buyer must advise the supplier. The production planner is in closer, more continuous contact with material requirements planning and frequently changing schedules than is the buyer. To improve the effectiveness of the planner/buyer activity, many companies have combined the two functions of buying and planning into a single job done by one person. **Planner/buyers** do the material planning for the items under their control, communicate the schedules to their suppliers, follow up, resolve problems, and work with other planners and the master scheduler when delivery problems arise. The planner/buyer is responsible for the following:

- Determining material requirements.
- Developing schedules.
- Issuing shop orders.
- Issuing material releases to suppliers.
- Establishing delivery priorities.
- Controlling orders in the factory and to suppliers.
- Handling all the activities associated with the buying and production scheduling functions.
- Maintaining close contact with supplier personnel.

Because the roles of production planning and buying are combined, there is a smoother flow of information and material between the supplier and the factory. The planner/buyer has a keener knowledge of factory needs than the buyer does and can better coordinate the material flow with suppliers. At the same time, the planner/buyer is better able to match material requirements with the supplier's manufacturing capabilities and constraints.

Contract buying Usually MRP generates frequent orders for relatively small quantities. This is particularly true for components that are ordered lot-for-lot. It can be costly, inefficient, and sometimes impossible to issue a new purchase order for every weekly requirement. The alternative is to develop a long-term contract with a supplier and to authorize releases against the contract. This can be in the form of a **blanket purchase order**, which provides a long-term commitment to the supplier, while actual shipments are based on short-term releases. Often suppliers are given a copy of the material requirements plan so they are aware of future demands. The buyer then issues a release against the schedule. This approach is efficient and cost-effective, and can be managed by the buyer/planner with close coordination and communication with the supplier.

Supplier responsiveness and reliability Because material requirements often change, suppliers must be able to react quickly to change. They must be highly flexible and reliable so they can react quickly to changes in schedules. Responsiveness and reliability are qualitative factors that must be taken into consideration when selecting suppliers. Long-term contracts ensure suppliers a given amount of business and commits them to allocating that amount of their capacity to the customer. Suppliers are more responsive

> ### HEALTHCARE SUPPLY AND DISTRIBUTION
>
> Healthcare is no different to the supply and demand process. It is all based on the need and current volume an institution has when taking care of patients. By working with suppliers and distributors, healthcare institutions can begin to forecast their needs based on trends or ordering patterns. The supplier/distributor will then provide the institution the inventory by following the defined replenishment process. That process typically follows:
>
> - Purchase request
> - Purchase order
> - Supplier/distributor delivering product
> - Receipt of product
> - Product replenishment to either a warehouse, store room, or par location
> - Par/cycle count
>
> This process can either happen manually or electronically depending on the organization's preference. Organizations will keep a certain amount of inventory in par locations and replenish from either a store room or warehouse.
>
> Submitted by:
>
> Luis A. Richard
> AVP, Supply Chain Management Health-Quest Systems, Inc.
> Poughkeepsie, New York

to customer needs and can react quickly to changes in schedules. Because customers know the capacity will be available when needed, they can delay ordering until they are more certain of their requirements.

Close relationship with suppliers Contract buying and the need for supplier **flexibility** and **reliability** mean the buyer/supplier relationship must be close and cooperative. There must be excellent two-way communication, cooperation, and teamwork. Both parties have to understand their own and the other's operations and problems.

The planner/buyer and the supplier counterpart, often the supplier's production planner, must work on a daily or weekly basis to ensure both parties are aware of any changes in material requirements or material availability.

Electronic data interchange Electronic data interchange (**EDI**) enables customers and suppliers to electronically exchange transaction information such as purchase orders, invoices, and material requirements planning information. This eliminates time-consuming paperwork and facilitates easy communication between planner/buyers and suppliers.

Vendor-managed inventory In recent years there has been an increase in the purchasing approach known as **vendor-managed inventory (VMI)**.

In this concept, a supplier maintains an inventory of certain items in the customer's facility. The supplier owns the inventory until the customer actually withdraws it for use, after which the customer pays for the amount consumed. The customer does not have to order any of the inventory, as the supplier is responsible for maintaining an adequate supply in the facility for customer use. This approach is most commonly used for lower-value products that have a relatively standard design, such as fasteners, standard electrical equipment, and so forth, although it can also be used for large-value items as well. VMI reduces the need for many small MRP order releases, and increases the collaboration of the supply chain.

e-procurement Many of the functions formerly performed manually or on paper have been eliminated in exchange for e-procurement. E-procurement encompasses a range of tools that reduces the paperwork and handling required for each purchase and transaction. This includes the automatic generation of documents, and the electronic transmission of purchasing-related documents to and from suppliers. This reduces the lead time between need recognition and receipt of an order, improves communication, and reduces costs and errors. It also allows purchasing personnel to spend more time on strategic value-added activities.

Global sourcing Technology has caused the world to shrink as far as communication, information sharing, available resources, and competition. Benefits of **global sourcing** may include access to additional products and processes, lower cost/price, quality, or even establishing a presence in a foreign market. In order to be successful, buyers must learn how to overcome any challenges, such as legal and cultural issues, language differences, logistical considerations, and currency risks.

8.7 ENVIRONMENTALLY RESPONSIBLE PURCHASING

The fifth category of purchasing objectives introduced at the beginning of this chapter is the objective for purchasing departments to minimize the impact their organization has on the environment, that is, **environmentally responsible purchasing**. Purchasing is responsible for managing waste products in most organizations since they have the following:

- First-hand knowledge of price trends for waste products.
- Contact with salespeople who are an excellent source of information as to possible uses of waste material.
- Familiarity with the company's own needs, or uses for materials within the organization.
- Knowledge of legislation involving the transportation and handling of environmentally sensitive materials.

Purchasing activities can have a direct impact on the reputation of an organization, so purchasing personnel should find and source from environmentally responsible suppliers who hold the same ethical and sustainability standards. Suppliers should be consulted when developing material specifications to reduce costs by cutting waste or recommending bulk storage and handling systems.

Several sustainability principles and guidelines can be applied to purchasing, including the following:

- Sourcing of environmentally friendly materials, components, and packaging.
- Keeping up to date with environmental regulations, including recycled material, proper disposal of hazardous waste and the reduction of ozone-depleting substances.
- Ensuring distributors, suppliers, and contractors adhere to the firm's sustainability standards, and monitoring their performance.
- Collaborating with suppliers and distributors to address sustainability issues that arise with materials and products.
- Working with suppliers to reduce or eliminate the amount of packaging material.
- Working with transportation partners to make use of reusable packaging, or redesign of packaging for better volume utilization.
- Considering transportation methods that use alternate fuels or aerodynamic vehicles.
- Coordinating inbound and outbound shipments to reduce fuel consumption and carbon emissions.

8.8 ETHICAL PROCUREMENT AND SOURCING

Ethical sourcing, as part of overall sustainability, has become an important aspect of supply chain management and purchasing. **Ethical standards** of behavior should be established among supply chain partners and all stakeholders. An organization's reputation and image with consumers can be irreparably damaged due to unethical behaviors, both internally and by their suppliers. The globalization of supply chains has made it more critical for companies to monitor the ethical behavior of their supply chain partners.

Ethical procurement encompasses several general ethical behaviors which are summarized below:

- **Fraud/dishonesty**—misrepresentation, deception, allocation of funds for personal use, disclosure of confidential information, or sharing competitive information with other suppliers.
- **Conflict of interest**—awarding business to an entity that a buyer has a financial or personal interest in.
- **Reciprocity**—giving preference to suppliers that are also customers.
- **Bribery and corruption**—kickbacks for contracts, insider information, bribes for buying goods and services, invoicing for work or products not delivered, or offering/accepting gifts or favors in exchange for business.

Ethical sourcing refers to sourcing products and services in a sustainable and responsible method, making sure that environmental standards are met, and that the employees are safe and treated fairly. Buyers should be cognizant of the following at both current and potential suppliers:

- Working conditions at supplier facilities, such as the use of child or forced labor, fair wages and benefits, working hours, health and safety issues, harassment or discrimination.
- Environmental impact of all materials and packaging used, and improper handling and disposal of hazardous materials.

Buyers must also determine if any of their purchased materials use **conflict minerals**. These are minerals mined under conditions of armed conflict or human rights abuses, and may be sold or traded by armed groups. This has especially been an issue in the high-tech and electronics manufacturing sector, which use minerals that may not be conflict-free such as tin, tantalum, tungsten, cassiterite, wolframite, coltan, and gold. Material traceability back to the origin of raw materials can ensure that sources being used are not involved in illegal extraction methods, smuggling, corruption, or human rights abuses.

Diversity is another element of ethical sourcing, and purchasing departments should have practices in place to give equal opportunities of bidding and winning contracts to economically disadvantaged businesses. These include minority-owned, women-owned, service-disabled, veteran-owned, etc.

Principles that purchasing organizations should use as best practices for ethical sourcing include the following:

- Monitoring distributors, suppliers, and contractors to ensure adherence to standards for ethics, health and safety, human rights, and child and forced labor.
- Collaborating with suppliers and distributors to address any social or ethical issues that arise.
- Allocating more contracts to minority, women-owned, and small businesses.
- Developing strategic buyer-seller relationships to deter unethical behavior.

8.9 SOME ORGANIZATIONAL IMPLICATIONS OF SUPPLY CHAIN MANAGEMENT

Organizations that move their perspective away from traditional purchasing toward supply chain management must recognize that their perspective toward managing the entire organization must also change. For instance, most organizations that have adopted a supply chain perspective find the following:

- Their cost focus has altered dramatically. Often decisions are not based on just product price but instead on total cost and value. This implies an integrated view of price, quality, serviceability, durability, and any other characteristic that the company places on total value. It also can include transportation, storage, and material handling costs. To accomplish this changed perspective, organizations have adopted techniques of process analysis, value stream analysis, and mutual value analysis between the company and its suppliers.

- Cross-functional teams are now used to plan and control the supply chain. These teams are made up of representatives from various departments, including production, quality control, engineering, finance, and purchasing. Cross-functional teams make decisions more quickly than traditional departmental organizations and also are likely to consider overall benefits to the company, not just benefits to the individual departments.
- The decision on whether the purchasing function should be centralized or decentralized can be impacted by supply chain management. Centralized purchasing allows organizations to be able to take advantage of volume discounts and maintain centralized control over suppliers, pricing, and so forth. However, the supply chain partnership concept often leads companies to consider a more decentralized approach, which can facilitate more of a relationship between purchasing and local suppliers. Integrated systems such as ERP can offer the benefits of centralized purchasing, while allowing purchasing to be located in multiple facilities. Both approaches have advantages, and the decision should be made based on the strategic plan of the company.
- Decision-making has changed from the "I say and you do" or a negotiated perspective with suppliers, to one of "Let's talk about the best way to handle this and make a mutually advantageous decision." This also implies supplier contracts extending into the long-term future.
- Information sharing has changed from simply giving out information about the order to the sharing of important information about the business itself, which requires mutual trust and cooperation between entities.
- Measurement systems look at all aspects of the supply chain and not just supplier performance.
- There is a growth in electronic business (e-business), or using the internet more for handling business information flows and transactions.
- The environment must be considered in the acquisition, storage, use, and disposal of all materials

Savings Can Be Substantial

There are many advantages associated with an effective supply chain perspective. Some of these savings include the following:

- More effective product specification, allowing for efficient product substitutions and product specifications focused on fitness of use.
- Better leveraging of volume discounts and supplier consolidation.
- Long-term contracts with efficient communication systems, significantly reducing the administrative cost of ordering and order tracking.
- More effective use of techniques such as electronic commerce, using credit cards for payments, and blanket ordering to lower payment costs.
- Reducing environmental costs by avoiding potentially hazardous materials and exercising the 3 Rs of reduce, reuse, and recycle.

SUMMARY

Purchasing has always been an important function in any company, especially manufacturers who use a lot of raw materials or materials that are difficult to obtain. Purchasing needs to continue to get the right products at the right time and at the best price but, the function is changing. The steps in the purchasing cycle are still necessary but many manual activities such as writing POs, getting information on products, and communicating with suppliers have been sped up through use of the internet and computerization. The reduction in routine clerical activity allows time to take a more strategic view of the organization and have an increasing impact on profits. Viewing the supply chain as an integrated function, outsourcing, and lean production are three management influences that have encouraged purchasing to improve their relations with suppliers and to take a more active role in the scheduling and flow of products. Purchasing also has the opportunity to take a lead role in reducing the environmental impact of a company by working with suppliers and the use of environmentally friendly materials. In addition, their role now includes monitoring the ethical sourcing of their products and services.

KEY TERMS

Blanket purchase order 203
Commodity 202
Conflict minerals 206
Electronic data interchange (EDI) 204
Engineering drawings 197
Enterprise resource planning (ERP) 192
Environmentally responsible purchasing 205
Ethical standards 205
Fixed cost 200
Flexibility 204
Global sourcing 205
Landed cost 198
Maintenance, repair, and operating (MRO) supplies 192
Make-or-buy cost analysis 191

Multiple sourcing 197
Offshore 191
Outsourcing 191
Planner/buyer 203
Purchase requisition 192
Reliability 204
Request for quote (RFQ) 193
Single sourcing 197
Sole sourcing 197
Specification 196
Total cost of ownership (TCO) 199
Value analysis 195
Variable cost 200
Vendor-managed inventory (VMI) 204

QUESTIONS

1. What are the five objectives of purchasing?
2. Describe outsourcing and when it should be considered.
3. List the seven steps in the purchasing cycle.
4. Describe the purposes, similarities, and differences among purchase requisitions, purchase orders, and requests for quotation.
5. What are the responsibilities of the purchasing department in follow-up?
6. Describe the duties of the receiving department upon receipt of goods.
7. Besides functional specifications, what other specifications must be determined? Why is each important?
8. Name two sources of specifications.
9. What is the difference between sole sourcing and single sourcing?
10. Describe the advantages and disadvantages of the following ways of describing functional requirements. Give examples of when each is used.
    a. By brand.
    b. By specification of physical and chemical characteristics, material and method of manufacture, and performance.
11. What are the advantages of using standard specifications?
12. Why is it important to select the right supplier and to maintain a relationship with that supplier?
13. Name and describe the three types of sourcing.
14. Describe the factors that should be used in selecting a supplier.
15. Type of product is a factor that influences the approach to negotiation. Name the four categories of products and state what room there is for negotiation.
16. What types of activities are included in environmentally responsible purchasing?
17. What are the best practices for ethical sourcing?
18. What are five savings that can result from adopting a supply chain management approach?
19. A company would like to reduce the amount of lead time in some of their soldered electronics. How can the purchasing department contribute to this endeavor?

PROBLEMS

8.1. If purchases were 45% of sales and other expenses were 45% of sales, what would be the increase in profit if, through better purchasing, the cost of purchases was reduced to 43% of sales?

8.2. If suppliers were to be rated on the following basis, what would be the ranking of the two suppliers listed?

Factor	Weight	Rating of Suppliers		Ranking of Suppliers	
		Supplier A	Supplier B	Supplier A	Supplier B
Function	7	6	9		
Cost	5	8	6		
Technical Assistance	4	5	8		
Credit Terms	1	7	5		

8.3. A company is negotiating with a potential supplier for the purchase of 100,000 widgets. The company estimates that the supplier's variable costs are $5 per unit and that the fixed costs, depreciation, overhead, and so on, are $50,000. The supplier quotes a price of $10 per unit. Calculate the estimated average cost per unit. Do you think $10 is too much to pay? Could the purchasing department negotiate a better price? How?

CASE STUDY 8.1

Let's Party!

"Let's party!" is still echoing in your head as you leave your Principles of Purchasing class. Again, you ask yourself, "Why did I ever let myself run for class president?" Most of the people in the class were good, level-headed individuals who enjoyed a good time and you enjoyed working with them. But a small group from your class, who were known on campus as The Rowdies, often bullied their way on decisions affecting class activities. The decision to have a year-end party was not surprising from The Rowdies, and class had ended with a chanting session of "Let's party." It sounded like a wrestling match to you. Fortunately, your professor had left the room early to let you discuss with the class the idea of some kind of year-end get-together.

The Rowdies had immediately suggested the Goat's Ear, a local hangout with not much to offer but overly loud music and poor-quality food and drink. The rest of your classmates had put forth some other suggestions, but no consensus on a location could be reached between the members of your executive committee or the rest of the class. If you went to the Goat's Ear, you suspected that most of the people in your class wouldn't attend, and even when you suggested more conventional locations, people couldn't agree because of factors such as the type of music played.

Since there were only two weeks left until the end of regular classes, you felt that you had to make arrangements in a hurry. It wasn't difficult to identify the most popular possible locations, but getting agreement from this group was going to be difficult.

One of your recent lectures was on supplier selection, and your professor had demonstrated the technique called the ranking or weighted-point method. It seemed simple enough in the lecture, and you had almost embarrassed yourself by asking the question "Why not just pick the least expensive supplier?" The thought occurred to you that there just might be some solution to your current problem in the professor's response, "One of the hardest things to do in any group, whether a business or a social club, is to get consensus on even the simplest choices."

Assignment

For this exercise, put yourself in the position of the class president described in this case and complete one of the two following exercises:

Exercise 1

1. Perform a supplier rating analysis for the situation. Include at least 10 factors and 4 possible locations.
2. Make the selection as indicated by the analysis.
3. Discuss why the analysis led to your selection in step 2 and whether you would change any of the criteria or weights.

Exercise 2

1. Prepare a presentation to be used in class to make a selection for a year-end get-together.
2. Lead a discussion to determine at least four possible locations and 10 factors.
3. Have the class agree on weighting factors for each criteria.
4. Perform the calculations and make the selection.
5. Discuss with the class why the analysis led to the selection in step 4 and whether you would change any of the criteria or weights.
6. Ask the class members whether they feel more in agreement with the decision after going through this process.

CASE STUDY 8.2

The Connery Company

When Juan Hernandez was first given the position of head buyer for the Connery Manufacturing Company, he visualized the job as merely an expansion of his old position as a commodity buyer. He had no formal training when he took the position, having been promoted to commodity buyer from his position as inventory clerk, which he had gotten directly out of high school.

The lack of formal training was not a problem when he first took the job. The Connery Company was small but growing, and the major concern of the purchasing department was to obtain adequate purchased material to support the production and the growth in sales. There was little done in the way of price negotiation. The reason was that there was little competition for their products and all costs could easily be passed along in the product price, leaving room for the healthy profits that have helped Connery grow so rapidly.

As is often the case in these types of situations the luxury of little competition and flexibility in pricing was fairly short-lived. The success of the products Connery produced attracted a lot of attention, and soon Connery found itself in a market with several strong competitors.

While they still had the advantage of some recognition in the market ("first mover" advantage, meaning that the first entrant into the market usually has a competitive advantage), that advantageous position was in grave danger of erosion. They also had an advantage in being further down the learning curve, and the quality of their product had always been quite good. The problem now was cost. Competition was driving down the prices and maintaining their delivery record at a lower price was rapidly becoming an important factor in stemming the tide of market share erosion.

These factors were one of the key reasons that Juan was promoted. He was recognized as the best and most experienced of all the buyers, and Lawanda Connery recognized the need to move the procurement activity from one of passive buying into an active and aggressive supply management group.

As a buyer, Juan's primary responsibility was to get the material they needed, when they needed it. He primarily was responsible for buying standard components and materials, so he had hundreds of catalogs from all possible suppliers of these standard commodities. When he needed to place an order, he would typically use the catalog price or the quoted price from a supplier as long as they could meet the delivery time he needed. He had little concern for transportation cost or even quality, since for these standard components the quality from all possible suppliers was roughly equivalent. There were a few cases in the past when quality did prove to be a problem, but the supplier could usually respond quickly with an appropriate replacement. Even though the supplier would typically give Connery credit for any rejected parts, changing schedules around the problem or carrying safety stock to protect against problems would both end up costing Connery more money.

Soon after the promotion to head buyer, Juan realized the job would be much more than merely an expansion of his old position. Lawanda told Juan she created the position of head buyer to move the company into a more cost-competitive condition. She wanted Juan to develop and implement a plan that would attempt to accomplish the following:

- Reduce purchased raw material inventory levels.
- Improve delivery speed and reliability of purchased material.
- Improve the quality performance of suppliers.
- Reduce the overall cost of purchased materials.

These actions were considered to be important if they were to reduce the overall cost and stay ahead of the pack on price competitiveness.

Juan now realized both the extent and the seriousness of the new position and his responsibility. The following give a little indication of the current position of the company:

Annual cost of goods sold	$14,827,527
Direct material cost	$ 8,517,323
Inventory (on balance sheet)	$ 2,352,117
Supplied parts transportation expense	$ 256,103
Number of suppliers	2872
Inventory holding cost	21% per year
Average total processing cost for products	3 hours, 27 minutes
Number of different designs for end product	72

Assignment

1. What additional information should Juan gather to help him develop his plan? Explain how you would use the information.
2. Using what information he has, make some logical assumptions for the additional information you decided he needed, and develop a plan for Juan.

CASE STUDY 8.3

Keltox Fabrication

NOTE: This case study reflects an actual situation that occurred in an actual company. Names of the company, all personnel, and type of product have been modified.

John Carson was getting really confused over what he should do. He had just recently been promoted to the supply chain manager position for Keltox Fabrication Company from his previous job as production control supervisor. He was attending only his second meeting of the plant manager's management team, and suddenly all the attention concerning a recently discussed serious problem was focused on him to analyze and possibly propose a solution to the problem. He knew that he had better be well prepared or this new job position might not last too long.

Background: A major product line for Keltox was to produce and sell new and replacement parts for gasoline and diesel automobiles and trucks of various sizes and uses.

Some of the products included, for example, water pumps, fuel pumps, and other small pumps for various uses such as windshield washer pumps.

The problem was first brought to his attention by the sales manager, Martina Moreno. Shortly after John has started this new job, Martina came to his office with the following statement: "John, welcome to the new job. I'm afraid that I have a problem that may be at least partially related to the supply chain, and therefore you may need to help handle somehow. Over the last few months both the on-time delivery of our products to our customers and the quality of some of them has been getting steadily worse. We need to fix this very soon or we'll start to lose business. See what you can do from the supply perspective to turn this around, and hopefully fast."

John felt he needed to find out what the causes of the problem were. The first stop was to the quality manager, but all he found out was that the problem with quality was all linked to the parts being shipped by one component supplier. The quality department did the best they could to try to remove any bad components, but their decision was sometimes overridden by the production manager, who told the quality people he would still "make the rejected components work in order to make shipments."

His second stop was to Randy Kellog, the production manager. Randy said he had plenty of capacity to produce what was scheduled, his production workers were all healthy and well trained, so that was not the problem. What he then said was that the seals, clips, and fasteners they needed to install on the products were the problem. Many of the supplied shipments to Keltox from the supplier of those parts were behind schedule and often had lots of quality problems. The fault, Randy concluded, was not an internal one. His work crews were doing the best they could with what they could get of the parts. He admitted he did force the use of some rejected parts, but that his workers could sometimes manipulate the bad part and still make the product work. He also admitted that the rework might sometime make the final product fail more easily and quickly, but "At least the customer has a product—with what we are getting from the supplier of good components there would be a lot of customer orders that would be unfillable. Isn't it better they get one that might be okay rather than none at all?"

Once he had that knowledge, he went to the buyers. It seemed that all the problems were coming from one supplier. The buyer (Reva Rao) assigned to work with that supplier was then able to give him a pretty good picture of what was happening. Reva explained that the supplier had been supplying Keltox for quite a few years, and during most of that time their performance was very good. He then filled John in about the recent discussions the Reva had with the supplier regarding the recent degrading of performance. With that knowledge, John felt he was able to probably able to appear knowledgeable during the management team meeting, but he still didn't know what to do.

The Management Team Meeting: The first person to bring the issue up was Martina, the sales manager. She pretty much told the meeting participants what she had already told John. When she was done, Lashanda Banks, the plant manager, looked at John and said, "Okay, John, tell us what the problem is and how do we fix it." John then summarized what he had learned from Reva (the buyer). It seems that all those problem parts were being supplied by a single supplier who had been Keltox's supplier for many years, and for most of that time had been very reliable in both delivery and quality. The recent decay in service was caused, according to Reva, when a very large company had approached the supplier with what could be a major contract for lots of work. The supplier was trying to "ramp up" attention and resources toward the potential new customer in the hopes that the preliminary "test" orders from the new and much larger customer could turn into a long-term contract. To summarize, John just stated "it appears we at Keltox are no longer considered to be very important to them for their future plans, especially compared to the work they could get from that new customer."

At that point some of the management team started to jump in with suggestions. The first was "Well then, why don't we just get a new supplier for those parts?" John was ready with the answer. "Those parts are quite specially designed and produced just for us. They are not the kind of part that you could buy on an open market. And because we have products that have so many different sizes and applications there are literally hundreds of part numbers being supplied by this one supplier. Even one of our product lines will sometimes have several different possible seals dependent on the environment where it will be used." Reva told me that soon after she became aware of what was happening, she started to talk to several other potential suppliers, but each of them told her the same story. They were very interested in our potential business, but could make no promises of price, delivery, or capability until they had the tooling used to make the parts and could therefore make some test runs. Because the parts are specialized, so is the tooling used to make it."

One of the managers then questioned "Why don't we arrange to get (or have one of the new potential suppliers get) a new set of tooling so we can switch?" The financial manager then broke in, "Hold on here—you all may not realize that we here at Keltox actually own that tooling being used by the supplier. The decision was made some years ago that we had to protect ourselves by owning that tooling. And don't think about getting a second set of tooling, because it was very expensive to get all that tooling made and at the time, we really had to stretch our credit line to do it. I'm not at all in favor of doing that again. We could send our component design and specifications to the potential new suppliers and have them make more tooling, but that would not only take lots of time but they would certainly want to recover the tooling cost by charging us a lot more for the components." "It's great that we own the tooling!" was the response from someone else—"then why don't we go to our current supplier and take our tooling and get it to one or more of the new potential suppliers?"

Again, Reva had given him the answer he needed here. "The problem is that the new potential suppliers can't really even say if the tooling would fit their machinery. There may need to be some major retrofitting done. Then also they have the problem of ramping up their capacity to meet our needs, and of course there is a learning curve involved for them even once they got the tooling and fitted it to their equipment. That means it is very difficult for them to be able to say when the would be completely 'up to speed' on the delivery. Once they got the tooling and agreed to a contract (for which there currently is no promise) it still could be several months before they are able to meet our quantity and quality needs." One last question then came up "then why don't we buy lots of extra parts from our current supplier before we take the tooling. That could keep us going until a new supplier is "up to speed." Again, John related to the group that Reva had thought of that but recognized a potential problem—"The supplier knows well our buying patterns over the last several years. They also know that we are not very happy with their recent performance. If we were to boost our order sizes much then they will likely figure out that we are planning on switching and cutting them off. We don't know for sure, but in that case, they may decide to place us on even a lower priority so they can pay even more attention to their big potential new customer. Their performance is likely to get even worse!"

Plant manager Banks ended the discussion by merely saying: "Okay, John, I think we understand the problem now. What I want you to do is come up with a comprehensive plan to deal with it by our next meeting. It's too important to not deal with this as soon as possible."

Assignment

Put yourself in John's position, then consider the following:

1. Are there any additional questions or data that you might want to deal with before you make a recommendation? What information and/or data would you try to discover, and how would you use it?
2. Try to come up with at least two plans to present to the next management team meeting. Be sure to list the good and bad points for each plan. What are the benefits and risks involved?
3. Select a plan and develop the specific recommendation and a specific order of actions.

CHAPTER NINE

FORECASTING AND DEMAND MANAGEMENT

9.1 INTRODUCTION

Forecasting is a prelude to planning. Before making plans, an estimate must be made of what conditions will exist over some future period. How estimates are made, and with what accuracy, is another matter, but little can be done without some form of estimation.

Why forecast? There are many circumstances and reasons, but forecasting is inevitable in developing plans to satisfy future demand. Most firms cannot wait until orders are actually received before they start to plan what to produce. Customers usually demand delivery in reasonable time, and manufacturers must anticipate future demand for products or services and plan to provide the capacity and resources to meet that demand. Firms that make standard products need to have saleable goods immediately available or at least to have materials and subassemblies available to shorten the delivery time. Firms that make to order cannot begin making a product before a customer places an order but must have the resources of labor and equipment available to meet demand.

Many factors influence the demand for a firm's products and services. Although it is not possible to identify all of them, or their effect on demand, it is helpful to consider some major factors:

- General business and economic conditions.
- Competitive factors.
- Market trends such as changing demand.
- The firm's own plans for advertising, promotion, pricing, and product changes.

9.2 DEMAND MANAGEMENT

The primary purpose of an organization is to serve the customer. By effectively serving customers, the company will often be successful, leading to what financial people identify as a primary purpose of maximizing company shareholder value. Marketing focuses on meeting customer needs, but operations, through materials management, must provide the resources. The coordination of plans by these two parties is demand management.

Demand management is the function of recognizing and managing all demands for products and/or services. It occurs in the short, medium, and long term. In the long term, demand projections are needed for strategic business planning of such things as facilities. In the medium term, the purpose of demand management is to project aggregate demand for production planning. In the short term, demand management is needed for items and is associated with master production scheduling. This text is most concerned with the latter.

If material and capacity resources are to be planned effectively, all sources of demand must be identified. These include domestic and foreign customers, other plants in the same corporation, branch warehouses, service parts and requirements, promotions, distribution inventory, and consigned inventory in customers' locations.

FIGURE 9.1 Demand management and the manufacturing planning and control system.

Demand management includes several major activities, all of which are primarily market driven:

- Identifying all product and service demand in the defined markets. This includes forecasting but also involves possible segmenting of markets, classifying customers, and identifying demand that does not add value and therefore should be ignored. This includes identifying customer desires for existing or possible new product, or service design and features.
- Identifying and understanding all aspects of the market that will potentially impact customer demand. This includes economic conditions and indicators, governmental laws and regulations, and sources of existing or potential competition, including possible new competitors.
- Synchronizing identified market demand with company capabilities.
- Setting priorities for demand when supply will not cover all demand.
- Making delivery promises. The concept of available-to-promise was discussed in Chapter 3.
- Interfacing between manufacturing planning and control and the marketplace. Figure 9.1 shows this relationship using a block diagram.
- Order processing.

In each of these cases, production (supply) is being planned to react to anticipated demand as shown by the forecast.

There are several other activities or components to demand management. They include the following:

- Setting and maintaining appropriate customer service levels.
- Planning for new product introductions and phase-out of obsolete inventory.
- Planning and managing interplant shipments and distribution requirements planning (discussed in Chapter 12).
- Establishing inventory target levels and maintaining them.
- Establishing performance metrics for demand and using them to evaluate performance.

A proactive approach to demand management (market driven) includes four major activities, as discussed in *Bricks Matter: The Role of Supply Chains in Building Market-Driven Differentiation* by Cecere and Chase:

- Sensing the demand.
- Shaping demand—based on strategic and market plans.
- Shifting demand—when appropriate, sales, marketing, and operations use communication, advertising, pricing, and promotion to shift demand in desirable patterns for the company. This implies encouraging demand for products or services that may be easiest to make or are the most profitable, while discouraging demand for products or services that may be difficult to make or provide little or no profit.
- Responding to demand.

Collaborative planning, forecasting, and replenishment As the concepts of supply chain continue to develop and mature, another approach to identify demand has been developed. Called **collaborative planning, forecasting, and replenishment (CPFR)**, the approach establishes a relationship between trading partners in a supply chain. They then create a joint business plan from which sales forecasts can be developed and communicated between the supply chain partners. This approach can be beneficial in obtaining more accurate demand information within the defined supply chain. This approach tends to be a closed-loop approach, where results are analyzed after plans are executed. The results from the analysis may then be used to evaluate and possibly improve the collaborative relationship.

Order processing   **Order processing** occurs when a customer's order is received. The product may be delivered from finished goods inventory or it may be made or assembled to order. If goods are sold from inventory, a sales order is produced authorizing the goods to be shipped from inventory. If the product is made or assembled to order, the sales department must write up a sales order specifying the product. This may be relatively simple if the product is assembled from standard components but can be a lengthy, complex process if the product requires extensive engineering. A copy of the sales order stating the terms and conditions of acceptance of the order is sent to the customer. Another copy, sent to the master planner, is authorization to go ahead and plan for manufacture. The master planner must know what to produce, how much to produce, and when to deliver. The sales order must be written in language that makes this information clear.

9.3 DEMAND FORECASTING

Forecasts differ depending on what is to be done. They must be made for the strategic plan, the strategic business plan, the sales and operation plan, and the master production schedule. As discussed in Chapter 2, the purpose, planning horizons, and level of detail vary for each.

The strategic plan and the business plan are concerned with overall markets and the direction of the economy over the next 2 to 10 years or more. Their purpose is to provide time to plan for those things that take long to change. For production, the strategic plan and the business plan should provide sufficient time for resource planning: plant expansion, capital equipment purchase, and anything requiring a long lead time to purchase. The level of detail is not high, and usually forecasts are in sales units, sales dollars, or capacity. Forecasts and planning will probably be reviewed quarterly or yearly.

Sales and operations planning is concerned with manufacturing activity for the next one to three years. For manufacturing, it means forecasting those items needed for production planning, such as budgets, labor planning, long lead time procurement items, and overall inventory levels. Forecasts are made for groups or families of products rather than specific end items. Forecasts and plans will probably be reviewed monthly.

Master production scheduling is concerned with production activity from the present to a few months ahead. Forecasts are made for individual items, as found on a master production schedule, individual item inventory levels, raw materials and component parts, labor planning, and so forth. Forecasts and plans will probably be reviewed weekly.

9.4 CHARACTERISTICS OF DEMAND

In this chapter, the term *demand* is used rather than *sales*. The difference is that sales implies what is actually sold, whereas demand shows market or customer requests. Sometimes demand cannot be satisfied, and sales will be less than demand.

Before discussing forecasting principles and techniques, it is best to look at some characteristics of demand that influence the forecast and the particular techniques used.

Demand Patterns

If historical data for demand are plotted against a time scale, they will show any shapes or consistent patterns that exist. A pattern is the general shape of a time series. Although some individual data points will not fall exactly on the pattern, they tend to cluster around it.

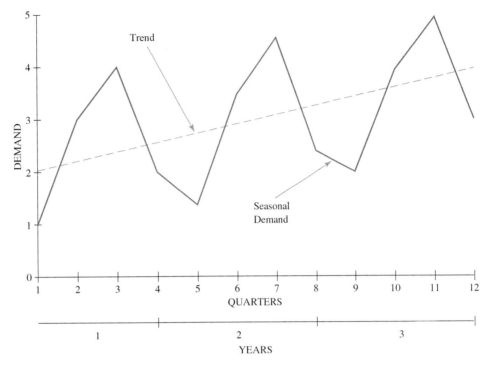

FIGURE 9.2 Demand over time.

Figure 9.2 shows a hypothetical historical demand pattern. The pattern shows that actual demand varies from period to period. There are four reasons for this: trend, seasonality, random variation, and cycle.

Trend Figure 9.2 shows that demand is increasing in a steady pattern of demand from year to year. This graph illustrates a linear trend, but there are different shapes, such as geometric or exponential. The **trend** can be level, having no change from period to period, or it can rise or fall.

Seasonality The demand pattern in Figure 9.2 shows each year's demand fluctuating depending on the time of year. This fluctuation may be the result of the weather, holiday seasons, or particular events that take place on a seasonal basis. **Seasonality** is usually thought of as occurring on a yearly basis, but it can also occur on a weekly or even daily basis. A restaurant's demand varies with the hour of the day, and supermarket sales vary with the day of the week.

Random variation **Random variation** occurs where many factors affect demand during specific periods and occur on a random basis. The variation may be small, with actual demand falling close to the pattern, or it may be large, with the points widely scattered. The pattern of variation can usually be measured, and this will be discussed in the section on tracking the forecast.

Cycle Over a span of several years and even decades, wavelike increases and decreases in the economy influence demand. However, forecasting of cycles is a job for economists and is beyond the scope of this text.

Stable Versus Dynamic

The shapes of the demand patterns for some products or services change over time, whereas others do not. Those that retain the same general shape are called **stable demand** and those that do not are called **dynamic demand**. Dynamic changes can affect the trend, seasonality, or randomness of the actual demand. The more stable the demand, the

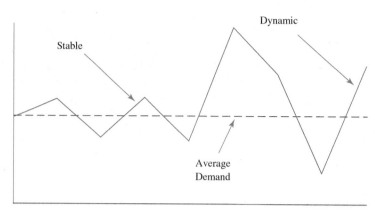

FIGURE 9.3 Stable and dynamic demand.

easier it is to forecast. Figure 9.3 shows a graphical representation of stable and dynamic demand. Notice the average demand is the same for both stable and dynamic patterns. It is usually the average demand that is forecast.

Dependent Versus Independent Demand

Chapter 4 discussed dependent and independent demand, where it was stated that demand for a product or service is independent when it is not related to the demand for any other product or service, or independent of internal activities of the firm. Dependent demand for a product or service occurs where the demand for the item is derived from that of a second item. Requirements for dependent demand items need not be forecast but are calculated from that of the independent demand item.

Only independent demand items need to be forecasted. These are usually end items or finished goods but should also include service parts and items supplied to other plants in the same company (intercompany transfers).

9.5 PRINCIPLES OF FORECASTING

Forecasts have four major characteristics or principles. An understanding of these will allow the more effective use of forecasts. They are simple and, to some extent, common sense.

1. **Forecasts are usually wrong.** Forecasts attempt to look into the unknown future and, except by sheer luck, will be wrong to some degree. Errors are inevitable and must be expected.

2. **Every forecast should include an estimate of error.** Since forecasts are expected to be wrong, the real question is "by how much?" Every forecast should include an estimate of error often expressed as a percentage (plus and minus) of the forecast or as a range between maximum and minimum values. Estimates of this error can be made statistically by studying the variability of demand about the average demand.

3. **Forecasts are more accurate for families or groups.** The behavior of individual items in a group is random even when the group has very stable characteristics. For example, the marks for individual students in a class are more difficult to forecast accurately than the class average. High marks average out with low marks. This means that forecasts are more accurate for large groups of items than for individual items in a group.

    For production planning, families or groups are based on the similarity of process and equipment used. For example, a firm forecasting the demand for knit socks as a product group might forecast men's socks as one group and women's as another since the markets are different. However, production of men's and women's ankle socks will be done on the same machines and knee socks on another. For production planning, the forecast should be for (a) men's and women's ankle socks and (b) men's and women's knee socks.

6. **Forecasts are more accurate for nearer time periods.** The near future holds less uncertainty than the far future. Most people are more confident in forecasting what they will be doing over the next week than a year from now. As someone once said, tomorrow is expected to be pretty much like today.

In the same way, demand for the near term is easier for a company to forecast than for a time in the distant future. This is extremely important for long lead time items and especially so if their demand is dynamic. Anything that can be done to reduce lead time will improve forecast accuracy.

9.6 COLLECTION AND PREPARATION OF DATA

Forecasts are usually based on historical data manipulated in some way using either judgment or a statistical technique. Thus, the forecast is only as good as the data on which it is based. To get good data, three principles of data collection are important.

1. **Record data in the same terms as needed for the forecast.** There is often a problem in determining the purpose of the forecast and what is to be forecast. There are three dimensions to this:

    a. If the purpose is to forecast demand on production, data based on demand, not shipments, are needed. Shipments show when goods were shipped and not necessarily when the customer wanted them. Thus, shipments do not necessarily give a true indication of demand.

    b. The forecast period, in weeks, months, or quarters, should be the same as the schedule period. If schedules are weekly, the forecast should be for the same time interval.

    c. The items forecasted should be the same as those controlled by manufacturing. For example, if there are a variety of options that can be supplied with a particular product, the demand for the product and for each option should be forecast.

    Suppose a firm makes a bicycle that comes in three frame sizes, three possible wheel sizes, a 3-, 5-, or 10-speed gear changer, and with or without deluxe trim. In all, there are 54 ($3 \times 3 \times 3 \times 2$) individual end items sold. If each were forecast, there would be 54 forecasts to make. A better approach is to forecast (a) total demand and (b) the percentage of the total that requires each frame size, wheel size, and so on. That way there need be only 12 forecasts: 3 frame sizes, 3 wheel sizes, 3 gear changers, 2 levels of trim, and the bike itself.

    In this example, the lead time to make the components would be relatively long in comparison to the lead time to assemble a bike. Manufacturing can make the components according to component forecast and can then assemble bikes according to customer orders. This would be ideal for situations where final assembly schedules are used as discussed in Chapter 3.

2. **Record the circumstances relating to the data.** Demand is influenced by particular events, and these should be recorded along with the demand data. For instance, artificial bumps in demand can be caused by sales promotions, price changes, changes in the weather, or a strike at a competitor's factory. It is vital that these factors be related to the demand history so they may be included or removed for future conditions.

3. **Record the demand separately for different customer groups.** Many firms distribute their goods through different channels of distribution, each having its own demand characteristics. For example, a firm may sell to a number of wholesalers that order relatively small quantities regularly and also sell to a major retailer that buys a large lot twice a year. Forecasts of average demand would be meaningless, and each set of demands should be forecast separately.

9.7 FORECASTING TECHNIQUES

There are many forecasting methods, but they can usually be classified into categories. Forecasting techniques generally may be qualitative or quantitative, and they can be based on extrinsic (external) or intrinsic (internal) factors (*APICS Dictionary*, 16th edition).

Qualitative Techniques

Qualitative forecasting techniques are projections based on judgment, intuition, and informed opinions. By their nature, they are subjective. Such techniques are used to forecast general business trends and the potential demand for large families of products over an extended period of time. As such, they are used mainly by senior management. Production and inventory forecasting is usually concerned with the demand for particular end items, and qualitative techniques are seldom appropriate.

When attempting to forecast the demand for a new product, there is no history on which to base a forecast. In these cases, the techniques of market research and historical analogy might be used. Market research is a systematic, formal, and conscious procedure for testing to determine customer opinion or intention. Historical analogy is based on a comparative analysis of the introduction and growth of similar products in the hope that the new product behaves in a similar fashion. Another method is to test-market a product.

There are several other methods of qualitative forecasting. An example of one such method is the Delphi method. This method uses a panel of experts who give their opinions on what is likely to happen.

Quantitative Techniques

Quantitative forecasting techniques are projections based on historical or numerical data, whether it be from inside or outside the organization.

Extrinsic Methods

Extrinsic forecasting methods are projections based on external (extrinsic) indicators that relate to the demand for a company's products. Examples of such data would be housing starts, birth rates, and disposable income. The theory is that the demand for a product group is directly proportional, or correlates, to activity in another field. Examples of correlation follow:

- Sales of bricks are proportional to housing starts.
- Sales of automobile tires are proportional to gasoline consumption.

Housing starts and gasoline consumption are called **economic indicators**. They describe economic conditions prevailing during a given time period. Because these indicators occur earlier in time than what they help to indicate, they are often called **leading indicators.** As an example, housing starts (often indicated by the requests for building permits) will lead to the need for housing materials, including roofing materials, electrical materials, and so on. Housing starts are, therefore, a good leading indicator for demand for many related housing materials. Some commonly used economic indicators are construction contract awards, automobile production, farm income, steel production, and gross national income. Data of this kind is compiled and published by various government departments, financial papers and magazines, trade associations, and banks.

The problem is to find an indicator that correlates with demand and one that preferably leads demand, that is, one that occurs before the demand does. For example, the number of construction contracts awarded in one period may determine the building material sold in the next period. When it is not possible to find a leading indicator, it may be possible to use a non-leading indicator for which the government or an organization forecasts. In a sense, it is basing a forecast on a forecast.

Extrinsic forecasting is most useful in forecasting the total demand for a firm's products or the demand for families of products. As such, it is used most often in business and production planning rather than the forecasting of individual end items.

Intrinsic Methods

Intrinsic forecasting methods use historical data to forecast. This data is usually recorded in the company and is readily available. Intrinsic forecasting techniques are based on the assumption that what happened in the past will happen in the future. This assumption has been likened to driving a car by looking out the rearview mirror. Although there is some obvious truth to this, it is also true that lacking any other crystal ball, the best guide to the future is what has happened in the past.

Since intrinsic techniques are so important, the next section will discuss some of the more important intrinsic techniques. They are often used as input to master production scheduling where end item forecasts are needed for the planning horizon of the plan.

9.8 SOME IMPORTANT INTRINSIC TECHNIQUES

Assume that the monthly demand for a particular item over the past year is as shown in Figure 9.4.

Suppose it is the end of December, and a forecast is needed for demand for January of the coming year. Several rules can be used:

- **Demand this month will be the same as last month.** January demand would be forecast at 84, the same as December. This may appear too simple, but if there is little change in demand month to month, it probably will be quite usable.
- **Demand this month will be the same as demand the same month last year.** Forecast demand would be 92, the same as January last year. This rule is adequate if demand is seasonal and there is little up or down trend.

Rules such as these, based on a single month or past period, are of limited use when there is much random fluctuation in demand. Usually, methods that average out history are better because they dampen out some effects of random variation.

As an example, the average of last year's demand can be used as an estimate for January demand. Such a simple average would not be responsive to trends or changes in level of demand. A better method would be to use a moving average.

Average demand This raises the question of what to forecast. As discussed previously, demand can fluctuate because of random variation. It is best to forecast the **average demand** rather than second-guess what the effect of random fluctuation will be. The second principle of forecasting discussed previously said that a forecast should include an estimate of error. This range can be estimated, so a forecast of average demand should be made, and the estimate of error applied to it.

Moving Averages

One simple way to forecast is to take the average demand for, say, the last three or six periods and use that figure as the forecast for the next period. At the end of the next period, the first period demand is dropped and the latest period demand added to determine a new average to be used as a forecast. This forecast would always be based on the average of the actual demand over the specified period.

For example, suppose it was decided to use a three-month moving average on the data shown in Figure 9.4. The forecast for January, based on the demand in October, November, and December, would be:

$$\frac{63 + 91 + 84}{3} = 79$$

January	92	July	84
February	83	August	81
March	66	September	75
April	74	October	63
May	75	November	91
June	84	December	84

FIGURE 9.4 A 12-month demand history.

Now suppose that January demand turned out to be 90 instead of 79. The forecast for February would be calculated as:

$$\frac{91 + 84 + 90}{3} = 88$$

Example Problem

Demand over the past three months has been 120, 135, and 114 units. Using a three-month moving average, calculate the forecast for the fourth month.

ANSWER

$$\text{Forecast for month 4} = \frac{120 + 135 + 114}{3} = \frac{369}{3} = 123$$

Actual demand for the fourth month turned out to be 129. Calculate the forecast for the fifth month.

$$\text{Forecast for month 5} = \frac{135 + 114 + 129}{3} = 126$$

In the previous discussion, the forecast for January was 79, and the forecast for February was 88. The forecast has risen, reflecting the higher January value and the dropping of the low October value. If a longer period, such as six months, is used, the forecast does not react as quickly. The fewer months included in the moving average, the more weight is given to the latest information, and the faster the forecast reacts to trends. However, the forecast will always lag behind a trend. For example, consider the following demand history for the past five periods:

Period	Demand
1	1000
2	2000
3	3000
4	4000
5	5000

There is a rising trend to demand. If a five-period moving average is used, the forecast for period 6 is $(1000 + 2000 + 3000 + 4000 + 5000) \div 5 = 3000$. It does not look very accurate since the forecast is lagging actual demand by a large amount. However, if a three-month moving average is used, the forecast is $(3000 + 4000 + 5000) \div 3 = 4000$. Not perfect, but somewhat better. The point is that a moving average always lags a trend, and the more periods included in the average, the greater the lag will be.

On the other hand, if there is no trend but actual demand fluctuates considerably due to random variation, a moving average based on a few periods reacts to the fluctuation rather than forecasts the average. Consider the following demand history:

Period	Demand
1	2000
2	5000
3	3000
4	1000
5	4000

The demand has no trend and is random. If a five-month moving average is used, the forecast for the next month is 3000. This reflects all the values. If a two-month average is taken, the forecasts for the third, fourth, fifth, and sixth months are as follows:

$$\text{Forecast for third month} = (2000 + 5000) \div 2 = 3500$$

$$\text{Forecast for fourth month} = (5000 + 3000) \div 2 = 4000$$

$$\text{Forecast for fifth month} = (3000 + 1000) \div 2 = 2000$$

$$\text{Forecast for sixth month} = (1000 + 4000) \div 2 = 2500$$

With a two-month moving average, the forecast reacts very quickly to the latest demand and thus is not stable.

Moving averages are best used for forecasting products with stable demand where there is little trend or seasonality. Moving averages are also useful to filter out random fluctuations. This has some common sense since periods of high demand are often followed by periods of low demand since the total market demand is usually constant and consumers often buy goods ahead of time due to sales or other outside influences, including the weather and holiday events. Buying early will lower future sales.

One drawback to using moving averages is the need to retain several periods of history for each item to be forecast. This will require a great deal of computer storage or clerical effort. Also, the calculations are cumbersome. A common forecasting technique, called **exponential smoothing**, gives the same results as a moving average but without the need to retain as much data and with easier calculations.

Exponential Smoothing

Using exponential smoothing, it is not necessary to keep months of history to get a moving average because the previously calculated forecast has already allowed for this history. Therefore, the forecast can be based on the old calculated forecast and the new data.

Using the data in Figure 9.4, suppose an average of the demand of the last six months (80 units) is used to forecast January demand. If at the end of January, actual demand is 90 units, July's demand is dropped and January's demand is used to determine the new forecast. However, if an average of the old forecast (80) and the actual demand for January (90) is taken, the new forecast, for February, is 85 units. This formula puts as much weight on the most recent month as on the old forecast (all previous months). If this does not seem suitable, less weight could be put on the latest actual demand and more weight on the old forecast. Perhaps putting only 10% of the weight on the latest month's demand and 90% of the weight on the old forecast would be better. In that case,

$$\text{February forecast} = 0.1(90) + 0.9(80) = 81$$

Notice that this forecast did not rise as much as the previous calculation in which the old forecast and the latest actual demand were given the same weight. One advantage to exponential smoothing is that the new data can be given any weight wanted.

The weight given to latest actual demand is called a **smoothing constant** and is represented by the Greek letter alpha (α). It is always expressed as a decimal and typically ranges from 0 to 0.3.

In general, the formula for calculating the new forecast is as follows:

$$\text{New forecast} = (\alpha)(\text{latest demand}) + (1 - \alpha)(\text{previous forecast})$$

Example Problem

The old forecast for May was 220, and the actual demand for May was 190. If alpha (α) is 0.15, calculate the forecast for June. If June demand turns out to be 218, calculate the forecast for July.

ANSWER

$$\text{June forecast} = (0.15)(190) + (1-0.15)(220) = 215.5$$

$$\text{July forecast} = (0.15)(218) + (0.85)(215.5) = 215.9$$

Exponential smoothing provides a routine method for regularly updating item forecasts. It works quite well when dealing with stable items. Generally, it has been found satisfactory for short-range forecasting. It is not satisfactory where the demand is low or intermittent.

Exponential smoothing will detect trends, although the forecast will lag actual demand if a definite trend exists. Figure 9.5 shows a graph of the exponentially smoothed forecast lagging the actual demand where a positive trend exists. Notice the forecast with the larger α follows actual demand more closely.

If a trend exists, it is possible to use a slightly more complex formula, called double exponential smoothing. This technique uses the same principles but notes whether each

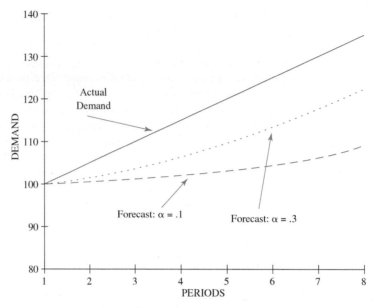

FIGURE 9.5 Exponential forecast where trend exists.

successive value of the forecast is moving up or down on a trend line. Double exponential smoothing is beyond the scope of this text.

A problem exists in selecting the "best" alpha factor. If a low factor such as 0.1 is used, the old forecast will be heavily weighted, and changing trends will not be picked up as quickly as might be desired. If a larger factor such as 0.4 is used, the forecast will react sharply to changes in demand and will be erratic if there is a sizable random fluctuation. A good way to get the best alpha factor is to use computer simulation. Using past actual demand, forecasts are made with different alpha factors to see which one best suits the historical demand pattern for particular products.

9.9 SEASONALITY

Many products have a seasonal or periodic demand pattern: skis, lawnmowers, and bathing suits are examples. Less obvious are products whose demand varies by the time of day, week, or month. Examples of these might be electric power usage during the day or grocery shopping during the week. Power usage peaks between 4:00 P.M. and 7:00 P.M. and supermarkets are most busy toward the end of the week or before certain holidays.

Seasonal Index

A useful indication of the degree of seasonal variation for a product is the **seasonal index**. This index is an estimate of how much the demand during the season will be above or below the average demand for the product. For example, swimsuit demand might average 100 per month, but in July the average could be 175, and in September it could be 35. The index for July demand would be 1.75, and for September it would be 0.35.

The formula for the seasonal index is as follows:

$$\text{Seasonal index} = \frac{\text{period average demand}}{\text{average demand for all periods}}$$

The period can be daily, weekly, monthly, or quarterly depending on the basis for the seasonality of demand.

The average demand for all periods is a value that averages out seasonality. This is called the *deseasonalized demand*. The previous equation can be rewritten as follows:

$$\text{Deseasonalized demand} = \frac{\text{period average demand}}{\text{seasonal index}}$$

Year	Quarter				
	1	2	3	4	Total
1	122	108	81	90	401
2	130	100	73	96	399
3	132	98	71	99	400
Average	128	102	75	95	400

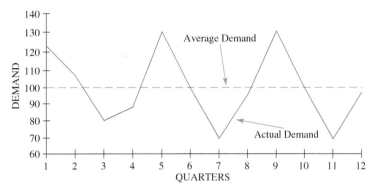

FIGURE 9.6 Seasonal sales history.

Example Problem

A product that is seasonally based on quarterly demand and the demand for the past three years is shown in Figure 9.6. There is no trend, but there is definite seasonality. Average quarterly demand is 100 units. Figure 9.6 also shows a graph of actual seasonal demand and average quarterly demand. The average demand shown is the historical average demand for all periods. Remember, this is a forecast of average demand, not seasonal demand.

Answer

The seasonal indices can now be calculated as follows:

$$\text{Seasonal index} = \frac{128}{100} = 1.28 \text{(quarter 1)}$$
$$= \frac{102}{100} = 1.02 \text{(quarter 2)}$$
$$= \frac{75}{100} = 0.75 \text{(quarter 3)}$$
$$= \frac{95}{100} = 0.95 \text{(quarter 4)}$$

Total of seasonal indices = 4.00

Note that the total of all the seasonal indices equals the number of periods. This is a good way to check whether the calculations are correct.

Seasonal Forecasts

The equation for developing seasonal indices is also used to forecast seasonal demand. If a company forecasts average demand for all periods, the seasonal indices can be used to calculate the seasonal forecasts. Changing the equation around results in:

$$\text{Seasonal demand} = (\text{seasonal index})(\text{average demand})$$

$$\text{Deseasonalized demand} = \frac{\text{actual seasonal demand}}{\text{seasonal index}}$$

Example Problem

The company in the previous problem forecasts an annual demand next year of 420 units. Calculate the forecast for quarterly sales.

ANSWER

$$\text{Expected quarter demand} = (\text{seasonal index})(\text{average quarterly demand})$$
$$\text{Expected first} - \text{quarter demand} = 1.28 \times 105 = 134.4 \text{ units}$$
$$\text{Expected second} - \text{quarter demand} = 1.02 \times 105 = 107.1 \text{ units}$$
$$\text{Expected third} - \text{quarter demand} = 0.75 \times 105 = 78.75 \text{ units}$$
$$\text{Expected fourth} - \text{quarter demand} = 0.95 \times 105 = \underline{99.75} \text{ units}$$
$$\text{Total forecast demand} = \qquad\qquad 420 \quad \text{units}$$

Deseasonalized Demand

Forecasts do not consider random variation. They are made for average demand, and seasonal demand is calculated from the average using seasonal indices. Figure 9.7 shows both actual demand and forecast average demand. The forecast average demand is also the deseasonalized demand. Historical data is of actual seasonal demand, and it must be deseasonalized before it can be used to develop a forecast of average demand.

Also, if comparisons are made between sales in different periods, they are meaningless unless deseasonalized data is used. For example, a company selling tennis rackets finds demand is usually largest in the summer. However, some people play indoor tennis, so there is demand in the winter months as well. If demand in January was 5200 units and in June was 24,000 units, how could January demand be compared to June demand to see which was the better demand month? If there is seasonality, comparison of actual demand would be meaningless. **Deseasonalized data** is needed to make a comparison.

The equation to calculate deseasonalized demand for each period is derived from the previous seasonal equation and is as follows:

$$\text{Deseasonalized demand} = \frac{\text{actual seasonal demand}}{\text{seasonal index}}$$

Example Problem

A company selling tennis rackets has a January demand of 5200 units and a June demand of 24,000 units. If the seasonal indices for January were 0.5 and for June were 2.5, calculate the deseasonalized January and June demand. How do the two months compare?

ANSWER

$$\text{Deseasonalized January demand} = 5200 \div 0.5 = 10,400 \text{ units}$$
$$\text{Deseasonalized June demand} = 24,000 \div 2.5 = 9600 \text{ units}$$

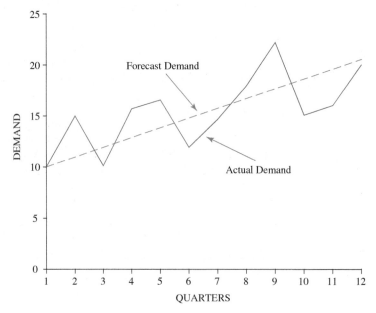

FIGURE 9.7 Seasonal demand.

June and January demand can now be compared. On a deseasonalized basis, January demand is greater than June demand.

Deseasonalized data must be used for forecasting. Forecasts are made for average demand, and the forecast for seasonal demand is calculated from the average demand using the appropriate season index.

The rules for forecasting with seasonality follow:

- Use only deseasonalized data to forecast.
- Forecast deseasonalized demand (base forecast), not seasonal demand.
- Calculate the seasonal forecast by applying the seasonal index to the base forecast.

Example Problem

A company uses exponential smoothing to forecast demand for its products. For April, the deseasonalized forecast was 1000, and the actual seasonal demand was 1250 units. The seasonal index for April is 1.2 and for May is 0.7. If α is 0.1, calculate the following:

a. The deseasonalized actual demand for April.

b. The deseasonalized May forecast.

c. The seasonal forecast for May.

ANSWER

a. Deseasonalized actual demand for April $= \dfrac{1250}{1.2} = 1042$

b. Deseasonalized May forecast $= \alpha(\text{latest actual}) + (1 - \alpha)$ (previous forecast)
$$= 0.1(1042) + 0.9(1000) = 1004$$

c. Seasonalized May forecast $= (\text{seasonal index})(\text{deseasonalized forecast})$
$$= 0.7(1004) = 703$$

9.10 TRACKING THE FORECAST

As noted in the discussion on the principles of forecasting, forecasts are usually wrong. There are several reasons for this, some of which are related to human involvement and others to the behavior of the economy and customers. If there were a method of determining how good a forecast is, forecasting methods could be improved and better estimates could be made accounting for the error. There is no point in continuing with a plan based on poor forecast data, so the forecast must be tracked. Tracking the forecast is the process of comparing actual demand with the forecast.

Forecast Error

Forecast error is the difference between actual demand and forecast demand. Error can occur in two ways: bias and random variation.

Bias Cumulative actual demand may not be the same as forecast. Consider the data in Figure 9.8. Actual demand varies from forecast, and over the six-month period, cumulative demand is 120 units greater than expected.

Bias exists when the cumulative actual demand varies from the cumulative forecast. This means the forecast average demand has been wrong. In the example in Figure 9.8, the forecast average demand was 100, but the actual average demand was $720 \div 6 = 120$ units. Figure 9.9 shows a graph of cumulative forecast and actual demand.

Bias is a systematic error in which the actual demand is consistently above or below the forecast demand. When bias exists, the forecast should be evaluated and possibly changed to improve its accuracy.

228 CHAPTER NINE

Month	Forecast		Actual	
	Monthly	Cumulative	Monthly	Cumulative
1	100	100	110	110
2	100	200	125	235
3	100	300	120	355
4	100	400	125	480
5	100	500	130	610
6	100	600	110	720
Total	600	600	720	720

FIGURE 9.8 Forecast and actual sales with bias.

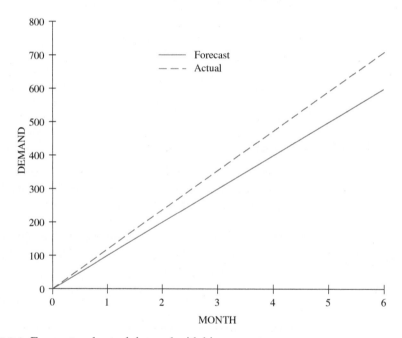

FIGURE 9.9 Forecast and actual demand with bias.

The purpose of tracking the forecast is to be able to react to forecast error by planning around it or by reducing it. When an unacceptably large error or bias is observed, it should be investigated to determine its cause.

Often there are exceptional one-time reasons for error. Examples are competitor actions, customer shutdown, large one-time orders, and sales promotions. These reasons relate to the discussion on collection and preparation of data and the need to record the circumstances relating to the data. On these occasions, the demand history must be adjusted to consider the exceptional circumstances.

Errors can also occur because of timing. For example, an early or late winter will affect the timing of demand for snow shovels although the cumulative demand will be the same.

Tracking cumulative demand will confirm timing errors or exceptional one-time events. The following example illustrates this. Note that in April the cumulative demand is back in a normal range.

Month	Forecast	Actual	Cumulative Forecast	Cumulative Actual
January	100	95	100	95
February	100	110	200	205
March*	100	155	300	360
April	100	45	400	405
May	100	90	500	495

*Customer foresaw a possible strike and stockpiled.

Random variation In a given period, actual demand will vary about the average demand. The variability will depend upon the demand pattern of the product. Some products will have a stable demand, and the variation will not be large. Others will be unstable and will have a large variation.

Consider the data in Figure 9.10, showing forecast and actual demand. Notice there is much random variation, but the average error is zero. This shows that the average forecast was correct and there was no bias. The data is plotted in Figure 9.11.

Mean Absolute Deviation

Forecast error must be measured before it can be used to revise the forecast or to help in planning. There are several ways to measure error. One method commonly used due to its ease of calculation is **mean absolute deviation (MAD).**

Consider the data on variability in Figure 9.10. Although the total error (variation) is zero, there is still considerable variation each month. Total error would be useless to measure the variation. One way to measure the variability is to calculate the total error ignoring the plus and minus signs and take the average. This is MAD:

- *mean* implies an average,
- *absolute* means without reference to plus and minus,
- *deviation* refers to the error:

$$\text{MAD} = \frac{\text{sum of absolute deviations}}{\text{number of observations}}$$

Month	Forecast	Actual	Variation (error)
1	100	105	5
2	100	94	−6
3	100	98	−2
4	100	104	4
5	100	103	3
6	100	96	−4
Total	600	600	0

FIGURE 9.10 Forecast and actual sales without bias.

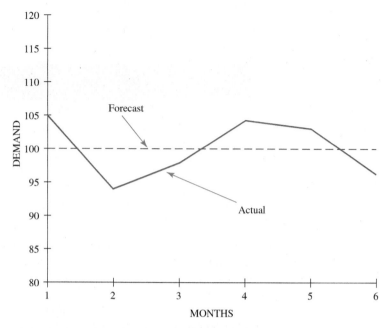

FIGURE 9.11 Forecast and actual sales without bias.

Example Problem

Given the data shown in Figure 9.10, calculate the MAD.

Answer

$$\text{Sum of absolute deviations} = 5 + 6 + 2 + 4 + 3 + 4 = 24$$

$$\text{MAD} = \frac{24}{6} = 4$$

Notice that if the deviations are not taken as absolute numbers, the result is quite different:

$$\text{Sum of deviations} = 5 + (-6) + (-2) + 4 + 3 + (-4) = 0$$

Clearly, this example shows that the actual deviations cannot be used as a good method to determine average forecast error, for in the above case it would show the "average" forecast error to be zero. The MAD is better for that. What the sum of the deviations being equal to zero does show is that there is no bias. In general, it can be said that any non-zero sum of deviations over time indicates forecast bias exists. Since the deviations are found by subtracting the forecast from the actual demand for a given period, a positive sum of deviations implies the sum of the actual demands are greater than the sum of the forecasts and implies the forecasting method is biased on the low side. A negative sum of deviations, on the other hand, implies the forecasting method is biased on the high side. This type of knowledge, when taken with the MAD for the data, can provide significant input into planning for safety inventory (discussed in more detail in Chapter 12).

Normal distribution The MAD measures the difference (error) between actual demand and forecast. Usually, actual demand is close to the forecast but sometimes is not. A graph of the number of times (frequency) actual demand is of a particular value produces a bell-shaped curve. This distribution is called a **normal distribution** and is shown in Figure 9.12. Chapter 12 gives a more detailed discussion of normal distributions and their characteristics.

There are two important characteristics to normal curves: the central tendency, or average, and the dispersion, or spread, of the distribution. In Figure 9.12, the central

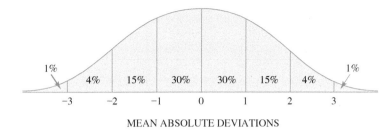

FIGURE 9.12 Normal distribution curve.

tendency is the forecast. The dispersion, the fatness or thinness of the normal curve, is measured by the standard deviation. The greater the dispersion, the larger the standard deviation. The MAD is an approximation of the standard deviation and is used because it is easy to calculate and apply.

From statistics it has been proven that the error will be within

± 1 MAD of the average about 60% of the time;
± 2 MAD of the average about 90% of the time;
± 3 MAD of the average about 98% of the time.

Mean absolute deviation has several uses. Some of the most important follow.

Tracking signal Bias exists when cumulative actual demand varies from forecast. The problem is in guessing whether the variance is due to random variation or bias. If the variation is due to random variation, the error will correct itself, and nothing should be done to adjust the forecast. However, if the error is due to bias, the forecast should be corrected. Using the MAD, judgment can be made about the reasonableness of the error. Under normal circumstances, the actual period demand will be within ±3 MAD of the average 98% of the time. If actual period demand varies from the forecast by more than 3 MAD, there is about 98% probability that the forecast is in error.

A **tracking signal** can be used to monitor the quality of the forecast. There are several procedures used, but one of the simpler ones is based on a comparison of the cumulative sum of the forecast errors (the cumulative bias) to the MAD. Following is the equation:

$$\text{Tracking signal} = \frac{\text{algebraic sum of forecast errors}}{\text{MAD}}$$

Example Problem

The forecast is 100 units a week. The actual demand for the past six weeks has been 105, 110, 103, 105, 107, and 115. If MAD is 7.5, calculate the sum of the forecast error and the tracking signal.

ANSWER

$$\text{Some of forecast error} = 5 + 10 + 3 + 5 + 7 + 15 = 45$$
$$\text{Tracking signal} = 45 \div 7.5 = 6$$

Example Problem

A company uses a trigger of ±4 to decide whether a forecast should be reviewed. Given the following history, determine in which period the forecast should be reviewed. MAD for the item is 2.

Period	Forecast	Actual	Deviation	Cumulative Deviation	Tracking Signal
				5	2.5
1	100	96			
2	100	98			
3	100	104			
4	100	110			

Answer

Period	Forecast	Actual	Deviation	Cumulative Deviation	Tracking Signal
				5	2.5
1	100	96	–4	1	0.5
2	100	98	–2	–1	–0.5
3	100	104	4	3	1.5
4	100	110	10	13	6.5

The forecast should be reviewed in period 4.

Contingency planning Suppose a forecast is made that demand for door slammers will be 100 units and that capacity for making them is 110 units. Mean absolute deviation of actual demand about the forecast historically has been calculated at 10 units. This means there is a 60% chance that actual demand will be between 90 and 110 units and a 40% chance that they will not. With this information, manufacturing management might be able to devise a contingency plan to cope with the possible extra demand.

Safety stock The data can be used as a basis for setting safety stock. This will be discussed in detail in Chapter 12.

P/D Ratio

Because of the inherent error in forecasts, companies that rely on them can run into a variety of problems. For example, the wrong material may be bought and perhaps processed into the wrong goods. A more reliable way of producing what is really needed is the use of the P/D ratio.

P, or **production lead time**, is the cumulative lead time for a product. It includes time for purchasing and arrival of raw materials, manufacturing, assembly, delivery, and sometimes the design of the product. Figure 1.1 on page 3 shows various times in different types of industries and is reproduced in Figure 9.13.

D, or **demand lead time**, is the customer's lead time. It is the time from when a customer places an order until the goods are delivered. It can be very short, as in a make-to-stock environment, or very long, as in an engineer-to-order company.

The traditional way to guard against inherent error in forecasting is to include safety stock in inventory. There is an added expense to the extra inventory carried "just in case." One other way is to make more accurate predictions. There are five ways to move in this direction.

Forecasting and Demand Management

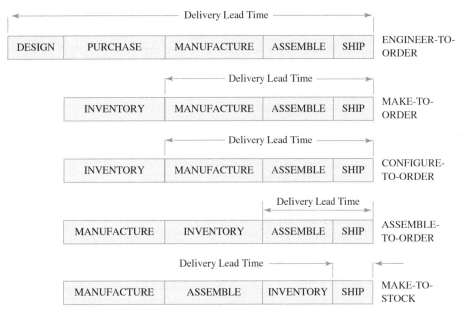

FIGURE 9.13 Manufacturing lead time and strategy.

1. **Reduce P time.** The longer the P time, the more chance there is for error. Ideally, P will be less than D.
2. **Force a match between P and D.** Moving in this direction can be done in two ways:
    a. Make the customer's D time equal to your P time. This is common with custom products when the manufacturer makes the product according to the customer's specification.
    b. Sell what you forecast. This will happen while controlling the market. One good example is the automobile market. It is common to offer special inducements toward the end of the automotive year in order to sell what the manufacturers have predicted.
3. **Simplify the product line.** The more variety in the product line, the more room for error.
4. **Standardize products and processes.** This means that customization must occur closer to final assembly. This technique is also known as postponement, as discussed in Chapter 1. The basic components are identical, or similar, for all components. Figure 9.14 shows this graphically.
5. **Forecast more accurately.** Make forecasts using a well-thought-out, well-controlled process.

If P is less than D, it implies the product can be produced in a shorter time period than the expected customer lead time. Production can be started with actual order information and still deliver within the customer lead time.

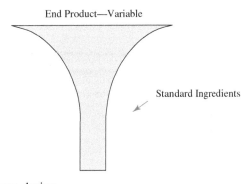

FIGURE 9.14 Mushroom design.

SUMMARY

Forecasting is an inexact science that is, nonetheless, an invaluable tool if the following are kept in mind:

- Forecasts should be tracked.
- There should be a measure of reasonableness of error.
- When actual demand exceeds the reasonableness of error, an investigation should be made to discover the cause of the error.
- If there is no apparent cause of error, the method of forecasting should be reviewed to see if there is a better way to forecast.

There are several methods used to forecast, including qualitative, quantitative, intrinsic, and extrinsic methods.

KEY TERMS

Average demand 221
Bias 227
Collaborative planning, forecasting, and replenishment (CPFR) 216
Demand lead time 232
Demand management 214
Deseasonalized data 226
Dynamic demand 217
Economic indicator 220
Exponential smoothing 223
Extrinsic forecasting method 220
Forecast 216
Forecast error 227
Forecasting 214
Intrinsic forecasting method 220

Leading indicator 220
Mean absolute deviation (MAD) 229
Moving average 223
Normal distribution 230
Order processing 216
Production lead time 232
Qualitative forecasting techniques 220
Quantitative forecasting techniques 220
Random variation 217
Seasonal index 224
Seasonality 217
Smoothing constant 223
Stable demand 217
Tracking signal 231
Trend 217

QUESTIONS

1. What is demand management? What functions does it include?
2. Why must we forecast?
3. What factors influence the demand for a firm's products?
4. Describe the purpose of forecasting for strategic business planning, sales and operations planning, and master production scheduling.
5. The text describes three characteristics of demand. Name and describe each.
6. Describe trend, seasonality, random variation, and cycle as applied to forecasting.
7. The text discusses four principles of forecasting. Name and describe each.
8. Name and describe the three principles of data collection.
9. Describe the characteristics and differences between qualitative, quantitative, extrinsic, and intrinsic forecasting techniques.
10. Describe and give the advantages and disadvantages of (a) moving averages and (b) exponential smoothing.

11. What is a seasonal index? How is it calculated?
12. What is meant by the term *deseasonalized demand*?
13. What is meant by the term *tracking the forecast*? In which two ways can forecasts go wrong?
14. What is bias error in forecasting? What are some of the causes?
15. What is random variation?
16. What is the mean absolute deviation (MAD)? Why is it useful in forecasting?
17. What action should be taken when unacceptable error is found in tracking a forecast?
18. What is the P/D ratio? How can it be improved?
19. How would a manufacturer with a P/D ratio less than 1 schedule production? How would this affect inventories?
20. What might it mean if a forecasting method has no bias yet has a large MAD?

PROBLEMS

9.1. Over the past three months, the demand for a product has been 255, 219, and 231. Calculate the three-month moving average forecast for month 4. If the actual demand in month 4 is 228, calculate the forecast for month 5.

   Answer. 235, 226

9.2. Given the following data, calculate the three-month moving average forecasts for months 4, 5, 6, and 7.

Month	Actual Demand	Forecast
1	60	
2	66	
3	40	
4	50	
5	78	
6	65	
7		

9.3. Monthly demand over the past 10 months is given in what follows.

   a. Graph the demand.

   b. What is your best guess for the demand for month 11?

   c. Using a three-month moving average, calculate the forecasts for months 4, 5, 6, 7, 8, 9, 10, and 11.

Month	Actual Demand	Forecast
1	102	
2	91	
3	95	
4	105	
5	94	
6	100	
7	106	
8	95	
9	105	
10	98	
11		

9.4. If the forecast for February was 122 and actual demand was 136, what would be the forecast for March if the smoothing constant (α) is 0.14? Use exponential smoothing for your calculation.

Answer. Forecast = 123.96 = 124

9.5. If the old forecast is 100 and the latest actual demand is 83, what is the exponentially smoothed forecast for the next period? Alpha is 0.2.

9.6. Using exponential smoothing, calculate the forecasts for months 2, 3, 4, 5, and 6. The smoothing constant is 0.2, and the old forecast for month 1 is 245.

Month	Actual Demand	Forecast Demand
1	260	
2	230	
3	225	
4	245	
5	250	
6		

9.7. Using exponential smoothing, calculate the forecasts for the same months as in problem 9.3c. The old average for month 3 was 96 and $\alpha = 0.4$. What is the difference between the two forecasts for month 11?

Month	Actual Demand	Forecast
1	102	
2	91	
3	95	
4	105	
5	94	
6	100	
7	106	
8	95	
9	105	
10	98	
11		

9.8. Weekly demand for an item averaged 100 units over the past year. Actual demand for the next eight weeks is shown in what follows:

a. Plot the data on graph paper.

b. Letting $\alpha = 0.25$, calculate the smoothed forecast for each week.

c. Comment on how well the forecast is tracking actual demand. Is it lagging or leading actual demand?

Week	Actual Demand	Forecast
1	103	100
2	112	
3	113	
4	120	
5	128	
6	131	
7	140	
8	142	
9		

9.9. If the average demand for the first quarter was 140 and the average demand for all quarters was 175, what is the seasonal index for the first quarter?

Answer.   Seasonal index = 0.80

9.10. Using the data in problem 9.9, if the forecast for next year is 1200, calculate the forecast for first quarterly demand next year.

Answer.   Forecast for first quarter = 240

9.11. The average demand for January has been 80, and the average annual demand has been 1800. Calculate the seasonal index for January. If the company forecasts annual demand next year at 2000 units, what is the forecast for January next year?

9.12. Given the following average demand for each month, calculate the seasonal indices for each month.

Month	Average Demand	Seasonal Index
January	30	
February	50	
March	85	
April	110	
May	125	
June	245	
July	255	
August	135	
September	100	
October	90	
November	50	
December	30	
Total		

Note that your answer, if done correctly, should have all the seasonal indices add up to the number of periods in the entire season, in this case 12.

9.13. Using the data in problem 9.12 and the seasonal indices you have calculated, calculate expected monthly demand if the annual forecast is 2000 units.

Month	Seasonal Index	Forecast
January		
February		
March		
April		
May		
June		
July		
August		
September		
October		
November		
December		

9.14. If the actual demand for April was 1480 units and the seasonal index was 2.5, what would be the deseasonalized April demand?

   Answer.   Deseasonalized demand = 592

9.15. Calculate the deseasonalized demands for the following:

Quarter	Actual Demand	Seasonal Index	Deseasonalized Demand
1	130	0.62	
2	170	1.04	
3	375	1.82	
4	90	0.52	
Total			

9.16. The old deseasonalized forecast is 100 units, $\alpha = 30$, and the actual demand for the last month was 180 units. If the seasonal index for the last month is 1.2 and the next month is 0.8, calculate:

   a. The deseasonalized actual demand for the last month.

   b. The deseasonalized forecast for next month using exponential smoothing.

   c. The forecast of actual demand for the next month.

   Answer.   a. Deseasonalized last month's demand = 150
              b. Deseasonalized forecast for next month = 115
              c. Forecast of seasonalized demand = 92

CHAPTER NINE

9.17. The Fast Track Ski Shoppe sells ski goggles during the four months of the ski season. Average demand follows:

Month	Average Past Demand	Seasonal Index	Forecast Demand Next Year
December	300		
January	400		
February	220		
March	130		
Total			

a. Calculate the deseasonalized sales and the seasonal index for each of the four months.

b. If next year's demand is forecast at 1200 pairs of goggles, what will be the forecast sales for each month?

9.18. Given the following forecast and actual demand, calculate the MAD.

Period	Forecast	Actual Demand	Absolute Deviation
1	110	90	
2	110	107	
3	110	122	
4	110	93	
5	110	90	
Total			

Answer. MAD = 14.4

9.19. For the following data, calculate the MAD.

Period	Forecast	Actual Demand	Absolute Deviation
1	100	106	
2	105	95	
3	110	85	
4	115	135	
5	120	105	
6	125	118	
Total	675	644	

9.20. A company uses a tracking signal trigger of ±4 to decide whether a forecast should be reviewed. Given the following history, determine in which period the forecast should be reviewed. MAD for the item is 15. Is there any previous indication that the forecast should be reviewed?

Period	Forecast	Actual	Deviation	Cumulative Deviation	Tracking Signal
1	100	110			
2	105	90			
3	110	85			
4	115	110			
5	120	105			
6	125	95			

CASE STUDY 9.1

Northcutt Bicycles: The Forecasting Problem

Jan Northcutt, owner of Northcutt Bicycles, started business in 2009. She noticed the quality of bikes she purchased for sale in her bicycle shop declining while the prices went up. She also found it more difficult to obtain the features she wanted on ordered bicycles without waiting for months. Her frustration turned to a determination to build her own bicycles to her particular customer specifications.

She began by buying all the necessary parts (frames, seats, tires, etc.) and assembling them in a rented garage using two helpers. As the word spread about her shop's responsiveness to options, delivery, and quality, however, the individual customer base grew to include other bicycle shops in the area. As her business grew and demanded more of her attention, she soon found it necessary to sell the bicycle shop itself and concentrate on the production of bicycles from a fairly large leased factory space.

As the business continued to grow, she backward integrated more and more processes into her operation, so that now she purchases less than 50% of the component value of the manufactured bicycles. This not only improves her control of production quality but also helps her control the costs of production and makes the final product more cost attractive to her customers.

The Current Situation

Jan considers herself a hands-on manager and has typically used her intuition and her knowledge of the market to anticipate production needs. Since one of her founding principles was rapid and reliable delivery to customer specification, she felt she needed to begin production of the basic parts for each particular style of bicycle well in advance of demand. In that way she could have the basic frame, wheels, and standard accessories started in production prior to the recognition of actual demand, leaving only the optional add-ons to assemble once the order came in. Her turnaround time for an order of less than half the industry average is considered a major strategic advantage, and she feels it is vital for her to maintain or even improve on response time if she is to maintain her successful operation.

As the customer base has grown, however, the number of customers Jan knows personally has shrunk significantly as a percentage of the total customer base for Northcutt Bicycles, and many of these new customers are expecting or even demanding very short response times, as that is what attracted them to Northcutt Bicycles in the first place. This condition, in addition to the volatility of overall demand, has put a strain on capacity planning. She finds that at times there is a lot of idle time (adding significantly to costs), whereas at other times the demand exceeds capacity and hurts customer response time. The production facility has therefore turned to trying to project demand for certain models and actually building a finished goods inventory of those models. This has not proven to be too satisfactory, as it has actually hurt costs and some response times. Reasons include the following:

- The finished goods inventory is often not the "right" inventory, meaning shortages for some goods and excessive inventory of others. This condition both hurts responsiveness and increases inventory costs.
- Often, to help maintain responsiveness, inventory is withdrawn from finished goods and reworked, adding to product cost.
- Reworking inventory uses valuable capacity for other customer orders, again resulting in poorer response times and/or increased costs due to expediting. Existing production orders and rework orders are both competing for vital equipment and resources during times of high demand, and scheduling has become a nightmare.

The inventory problem has grown to the point that additional storage space is needed, and that is a cost that Jan would like to avoid if possible.

Another problem Jan faces is the volatility of demand for bicycles. Since she is worried about unproductive idle time and yet does not wish to lay off her workers during times of low demand, she has allowed them to continue to work steadily and build finished goods. This makes the problem of building the "right" finished goods even more important, especially given the tight availability of storage space.

Past Demand

The following shows the monthly demand for one major product line: the standard 26-inch 10-speed street bicycle. Although it is only one of Jan's products, it is representative of most of the major product lines currently being produced by Northcutt Bicycles. If Jan can find a way to use this data to more constructively understand her demand, she feels she can probably use the same methodologies to project demand for other major product families. Such knowledge can allow her, she feels, to plan more effectively and continue to be responsive while still controlling costs.

Month	ACTUAL DEMAND			
	2011	2012	2013	2014
January	437	712	613	701
February	605	732	984	1291
March	722	829	812	1162
April	893	992	1218	1088
May	901	1148	1187	1497
June	1311	1552	1430	1781
July	1055	927	1392	1843
August	975	1284	1481	839
September	822	1118	940	1273
October	893	737	994	912
November	599	983	807	996
December	608	872	527	792

Assignment

1. Plot the data and describe what you see. What does it mean and how would you use the information from the plot to help you develop a forecast?

2. Use at least two different methodologies to develop as accurate a forecast as possible for the demand. Use each of those methods to project the next four months demand.

3. Which method from question 2 is "better"? How do you know that?

4. How, if at all, could we use Jan's knowledge of the market to improve the forecast? Would it be better to forecast in quarterly increments instead of monthly? Why or why not?

5. Are there other possible approaches that might improve Jan's operation and situation? What would they be and how could they help?

6. Has Jan's operation grown too large for her to control well? Why or why not? What would you suggest she do? What additional information would you suggest she look for to help her situation?

CASE STUDY 9.2

Hatcher Gear Company

Jaya Mehta found herself in a real dilemma when the sales and marketing department presented her with the annual sales demand forecast for one of the gear lines. Jaya, as materials manager for Hatcher Gears, was responsible for taking the forecast and translating it into projected production needs and, as part of that, requirements for raw material purchasing needs—both in quantities and dates. That fairly routine task went fine until she came to the V27 family of gears.

The V27 family of gears was used by customers in highly stressful applications, and as a result they needed to be made from a highly specialized steel, made with a complex mixture of chemicals. The steel mill made the steel reluctantly, since it required shutting down a furnace and completely cleaning it out in order to avoid contamination. The time and effort to make the furnace ready, in combination with the costly chemicals used to make the steel, meant that the steel was extremely expensive for Hatcher to buy. In addition, the steel company had told Hatcher Gear that they would make only one batch of the steel and only once a year. They had their own annual production plans to execute, and the special steel was simply too disruptive to their production for Hatcher Gears to request any additional steel beyond the one batch during the year. Since Hatcher Gears was the only customer that needed that steel, the steel company required Hatcher to buy all the steel that was made in the batch. The steel company had no desire to maintain a very expensive inventory for Hatcher Gear. As it was, the steel company had reluctantly kept Hatcher Gears as a customer, and his recent attempts to find another steel company willing to make the steel was met with rapid and emphatic refusals.

Those facts meant that Jaya must figure out how much steel to buy as accurately as she could—whatever amount she bought had to last her for the remainder of the year. If she bought too little, she might not be able to supply some of the customers for the gears, and that could be disastrous for customer service, and her boss (the general manager) would certainly hold her responsible. In fact, just recently the general manager had again emphasized that the customers for V27 gears were important enough that he wanted to have enough of the steel to be able to supply those customers on time at least 97% of the time, even though the inventory to do that was certainly very expensive. All of their customers for the V27 gears used Hatcher as their single source of supply. But buying too much would also be bad. Jaya, being materials manager, essentially "owned" the inventory in the facility, meaning that she was held responsible for all raw, in-process inventory, and finished goods inventory, including having to answer for the cost of holding that inventory. Having too much of that expensive steel over the year could get her in a lot of trouble with the chief financial officer and the general manager. While the customers for the gears knew the steel was expensive and they were therefore willing to pay a higher price, they were not willing to accept a price increase to help Hatcher pay for excessive inventory. They expected Hatcher to manage that.

Knowing all that, Jaya's dilemma was based on looking at the annual sales demand forecast developed by marketing for the V27 gear family of 16,000 gears. When she got that forecast for 16,000, she looked at the sales for the V27 over the last 10 years:

Ten years ago—9733
Nine years ago—10,115
Eight years ago—9814
Seven years ago—10,033
Six years ago—10,077
Five years ago—9782
Four years ago—10,145
Three years ago—10,097
Two years ago—9924
Last year—9897

Jaya decided to call the director of sales and marketing (Phil Johnson) and ask about the forecast. This is how the conversation went:

JAYA : "I wanted to ask about the demand forecast for the V27 gear line. You said it was 16,000. How did you come up with that number?"
PHIL: "Because that is what we plan to sell."
JAYA: "Do you have or expect to have any new customers for the V27 line?"
PHIL: "No."
JAYA: "Do any of your customers have any new applications that use the V27 gears?"
PHIL: "Not that I know of."
JAYA: "Do any of your customers have or expect to have new customers for their products that use our V27 gears?"
PHIL: "Not that I know of."
JAYA: "Are any of your customers experiencing growth in their products that use V27 gears?"
PHIL: "Not that I know of."
JAYA: "Well then why do you think you will sell 16,000."
PHIL: "Because that is what we plan to sell—and you hadn't better disappoint any of our customers or be late with any of their orders!"

Jaya hung up the phone and seriously wondered what she was to do. If she ended up with too much inventory or disappointed any customers, she knew the general manager would blame her. She knew it could be a big mistake to try to put the blame back on the director of sales and marketing because not only was the director of sales and marketing at a higher level in the company, but he was also a personal friend of the general manager. Jaya knew they often played golf together on weekends and that their families frequently attended social events together.

Assignment

1. Try to help Jaya out with a short-term solution. Help him come up with how much steel to buy—enough to cover production of how many V27 gears over the next year? After you come up with a number, justify your solution. Be as comprehensive as possible in the justification—include all possible options, discussing why you reject those you reject and why you accepted your recommendation.

2. Develop a recommendation for Jaya and the Hatcher Gear Company to use in the long run, specifically to try to minimize the current dilemma from occurring again.

CHAPTER TEN

INVENTORY FUNDAMENTALS

10.1 INTRODUCTION

Inventories are materials and supplies that a business or institution carries either for sale or to provide inputs or supplies to the production process. All businesses and institutions require inventories and they are often a substantial part of total assets.

Financially, inventories are very important to manufacturing companies. On the balance sheet, they usually represent from 20 to 60% of total assets. As inventories are used by consumption into other goods or through sales, their value is converted into cash, which improves cash flow and return on investment. There is a cost for carrying inventory which increases operating costs and decreases profits. Therefore, good inventory management is essential.

Inventory management is responsible for planning and controlling inventory from the raw material phase to delivery to the customer. Since inventory either results from production or supports it for service, the management of it must be coordinated. Inventory must be considered at each of the planning levels and is thus part of production planning, master production scheduling, and material requirements planning. Production planning is concerned with overall inventory, master planning with end items, and material requirements planning with component parts and raw material.

10.2 AGGREGATE INVENTORY MANAGEMENT

Aggregate inventory management deals with managing inventories according to their classification (raw material, work in process, and finished goods) and the function they perform rather than at the individual item level. It is financially oriented and is concerned with the costs and benefits of carrying the different classifications of inventory. As such, aggregate inventory management involves:

- Flow and kinds of inventory needed.
- Supply and demand patterns.
- Functions that inventories perform.
- Objectives of inventory management.
- Costs associated with inventories.

10.3 ITEM INVENTORY MANAGEMENT

Inventory is managed not only at the aggregate level but also at the item level. Management must establish decision rules about inventory items so those responsible for inventory control can do their job effectively. These rules include the following:

- Which individual inventory items are most important.
- How individual items are to be controlled.
- How much to order at one time.
- When to place an order.

This chapter will study aggregate inventory management and some factors influencing inventory management decisions, which include the following:

- Types of inventory based on the flow of material.
- Supply and demand patterns.
- Functions performed by inventory.
- Objectives of inventory management.
- Inventory costs.

Finally, this chapter will conclude with a study of the first two decisions, deciding the importance of individual end items and how they are controlled. Subsequent chapters will discuss the question of how much stock to order at one time and when to place orders.

10.4 INVENTORY AND THE FLOW OF MATERIAL

There are many ways to classify inventories. One often-used classification is related to the flow of materials into, through, and out of a manufacturing organization, as shown in Figure 10.1.

- **Raw materials.** These are purchased items received that have not entered the production process. They include purchased materials, component parts, and subassemblies.
- **Work in process (WIP).** Raw materials that have entered the manufacturing process and are being worked on or waiting to be worked on.
- **Finished goods.** The finished products of the production process that are ready to be sold as completed items. They may be held at a factory or central warehouse or at various points in the distribution or retail system.
- **Distribution inventory.** Finished goods located in the distribution system.

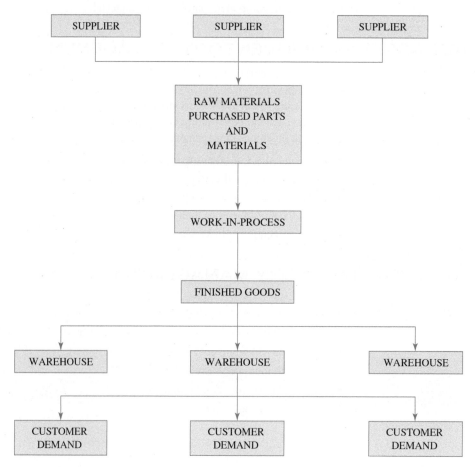

FIGURE 10.1 Inventories and the flow of materials.

- **Maintenance, repair, and operating (MRO) supplies.** Items used in production that do not become part of the product. These include hand tools, spare parts, lubricants, and cleaning supplies.

Classification of an item into a particular inventory depends on the production environment. For instance, sheet steel or tires are finished goods to the supplier but are raw materials and component parts to the car manufacturer.

10.5 SUPPLY AND DEMAND PATTERNS

If supply met demand exactly, there would be little need for inventory. Goods could be made at the same rate as demand, and no inventory would build up. For this situation to exist, demand must be predictable, stable, and relatively constant over a long time period.

If this is so, manufacturing can produce goods on a line-flow basis, matching production to demand. Using this system, raw materials are fed to production as required, work flow from one work center to another is balanced so little work in process inventory is required, and goods are delivered to the customer at the rate the customer needs them. Flow manufacturing systems were discussed in Chapter 1. Because the variety of products they can make is so limited, demand has to be large enough to justify economically setting up the system. These systems are characteristic of lean manufacturing, which will be discussed in Chapter 16.

Demand for most products is neither sufficient nor constant enough to warrant setting up a line-flow system, and these products are usually made in lots or batches. Work centers are organized by function, for example, all machining tools in one area, all welding in another, and assembly in another. Work moves in lots from one work center to another as required by the routing. By the nature of the process, inventory will build up in raw materials, work in process, and finished goods.

10.6 FUNCTIONS OF INVENTORIES

In intermittent manufacturing, the basic purpose of inventory is to decouple supply and demand. Inventory serves as a buffer between:

- Supply and demand.
- Customer demand and finished goods.
- Finished goods and component availability.
- Requirements for an operation and the output from the preceding operation.
- Parts and materials to begin production and the suppliers of materials.

Based on this, inventories can be classified according to the function they perform.

Anticipation Inventory

Anticipation inventories are built up in anticipation of future demand. For example, they are created ahead of a peak selling season, a promotion program, vacation shutdown, or possibly the threat of a strike. They are built up to help level production and to reduce the costs of changing production rates.

Fluctuation Inventory (Safety Stock)

Fluctuation inventory is held to cover random unpredictable fluctuations in supply and demand or lead time. If demand or lead time is greater than the forecast, a **stockout** will occur. **Safety stock** is carried to protect against this possibility. Its purpose is to prevent disruptions in manufacturing or deliveries to customers. Safety stock is also called buffer stock or reserve stock.

Lot-Size Inventory

Items purchased or manufactured in quantities greater than needed immediately create **lot-size inventory**. This is to take advantage of quantity discounts, to reduce shipping and setup costs, and in cases where it is impossible to make or purchase items at the same rate that they will be used or sold. Lot-size inventory is sometimes called **cycle stock**. It is the portion of inventory that depletes gradually as customers' orders come in and is replenished cyclically when suppliers' orders are received.

Transportation Inventory

Transportation inventory exists because of the time needed to move goods from one location to another, such as from a plant to a distribution center or a customer. They are sometimes called **pipeline** or **movement inventory**. The average amount of inventory in transit is

$$I = \frac{tA}{365}$$

where I is the average annual inventory in transit, t is transit time in days, and A is annual demand. Notice that the transit inventory does not depend upon the shipment size but on the transit time and the annual demand. The only way to reduce the inventory in transit, and its cost, is to reduce the transit time. In addition, since there is a direct relationship between the amount of inventory in a process and the throughput time (flow time) to move inventory through a process, there is both a cost and time advantage to reduce the inventory in transit. This concept will be developed more completely in Chapter 11.

Example Problem

Delivery of goods from a supplier is in transit for 10 days. If the annual demand is 5200 units, what is the average annual inventory in transit?

Answer

$$I = \frac{10 \times 5200}{365} = 142.5 \text{ units}$$

The problem can be solved in the same way using dollars instead of units.

Hedge Inventory

Some products, such as minerals and commodities, grains or animal products, are traded on a worldwide market. The price for these products fluctuates according to world supply and demand. If buyers expect prices to rise, they can purchase **hedge inventory** when prices are low. Hedging inventory is complex and beyond the scope of this text.

Maintenance, Repair, and Operating (MRO) Supplies

MROs are items used to support general operations and maintenance but do not become directly part of a product. They include maintenance supplies, equipment spare parts, gloves, and consumables such as cleaning compounds and lubricants. They are not part of the product structure.

In most cases materials planners and MRP do not create orders for the purchase of MRO material. Orders are often generated directly from the maintenance or engineering functions. Some MRO material inventories are maintained based on probability of need, especially in the case of frequently used materials such as cleaning supplies. Materials used for maintenance can often be determined from the preventive maintenance schedule for the equipment. This is especially important for maintenance material that may be too expensive to justify keeping inventory on hand at all times. Some repair material, even if expensive to keep in inventory, may need to be held in inventory in cases where the material may have a long lead time or if the lack of the material may mean a critical piece of equipment would need to remain idle until the material is available. Buffer stocks can be held, established in much the same way as production inventory (discussed in Chapter 12). Statistical analysis for critical maintenance material can be done to establish mean time between failures (MTBF) of the part. In that way, maintenance can be scheduled specifically to minimize the probability of a breakdown of the equipment, and the purchase of the parts can similarly be synchronized with the maintenance schedule to minimize inventory holding costs. This is very much like the maintenance schedules that automobile companies provide with the purchase of an automobile. The owner is not obligated to follow the maintenance schedule, but ignoring recommendations will increase the probability of a costly and inconvenient breakdown.

Location of MRO material depends on the needs and policies of the company. In some cases, MRO material is stocked in the same secure warehouse with production material, but sometimes there is a separate MRO stocking location. In either case, much as with production material, MRO material often represents a substantial financial cost, and location and count accuracies are often just as important as they are with production material.

10.7 OBJECTIVES OF INVENTORY MANAGEMENT

A firm wishing to maximize profit will have at least the following objectives:

- Maximum customer service.
- Low-cost plant operation.
- Minimum inventory investment.

Customer Service

In broad terms, customer service is the ability of a company to satisfy the needs of customers. In inventory management, the term is used to describe the availability of items when needed and is a measure of inventory management effectiveness. The customer can be a purchaser, a distributor, another plant in the organization, or the work center where the next operation is to be performed.

There are many different ways to measure customer service, each with its strengths and weaknesses, and there is no one best measurement. Some measures are percentage of orders shipped on schedule, percentage of line items shipped on schedule, and days out of stock.

Inventories help to maximize customer service by protecting against uncertainty. If a company could forecast exactly what customers want and when, they could plan to meet demand with no uncertainty. However, demand and the lead time to get an item are often uncertain, possibly resulting in stockouts and customer dissatisfaction. For these reasons, it may be necessary to carry extra inventory to protect against uncertainty. As mentioned, this inventory is called safety stock and will be discussed in more detail in Chapter 12.

Operating Efficiency

Inventories help make a manufacturing operation more productive in four ways:

1. Inventories allow operations with different rates of production to operate separately and more economically. If two or more operations in a sequence have different rates of output and are to be operated efficiently, inventories must build up between them. The inventory purposely used to separate operations to improve operational efficiency is often called **decoupling inventory**.

2. Chapter 2 discussed production planning for seasonal products in which demand is nonuniform throughout the year. One strategy discussed was to level production and build anticipation inventory. Production should be leveled and anticipation inventory built for sale in the peak periods. This would result in the following:
    - Lower overtime costs.
    - Lower hiring and firing costs.
    - Lower training costs.
    - Lower subcontracting costs.
    - Lower capacity required.

By leveling production, manufacturing can continually produce an amount equal to the average demand. The advantage of this strategy is that the costs of changing production levels are avoided. Figure 10.2 shows this strategy.

3. Inventories allow manufacturing to run longer production runs, which result in the following:

- **Lower setup costs per item.** The total cost to make a lot or batch depends upon the setup costs and the run costs. The setup costs are fixed, but the run costs vary with the number produced. If larger lots are run, the setup costs are absorbed over a larger number, and the average (unit) cost is lower.
- **An increase in production capacity due to production resources being used at a greater portion of the time for processing as opposed to setup.** Capacity at a work center is taken up by setup and by run time. Output occurs only when an item is being worked on and not when setup is taking place. If larger quantities are produced at one time, there are fewer setups required to produce a given annual output, and thus more time is available for producing goods. This is most important with bottleneck resources. Time lost on setup on these resources is lost throughput (total production) and lost capacity.

4. Inventories allow manufacturing to purchase in larger quantities, which results in lower ordering costs per unit and quantity discounts.

But all of this is at a price. The problem is to balance inventory investment with the following:

1. **Customer service.** The lower the inventory, the higher the likelihood of a stockout and the lower the level of customer service. The higher the inventory level, the higher customer service will be.
2. **Costs associated with changing production levels.** Excess equipment capacity, overtime, hiring, training, and layoff costs will all be higher if production fluctuates with demand.
3. **Cost of placing orders.** Lower inventories can be achieved by ordering smaller quantities more often, but this practice results in higher annual ordering costs.
4. **Transportation costs.** Goods moved in small quantities cost more to move per unit than those moved in large quantities. However, moving large lots implies higher inventory.

If inventory is carried, there has to be a benefit that exceeds the costs of carrying that inventory. The only good reason for carrying inventory beyond current needs is if it costs less to carry it than not. This being so, the discussion will now turn to the costs associated with inventory.

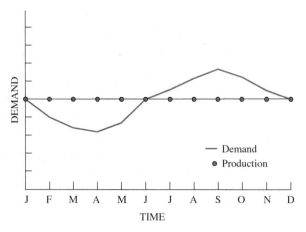

FIGURE 10.2 Operation leveling.

10.8 INVENTORY COSTS

The following costs are used for inventory management decisions:

- Unit cost.
- Carrying costs.
- Ordering costs.
- Stockout costs.
- Capacity-related costs.

Unit Cost

Unit cost is the price paid for a purchased item, which consists of the cost of the item and any other direct costs associated in getting the item into the plant. These could include such things as transportation, custom duties, and insurance. The inclusive cost is called the **landed cost**. For an item manufactured in-house, the cost includes direct material, direct labor, and factory overhead.

Carrying Costs

Carrying costs (sometimes called holding costs) include all expenses incurred by the firm because of the volume of inventory carried. As inventory increases, so do these costs. They can be broken down into three categories:

1. **Capital costs.** Money invested in inventory is not available for other uses and as such represents a lost opportunity cost. The minimum cost would be the interest lost by not investing the money at the prevailing interest rate, and it may be much higher depending on investment opportunities for the firm.
2. **Storage costs.** Storing inventory requires space, material handlers, and equipment. As inventory increases, so do these costs.
3. **Risk costs.** The risks in carrying inventory are as follows:
   a. Obsolescence—Loss of product value resulting from a model or style change or technological development.
   b. Damage—Inventory damaged while being held or moved.
   c. Pilferage—Goods lost, strayed, or stolen.
   d. Deterioration—Inventory that rots or dissipates in storage or whose shelf life is limited.

What does it cost to carry inventory? Actual figures vary from industry to industry and company to company. Capital costs may vary depending upon interest rates, the credit rating of the firm, and the opportunities the firm may have for investment. Storage costs vary with location and type of storage needed. Risk costs can be very low or can be close to 100% of the value of the item for perishable goods. The carrying cost is usually defined as a percentage of the dollar value of inventory per unit of time (usually one year). In some manufacturing industries a realistic number is typically 20–30%. However, in those industries dealing with fashion, technology, or short-term products, the possibility of obsolescence is high, and the cost of carrying such items is greater.

Example Problem

A company carries an average annual inventory of $2,000,000. If it estimates the cost of capital is 10%, storage costs are 7%, and risk costs are 6%, what does it cost per year to carry this inventory?

ANSWER

$$\text{Total cost of carrying inventory} = 10\% + 7\% + 6\% = 23\%$$
$$\text{Annual cost of carrying inventory} = 0.23 \times 2,000,000 = \$460,00$$

Ordering Costs

Ordering costs are those costs that are associated with placing an order either with the factory or with a supplier. The cost of placing an order does not depend upon the quantity ordered. Whether a lot of 10 or 100 is ordered, the costs associated with placing the order are essentially the same. However, the annual cost of ordering depends upon the number of orders placed in a year.

Ordering costs include the following:

- **Production control costs.** The costs incurred are those of issuing and closing orders, scheduling, loading, dispatching, and expediting. The annual cost and effort expended in production control depends on the number of orders placed, not on the quantity ordered. The fewer orders per year, the less cost.
- **Setup and teardown costs.** Every time an order is issued, work centers have to set up to run the order and tear down the setup at the end of the run. These costs do not depend upon the quantity ordered but on the number of orders placed per year.
- **Lost capacity cost.** Every time an order is placed at a work center, the time taken to set up is lost as productive output time. This represents a loss of capacity and is directly related to the number of orders placed. It is particularly important and costly with bottleneck work centers.
- **Purchase order cost.** Every time a purchase order is placed, costs are incurred to place the order. These costs include order preparation, follow-up, expediting, receiving, authorizing payment, and the accounting cost of receiving and paying the invoice.
- **Movement or transportation cost.** When an order is placed, material for the order has to be moved from operation to operation.

The annual cost of ordering depends upon the number of orders placed in a year. This can be reduced by ordering more at one time, resulting in the placing of fewer orders. However, this drives up the inventory level and the annual cost of carrying inventory.

Example Problem

Given the following annual costs, calculate the average cost of placing one order.

Production control salaries = $60,000
Supplies and operating expenses for production control department = $15,000
Cost of setting up work centers for an order = $120
Orders placed each year = 2000

ANSWER

$$\text{Average cost} = \frac{\text{fixed costs}}{\text{number of orders}} + \text{variable cost}$$

$$= \frac{\$60,000 + \$15,000}{2000} + \$120 = \$157.50$$

Stockout Costs

If demand during the lead time exceeds forecast, a stockout can be expected. A stockout can potentially be expensive because of backorder costs, lost sales, and possibly lost customers. **Stockouts costs** can be reduced by carrying extra inventory to protect against those times when the demand during lead time is greater than forecast.

Capacity-Related Costs

When output levels must be changed, there may be costs for overtime, hiring, training, extra shifts, and layoffs. These **capacity-related costs** can be avoided by leveling production, that is, by producing items in slack periods for sale in peak periods. However, this builds inventory in the slack periods.

Example Problem

A company makes and sells a seasonal product. Based on a sales forecast of 2000, 3000, 6000, and 5000 per quarter, calculate a level production plan, quarterly ending inventory, and average quarterly inventory.

If inventory carrying costs are $3 per unit per quarter, what is the annual cost of carrying inventory? Opening and ending inventories are zero.

ANSWER

		Quarter 1	Quarter 2	Quarter 3	Quarter 4	Total
Forecast Demand		2000	3000	6000	5000	16,000
Production		4000	4000	4000	4000	16,000
Ending Inventory	0	2000	3000	1000	0	
Average Inventory		1000	2500	2000	500	
Inventory Cost (dollars)		3000	7500	6000	1500	18,000

10.9 FINANCIAL STATEMENTS AND INVENTORY

The two major financial statements are the balance sheet and the income statement. The balance sheet shows assets, liabilities, and owners' equity. The income statement shows the revenues earned and the expenses incurred in achieving that revenue.

Balance Sheet

An **asset** is something that has value and is expected to benefit the future operation of the business. Assets may be tangible, such as cash, inventory, machinery, and buildings, or may be intangible, such as accounts receivable or a patent.

Liabilities are obligations or amounts owed by a company. Accounts payable, wages payable, and long-term debt are examples of liabilities.

Owners' equity is the difference between assets and liabilities. After all the liabilities are paid, it represents what is left for the owners of the business. Owners' equity is created either by the owners investing money in the business or through the operation of the business when it earns a profit. It is decreased when owners take money out of the business or when the business loses money.

The accounting equation is the relationship between assets, liabilities, and owners' equity. The equation is expressed by the balance sheet equation:

$$\text{Assets} = \text{liabilities} + \text{owners' equity}$$

This is a basic accounting equation. Given two of the values, the third can always be found.

Example Problem

If the owners' equity is $1000 and liabilities are $800, what are the assets?
If the assets are $1000 and liabilities are $600, what is the owners' equity?

ANSWER

Assets = Liabilities + owners' equity
Assets = $800 + $1000 = $1800

Owners' equity = assets − liabilities
    = $1000 − $600 = $400

Balance sheet The **balance sheet** usually has the assets on the left side and the liabilities and owners' equity on the right side, as shown.

Assets		Liabilities	
Cash	$100,000	Notes payable	$5000
Accounts receivable	$300,000	Accounts payable	$20,000
Inventory	$500,000	Long-term debt	$500,000
Fixed assets	$1,000,000	Total liabilities	$525,000
		Owners' equity	
		Capital	$1,000,000
		Retained earnings	$375,000
Total assets	$1,900,000	Total liabilities and owners' equity	$1,900,000

Capital is the amount of money the owners have invested in the company.

Retained earnings are increased by the revenues a company makes and decreased by the expenses incurred. The summary of revenues and expenses is shown on the income statement.

Income Statement

Income (profit) The primary purpose of a business is to increase the owners' equity by making a **profit**. For this reason, owners' equity is broken down into a series of accounts, called revenue accounts, which show what increased owners' equity, and expense accounts, which show what decreased owners' equity.

$$\text{Income} = \text{revenue} - \text{expenses}$$

Revenue comes from the sale of goods or services. Payment is sometimes immediate in the form of cash, but often is made as a promise to pay at a later date, called an accounts receivable.

Expenses are the costs incurred in the process of making revenue. They are usually categorized into the cost of goods sold and general and administrative expenses.

Cost of goods sold (COGS) are costs that are incurred to make the product. They include direct labor, direct material, and factory overhead. Factory overhead is all other factory costs except direct labor and direct material, and is allocated to a product using some kind of percentage, such as units produced.

General and administrative expenses (G&A) include all other costs in running a business. Examples of these are advertising, insurance, property taxes, and wages and benefits other than direct material, direct labor, and factory overhead costs.

The following is an example of an **income statement**.

Revenue		$1,000,000
Cost of goods sold		
Direct labor	$200,000	
Direct material	400,000	
Factory overhead	200,000	$800,000
Gross margin (profit)		$200,000
General and administrative expenses		$100,000
Net income (profit)		$100,000

Example Problem

Given the following data, calculate the gross margin and the net income.

Revenue	= $1,500,000
Direct labor	= $300,000
Direct material	= $500,000
Factory overhead	= $400,000
General and administrative expenses	= $150,000

How much would profits increase if, through better materials management, material costs are reduced by $50,000?

Revenue		$1,500,000
Cost of goods sold		
Direct labor	$300,000	
Direct material	500,000	
Overhead	400,000	$1,200,000
Gross margin (gross profit)		$300,000
General and administrative expenses		$150,000
Net income (profit)		$150,000

If material costs are reduced by $50,000, income increases by $50,000. Materials management can have a direct impact on the bottom line—net income.

Cash Flow Management

When inventory is purchased as raw material, it is recorded as an asset. When it enters production, it is recorded as work in process inventory, and as it is processed, its value increases by the amount of direct labor applied to it and the overhead attributed to its processing. The material is said to absorb overhead. Goods do not become revenue until they are sold. However, the expenses incurred in producing the goods must be paid for. This raises another financial issue: businesses must have the cash to pay their bills. Cash is generated by sales, and the flow of cash into a business must be sufficient to pay bills as they become due. Businesses develop financial statements showing the cash flow into and out of the business. Any shortfall of cash must be provided for by borrowing or in some other way. This is called **cash flow management**.

Cash-to-cash cycle time **Cash-to-cash cycle time** is defined in the *APICS Dictionary*, 16th edition, as "an indicator of how efficiently a company manages its assets to improve cash flow." Cash-to-cash cycle time is a means of measuring the length of time between the purchase of materials to the receipt of payment by the customer for the sale of products or services that used those materials. It is calculated using the following formula:

Cash-to-cash cycle time = inventory days + accounts receivable days − accounts payable days

For example, if the average days material stays in inventory is 60, the average accounts receivable outstanding is 45 days, and the average accounts payable days is 50, then

Cash to cash cycle time = 60 + 45 − 50 = 55

Return on Investment (ROI)

Another measurement often used to evaluate the financial efficiency of an investment is the **return on investment**, or ROI. There are several possible variations for calculating ROI, but one general approach is the following simple formula:

$$\text{Return on Investment} = \frac{\text{Net profit from investment}}{\text{Investment cost}}$$

The net profit from the investment is, of course, equal to the financial gain from the investment minus the investment cost.

Financial Inventory Performance Measures

As discussed previously in this chapter, from a financial point of view, inventory is an asset and represents money that is tied up and cannot be used for other purposes. Because inventory has a carrying cost—the costs of capital, storage, and risk—finance wants as little inventory as possible and needs some measure of the level of inventory. Total inventory investment is one measure, but in itself does not relate to sales. Two measures that do relate to sales are the inventory turns ratio and days of supply.

Inventory turns Ideally, a manufacturer carries no inventory. This is impractical, since inventory is needed to support manufacturing and often to supply customers. How much inventory is enough? There is no one correct answer. A convenient measure of how effectively inventories are being used is the **inventory turns** ratio:

$$\text{Inventory turns} = \frac{\text{annual cost of goods sold}}{\text{average inventory in dollars}}$$

The calculation of average inventory can be complicated and is a subject for cost accounting. In this text it will be taken as a given, although a simple formula under the assumption of relatively constant demand and replenishment is often given as:

$$\text{Average inventory} = (\text{inventory at beginning of period} + \text{inventory at end of period})/2$$

In some cases, inventory turns are calculated using the inventory position on the balance sheet as the value of the average inventory.

For example, if the annual cost of goods sold is $1 million and the average inventory is $500,000, then

$$\text{Inventory turns} = \frac{\$1,000,000}{\$500,000} = 2$$

What does this mean? At the very least, it means that with $500,000 of inventory a company is able to generate $1 million in sales. If, through better materials management, the firm is able to increase its turnover ratio to 10, the same sales are generated with only $100,000 of average inventory. If the annual cost of carrying inventory is 25% of the inventory value, the reduction of $400,000 in inventory results in a cost reduction (and profit increase) of $100,000.

Inventory turns is also a convenient method to estimate the average length of time inventory stays within the operation. For example, if inventory turnover is four times per year, then that implies there is an average of 3 months of inventory in the operation (12 months/4 turns). This concept is somewhat similar to the concept of cycle time or throughput time (see Chapter 6). Another useful concept relating to inventory and time is **inventory velocity**, which essentially means the time from receipt of the inventory as raw material until it is sold as finished goods.

Example Problem

a. What will be the inventory turns ratio if the annual cost of goods sold is $24 million a year and the average inventory is $6 million?

Answer

$$\text{Inventory turns} = \frac{\text{annual cost of goods sold}}{\text{average inventory in dollars}}$$

$$= \frac{24,000,000}{6,000,000} = 4$$

b. What would be the reduction in inventory if inventory turns were increased to 12 times per year?

ANSWER

$$\text{Average inventory} = \frac{\text{annual cost of goods sold}}{\text{inventory turns}}$$
$$= \frac{24{,}000{,}000}{12}$$
$$= \$2{,}000{,}000$$
$$\text{Reduction in inventory} = 6{,}000{,}000 - 2{,}000{,}000 = \$4{,}000{,}000$$

c. If the cost of carrying inventory is 25% of the average inventory, what will the savings be?

ANSWER

$$\text{Reduction in inventory} = \$4{,}000{,}000$$
$$\text{Savings} = \$4{,}000{,}000 \times 0.25 = \$1{,}000{,}000$$

Days of supply Companies need to know how much inventory they have in order to make effective decisions at both management and operational levels. One of the easiest tools to use and understand is **days of supply**. It is essentially a measure of how long in days, at the rate of anticipated demand, it will take for the current inventory level to reach zero or a set safety stock level. This is useful for deciding when to order and also to balance item inventory levels to have equal days of supply. On the next order cycle, items with an equal time-value of inventory can be ordered at the same time. Days of supply is also one of the most easily understood measures and avoids the calculations of other measures such as inventory turns. The equation to calculate the days of supply is

$$\text{Days of supply} = \frac{\text{inventory on hand}}{\text{average daily usage}}$$

This measure can be fairly easily converted to weeks of inventory on hand as a useful look at total inventory that may have more relevance to people than just looking at the total inventory count.

With the growth of the supply chain concept, the days of supply figures for inventory can be established for each part of the supply chain. This concept, often called *inventory profiling*, can be useful to see where inventory may be amassed or where performance problems exist. The actual inventory profile can be compared to a planned profile or even an optimal profile to establish plans for corrective actions.

Example Problem

A company has 9000 units on hand and the annual usage is 48,000 units. There are 240 working days in the year. What is the days of supply?

ANSWER

$$\text{Average daily usage} = \frac{48{,}000}{240} = 200 \text{ units}$$
$$\text{Days of supply} = \frac{\text{inventory on hand}}{\text{average daily usage}} = \frac{9000}{200} = 45 \text{ days}$$

In general, the ability to operate a business effectively with less inventory can have a positive impact on the financial side of the business. For example, since inventory appears as an asset, the business would likely see an increase in cash (also an asset), particularly if the company uses its own cash to purchase the inventory. Cash, of course, is a much more flexible asset to have than is inventory. In addition, if the company has its cash tied up in inventory, there is often an opportunity cost associated, meaning that the cash tied up in inventory cannot be used to invest in other income-generating methods. If, on the other hand, the company does not use its own cash to acquire inventory, they will typically have to borrow the money to buy the inventory, decreasing their profit because of the need to pay interest charges on the borrowed money. Another inventory impact on income is the cost to handle, store, move, and risk damage to inventory. This often is included in overhead expenses, which also reduce profit as they grow. These financial impacts are one, but only one, of several reasons that companies all over the world have embraced the concepts of lean management to reduce inventory, discussed in much more detail in Chapter 16.

Methods of Evaluating Inventory

There are five methods accounting uses to cost inventory. Each has implications for the value placed on inventory. If there is little change in the price of an item, any of the five ways will produce about the same results. However, in rising or falling prices, there can be a pronounced difference. There is no relationship with the actual physical movement of actual items in any of the methods. Whatever method is used is only to account for usage and the accounting value.

First in, first out (FIFO) The **first in, first out (FIFO)** method assumes that the oldest (first) item in stock is used first. In rising prices, replacement is at a higher price than the assumed cost. This method does not reflect current prices, and replacement will be understated. The reverse is true in a falling price market.

Last in, first out (LIFO) The **last in, first out (LIFO)** method assumes the newest (last) item in stock is the first used. In rising prices, replacement is at the current price. In a falling price market existing inventory is overvalued. However, the company is left with an inventory that may be grossly understated in value.

Average cost This method assumes an average of all prices paid for the article. During a period of changing prices (rising or falling), **average cost** can be skewed from the actual cost.

Standard cost This method uses a predetermined cost that includes direct material, direct labor, and overhead. Any difference between the standard cost and actual cost is stated as a variance. **Standard costs** are typically changed once a year by analyzing actual costs expended and setting a new standard for raw materials and finished goods.

Actual cost This method uses the **actual costs** paid for each purchase or production run. This results in a layered cost in accounting for those items in inventory that have been purchased or produced at different times. Lot control, as discussed in Chapter 6, is often required in order to identify the items costed differently.

10.10 ABC INVENTORY CONTROL

Control of inventory is exercised by controlling individual items, which are called **stock keeping units (SKUs)**. In controlling inventory, four questions must be answered:

1. What is the importance of the inventory item?
2. How are they to be controlled?
3. How much should be ordered at one time?
4. When should an order be placed?

ABC classification answers the first two questions by determining the importance of items and thus allowing different levels of control based on the relative importance of items.

Most companies carry a large number of items in stock. To have better control at a reasonable cost, it is helpful to classify the items according to their importance. Usually this is based on annual dollar usage, but other criteria may be used. For example, if an item is particularly difficult to obtain and has a long replenishment lead time, that item may be placed in a more important classification even though the annual dollar usage may be relatively small.

The ABC principle is based on the observation that a small number of items often dominate the results achieved in any situation. This observation was first made by an Italian economist, Vilfredo Pareto, and is called **Pareto's law**. As applied to inventories, it is usually found that the relationship between the percentage of items and the percentage of annual dollar usage follows a pattern in which three groups can be defined:

Group A   About 20% of the items account for about 80% of the dollar usage.

Group B   About 30% of the items account for about 15% of the dollar usage.

Group C   About 50% of the items account for about 5% of the dollar usage.

The percentages are approximate and should not be taken as absolute. This type of distribution can be used to help control inventory.

Steps in Performing an ABC Analysis

1. Establish the item characteristics that influence the results of inventory management. This is usually annual dollar usage but may be other criteria, such as scarcity of material, very long replenishment lead times, short effective shelf life, or quality issues.
2. Classify items into groups based on the established criteria.
3. Apply a degree of control in proportion to the importance of the group.

The factors affecting the importance of an item include annual dollar usage, unit cost, and scarcity of material. For simplicity, only annual dollar usage is used in this text. The procedure for classifying by annual dollar usage is as follows:

1. Determine the annual usage for each item.
2. Multiply the annual usage of each item by its cost to get its total annual dollar usage.
3. List the items according to their annual dollar usage.
4. Calculate the cumulative annual dollar usage and the cumulative percentage of items.
5. Examine the annual usage distribution and group the items into A, B, and C groups based on percentage of annual usage.

Example Problem

A company manufactures a line of 10 items. The usage and unit cost are shown in the following table, along with the annual dollar usage. The latter is obtained by multiplying the unit usage by the unit cost.

a. Calculate the annual dollar usage for each item.

b. List the items according to their annual dollar usage.

c. Calculate the cumulative annual dollar usage and the cumulative percentage of items.

d. Group items into an A, B, and C classification.

ANSWER

a. Calculate the annual dollar usage for each item.

Part Number	Unit Usage	Unit Cost $	Annual $ Usage
1	1100	2	2200
2	600	40	24,000
3	100	4	400
4	1300	1	1300
5	100	60	6000
6	10	25	250
7	100	2	200
8	1500	2	3000
9	200	2	400
10	500	1	500
Total	5510		$38,250

b. **b**, **c**, and **d**.

Part Number	Annual $ Usage	Cumulative $ Usage	Cumulative % $ Usage	Cumulative % of Items	Class
2	24,000	24,000	62.75	10	A
5	6000	30,000	78.43	20	A
8	3000	33,000	86.27	30	B
1	2200	35,200	92.03	40	B
4	1300	36,500	95.42	50	B
10	500	37,000	96.73	60	C
9	400	37,400	97.78	70	C
3	400	37,800	98.82	80	C
6	250	38,050	99.48	90	C
7	200	38,250	100.00	100	C

The percentage of value and the percentage of items are often shown as a graph, as in Figure 10.3.

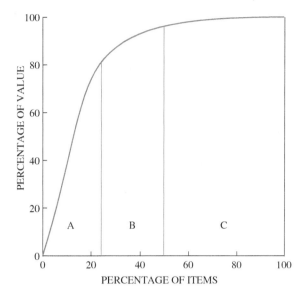

FIGURE 10.3 ABC curve: Percentage of value versus percentage of items.

Control Based on ABC Classification

Using the ABC approach, there are two general rules to follow:

1. **Have plenty of low-value items in supply.** C items represent about 50% of the items but account for only about 5% of the total inventory value. Carrying extra stock of C items adds little to the total value of the inventory. C items are really only important if there is a shortage of one of them, when they become extremely important, so a supply should always be on hand. For example, order a year's supply at a time and carry plenty of safety stock. That way, there is only one time a year when a stockout is even possible.

2. **Use the money and control effort saved to reduce the inventory of high-value items. A items** represent about 20% of the items and account for about 80% of the value. They are extremely important and deserve the tightest control and the most frequent review.

Different controls used with different classifications might be the following:

- **A items: high priority.** Tight control including complete accurate records, regular and frequent review by management, frequent review of demand forecasts, and close follow-up and expediting to reduce lead time.
- **B items: medium priority.** Normal controls with good records, regular attention, and normal processing.
- **C items: lowest priority.** Simplest possible controls, simple or no records, such as a two-bin system or periodic review system. Order large quantities and carry safety stock.

SUMMARY

There are benefits as well as costs to having inventory. The problem is to balance the cost of carrying inventory with the following:

- **Customer service.** The lower the inventory level, the higher the likelihood of a stockout and the potential cost of back orders, lost sales, and lost customers. The higher the inventory level, the higher the level of customer service.
- **Operating efficiency.** Inventories decouple one operation from another and allow manufacturing to operate more efficiently. They allow leveling production and avoid the costs of changing production levels. Carrying inventory allows longer production runs and reduces the number of setups. Finally, inventories let manufacturing purchase in larger quantities. ABC classification prioritizes individual items so that inventory and costs can be better controlled.

- **Cost of placing orders.** Inventory can be reduced by ordering less each time an order is placed. However, this increases the annual cost of ordering.
- **Transportation and handling costs.** The more often goods have to be moved and the smaller the quantities moved, the greater the transportation and material handling costs.
- Inventory management is influenced by several factors:
- The classification of the inventory, whether raw material, work in process, or finished goods.
- The functions that inventory serves: anticipation, fluctuation, lot size, or transportation.
- Supply and demand patterns.
- The costs associated with carrying (or not carrying) inventory.

Besides managing inventory at the aggregate level, it must also be managed at the item level. Management needs to establish decision rules about inventory items so inventory control personnel can do their job effectively.

KEY TERMS

ABC classification 258	Inventory velocity 256
Actual costs 258	Landed cost 251
Anticipation inventories 247	Last in, first out (LIFO) 258
Assets 253	Liabilities 253
Average cost 258	Lot-size inventory 248
Balance sheet 254	Maintenance, repair, and operating (MRO) supplies 247
Capacity-related costs 252	
Carrying cost 251	Movement inventory 248
Cash flow management 255	Ordering costs 252
Cash-to-cash cycle time 255	Owners' equity 253
Cost of goods sold (COGS) 255	Pareto's law 259
Cycle stock 248	Pipeline inventory 248
Days of supply 257	Profit 254
Decoupling inventory 249	Raw materials 246
Distribution inventory 246	Return on investment (ROI) 255
Expenses 254	Revenue 254
Finished goods 246	Safety stock 247
First in, first out (FIFO) 258	Standard costs 258
Fluctuation inventory 247	Stock keeping unit (SKU) 258
General and administrative expenses (G&A) 254	Stockout 247
	Stockout costs 252
Hedge inventory 248	Transportation inventory 248
Income statement 254	Unit cost 251
Inventory turns 256	Work in process (WIP) 246

QUESTIONS

1. What are inventories? Why are they important to manufacturing companies?
2. What are the responsibilities of inventory management?
3. What is aggregate inventory management? With what is it concerned?
4. What are decision rules? Why are they necessary?
5. According to the flow of material, what are the four classifications of inventories?

6. Why is less inventory needed in a line-flow manufacturing system than in lot or batch manufacturing?
7. What is the basic purpose of inventories? In what five areas do they provide a buffer?
8. Describe the function and purpose of the following kinds of inventories:
   a. Anticipation.
   b. Fluctuation.
   c. Lot size.
   d. Transportation.
9. Describe how inventories influence each of the following:
   a. Customer service.
   b. Plant operations.
10. What are the five costs associated with inventory?
11. Name and describe the categories of inventory-carrying costs.
12. Name and describe the categories of ordering costs found in a factory.
13. What are stockout costs and capacity-related costs? What is their relationship to inventories?
14. What are the balance sheet equation and the income statement equation?
15. What is the purpose of cash flow management?
16. How is cash-to-cash cycle time calculated?
17. What do inventory turns and days of supply measure?
18. What is the basic premise of ABC analysis? What are the steps for performing an ABC inventory analysis?
19. What are the five steps in the procedure for classifying inventory by annual dollar usage?
20. Using the items in your own kitchen, classify them in to A, B, and C categories. Are you practicing good control based on ABC classification? How?
21. What is the difference between FIFO and LIFO?
22. During times of inflation why does the value of inventory change between FIFO and LIFO evaluations?

PROBLEMS

10.1. If the transit time is 11 days and the annual demand for an item is 10,000 units, what is the average annual inventory in transit?

Answer. 301.4 units

10.2. A company is using a carrier to deliver goods to a major customer. The annual demand is $2,500,000, and the average transit time is 10 days. Another carrier promises to deliver in seven days. What is the reduction in transit inventory?

10.3. Given the following percentage costs of carrying inventory, calculate the annual carrying cost if the average inventory is $1 million. Capital costs are 12%, storage costs are 7%, and risk costs are 8%.

Answer. $270,000

10.4. A florist carries an average inventory of $12,000 in cut flowers. The flowers require special storage and are highly perishable. The florist estimates capital costs at 10%, storage costs at 23%, and risk costs at 55%. What is the annual carrying cost?

10.5. Annual purchasing salaries are $65,000, operating expenses for the purchasing department are $25,000, and inspecting and receiving costs are $22 per order. If the purchasing department places 9000 orders a year, what is the average cost of ordering? What is the annual cost of ordering?

Answer. Average ordering cost = $3

Average ordering cost = $288,000

10.6. An importer operates a small warehouse that has the following annual costs. Wages for purchasing are $45,000, purchasing expenses are $30,000, customs and brokerage costs are $30 per order, the cost of financing the inventory is 8%, storage costs are 9%, and the risk costs are 10%. The average inventory is $250,000, and 6000 orders are placed in a year. What are the annual ordering and carrying costs?

10.7. A company manufactures and sells a seasonal product. Based on the sales forecast that follows, calculate a level production plan, quarterly ending inventories, and average quarterly inventories. Assume that the average quarterly inventory is the average of the starting and ending inventory for the quarter. If inventory carrying costs are $3 per unit per quarter, what is the annual cost of carrying this anticipation inventory? Opening and ending inventories are zero.

Answer. Annual inventory costs = $6000

	Quarter 1	Quarter 2	Quarter 3	Quarter 4	Totals
Sales	1000	2000	3000	2000	
Production					
Ending Inventory					
Average Inventory					
Inventory Cost					

10.8. Given the following data, calculate a level production plan, quarterly ending inventory, and average quarterly inventory. If inventory carrying costs are $6 per unit per quarter, what is the annual carrying cost? Opening and ending inventories are zero.

	Quarter 1	Quarter 2	Quarter 3	Quarter 4	Totals
Forecast Demand	5000	7000	8000	10,000	
Production					
Ending Inventory					
Average Inventory					
Inventory Cost					

If the company always carries 100 units of safety stock, what is the annual cost of carrying it?

10.9. Given the following data, calculate a level production plan, quarterly ending inventory, and average quarterly inventory. If inventory carrying costs are $3 per unit per quarter, what is the annual carrying cost? Opening and ending inventories are zero.

	Quarter 1	Quarter 2	Quarter 3	Quarter 4	Totals
Forecast Demand	3000	4000	6500	6500	
Production					
Ending Inventory					
Average Inventory					
Inventory Cost					

10.10. If the assets are $2,000,000 and liabilities are $1,600,000, what is the owners' equity?

Answer. $400,000

10.11. If the liabilities are $4,000,000 and the owners' equity is $1,300,000, what are the assets worth?

10.12. Given the following data, calculate the gross margin and the net income.

$$\text{Revenue} = \$3,000,000$$
$$\text{Direct labor} = \$700,000$$
$$\text{Direct material} = \$900,000$$
$$\text{Factory overhead} = \$800,000$$
$$\text{General and administrative expense} = \$350,000$$
$$\text{Gross margin} = \$600,000$$
$$\text{Net income} = \$250,000$$

Answer. Gross margin = $600,000

Net income = $250,000

10.13. In question 10.12, how much would profit increase if the materials costs are reduced by $200,000?

10.14. If the annual cost of goods sold is $12,000,000 and the average inventory is $2,500,000,

a. What is the inventory turns ratio?

b. What would be the reduction in average inventory if, through better materials management, inventory turns were increased to 10 times per year?

c. If the cost of carrying inventory is 20% of the average inventory, what is the annual savings?

Answer. a. 4.8
b. $1,300,000
c. $260,000

10.15. If the annual cost of goods sold is $30,000,000 and the average inventory is $5,000,000,

a. What is the inventory turns ratio?

b. What would be the reduction in average inventory if, through better materials management, inventory turns were increased to 10 times per year?

c. If the cost of carrying inventory is 25% of the average inventory, what is the annual savings?

10.16. A company has 900 units on hand and the annual usage is 7200 units. There are 240 working days in the year. What is the days of supply?

Answer. 30 days

10.17. Over the past year, a company has sold the following 10 items. The following table shows the annual sales in units and the cost of each item.

a. Calculate the annual dollar usage of each item.

b. List the items according to their total annual dollar usage.

c. Calculate the cumulative annual dollar usage and the cumulative percentage of items.

d. Group the items into A, B, and C groups based on percentage of annual dollar usage.

Part Number	Annual Unit Usage	Unit Cost $	Annual $ Usage
1	21,000	1	
2	5000	40	
3	1600	3	
4	12,000	1	
5	1000	100	
6	50	50	
7	800	2	
8	10,000	3	
9	4000	1	
10	5000	1	

Answer. A items 2, 5

B items 8, 1, 4

C items 10, 3, 9, 6, 7

10.18. Analyze the following data to produce an ABC classification based on annual dollar usage.

Part Number	Annual Unit Usage	Unit Cost $	Annual $ Usage
1	200	10	
2	17,000	4	
3	60,000	6	
4	15,000	15	
5	1500	10	
6	120	50	
7	25,000	2	
8	700	3	
9	25,000	1	
10	7500	1	

10.19. a. The Ajex Company has several of items they store in inventory. The table below shows the annual demand, and the unit cost. Develop a logical classification for the inventory based on ABC:

Item	Unit Cost	Annual Usage (# units)
C34	$12	4000
B99	$23	8000
V94	$19	5500
H64	$41	1200
P77	$72	400
Y12	$62	1100
R74	$33	1440

b. One additional item, the M22, has a very low usage (300 per year) and a low unit cost ($3 per unit), but has a very long lead time and is often difficult to obtain. How should that item be handled and why?

CASE STUDY 10.1

Randy Smith, Inventory Control Manager

Randy Smith was very proud of his new position as inventory control manager for the Johnson Trinket Company. His primary responsibility had been fairly clearly defined: Maintain an inventory level in the warehouse that ensures that production will not run out of stock, yet also maintain an inventory level that will minimize inventory holding and control costs. Since Randy had recently had a course in materials management, he knew an approach that should help him. He decided to make a list of inventory items in one small section of the warehouse to see if he could develop a good plan for inventory control. If it worked in the one small section, he could expand it to the rest of the more than 30,000 part numbers in the warehouse.

The following is the data Randy compiled:

Part Number	Part Unit Value in $	Quantity Currently in Inventory	Average Annual Usage
1234	$2.50	300	3000
1235	$0.20	550	900
1236	$15.00	400	1000
1237	$0.75	50	7900
1238	$7.60	180	2800
1239	$4.40	20	5000
1240	$1.80	200	1800
1241	$0.05	10	1200
1242	$17.20	950	2000
1243	$9.00	160	2500
1244	$3.20	430	7000

(Continued)

Part Number	Part Unit Value in $	Quantity Currently in Inventory	Average Annual Usage
1245	$0.30	500	10000
1246	$1.10	25	7500
1247	$8.10	60	2100
1248	$5.00	390	4000
1249	$0.90	830	6500
1250	$6.00	700	3100
1251	$2.20	80	6000
1252	$1.20	480	4500
1253	$5.90	230	900

When Randy scanned the list, he noticed several things that disturbed him, and he asked one of the experienced inventory clerks. The following list summarizes the issues that concerned Randy, and the explanation from the clerk:

- Part number 1236, a very expensive part with almost half-a-years' worth of inventory. This part is used for a product that has very cyclical demand, and the busy time of the year is about to start.
- Part number 1241 is very inexpensive, yet the inventory is very small. This part has a supplier with an erratic delivery history, and the part also has a very long lead time. A lot of 150 has been on order for some time, and is now several days past due.
- Part number 1242, like 1236, is expensive with almost half a year worth of inventory—this part is shipped to a location on the other side of the country and is being accumulated into a large lot to save shipping costs.
- Part number 1246 is not too expensive, with a low inventory. This part is produced in-house, and has a quality tolerance that the older equipment, which was used to produce it, had a difficult time meeting. The last batch was rejected by the quality department.
- Part number 1253 is moderately expensive with a large inventory compared to usage. This part was subject to a recent quality audit, and almost 150 of the items were rejected as a result of that audit.

Once Randy understood some of the issues, suddenly he did not feel quite as confident that he had the best approach in mind to control the inventory to meet the expectations of his boss.

Assignment

1. Use the information above to evaluate the current situation.
2. Given your evaluation, try to develop an integrated inventory control policy that Randy should consider.
3. Is there other information that you would like to see that might help you to make a more effective policy? If so, what would that information be and how would you use it to help you?

CHAPTER ELEVEN

ORDER QUANTITIES

11.1 INTRODUCTION

The objectives of inventory management are to provide the required level of customer service and to reduce the sum of all costs involved. To achieve these objectives, two basic questions must be answered:

1. How much should be ordered at one time?
2. When should an order be placed?

Management must establish decision rules to answer these questions so inventory management personnel know when to order and how much. Lacking any better knowledge, decision rules are often made based on what seems reasonable. Unfortunately, such rules do not always produce the best results.

This chapter will examine methods of answering the first question, and the next chapter will deal with the second question. First, it must be determined what is being ordered and controlled.

Stock Keeping Unit

Control is exercised through individual items in a particular inventory. These are called **stock keeping units (SKUs)**. Two white shirts in the same inventory but of different sizes or styles would be two different SKUs. The same shirt in two different inventories would be two different SKUs.

Lot-Size Decision Rules

The *APICS Dictionary*, 16th edition, defines a lot, or batch, as "a quantity produced together and sharing the same production costs and specifications." Following are some common decision rules for determining what lot size to order at one time.

Lot-for-lot The **lot-for-lot** rule says to order exactly what is needed: no more, no less. Another way to think of this is that it really assumes a basic lot size of one, allowing the order quantity to change whenever requirements change. This technique requires time-phased information such as that provided by a material requirements plan or a master production schedule. Since items are ordered only when needed, this system creates no unused lot-size inventory. Because of this, it is often the preferred method for planning "A" items (see Chapter 10) and is also used (or is a target quantity to be used) in a just-in-time or lean environment.

Fixed order quantity **Fixed order quantity** rules specify the number of units to be ordered each time an order is placed for an individual item or SKU. The quantity is sometimes arbitrary, such as 200 units at a time, but is sometimes based on an economic order size calculation, the size of a container, or the size of a package the material comes in. The advantage to this type of rule is that it is easily understood. The disadvantage is that it does not minimize the costs involved.

A variation on the fixed order quantity system is the **min–max system**. In this system, an order is placed when the quantity available falls below the order point (discussed in Chapter 12). The quantity ordered is the difference between the actual quantity available at the time of order and the maximum. For example, if the order point is 100 units, the maximum is 300 units, and the quantity actually available when the order is placed is 75, the order quantity is 225 units. If the quantity actually available is 80 units, an order for 220 units is placed.

One commonly used method of calculating the quantity to order is the economic order quantity, which is discussed in the next section.

Period order quantity Rather than ordering a fixed quantity, inventory management can order enough to satisfy future demand for a given period of time. The question is how many periods should be covered? The answer is given later in this chapter in the discussion on the **period order quantity** system.

Costs

As shown in Chapter 10, the cost of ordering and the cost of carrying inventory both depend on the quantity ordered. Ideally, the ordering decision rules used will minimize the sum of these two costs. The best-known system is the economic order quantity.

11.2 ECONOMIC ORDER QUANTITY

Assumptions

The assumptions on which the **economic order quantity (EOQ)** is based are as follows:

1. Demand is relatively constant and is known.
2. The item is produced or purchased in lots or batches and not continuously.
3. Order preparation costs and inventory carrying costs are constant and known.
4. Replacement occurs all at once.

These assumptions are usually valid for finished goods whose demand is independent and fairly uniform. However, there are many situations where the assumptions are not valid and the EOQ concept is of no use. For instance, there is no reason to calculate the EOQ for make-to-order items in which the customer specifies the order quantity, the shelf life of the product is short, or the length of the run is limited by tool life or raw material batch size. In material requirements planning, the lot-for-lot decision rule is often used, but there are also several rules used that are variations of the EOQ.

Development of the EOQ Formula

Under the assumptions given, the quantity of an item in inventory decreases at a uniform rate. Suppose for a particular item that the order quantity is 200 units and the usage rate is 100 units a week. Figure 11.1 shows how inventory would behave.

The vertical lines represent stock arriving all at once as the stock on hand reaches zero. The quantity of units in inventory then increases instantaneously by Q, the quantity

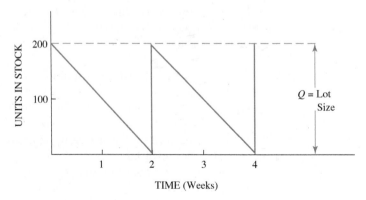

FIGURE 11.1 Inventory on hand over time.

ordered. This is an accurate representation of the arrival of purchased parts or manufactured parts where all parts are received at once.

From the preceding,

$$\text{Average lot size inventory} = \frac{\text{order quantity}}{2} = \frac{200}{2} = 100 \text{ units}$$

$$\text{Number of orders per year} = \frac{\text{annual demand}}{\text{order quantity}} = \frac{100 \times 52}{200}$$

$$= 26 \text{ times per year}$$

Example Problem

The annual demand for an SKU is 10,075 units, and it is ordered in quantities of 650 units. Calculate the average inventory and the number of orders placed per year.

ANSWER

$$\text{Average cycle inventory} = \frac{\text{order quantity}}{2} = \frac{650}{2} = 325 \text{ units}$$

$$\text{Number of orders per year} = \frac{\text{annual demand}}{\text{order quantity}} = \frac{10,075}{650} = 15.5$$

Notice in the example problem that the number of orders per year is rounded neither up nor down. It is an average figure, and the actual number of orders per year will vary from year to year but will average to the calculated figure. In the example, 16 orders will be placed in one year and 15 in the second.

Relevant costs The relevant costs are as follows:

- Annual cost of placing orders.
- Annual cost of carrying inventory.

As the order quantity increases, the average inventory and the annual cost of carrying inventory increase, but the number of orders per year and the ordering cost decrease. It is a bit like a seesaw where one cost can be reduced only at the expense of increasing the other. The key is to find the particular order quantity in which the total cost of carrying inventory and the cost of ordering will be a minimum.

Let

A = annual usage in units

S = ordering cost in dollars per order

i = annual carrying cost rate as a decimal or a percentage

c = unit cost in dollars

Q = order quantity in units

Then,

Annual ordering cost = number of orders × costs per order

$$= \frac{A}{Q} \times S$$

Annual carrying cost = average inventory × cost of carrying 1 unit for 1 year

= average inventory × unit cost × carrying cost

$$= \frac{Q}{2} \times c \times i$$

Total annual costs = annual ordering costs + annual carrying costs

$$= \frac{A}{Q} \times S + \frac{Q}{2} \times c \times i$$

Example Problem

The annual demand is 10,000 units, the ordering cost is $30 per order, the carrying cost is 20%, and the unit cost is $15. The order quantity is 600 units. Calculate:

a. Annual ordering cost.
b. Annual carrying cost.
c. Total annual cost.

ANSWER

$$A = 10{,}000 \text{ units}$$
$$S = \$30$$
$$i = 0.20$$
$$c = \$15$$
$$Q = 600 \text{ units}$$

a. Annual ordering cost $= \dfrac{A}{Q} \times S = \dfrac{10{,}000}{600} \times 30 = \$500$

b. Annual carrying cost $= \dfrac{Q}{2} \times c \times i = \dfrac{600}{2} \times 15 \times 0.2 = \$900$

c. Total annual cost $= \$1400$

Ideally, the total cost will be a minimum. For any situation in which the annual demand (A), the cost of ordering (S), and the cost of carrying inventory (i) are given, the total cost will depend upon the order quantity (Q).

Trial-and-Error Method

Consider the following example:

A hardware supply distributor carries boxes of 3-inch bolts in stock. The annual usage is 1000 boxes, and demand is relatively constant throughout the year. Ordering costs are $20 per order, and the cost of carrying inventory is estimated to be 20%. The cost per unit is $5.

Let:

$A = 1000$ units

$S = \$20$ per order

$c = \$5$ per unit

$i = 20\% = 0.20$

Then:

Annual ordering cost $= \dfrac{A}{Q} \times S = \dfrac{1000}{Q} \times 20$

Annual carrying cost $= \dfrac{Q}{2} \times c \times i = \dfrac{Q}{2} \times 5 \times 0.20$

Total annual cost = annual ordering cost + annual carrying cost

Figure 11.2 shows a tabulation of the costs for different order quantities. The results from the table in Figure 11.2 are represented on the graph in Figure 11.3.

Figures 11.2 and 11.3 show the following important facts:

1. There is an order quantity in which the sum of the ordering costs and carrying costs is a minimum.
2. This EOQ occurs when the cost of ordering equals the cost of carrying.
3. The total cost varies little for a wide range of lot sizes about EOQ.

The last point is very important for two reasons. First, it is usually difficult to determine accurately the cost of carrying inventory and the cost of ordering. Since the total cost is relatively flat around the EOQ, it is not critical to have exact values. Good

Order Quantity (Q)	Ordering Costs (AS/Q)	Carrying Costs (Qci/2)	Total Cost
50	$400	$25	$425
100	200	50	250
150	133	75	208
200	100	100	200
250	80	125	205
300	67	150	217
350	57	175	232
400	50	200	250

FIGURE 11.2 Costs for different lot sizes.

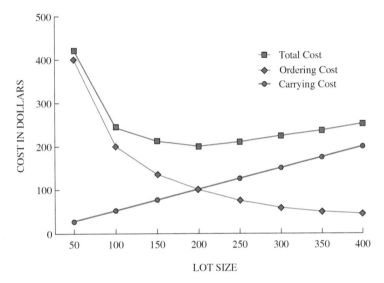

FIGURE 11.3 Cost versus lot size.

approximations are sufficient. Second, parts are often ordered in convenient packages such as pallet loads, cases, or dozens, and it is adequate to pick the package quantity closest to the EOQ.

EOQ Formula

The previous section showed that the EOQ occurred at an order quantity in which the ordering costs equal the carrying costs. If these two costs are equal, the following formula can be derived:

$$\text{Carrying costs} = \text{ordering costs}$$

$$\frac{Qic}{2} = \frac{AS}{Q}$$

Solving for Q gives

$$Q^2 = \frac{2AS}{ic}$$

$$Q = \sqrt{\frac{2AS}{ic}}$$

This value for the order quantity is the EOQ. Using the formula to calculate the EOQ in the preceding example yields:

$$\text{EOQ} = \sqrt{\frac{2AS}{ic}} = \sqrt{\frac{2 \times 1000 \times 20}{0.20 \times 5}} = 200 \text{ units}$$

How to Reduce Lot Size

The EOQ formula has four variables. The EOQ will increase as the annual demand (A) and the cost of ordering (S) increase, and it will decrease as the cost of carrying inventory (i) and the unit cost (c) increase.

The annual demand (A) is a condition of the marketplace and is beyond the control of manufacturing. The cost of carrying inventory (i) is determined by the product itself and the cost of money to the company. As such, it is beyond the control of manufacturing.

The unit cost (c) is either the purchase cost of the SKU or the cost of manufacturing the item. Ideally, both costs should be as low as possible. In any event, as the unit cost decreases, the EOQ increases.

The cost of ordering (S) is either the cost of placing a purchase order or the cost of placing a manufacturing order. The cost of placing a manufacturing order is made up from production control costs and setup costs. Anything that can be done to reduce these costs reduces the EOQ.

Lean production emphasizes reduction of setup time (ordering cost). There are several reasons why this is desirable, and the reduction of order quantities is one. Chapter 16 discusses these issues further.

11.3 VARIATIONS OF THE EOQ MODEL

There are several modifications that can be made to the basic EOQ model to fit particular circumstances. Two that are often used are the monetary unit lot-size model and the non-instantaneous receipt model.

Monetary Unit Lot Size

The EOQ can be calculated in monetary units rather than physical units. The same EOQ formula given in the preceding section can be used, but the annual usage changes from units to dollars.

A_D = annual usage in dollars

S = ordering costs in dollars

i = carrying cost rate as a decimal or a percent

Because the annual usage is expressed in dollars, the unit cost is not needed in the modified EOQ equation.

The EOQ in dollars is:

$$EOQ = \sqrt{\frac{2 A_D S}{i}}$$

Example Problem

An item has an annual demand of $5000, preparation costs of $20 per order, and a carrying cost of 20%. What is the EOQ in dollars?

A_D = $5000
S = $20
i = 20% = 0.20

$$EOQ = \sqrt{\frac{2 A_D S}{i}} = \sqrt{\frac{2 \times \$5000 \times \$20}{0.2}} = \$1000$$

Non-instantaneous Receipt Model

In some cases, when a replenishment order is made, the order is not all received at one time. The most common reason for this is that the ordered material is being produced over an extended period of time, yet material is received for the order as it is being produced. In this case, the EOQ is modified to reflect the rate of production as related to rate of demand:

$$EOQ = \sqrt{\frac{2AS}{ic}\left(\frac{p}{p-d}\right)}$$

where
- p = the rate of production (in units per day) and
- d = the rate of demand (in units per day)

The units of time for p and d (shown above as units per day) can be any unit of time as long as that unit of time is the same for both.

Example Problem

An item has a setup cost for production of $500 per order, and the inventory carrying costs for the item is $12 per year. The demand for the item is constant at 11 units per day. The production rate is 50 units per day while the item is being produced. What is the non-instantaneous EOQ?

Annual demand (A) = (11 units per day) × (365 days per year)
= 4015 units per year

$$EOQ = \sqrt{\frac{2(4015)\,500}{12}\left(\frac{50}{50-11}\right)}$$
$$= \sqrt{334{,}583.3(1.3)}$$
$$= 655 \text{ units}$$

11.4 QUANTITY DISCOUNTS

When material is purchased, suppliers often give a discount on orders over a certain size. This can be done because larger orders reduce the supplier's costs; to get larger orders, suppliers are willing to offer volume discounts. The buyer must decide whether to accept the discount, and in doing so must consider the relevant costs:

- Purchase cost.
- Ordering costs.
- Carrying costs.

Example Problem

An item has an annual demand of 25,000 units, a unit cost of $10, an order preparation cost of $10, and a carrying cost of 20%. It is ordered on the basis of an EOQ, but the supplier has offered a discount of 2% on orders of $10,000 or more. Should the offer be accepted?

ANSWER

A_D = 25,000 × $10 = $250,000
S = $10
i = 20%

$$EOQ = \sqrt{\frac{2 \times 250{,}000 \times 10}{0.2}} = \$5000$$

Discounted order quantity = $10,000 × 0.98 = $9800

	No Discount	Discount Lot Size
Unit Price	$10	$9.80
Lot Size	$5000	$9800
Average Lot-Size Inventory ($Qc \div 2$)	$2500	$4900
Number of Orders per Year	50	25
Purchase Cost	$250,000	$245,000
Inventory Carrying Cost (20%)	500	980
Order Preparation Cost ($10 each)	500	250
Total Cost	$251,000	$246,230

From the preceding example problem, it can be seen that taking the discount results in the following:

- There is a saving in purchase cost.
- Ordering costs are reduced because fewer orders are placed since larger quantities are being ordered.
- Inventory carrying costs rise because of the larger order quantity.

The buyer must weigh the first two against the last and decide what to do. What counts is the total cost. Depending on the figures, it may or may not be best to take the discount. When considering a **quantity discount**, the total costs of using the discount are usually compared with the total costs of using the EOQ method.

11.5 ORDER QUANTITIES FOR FAMILIES OF PRODUCT WHEN COSTS ARE NOT KNOWN

The EOQ formula depends upon the cost of ordering and the cost of carrying inventory. In practice, these costs are not necessarily known or easy to determine. However, the formula can still be used to an advantage when applied to a family of items.

For a family of items, the ordering costs and the carrying costs are generally the same for each item. For instance, if ordering hardware items—nuts, bolts, screws, nails, and so on—the carrying costs would be virtually the same (storage, capital, and risk costs) and the cost of placing an order with the supplier would be the same for each item. In cases such as this, the cost of placing an order (S) is the same for all items in the family as is cost of carrying inventory (i).

Now

$$Q = \sqrt{\frac{2 A_D S}{i}}$$

where A (annual demand) is in *dollars*.

Since S is the same for all the items and i is the same for all items, the ratio $2S \div i$ must be the same for all items in the family. For convenience, let

$$K = \sqrt{\frac{2S}{i}}$$

Then, $$Q = K\sqrt{A_D}$$

Also, $$Q = \frac{\text{annual demand}}{\text{orders per year}} = \frac{A_D}{N}$$

Therefore, $$K\sqrt{A_D} = \frac{A_D}{N}$$

$$K = \frac{\sqrt{A_D}}{N}$$

Example Problem

Suppose there were a family of items for which the decision rule was to order each item four times a year. Since the cost of ordering (S) and the cost of carrying inventory (i) are not known, ordering four times a year is not based on an EOQ. Can we come up with a better decision rule even if the EOQ cannot be calculated?

Item	Annual Usage	Orders per Year	Present Lot Size	$\sqrt{A_D}$	$K = \frac{\sqrt{A_D}}{N}$
A	$10,000	4	$2500	$100	25
B	400	4	100	20	5
C	144	4	36	12	3
		12	$2636	$132	33
Average inventory =			$1318		

The sum of all the lots is $2636. Since the average inventory is equal to half the order quantity, the average inventory is $2636 ÷ 2 = $1318.

Since this is a family of items where the preparation costs are the same and the carrying costs are the same, the values for $K = (2S \div i)^{1/2}$ should be the same for all items. The preceding calculations show that they are not. The correct value for K is not known, but a better value would be the average of all the values:

$$K = \frac{\sum \sqrt{A_D}}{\sum N}$$

$$= \frac{132}{12}$$

$$= 11$$

This value of K can be used to recalculate the order quantities for each item.

Item	Annual Usage	Present Orders per Year	Present Lot Size	$\sqrt{A_D}$	New Lot Size $= K\sqrt{A_D}$	New Orders per Year $N = A_D/Q$
A	$10,000	4	$2500	$100	$1100	9.09
B	400	4	100	20	220	1.82
C	144	4	36	2	132	1.09
	$10,544	12	$2636	$132	$1452	12.00
Average inventory			$1318		$726	

Item A: New lot size $= K\sqrt{A_D} = 11 \times 100 = \$1100$

$$\text{New orders per year} = \frac{A_D}{Q} = \frac{10,000}{1100} = 9.09$$

The average inventory has been reduced from $1318 to $726 while the number of orders per year (12) remains the same. Thus, the total costs associated with inventory have been reduced.

11.6 PERIOD ORDER QUANTITY

The EOQ attempts to minimize the total cost of ordering and carrying inventory and is based on the assumption that demand is uniform. Often demand is not uniform, particularly in material requirements planning (MRP), and using the EOQ does not produce a minimum cost. The order quantity will often exceed the demand expected for the next few periods, which may result in inventory carried over periods of no demand to avoid the cost of placing another order. However, the quantity ordered may not be exactly enough to cover the demand in the next few periods resulting in the placement of another order even while a remnant of the last order is available. Keeping inventory to avoid placing orders is a good idea but not keeping enough inventory to avoid placing orders defeats the goal of EOQ. A change in the application is required and this is demonstrated in the period order quantity method.

The period order quantity (POQ) lot-size rule is based on the same theory as the EOQ. It uses the EOQ formula to calculate an economic *time between orders*. Note that POQ does not calculate a quantity but actually calculates the number of periods that are to be covered. The POQ is calculated by dividing the EOQ by the demand rate. This produces a time interval for which orders are placed. Instead of ordering the same quantity (EOQ), orders are placed to satisfy requirements for the calculated time interval. The number of orders placed in a year is approximately the same as for an EOQ, but the amount ordered each time varies. Thus, the ordering cost is the same as it would be using the EOQ but because the order quantities are determined by actual demand, the carrying cost is reduced.

$$\text{Period order quantity} = \frac{\text{EOQ}}{\text{average weekly usage}}$$

Note that although POQ calculates an order interval say of two weeks, this does not mean that there is an order placed every two weeks. It actually means that when an order needs to be placed, then enough inventory is ordered to cover the next two weeks. This is especially important in discontinuous demand, as shown in the following example.

Example Problem

The EOQ for an item is 2800 units, and the annual usage is 52,000 units. What is the POQ?

ANSWER

$$\text{Average weekly usage} = 52{,}000 \div 52 = 1000 \text{ per week}$$

$$\text{Period order quantity} = \frac{\text{EOQ}}{\text{average weekly usage}}$$

$$= \frac{2800}{1000} = 2.8 \text{ weeks} \longrightarrow 3 \text{ weeks}$$

When an order is placed, it will cover the requirements for the next three weeks. Notice the calculation is approximate. Precision is not important.

Example Problem

Given the following MRP record and an EOQ of 250 units, calculate the planned order receipts using the EOQ. Next, calculate the period order quantities and the planned order receipts. In both cases, calculate the ending inventory and the total inventory carried over the 10 weeks.

Week	1	2	3	4	5	6	7	8	9	10	Total
Net Requirements	100	50	150		75	200	55	80	150	30	890
Planned Order Receipt											

ANSWER

EOQ = 250 units

Week	1	2	3	4	5	6	7	8	9	10	Total
Net Requirements	100	50	150		75	200	55	80	150	30	890
Planned Order Receipt	250		250			250			250		
Ending Inventory	150	100	200	200	125	175	120	40	140	110	1360

Period order quantity:

Weekly average demand = 890 ÷ 10 = 89 units

POQ = 250 ÷ 89 = 2.81 → 3 weeks

Week	1	2	3	4	5	6	7	8	9	10	Total
Net Requirements	100	50	150		75	200	55	80	150	30	890
Planned Order Receipt	300				330			260			
Ending Inventory	200	150	0	0	255	55	0	180	30	0	870

Notice in the example problem the total inventory is reduced from 1360 to 870 units over the 10-week period. Note also that an order is not placed in week 4 since there is no demand in period 4 and that the period before each order has zero inventory having used up the previous order entirely.

Practical Considerations When Using the EOQ

Lumpy demand The EOQ assumes that demand is uniform and replenishment occurs all at once. When this is not true, the EOQ will not produce the best results. It is better to use the POQ.

Anticipation inventory Demand is not uniform, and stock must be built ahead of periods of high demand. It is better to plan a buildup of inventory based on capacity and future demand.

Minimum order quantities Some suppliers require a minimum order. This minimum may be based on the total order rather than on individual items. Often these are C items where the rule is to order plenty, not an EOQ.

Transportation inventory As will be discussed in Chapter 14, transportation carriers give rates based on the amount shipped. A full load costs less per ton to ship than a partial load. This is similar to the price break given by suppliers for large quantities. The same type of analysis can be used.

Multiple order quantities Sometimes order size is constrained by package size. For example, a supplier may ship only in skid-load lots. In these cases, the unit used should be multiples of the minimum package size.

Order quantities and lean production As will be discussed in much more detail in Chapter 16, lean production has a profound effect on the amount of inventory to be produced at one time. The replenishment quantity of an item is adjusted to match the demand of the next operation in the supply chain. This adjustment leads to smaller lot sizes and is often determined by the frequency of shipments to a customer or the size of an easily moved container rather than by calculation.

What happens if the EOQ assumptions are not valid? Earlier in the chapter we listed the key assumptions for using the EOQ. To repeat, those assumptions were:

1. Demand is relatively constant and is known.
2. The item is produced or purchased in lots or batches and not continuously.
3. Order preparation costs and inventory carrying costs are constant and known.
4. Replacement occurs all at once.

Often those assumptions are not totally true in a given situation. An examination of the total cost curve for the EOQ (Figure 11.3) provides some insight to realize that the EOQ formula will still be useful as a close assumption to an optimum lot size even when the environment does not precisely match the assumptions. Notice the total cost curve is quite flat in the region on both sides of the EOQ value of the curve. That implies that the total cost will rise very little until the environmental assumptions are off by a fairly large extent. In those cases, other options for order quantities should be evaluated.

SUMMARY

The economic order quantity (EOQ) is based on the assumption that demand is relatively uniform. This is appropriate for some inventories, and the EOQ formula can be used with reasonable results. One problem in using the EOQ formula is in determining the cost of ordering and the cost of carrying inventory. Since the total cost curve is flat at the bottom, good guesses very often will produce an order quantity that is economical. It has also been demonstrated that the EOQ concept can be used effectively with groups of items when the costs of carrying and ordering are not known.

The two costs influenced by the order quantity are the cost of ordering and the cost of carrying inventory. All methods of calculating order quantities attempt to minimize the sum of these two costs. The period order quantity does this. It has the advantage over the EOQ in that it is better for lumpy demand because it looks forward to see what is actually needed.

KEY TERMS

Economic order quantity (EOQ) 270
Fixed order quantity 269
Lot-for-lot 269
Min–max system 269

Period order quantity 270
Quantity discount 276
Stock keeping unit (SKU) 269

QUESTIONS

1. What are the two basic questions in inventory management that are discussed in the text?
2. What are decision rules? What is their purpose?
3. What is an SKU?
4. What is the lot-for-lot decision rule? What is its advantage? Where would it be used?
5. What are the four assumptions on which economic order quantities are based? For what kind of items are these assumptions valid? When are they not?
6. Under the assumptions on which EOQs are based, what are the formulas for average lot size and the number of orders per year?
7. What are the relevant costs associated with the two formulas? As the order quantities increase, what happens to each cost? What is the objective in establishing a fixed order quantity?
8. Define each of the following in your own words and as a formula:
   a. Annual ordering cost.
   b. Annual carrying cost.
   c. Total annual cost.
9. What is the EOQ formula? Define each term and give the units used. How do the units change when monetary units are used?

10. What are the relevant costs to be considered when deciding whether to take a quantity discount? On what basis should the decision be made?

11. What is the POQ? How is it established? When can it be used?

12. How do each of the following influence inventory lot-size decisions?
    a. Lumpy demand.
    b. Minimum orders.
    c. Transportation costs.
    d. Multiples.

13. A company working toward lean will have smaller lot sizes when compared to using traditional methods. Discuss how this will affect the costs associated with inventory. What are the controllable and the uncontrollable costs?

PROBLEMS

11.1. An SKU costing $10 is ordered in quantities of 500 units, annual demand is 5200 units, carrying costs are 20%, and the cost of placing an order is $50. Calculate the following:

   a. Average inventory.
   b. Number of orders placed per year.
   c. Annual inventory carrying cost.
   d. Annual ordering cost.
   e. Annual total cost.

   Answer. a. 250 units
   b. 10.4 orders per year
   c. Inventory carrying cost = $500
   d. Annual ordering cost = $520
   e. Annual total cost = $1020

11.2. If the order quantity is increased to 1200 units, recalculate problems 11.1a to 11.1e and compare the results.

11.3. A company decides to establish an EOQ for an item. The annual demand is 500,000 units, each costing $9, ordering costs are $35 per order, and inventory carrying costs are 24%. Calculate the following:

   a. The EOQ in units.
   b. Number of orders per year.
   c. Cost of ordering, cost of carrying inventory, and total cost.

   Answer. a. EOQ = 4025 units
   b. Number of orders per year = 124
   c. Annual cost of ordering = $4348
   d. Annual cost of carrying = $4347
   e. Annual total cost = $8695

11.4. A company wishes to establish an EOQ for an item for which the annual demand is $800,000, the ordering cost is $32, and the cost of carrying inventory is 20%. Calculate the following:

   a. The EOQ in dollars.
   b. Number of orders per year.
   c. Cost of ordering, cost of carrying inventory, and total cost.
   d. How do the costs of carrying inventory compare with the costs of ordering?

Answer. a. EOQ = $16,000
b. Number of orders per year = 50
c. Annual cost of ordering = $1600
Annual cost of carrying = $1600
Annual total cost = $3200
d. The costs of carrying equal the costs of ordering since the order quantity used was the EOQ.

11.5. An SKU has an annual demand of 20,000 units, each costing $15, ordering costs are $90 per order, and the cost of carrying inventory is 23%. Calculate the EOQ in units and then convert to dollars.

11.6. A company is presently ordering on the basis of an EOQ. The demand is 10,000 units a year, unit cost is $10, ordering cost is $30, and the cost of carrying inventory is 20%. The supplier offers a discount of 3% on orders of 1000 units or more. What will be the saving (loss) of accepting the discount?

Answer. Savings = $2825.45

11.7. Refer to problem 11.3. The supplier offers a 2% discount on orders of 6000 units. Calculate the purchase cost, the cost of ordering, the cost of carrying, and the total cost if orders of 6000 are placed. Compare the results and calculate the savings if the discount is taken.

11.8. The local fire department uses 10,000 alkaline flashlight batteries per year, which cost $4 each. The cost of ordering batteries is estimated to be $50. The current interest rate suggested by the city council is 25%. The sales rep has recently suggested that you could get a discount of 2% for orders of 2000 batteries at a time. Should you take advantage of this special offer?

11.9. Calculate the new lot size for the following if $K = 5$

Item	Annual Demand	$\sqrt{A_D}$	New Lot Size
1	2500		
2	900		
3	121		

Answer. Item 1 250
Item 2 150
Item 3 55

11.10. Calculate K for the following data:

Item	Annual Demand	Orders per Year	$\sqrt{A_D}$
11	$14,400	5	
2	4900	5	
3	1600	5	
Total			

Answer. $K = 15.33$

11.11. A company manufactures three sizes of lightning rods. Ordering costs and carrying costs are not known, but it is known that they are the same for each size. Each size is produced six times per year. If the demand for each size is as follows, calculate order quantities to minimize inventories and maintain the same total number of runs. Calculate the old and new average inventories. Is there any change in the number of orders per year?

Item	Annual Usage	Present Orders per Year	Present Lot Size	$\sqrt{A_D}$	New Lot Size $= K\sqrt{A_D}$	New Orders per Year $N = A_D/Q$
1	$22,500	6				
2	$7569	6				
3	$1600	6				
Total						
Average Inventory						

Answer. Average inventory with present lot sizes = $2639.09
Average inventory with new lot sizes = $2131.52

11.12. A company manufactures five sizes of screwdrivers. Ordering costs and carrying costs are not known, but it is known that they are the same for each size. At present, each size is produced four times per year. If the demand for each size is as follows, calculate order quantities to minimize inventories and maintain the same total number of runs. Calculate the old and new average inventories. Is there any change in the number of orders per year?

Item	Annual Usage	Present Orders per Year	Present Lot Size	$\sqrt{A_D}$	New Lot Size $= K\sqrt{A_D}$	New Orders per Year $N = A_D/Q$
1	$12,100	6				
2	$8100	6				
3	$3600	6				
4	$1600					
5	$225					
Total						
Average Inventory						

11.13. The EOQ for an item is 700 units, and the annual usage is 2600 units. What is the POQ?

Answer. POQ = 14 weeks

284 CHAPTER ELEVEN

11.14. Given the following net requirements, calculate the planned order receipts based on the POQ. The EOQ is 250 units, and the annual demand is 4200 units.

Week	1	2	3	4	5	6	7	8	
Net Requirements	100	85	90	0	85	80	90	100	630
Planned Order Receipts									

Answer. POQ = 3 weeks
Planned order period 1 = 275 units
Planned order period 5 = 255 units
Planned order period 8 = 100 units

11.15. Given the following MRP record and an EOQ of 200 units, calculate the planned order receipts using the EOQ. Next, calculate the period order quantities and the planned order receipts. In both cases, calculate the ending inventory and the total inventory carried over the 10 weeks.

Week	1	2	3	4	5	6	7	8	9	10	Total
Net Requirements	70	75	60	40	60	80	70	45	20	80	
Planned Order Receipt											
Ending Inventory											

Answer. POQ = 3 weeks
EOQ total ending inventory = 970 units
POQ total ending inventory = 500 units

11.16. An item has a weekly demand of 240 units throughout the year. The item has a unit value of $42 and the company uses 20% of the item value for the annual inventory cost. When ordered, the setup cost to produce an order is $600, and the production process is able to produce 500 per week and deliver them weekly as produced. What is the EOQ?

CASE STUDY 11.1

Jack's Hardware

Several years ago, Jack Adams decided he had been working as a skilled building contractor long enough. He wanted to establish himself in a more predictable and pleasant work environment. He realized that there was an opportunity in the medium-sized village where he lived, in that there was no local hardware store. People from the village would have to travel by car more than 30 minutes to find the nearest hardware store, and Jack figured many in his village (and surrounding countryside) would rather come to a local store for their needs. His knowledge of hardware gained from his years as a contractor would serve him well in the hardware business.

The venture turned out to be quite successful and his business grew. The problem he was currently facing is that the village was also growing and at least one large hardware chain company was starting to look at the village as a possible opportunity for a new store. Jack realized that if he had competition in town, he needed to evaluate his own business.

Jack had never paid a lot of attention to what he considered the "small details" of his cost structure. He knew that his prices were somewhat higher than what the chain hardware stores in the nearby city charged, but he knew that his customer still came to him because he was in their village and also because they knew that Jack had lots of knowledge about hardware and could almost always answer their questions. His customers were loyal, but he also knew that if another store opened in the village with lower prices that he would certainly lose at least some of his business. Since those chains could order in bulk and get quantity discounts, they could charge less for their products.

One of Jack's clerks had recently had experience in a warehouse setting and suggested to Jack that he might want to evaluate his inventory policies. If he could save some money with inventory, he might be able to lower his prices enough to stay reasonably price competitive, especially in view of his great reputation for customer service.

Jack had a good friend who had a daughter who was a business student studying operations management at a well-known university, so Jack asked if his friend if his daughter might be able to help. She agreed, and asked Jack to provide some basic data about two of his more important products. She said she would evaluate options for Jack, and if it turned out successful, he could potentially expand the approach to more of his items.

Here is the data that Jack provided for the two items:

ITEM A
Item cost = $2.80 each
Order cost = $15.00 per order
Inventory holding cost = 18% per year
Average demand per year = 1000
Currently (at the recommendation of the supplier) the amount ordered per order = 100

ITEM B
Item cost = $12.50 each
Order cost = $22.50 per order
Inventory holding cost = 18% per year
Average demand per year = 750
Currently (at the recommendation of the supplier) amount ordered per order = 60
Actual demand over the last 20 weeks:

ITEM A	Week	ITEM B
24	1	12
15	2	9
21	3	17
11	4	20
8	5	18
26	6	6
18	7	12
30	8	16
14	9	4
21	10	13

ITEM A	Week	ITEM B
28	11	5
7	12	19
19	13	22
23	14	8
12	15	10
29	16	11
17	17	16
9	18	13
22	19	8
13	20	19

The current inventory (at the beginning of week 1) for Item A is 30, and for Item B is 25.

The student doing the evaluation said she wanted to evaluate using and economic order quantity (EOQ) and a periodic order quantity (POQ) and compare both to the existing policy using the same criteria (the only difference being the order size). The current policy on reordering items was to order when the inventory level dropped to one week's average usage (the supplier was able to respond to all orders from Jack in about one week), but just in case that week's demand was a little higher than average, he wanted to add a bit of a buffer of 15% weekly demand. He didn't want to lose a sale if he ran out before his new order from the supplier came in. In the past he didn't care so much because most of his customers could wait for a day or two for a part that was out of stock, but Jack knew that with a potential new competitor coming to town that many of his customers might go to the new store if Jack was out of stock, and he thought that even though the 15% extra might cost him a bit more in inventory cost it was likely worth it to prevent lost sales and perhaps lost customer loyalty.

The student was a bit concerned about evaluating the POQ situation because it meant the inventory would not be monitored on a regular basis until the end of the defined -period. While that was an advantage for saving constant inventory counts, it might mean that if the demand was unusually high, they could run out of stock without even knowing it was low. Given that the items are sold on demand, there is no opportunity to "look ahead" at demand.

For the two items in question, the average weekly usage is as follows:

The average weekly usage for A = 1000/52 = 19.2 or 19 units
The average weekly usage for B = 750/52 = 14.4 or 14 units

Applying the 15% "buffer," the critical inventory values are 22 units for Item A and 16 units for Item B. Applying those values for the current lot size and for the EOQ would be simple—whenever the inventory equaled or went lower than the critical value then a replacement order would be activated.

The POQ situation was a bit different. For convenience, the student decided that when the end of the period came that they should order based on the following approach:

$$\text{Order size} = \text{EOQ} + (\text{critical inventory value}) - (\text{existing inventory at time of reorder}).$$

She decided this since the POQ is based on the EOQ, she did not want to order too much (in case there was lots of inventory at the time of review) yet she also wanted to try to keep the inventory as close to the "critical value" as possible by the end of the review period.

Assignment

1. Calculate the EOQ for both Item A and B.
2. Calculate the ordering period for each item using period order quantity (POQ) based on the EOQ.
3. For the 20 weeks of data (and the current inventory) calculate the order receipts using the current order size, the EOQ, and the POQ. Calculate the ending inventory for each week and then the total inventory carried over the 20 weeks. For the POQ calculation, assume that week 1 is the review period for calculating the order.
4. Based on the data and calculations, calculate the total cost (order cost and inventory holding cost) for each order quantity over the 20 weeks given.
5. Since the POQ is calculated based on the EOQ, why do you think the results on your charts are different from each other?
6. Based on the full 52 weeks of the year and the fact that those two items represent about 4% of the total of similar items in the store, project the potential saving using what appears to be the best approach.
7. Can you think of a better approach to use in this situation?

CHAPTER TWELVE

INDEPENDENT DEMAND ORDERING SYSTEMS

12.1 INTRODUCTION

The concept of an economic order quantity, covered in Chapter 11, addresses the question of how much to order at one time. Another important question is when to place a replenishment order. If stock is not reordered soon enough, there will be a stockout and a potential loss in customer service. However, stock ordered earlier than needed will create extra inventory. The problem is how to balance the costs of carrying extra inventory against the costs of a stockout.

No matter what the items are, some rules for reordering are needed and can be as simple as order when needed, order every month, or order when stock falls to a predetermined level. Rules are used in all parts of daily life, and they vary depending on the significance of the item. People use some intuitive rules to make up their weekly shopping list. Purchase enough food for a week, buy fuel when the tank reaches ¼ full, order holiday gifts in time for giving, purchase new clothes to begin school, and so on.

In industry there are many inventories that involve a large investment and have high stockout costs. Controlling these inventories requires effective reorder systems. Three basic systems are used to determine when to order:

- Order point system.
- Periodic review system.
- Material requirements planning (MRP).

The first two are for independent demand items; the last is for dependent demand items.

12.2 ORDER POINT SYSTEM

When the quantity of an item on hand in inventory falls to a predetermined level, called an **order point** or **reorder point (ROP)**, an order is placed. The quantity ordered is usually precalculated and based on economic order quantity concepts.

Using this technique, an order must be placed when there is enough stock on hand to satisfy demand from the time the order is placed until the new stock arrives, i.e., the **lead time**. Assume that the average demand for a particular item is 100 units a week and the lead time is four weeks. If an order is placed when there are 400 units on hand, on average there will be enough stock on hand to last until the new stock arrives. However, demand during any one lead time period probably varies from the average—sometimes more and sometimes less than the 400. Statistically, half the time the demand is greater than average, and there is a stockout; half the time the demand is less than average, and there is extra stock. If it is necessary to provide some protection against a stockout, safety stock can be added. The item is ordered when the quantity on hand falls to a level equal to the demand during the lead time plus the safety stock:

$$OP = DDLT + SS$$

where

$\quad$ OP $\quad$ = order point
$\quad$ DDLT = demand during the lead time
$\quad$ DDLT = safety stock

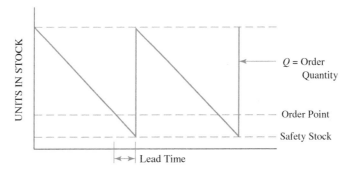

FIGURE 12.1 Quantity on hand versus time: Independent demand item.

It is important to note that it is the *demand (and the variation in demand) during the lead time that is important*. The only time a stockout is possible is during the lead time. If demand during the lead time is greater than expected, there will be a stockout unless sufficient safety stock is carried.

Example Problem

Demand is 200 units a week, the lead time is three weeks, and safety stock is 300 units. Calculate the order point.

ANSWER

$$\begin{aligned} \text{OP} &= \text{DDLT} + \text{SS} \\ &= 200 \times 3 + 300 \\ &= 900 \text{ units} \end{aligned}$$

Figure 12.1 shows the relationship between safety stock, lead time, order quantity, and order point. With the order point system:

- Order quantities are usually fixed.
- The order point is determined by the average demand during the lead time. If the average demand or the lead time changes and there is no corresponding change in the order point, effectively there has been a change in safety stock.
- The intervals between replenishment are not constant but vary depending on the actual demand during the reorder cycle.
- The average inventory for a period is equal to the opening inventory plus the ending inventory, divided by 2.

$$\begin{aligned} \text{Period opening inventory} &= \text{order quantity plus safety stock} \\ \text{Period ending inventory} &= \text{safety stock} \\ \text{Average inventory} &= \frac{\text{order quantity} + \text{safety stock} + \text{safety stock}}{2} \\ &= \frac{\text{order quantity}}{2} + \text{safety stock} \\ &= \frac{Q}{2} + \text{SS} \end{aligned}$$

Example Problem

Order quantity is 1000 units and safety stock (SS) is 300 units. What is the average inventory?

ANSWER

$$\begin{aligned} \text{Average inventory} &= \frac{Q}{2} + \text{SS} \\ &= \frac{1000}{2} + 300 \\ &= 800 \text{ units} \end{aligned}$$

Determining the order point depends on the demand during the lead time and the safety stock required.

Methods of estimating the demand during the lead time were discussed in Chapter 9. In this chapter, we discuss the factors to consider when determining safety stock.

12.3 DETERMINING SAFETY STOCK

Safety stock is intended to protect against uncertainty in supply and demand. Uncertainty may occur in two ways: quantity uncertainty and timing uncertainty. Quantity uncertainty occurs when the amount of supply or demand varies; for example, if the demand is greater or less than expected in a given period. Timing uncertainty occurs when the time of receipt of supply or demand differs from that expected. A customer or a supplier may change a delivery date, for instance.

There are two ways to protect against uncertainty: carry extra inventory, called safety stock, or order early, called safety lead time. **Safety stock** is a calculated extra amount of stock carried in inventory and is generally used to protect against quantity uncertainty. **Safety lead time** is used to protect against uncertainty in delivery lead time by planning order releases and order receipts earlier than required. Both safety stock and safety lead time result in extra inventory, but the methods of calculation are different.

Safety stock is the most common way of buffering against uncertainty and is the method described in this text. The safety stock required depends on the following:

- Variability of demand during the lead time. The higher the variability, the greater the need for safety stock.
- Frequency of reorder. If orders are placed frequently, changes to the demand or variability of the demand will be detected earlier.
- Service level desired. Higher service levels require more inventory, to accommodate periods of increased demand.
- Length of the lead time. The longer the lead time, the more safety stock has to be carried to provide a specified service level. This is one reason it is important to reduce lead times as much as possible.

Variation in Demand During Lead Time

Chapter 9 discussed forecast error and it was determined that actual demand varies from forecast for two reasons: bias error in forecasting the average demand and random variations in demand about the average. It is the latter for which safety stock should be determined.

Suppose two items, A and B, have a 10-week sales history, as shown in Figure 12.2. Average demand over the lead time of one week is 1000 per week for both

Week	Item A	Item B
1	1200	400
2	1000	600
3	800	1600
4	900	1300
5	1400	200
6	1100	1100
7	1100	1500
8	700	800
9	1000	1400
10	800	1100
Total	10,000	10,000
Average	1000	1000

FIGURE 12.2 Actual demand for two items.

items. However, the weekly demand for item A has a range from 700 to 1400 units a week and for item B the range is from 200 to 1600 units per week. The demand for B is more erratic than that for A. If the order point is 1200 units for both items, there will be one stockout for A and four for B, which will occur when the demand during any one week exceeds 1200 units. If the same service level is to be provided (the same chance of stockout for all items), some method of estimating the randomness of item demand is needed.

Variation in Demand About the Average

Over the past 100 weeks a history of weekly demand for a particular item shows an average demand of 1000 units. As expected, most of the demands are around 1000; a smaller number would be farther away from the average, and still fewer would be farthest away. If the weekly demands are classified into groups or ranges about the average, a picture of the distribution of demand about the average appears. Suppose the demand is distributed as follows:

Weekly Demand	Number of Weeks
725–774	2
775–824	3
825–874	7
875–924	12
925–974	17
975–1024	20
1025–1074	17
1075–1124	12
1125–1174	7
1175–1224	3
1225–1274	2

These data are plotted to give the results shown in Figure 12.3. This type of chart is a **histogram**.

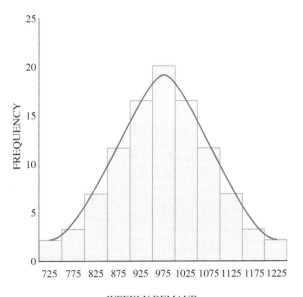

FIGURE 12.3 Histogram of actual demand.

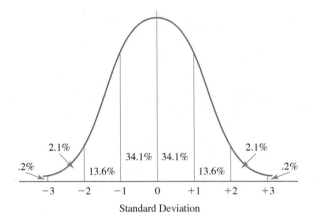

FIGURE 12.4 Normal distribution.

Normal distribution Everything in life varies, even identical twins in some respects. The pattern of demand distribution about the average will differ for different products and markets. Some method is needed to describe the distribution—its shape, center, and spread.

The shape of the histogram in Figure 12.3 indicates that although there is variation in the distribution, it follows a definite pattern, as shown by the smooth curve. Such a natural pattern shows predictability. As long as the demand conditions remain the same, the pattern can be expected to remain very much the same. If the demand is erratic, so is the demand pattern, making it difficult to predict future demand with any accuracy. Fortunately, most demand patterns are stable and predictable.

The most common predictable pattern is similar to the one outlined by the histogram in Figure 12.3 and is called a *normal curve*, or *bell curve*, because its shape resembles a bell. The shape of a perfectly normal distribution is shown in Figure 12.4.

The **normal distribution** has most of the values clustered near a central point with progressively fewer results occurring away from the center. It is symmetrical about this central point in that it spreads out evenly on both sides.

The normal curve is described by two characteristics. One relates to its central tendency, or average, and the other to the variation, or dispersion, of the actual values about the average.

Average or mean The average or **mean** value is at the high point of the curve. It is the central tendency of the distribution. The symbol for the mean is $\bar{x}$ (pronounced "x bar"). It is calculated by adding the data and dividing by the total number of data. In mathematical terms, it can be written as:

$$\bar{x} = \frac{\sum x}{n}$$

where x stands for the individual data [(in this case, the individual demands) $\sum$ (capital Greek letter sigma) is the summation sign and n is the number of data (demands)].

Example Problem
Given the following actual demands for a 10-week period, calculate the average ($\bar{x}$) of the distribution.

Period	Actual Demand
1	1200
2	1000
3	800
4	900
5	1400
6	1100
7	1100
8	700
9	1000
10	800
Total	10,000

ANSWER

$$\text{Average Demand} = \bar{x} = \frac{\sum x}{n} = \frac{10,000}{10} = 1000 \text{ units}$$

Dispersion The variation, or **dispersion**, of actual demands about the average refers to how closely the individual values cluster around the mean or average. It can be measured in several ways:

- As a range of the maximum minus the minimum value.
- As the **mean absolute deviation (MAD)**, which is a measure of the average forecast error. (Calculation of MAD was discussed in Chapter 9.)
- As a standard deviation.

Standard Deviation (σ)

The **standard deviation** is a statistical value that measures how closely the individual values cluster about the average. It is represented by the Greek letter sigma (σ). The standard deviation is calculated as follows:

1. Calculate the deviation for each period by subtracting the actual demand from the forecast demand.
2. Square each deviation.
3. Add the squares of the deviations.
4. Divide the value in step 3 by the number of periods to determine the average of the squared deviations.
5. Calculate the square root of the value calculated in step 4. This is the standard deviation (σ).

6. Most calculators and spreadsheet applications have a statistical function that can calculate sigma (σ) directly and have two variations of sigma: σ_n, which is based on a population and σ_{n-1}, which is based on a sample estimating the population. Sigma is used based on $n(\sigma_n)$. Further explanation of this difference is unfortunately beyond the scope of this text.

Note that previous versions of this text have used MAD as the measure of dispersion since it is easier to calculate manually. However, standard deviation is a more widely accepted technique, which will be used in the calculation of safety stocks. One standard deviation is approximately equal to 1.25 MADs.

It is also important to note that in the discussion of safety stock the deviations in demand are still calculated for the same time intervals as the lead time. If the lead time is one week, then the variation in demand over a one-week period is needed to determine the safety stock. Differences in the lead time and forecast interval will be discussed later in this chapter.

Example Problem

Given the data from the previous example problem, calculate the standard deviation (σ).

Period	Forecast Demand	Actual Demand	Deviation	Deviation Squared
1	1000	1200	200	40,000
2	1000	1000	0	0
3	1000	800	−200	40,000
4	1000	900	−100	10,000
5	1000	1400	400	160,000
6	1000	1100	100	10,000
7	1000	1100	100	10,000
8	1000	700	−300	90,000
9	1000	1000	0	0
10	1000	800	−200	40,000
Total	10,000	10,000	0	400,000

ANSWER

$$\text{Average of the squares of the deviation} = 400{,}000 \div 10 = 40{,}000$$
$$\text{Sigma} = \sqrt{40{,}000} = 200 \text{ units}$$

From statistics, it can be determined that:

The actual demand will be within ± 1 sigma (± 200 units) of the forecast average approximately 68% of the time.

The actual demand will be within ± 2 sigma (± 400 units) of the forecast average approximately 95% of the time.

The actual demand will be within ± 3 sigma (± 600 units) of the forecast average approximately 99.7% of the time.

Determining the Safety Stock and Order Point

Now that the standard deviation has been calculated, how much safety stock is needed can be determined.

One property of the normal curve is that it is symmetrical about the average. This means that half the time the actual demand is less than the average and half the time it is greater. Safety stocks are needed to cover only those periods in which the demand during the lead time is greater than the average. Thus, a service level of 50% can be attained with no safety stock. If a higher service level is needed, safety stock must be provided to protect against those times when the actual demand is greater than the average.

As stated, statistics have shown that the error is within ±1 sigma of the forecast about 68% of the time (34% of the time less and 34% of the time greater than the forecast).

Suppose the standard deviation of demand during the lead time is 100 units and this amount is carried as safety stock. This much safety stock provides protection against stockout for the 34% of the time that actual demand is greater than expected. In total, there is enough safety stock to provide protection for the 84% of the time (50% + 34% = 84%) that a stockout is possible.

The service level is a statement of the percentage of time there is no stockout. But what exactly is meant by supplying the customer 84% of the time? It means being able to supply when a stockout is possible, and a stockout is possible only during the time interval between when an order is to be placed and when the replenishment is received. If an order is placed 100 times a year, there are 100 chances of a stockout. With safety stock equivalent to one standard deviation, on the average one would expect no stockouts about 84 of the 100 times.

It should be noted that there are other definitions and interpretations of the concept of service level, but a complete treatment of additional interpretations will yield fairly small variations in the safety stock and are beyond the scope of this text.

Example Problem

Using the figures in the last example problem, in which the standard deviation was calculated as 200 units, and assuming the lead time to be one period,

a. Calculate the safety stock and the order point for an 84% service level.

b. If a safety stock equal to two standard deviations is carried, calculate the safety stock and the order point.

ANSWER

a. Safety stock = 1 sigma
    = 1 × 200
    = 200 units
   Order point = DDLT + SS
    = 1000 + 200 = 1200 units

where DDLT and SS are as defined previously. With this order point and level of safety stock, on the average there are no stockouts 84% of the time when a stockout is possible.

b. SS = 2 × 200
    = 400 units
   OP = DDLT + SS
    = 1000 + 400
    = 1400 units

Similar to using the demand during lead time approach to calculating how much inventory to keep on hand, **days of supply**, as discussed in Chapter 10, can also be used.

$$\text{Days of supply} = \frac{\text{inventory on hand}}{\text{average daily usage}}$$

Service Level (%)	Safety Factor
50	0.00
75	0.67
80	0.84
85	1.04
90	1.28
94	1.56
95	1.65
96	1.75
97	1.88
98	2.05
99	2.33
99.86	3.00
99.99	4.00

FIGURE 12.5 Table of safety factors.

A decision would be made by the inventory management team on how many days of product or raw materials should be stocked to hedge against uncertainty, and this is then used to set safety stock levels. This can be a static number, using average daily usage, or a more dynamic number, using the demand driven through the enterprise resource planning (ERP) system. The latter allows safety stock for an item to decrease or increase based on future demand, and prevents the overstocking of inventory when demand is decreasing.

Safety factor The service level is directly related to the number of standard deviations provided as safety stock and is usually called the **safety factor**. In earlier discussions, the normal curve has been used to determine the service level produced by one, two, or three standard deviations. Safety factors are the inverse of this procedure, wherein one can determine a desired service level and look up the corresponding number of standard deviations required.

Figure 12.5 shows safety factors for various service levels. Note that the service level is the percentage of order cycles without a stockout. For values not shown on the table, a close approximation of the safety factor can be made by interpolating the factors given. For example, to find the safety factor for a desired service level of 77%, calculate the average safety factor for a 75% service level (0.67) and an 80% service level (0.84). The safety factor for a 77% service level would be approximately

$$\frac{0.67 + 0.84}{2} = 0.76$$

Example Problem

If the standard deviation is 200 units, what safety stock should be carried to provide a service level of 90%? If the expected demand during the lead time is 1500 units, what is the order point?

Answer

From Figure 12.5, the safety factor for a service level of 90% is 1.28. Therefore,

$$\begin{aligned}
\text{Safety stock} &= \sigma \times \text{safety factor} \\
&= 200 \times 1.28 \\
&= 256 \text{ units} \\
\text{Order point} &= \text{DDLT} + \text{SS} \\
&= 1500 + 256 \\
&= 1756 \text{ units}
\end{aligned}$$

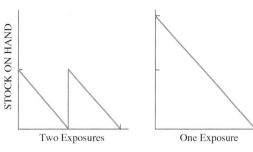

FIGURE 12.6 Exposures to stockout.

12.4 DETERMINING SERVICE LEVELS

Theoretically, a company wants to carry enough safety stock on hand so the cost of carrying the extra inventory plus the cost of stockouts is a minimum. Stockouts cost money for the following reasons:

- Backorder costs.
- Lost sales, present or future.
- Lost customers.

The cost of a stockout varies depending on the item, the market served, the customer, and competition. In some markets, customer service is a major competitive tool, and a stockout can be very expensive. In others, it may not be a major consideration. Stockout costs are difficult to establish due to the intangible nature of lost sales or customers. Usually, the decision about what the service level should be is a senior management decision and is part of the company's corporate and marketing strategy. As such, it is beyond the scope of this text.

The only time it is possible for a stockout to occur is when stock is running low, and this happens every time an order is to be placed. Therefore, the chances of a stockout are directly proportional to the frequency of reorder. The more often stock is reordered, the more often there is a chance of a stockout. Figure 12.6 shows the effect of the order quantity on the number of exposures per year. Note also that when the order quantity is increased, exposure to stockout decreases. The safety stock needed decreases, but because of the larger order quantity, the average inventory increases. When a company pursues a lean production approach, the tactics to reduce average inventory tends to drive the order quantity down significantly as a result of reducing order cost. While this will often increase the exposures to stockout by a large number, other tactics are used to minimize the risks associated with those stockouts. Lean production is discussed more completely in Chapter 16.

It is the responsibility of management to determine the number of stockouts per year that are tolerable. Then the service level, safety stock, and order point can be calculated.

Example Problem

Suppose management stated that it could tolerate only one stockout per year for a specific item.

For this particular item, the annual demand is 52,000 units, it is ordered in quantities of 2600, and the standard deviation of demand during the lead time is 100 units. The lead time is one week. Calculate:

a. Number of orders per year.

b. Service level.

c. Safety stock.

d. Order point.

Answer

a. Number of orders per year $= \dfrac{\text{annual demand}}{\text{order quantity}}$

$= \dfrac{52{,}000}{2600} = 20$ times per year

b. Since one stockout per year is tolerable, there must be no stockouts 19 (20 − 1) times per year.

$$\text{Service level} = \frac{20 - 1}{20} = 95\%$$

c. From Figure 12.5

$$\text{Safety factor} = 1.65$$
$$\text{Safety stock} = \text{safety factor} \times \sigma$$
$$= 1.65 \times 100 = 165 \text{ units}$$

d. Demand during lead time (1 week) $= \frac{(52,000)}{52} = 1000$ units

$$\text{Order point} = \text{demand during lead time} + \text{SS}$$
$$= 1000 + 165 = 1165 \text{ units}$$

12.5 DIFFERENT FORECAST AND LEAD TIME INTERVALS

Usually, there are many items in an inventory, each with different lead times. Records of actual demand and forecasts are normally made on a weekly or monthly basis for all items regardless of what the individual lead times are. It is almost impossible to measure the variation in demand about the average for each of the lead times, so some method of adjusting standard deviation for the different time intervals is needed.

If the lead time is zero, the standard deviation of demand is zero. As the lead time increases, the standard deviation increases. However, it will not increase in direct proportion to the increase in time. For example, if the standard deviation is 100 for a lead time of one week, then for a lead time of four weeks it will not be 400, since it is very unlikely that the deviation would be high for four weeks in a row. As the time interval increases, there is a smoothing or canceling effect, and the longer the time interval, the more smoothing takes place.

The following adjustment can be made to the standard deviation or the safety stock to compensate for differences between lead time interval (LTI) and forecast interval (FI). It states that the standard deviation changes as the square root of the change in the interval. Although not exact, the formula gives a good approximation.

$$\text{sigma for LTI} = (\text{sigma for FI})\sqrt{\frac{\text{LTI}}{\text{FI}}}$$

Example Problem

The forecast interval is four weeks, the lead time interval is two weeks, and sigma for the forecast interval is 150 units. Calculate the standard deviation for the lead time interval.

Answer

$$\text{sigma for LTI} = 150\sqrt{\frac{2}{4}} = 150 \times 0.707 = 106 \text{ units}$$

The preceding relationship is also useful where there is a change in the LTI. Now it is probably more convenient to work directly with the safety stock rather than the standard deviation. The relationship is as follows:

$$\text{New safety stock} = \text{old safety stock}\sqrt{\frac{\text{new interval}}{\text{old interval}}}$$

Example Problem

The safety stock for an item is 150 units, and the lead time is two weeks. If the lead time changes to three weeks, calculate the new safety stock.

Answer

$$\text{SS (new)} = 150\sqrt{\frac{3}{2}}$$
$$= 150 \times 1.22 = 183 \text{ units}$$

It should also be noted that as the lead time is decreased, the amount of safety stock is also decreased. This is one of the many direct benefits of just-in-time and lean production discussed in Chapter 16.

12.6 DETERMINING WHEN THE ORDER POINT IS REACHED

There must be some method to show when the quantity of an item on hand has reached the order point. In practice, there are many systems, but they all are inclined to be variations or extensions of three basic methods: the two-bin inventory system, kanbans, and perpetual inventory records.

Two-Bin Inventory System

A quantity of an item equal to the order point quantity is set aside, frequently in a separate or second bin, and not touched until all the main stock is used up. When the stock in the second bin needs to be used, the production control or purchasing department is notified and a replenishment order is placed. There are variations on this system, such as the red tag system, where a tag is placed in the stock at a point equal to the order point.

The **two-bin inventory system** is a simple way of keeping control of C items. Because they are of low value, it is best to spend the minimum amount of time and money controlling them. However, they do need to be managed, and someone should be assigned to ensure that when the reserve stock is reached an order must be placed. When it is out of stock, a C item can become an A item.

Kanbans

The use of **kanbans** is a simple technique that signals the need for more product. It normally consists of a card or ticket that has information on the item and the quantity to be produced. It avoids the need for formal record keeping and, like the two-bin system, makes a visual signal of the need for more product when the inventory falls below a preset level. It is used to replenish all items and not just low-value C items.

Bookstores frequently use this system. A tag or card is placed in a book that is in a stack in a position equivalent to the order point. When a customer takes that book to the checkout, the store is effectively notified that it is time to reorder that title.

Kanbans are discussed in detail in Chapter 16.

Perpetual Inventory Record

A **perpetual inventory record** is a continual account of inventory transactions as they occur. At any instant, it holds an up-to-date record of transactions. At a minimum, it contains the balance on hand, but it may also contain the quantity on order but not received, the quantity allocated but not issued, and the available balance. The accuracy of the record depends upon the speed with which transactions are recorded and the accuracy of the input. Because manual systems rely on the input of humans, they are more likely to have slow response and inaccuracies. Computer-based inventory systems such as ERP have a higher transaction speed and reduce the possibility of human error.

An inventory record contains variable and permanent information. Figure 12.7 shows an example of a perpetual inventory record.

Permanent or static information that does not change frequently is shown at the top of Figure 12.7. Any alteration is usually the result of an engineering change, manufacturing process change, or inventory management change. It includes data such as the following:

- Part number, name, and description.
- Order point.
- Order quantity.
- Lead time.
- Safety stock.

426254 SCREW					
ORDER POINT = 100 ORDER QUANTITY = 500			LEAD TIME = 2 WEEKS SAFETY STOCK = 50		
DATE	ORDERED	RECEIVED	ISSUED	ON HAND	ALLOCATED	AVAILABLE
01				500		500
02				500	400	100
03	500			500		100
04			400	100		100
05		500		600		600

FIGURE 12.7 Perpetual inventory record.

Variable or dynamic information changes with each transaction and includes the following:

- Quantities ordered: dates, order numbers, and quantities.
- Quantities received: dates, order numbers, and quantities.
- Quantities issued: dates, order numbers, and quantities.
- Balance on hand.
- Allocated: dates, order numbers, and quantities.
- Available balance.

The information depends on the needs of the company and the particular circumstances.

12.7 PERIODIC REVIEW SYSTEM

In the order point system, an order is placed when the quantity on hand falls to a predetermined level called the order point. The quantity ordered is usually predetermined on some basis such as the economic order quantity. The interval between orders varies depending on the demand during any particular cycle. In most retail environments, such as supermarkets or big-box stores, there are thousands of items, and each item could be reaching its order point at any time. It is impractical to place many individual orders and it is anticipated that many items will be delivered on one truck at a time. In this way, the ordering cost, which includes the price of transportation, is spread over many small items. The periodic review system makes the timing of each order a regular interval.

Using the **periodic review system**, the quantity on hand of a particular item is determined at specified, fixed-time intervals, and an order is placed. Figure 12.8 illustrates this system.

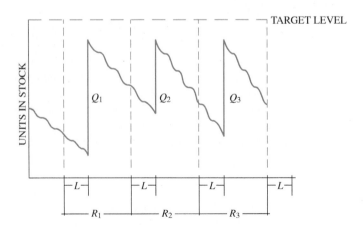

FIGURE 12.8 Periodic review system: Units in stock versus time.

Figure 12.8 shows that the review intervals (t_1, t_2, and t_3) are equal and that Q_1, Q_2, and Q_3 are not necessarily the same. Thus, the review period is fixed, and the order quantity is allowed to vary. The quantity on hand plus the quantity ordered must be sufficient to last until the next shipment is received. That is, the quantity on hand plus the quantity ordered must equal the sum of the demand during the lead time plus the demand during the review period plus the safety stock.

Target or Maximum Inventory Level

In the periodic review technique, the variable is the order quantity. The objective is to restock the inventory to a **target inventory level** or **maximum inventory**, which is the quantity equal to the demand during the lead time plus the demand during the review period plus safety stock:

$$T = D(R + L) + SS$$

where

T = target (maximum) inventory level
D = demand per unit of time
L = lead time duration
R = review period duration
SS = safety stock

Note that, as illustrated in Figure 12.8, inventory rarely actually reaches the target or maximum level inventory since product will continue to be consumed during the lead time interval.

The order quantity is equal to the maximum inventory level minus the quantity on hand at the review period:

$$Q = T - I$$

where

Q = order quantity
I = inventory on hand

The periodic review system is useful for the following:

- Where there are many small issues from inventory, and posting transactions to inventory records are very expensive. Supermarkets and retailers are in this category.
- Where ordering costs per item need to be kept small. This occurs when many different items are ordered from one source. A regional distribution center may order most or all of its stock from a central warehouse in regular, large, multi-item shipments.
- Where many items are ordered together to make up a production run or fill a truckload. A good example of this is a regional distribution center that orders a truckload once a week from a central warehouse.

Example Problem

A hardware company stocks nuts and bolts and orders them from a local supplier once every two weeks (10 working days). Lead time is two days. The company has determined that the average demand for ½-inch bolts is 150 per week (five working days), and it wants to keep a safety stock of three days' supply on hand. An order is to be placed this week, and stock on hand is 130 bolts.

a. What is the target level?

b. How many ½-inch bolts should be ordered this time?

ANSWER

Let

D = demand per unit of time = $150 \div 5$ = 30 per working day
L = lead time duration = 2 days
R = review period duration = 10 days
SS = safety stock = 3 days' supply = 3×30 = 90 units
I = inventory on hand = 130 units

Then,

a. Target level $T = D(R + L) + SS$
$= 30(10 + 2) + 90$
$= 450$ units

b. Order quantity $Q = T - I$
$= 450 - 130 = 320$ units

12.8 DISTRIBUTION INVENTORY

Distribution inventory includes all the finished goods held anywhere in the distribution system. The purpose of holding inventory in distribution centers is to improve customer service by locating stock near the customer and to reduce transportation costs by allowing the manufacturer to ship full loads rather than partial loads over long distances. This will be studied in Chapter 14.

The objectives of distribution inventory management are to provide the required level of customer service, to minimize the costs of transportation and handling, and to be able to interact with the factory to minimize scheduling problems.

Distribution systems vary considerably, but in general they have a central supply facility that is supported by a factory, a number of distribution centers, and, finally, customers. Figure 12.9 shows a representation of such a system. The customers may be the final consumer or some intermediary in the distribution chain.

Unless a firm delivers directly from factory to customer, demand on the factory is created by central supply. In turn, demand on central supply is created by the distribution centers. This can have severe repercussions on the pattern of demand on central supply and the factory. Although the demand from customers may be relatively uniform, the demand on central supply is not, because it is dependent on when the distribution centers place replenishment orders. In turn, the demand on the factory depends on when central supply places orders. Figure 12.10 shows the distribution inventory process.

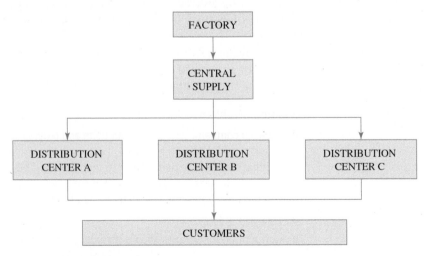

FIGURE 12.9 A distribution system.

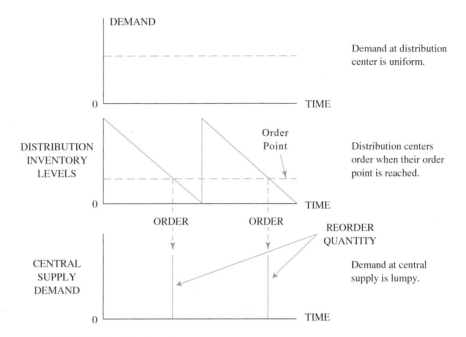

FIGURE 12.10 Distribution inventory.

Notice in Figure 12.10 how continuous, steady demand at the distribution center has changed, due to lot sizing into discontinuous or lumpy demand at the central supply. The distribution system is the factory's customer, and the way the distribution system interfaces with the factory has a significant effect on the efficiency of factory operations.

Distribution inventory management systems can be classified into decentralized, centralized, and distribution requirements planning.

Decentralized Inventory Control

In **decentralized inventory control**, each distribution center first determines what it needs and when, and then places orders to central supply. Each center orders on its own to service local demand without regard for the needs of other centers, available inventory at central supply, or the production schedule of the factory.

The advantage of the decentralized system is that each center can operate on its own and thus reduce communication and coordination expense. The disadvantage is the lack of coordination and the effect this may have on overall inventories, customer service, and factory schedules. Because of these deficiencies, some distribution systems have moved toward more central control.

A number of ordering systems can be used at each distribution center, including the order point and periodic review systems. The decentralized system is sometimes called the pull system because orders are placed on central supply and pulled through the system.

Centralized Inventory Control

In **centralized inventory control**, all forecasting and order decisions are made centrally. Stock is pushed out into the system from central supply. Distribution centers have no say about what they receive.

Different ordering systems can be used, but generally an attempt is made to replace the stock that has been sold and to provide for special situations such as seasonality or sales promotions. These systems attempt to balance the available inventory with the needs of each distribution center.

The advantage of these systems is the coordination between factory, central supply, and distribution center needs. The disadvantage is the inability to react to local demand, thus lowering the level of service.

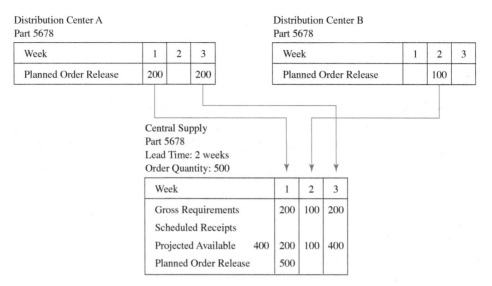

FIGURE 12.11 Distribution requirements planning.

Distribution Requirements Planning

Distribution requirements planning (DRP) is a system that forecasts when the various demands will be made by the system on central supply. This gives central supply and the factory the opportunity to plan for the goods that will actually be needed and when. It is able both to respond to local customer demand and coordinate planning and control.

The system translates the logic of material requirements planning to the distribution system. Planned order releases from the various distribution centers become the input to the material plan of central supply. The planned order releases from central supply become the forecast of demand for the factory master production schedule. Figure 12.11 shows the system schematically. The records shown are all for the same part number.

Example Problem

A company making lawnmowers has a central supply attached to its factory and two distribution centers. Distribution center A forecasts demand at 25, 30, 55, 50, and 30 units over the next five weeks and has 100 lawnmowers in transit that are due in week 2. The transit time is two weeks, the order quantity is 100 units, and there are 50 units on hand. Distribution center B forecasts demand at 95, 85, 100, 70, and 50 over the next five weeks. Transit time is one week, the order quantity is 200 units, and there are 100 units on hand. The central warehouse has a lead time of two weeks, the order quantity is 500 units, and there are 400 on hand. Calculate the gross requirements, projected available, and planned order releases for the two distribution centers, and the gross requirements, projected available, and planned order releases for the central warehouse.

Answer

Distribution Center A

Transit Time: 2 weeks

Order Quantity: 100 units

Week	1	2	3	4	5
Gross Requirements	25	30	55	50	30
In Transit		100			
Projected Available 50	25	95	40	90	60
Planned Order Release		100			

Distribution Center B
Transit Time: 1 week
Order Quantity: 200 units

Week	1	2	3	4	5
Gross Requirements	95	85	100	70	50
In Transit					
Projected Available 100	5	120	20	150	100
Planned Order Release	200		200		

Central Supply
Lead Time: 2 weeks
Order Quantity: 500 units

Week	1	2	3	4	5
Gross Requirements	200	100	200		
Scheduled Receipts					
Projected Available 400	200	100	400		
Planned Order Release	500				

SUMMARY

This chapter addresses the problem of when to order inventory items to fulfill customer and production demand. Various methods for determining order timing exist, such as order point or periodic review. Ordering systems need to ensure orders are placed in time to avoid running out of stock. Statistical applications allow a prediction of the demand during the critical lead time to establish safety stocks. The order point uses this safety stock plus anticipated demand during the lead time to ensure acceptable service to the customer. Distribution systems can take advantage of MRP-type logic to optimize order quantities all the way through the supply chain. Inventory and ordering systems are all affected by the lead time.

KEY TERMS

Centralized inventory control 303
Days of supply 295
Decentralized inventory control 303
Dispersion 293
Distribution requirements planning (DRP) 304
Histogram 291

Kanban 299
Lead time 288
Maximum inventory 301
Mean 292
Mean absolute deviation (MAD) 293
Normal distribution 292
Order point 288

Periodic review system 300
Perpetual inventory record 299
Reorder point (ROP) 288
Safety factor 296
Safety lead time 290

Safety stock 290
Standard deviation 293
Target inventory level 301
Two-bin inventory system 299

QUESTIONS

1. What are independent demand items? What two basic ordering systems are used for these items? What are dependent demand items? What system should be used to order these items?
2. a. Using the order point system, when must an order be placed?
   b. Why is safety stock carried?
   c. What is the formula for the order point?
   d. On what two things does the order point depend?
   e. Why is the demand during the lead time important?
3. What are four characteristics of the order point system?
4. What are the four factors that can influence the amount of safety stock that should be carried? How does the length of the lead time affect the safety stock carried?
5. What is a normal distribution? What two characteristics define it? Why is it important in determining safety stock?
6. What is the standard deviation of demand during the lead time? If the standard deviation for the lead time interval is 100 units, what percentage of the time would the actual demand be equal to ± 100 units? To ± 200 units? To ± 300 units?
7. What is service level?
8. What are the three categories of stockout costs? What do these costs depend upon in any company?
9. Why does the service level depend upon the number of orders per year?
10. If the lead time increases from one to four weeks, will the standard deviation of demand during the lead time increase four times? If not, why not?
11. Describe the two-bin inventory system.
12. What kinds of information are shown on a perpetual inventory record?
13. What are the differences between the order point system and the periodic review system regarding when orders are placed and the quantity ordered at any one time?
14. Define how the target inventory level is used in the periodic review system.
15. Describe how changes in demand will affect the order quantity when using the period review system.
16. What are the objectives of distribution inventory management?
17. If a factory does not supply the customer directly, from where does demand on the factory come? Is it independent or dependent demand?
18. Describe distribution requirements planning (DRP).
19. Grocery stores are an example of well-controlled inventory and replenishment systems. Describe in your own words examples of safety stock, the two-bin system, and the periodic review system, including target level, review period, replenishment period, and order quantity.
20. Describe the advantages and disadvantages of a decentralized distribution system.
21. Describe how the pattern of continuous demand changes from a distribution center to central supply.

PROBLEMS

12.1. For a particular SKU, the lead time is four weeks, the average demand is 150 units per week, and safety stock is 100 units. What is the average inventory if 1600 units are ordered at one time? What is the order point?

Answer. Average inventory = 900 units

Order point = 700 units

12.2. For a particular SKU, the lead time is seven weeks, the average demand is 80 units a week, and safety stock is 200 units. What is the average inventory if 10 weeks' supply is ordered at one time? What is the order point?

12.3. Given the following data, calculate the average x of the distribution and the standard deviation (σ).

Period	Actual Demand	Deviation	Deviation Squared
1	500		
2	600		
3	425		
4	450		
5	600		
6	575		
7	375		
8	475		
9	525		
10	475		
Total			

Answer. Average demand $\bar{x}$ = 500 units
Sigma = 68.92 units

12.4. Given the following data, calculate the average demand and the standard deviation.

Period	Actual Demand	Deviation	Deviation Squared
1	1700		
2	2100		
3	1900		
4	2200		
5	2000		
6	1800		
7	2100		
8	2300		
9	2100		
10	1800		
Total			

12.5. If sigma is 130 units, and the demand during the lead time is 250 units, calculate the safety stock and order point for:

a. A 50% service level.

b. An 85% service level.

Use the table in Figure 12.5 as help to calculate your answer.

Answer. a. Safety stock = zero

Order point = 250 units

b. Safety stock = 135 units

Order point = 385 units

12.6. The standard deviation of demand during the lead time is 80 units.

a. Calculate the safety stock required for the following service levels: 75%, 80%, 85%, 90%, 95%, and 99.99%.

b. Calculate the *change* in safety stock required to increase the service levels from 75 to 80%, 80 to 85%, 85 to 90%, 90 to 95%, and 95 to 99.99%. What conclusion do you reach?

12.7. For an SKU, the standard deviation of demand during the lead time is 150 units, the annual demand is 10,000 units, and the order quantity is 750 units. Management says it will tolerate only one stockout per year. What safety stock should be carried? What is the average inventory? If the lead time is three weeks, what is the order point?

Answer. Safety stock 213 units

Average inventory = 588 units

Order point 790 units

12.8. A company stocks an SKU with a weekly demand of 210 units and a lead time of three weeks. Management will tolerate one stockout per year. If sigma for the lead time is 175 and the order quantity is 900 units, what is the safety stock, the average inventory, and the order point?

12.9. A company stocks an SKU with a weekly demand of 600 units and a lead time of four weeks. Management will tolerate one stockout per year. If sigma for the lead time is 100 and the order quantity is 2500 units, what is the safety stock, the average inventory, and the order point?

12.10. If the standard deviation is calculated from weekly demand data at 100 units, what is the equivalent sigma for a three-week lead time?

Answer. 173 units

12.11. If the safety stock for an item is 200 units and the lead time is three weeks, what should the safety stock become if the lead time is extended to five weeks?

Answer. 258 units

12.12. If the weekly standard deviation is 140 units, what is it if the lead time is three weeks?

12.13. The safety stock on an SKU is set at 240 units. The supplier says it can reduce the lead time from eight to five weeks. What should be the new safety stock?

Answer. 190 units

12.14. The safety stock on an SKU is set at 500 units. The supplier says it has to increase the lead time from four to five weeks. What should be the new safety stock?

12.15. Management has stated that it will tolerate one stockout per year. The forecast of annual demand for a particular SKU is 100,000 units, and it is ordered in quantities of 10,000 units. The lead time is two weeks. Sales history for the past 10 weeks follows. Calculate:

a. Sigma for the demand history time interval.

b. Sigma for the lead time interval.

c. The service level.

d. The safety stock required for this service level.

e. The order point.

Week	Actual Demand	Deviation	Deviation Squared
1	2100		
2	1600		
3	2700		
4	1400		
5	1800		
6	2400		
7	2100		
8	1600		
9	2100		
10	2200		
Total			

12.16. If in problem 12.15, management said that it is considering increasing the service level to one stockout every two years, what would the new safety stock be? If the cost of carrying inventory on this item is $10 per unit per year, what is the cost of increasing the inventory from one stockout per year to one every two years?

12.17. The annual demand for an item is 10,000 units, the order quantity is 250, and the service level is 90%. Calculate the probable number of stockouts per year.

Answer. 4 stockouts per year

12.18. A company that manufactures stoves has one plant and two distribution centers (DCs). Given the following information for the two DCs, calculate the gross requirements, projected available, and planned order releases for the two DCs and the gross requirements, projected available, and planned order releases for the central warehouse.

Distribution Center A
Transit Time: 2 weeks
Order Quantity: 100 units

Week	1	2	3	4	5
Gross Requirements	50	50	85	50	110
In Transit		100			
Projected Available 75					
Planned Order Release					

Distribution Center B
Transit Time: 1 week
Order Quantity: 200 units

Week	1	2	3	4	5
Gross Requirements	120	110	115	100	105
In Transit	200				
Projected Available 50					
Planned Order Release					

Center Supply
Lead Time: 2 weeks
Order Quantity: 500 units

Week	1	2	3	4	5
Gross Requirements					
Scheduled Receipts					
Projected Available 400					
Planned Order Release					

Answer. Planned order release from central supply: 500 in week 2.

12.19. A company that manufactures snow shovels has one plant and two distribution centers (DCs). Given the following information for the two DCs, calculate the gross requirements, projected available, and planned order releases for the two DCs and the gross requirements, projected available, and planned order releases for the central warehouse.

Distribution Center A

Transit Time: 2 weeks

Order Quantity: 500 units

Week	1	2	3	4	5
Gross Requirements	300	200	150	275	300
In Transit	500				
Projected Available 200					
Planned Order Release					

Distribution Center B

Transit Time: 2 weeks

Order Quantity: 200 units

Week	1	2	3	4	5
Gross Requirements	50	75	100	125	150
In Transit					
Projected Available 150					
Planned Order Release					

Central Supply

Lead Time: 1 week

Order Quantity: 600 units

Week	1	2	3	4	5
Gross Requirements					
Scheduled Receipts					
Projected Available 400					
Planned Order Release					

12.20. A firm orders a number of items from a regional warehouse every two weeks. Delivery takes one week. Average demand is 300 units per week, and safety stock is held at two weeks' supply.

a. Calculate the target level.

b. If 600 units are on hand, how many should be ordered?

Answer. Target level = 1500 units

Order quantity = 900 units

12.21. A regional warehouse orders items once a week from a central warehouse. The truck arrives three days after the order is placed. The warehouse operates five days a week. For a particular brand and size of chicken soup, the demand is fairly steady at 20 cases per day. Safety stock is set at two days' supply.

a. What is the target level?

b. If the quantity on hand is 90 cases, how many should be ordered?

12.22. A small hardware store uses a periodic review system to control inventory and calculate order quantities. Inventory is reviewed every two weeks and there is a one-week lead time for delivery. Safety stock is set at one week's supply. The following table shows the products and the current on-hand inventory. Calculate the target level and order quantity for each of the items. Use 50 weeks per year for your calculations.

Item	Annual Demand	Target Level	On-Hand	Order Quantity
Nut – 6 mm	500		22	
Nut – 8 mm	750		54	
Bolt – 6 mm	200		0	
Bolt – 8 mm	100		6	
Screw #8 – 30 mm	250		12	
Screw #8 – 40 mm	200		8	
Washer – 8 mm	380		20	
Washer – 10 mm	100		5	
Pin – Split	400		40	

12.23. Using the data from problem 12.22, calculate the order quantities that would result if the store switched to a weekly review of inventory. What effect does more frequent ordering have on order quantities?

CASE STUDY 12.1

Carl's Computers

There was no question about Carl's genius. In the late 1980's he decided to enter the competitive nightmare that the personal computer business had become. Although on the surface that appeared to be a rather non-genius-like move, the genius came in the unique designs and features that he developed for his computer. He also figured a way to promise delivery in only two days for the local and regional market. Other computer makers also had rapid production and delivery, but they were national competitors, and the delivery time from distant locations generally made Carl able to outcompete them on delivery.

Carl soon had a loyal following, especially among the many small businesses in the area. Not only could Carl deliver quickly, but he also had very rapid service to deal with any technical problems. That service feature became critical for the local businesses whose very livelihood depended on the computers, and soon that rapid service capability became more important than the initial product delivery. Since most of these businesses were fairly small, they could not afford to have their own in-house computer experts, so they depended heavily on Carl.

The Current Situation

All was not totally rosy at Carl's Computers, however. Recently they had hired Rosa Chang for the newly developed position of inventory manager for aftermarket service. In the first week Rosa got a good idea of the challenges facing her after she interviewed several of the people at Carl's.

> DEONTE SMITH, CUSTOMER SERVICE MANAGER: "I'm not sure what you need to do, but whatever it is needs to be done fast! At this point our main competitive edge other than product delivery is service response, and I'm always hearing that we can't get a unit in the field serviced because some critical part is missing. Both the customers and the field service people are complaining about it. They make a service call, find out they need a certain part, but in many cases we're out of the part. The customers tend to be fairly loyal, but their patience is wearing thin—our policy is to provide at least a 98% customer service level, and we're not even close. That's not the only problem, though. Since our service is declining, the customers are looking more closely at our prices. I'd like to cut them a break, but our financial people tell us our margins are already too thin, and get this—one major reason is that our inventory and associated inventory costs are too high! It looks to me as if we have a very large amount of the wrong stuff here. I don't know that for sure, but I sure hope you can find a solution, and fast!"
>
> ELLEN BEDROSIAN, CHIEF ENGINEER: "Boy, am I glad you're here! The inventory problems are killing us in engineering. Carl's has always been known for unique designs, and we've been trying hard to keep ahead of the competitive curve on that issue. The problem is that most of the time when we push hard to get a new design out, the inventory and financial people tell us we have to wait. It seems like they always have too much of the old design inventory around, and the financial 'hit' to make it immediately obsolete would be too severe. We're told that as soon as we announce a new design, many of our customers would want it, so that tends to make most existing old design material—even for service—obsolete. We try to tell the service inventory people when we have a new design coming so they can use up the old material, but somehow it never seems to work out."
>
> CARLOS PEREZ, PURCHASING MANAGER: "Well, Rosa, I wish you luck—you'll need it. I'm getting pressure from so many directions, sometimes I don't know how to respond. First, the financial people are always telling me to cut or control costs. The engineers then are always coming out with new designs, most of which represent purchased parts. A lot of our time is spent working with suppliers on the new designs, while trying to get them to have very rapid delivery with low prices. Although most can live with that, where we really jerk them around is with the changes in orders. One minute our field service people tell us they've run out of

something and they need delivery immediately. In many cases they don't even have an order for that part on the books. The next thing you know they want us to cancel an order for something that only a day before they said was critical. Our buyers and suppliers are good, but they're not miracle workers and they can't do everything at once. Some of our suppliers are even threatening to refuse our business if we don't get our act together. We've tried to offer solutions for the field service people, but nothing seems to work. Maybe they just don't care."

CHANDRA DAS, CHIEF FINANCIAL OFFICER: "If you can help us with this inventory problem, you'll be well worth your salary, and then some! Here we are being competitively crunched for price, delivery, and efficient service, and our service inventory costs seem to have gone completely out of control. The total inventory has climbed more than 200% in the last two years, while our service revenues have only grown 15%. On top of that, we have had an increase in obsolete material write-off of 80% in that same two-year period. In addition, significant inventory-related costs have come from expediting. Premium freight shipments, such as flying in parts, caused by critical part shortages cost us over $67,000 last year alone. Do you realize that represents almost 20% of our gross profit margin from the service business? With our interest rates, warehousing, and obsolete inventory costs, we recognize a 23% inventory holding cost. Given our huge inventory level, that takes another big bite out of profits. All this suggests to me we need to get control of the situation or we may find ourselves out of business!"

JAMAYA KNOWLES, FIELD SERVICE SUPERVISOR: "Until they hired you, the other production supervisor and I had been in charge of inventory. I hate to discourage you, but it looks like an impossible job. The purchasing people bought a bunch of standard-size bins, and they told us that as soon as we had a week's average part usage for each part, we should order more—specifically, 'enough to fill up the bin.' Since most of their lead times were a week or less, it sure made sense. All the records were kept on computer; therefore, the computer could be programmed to tell us when we had only the week's supply. It made great sense to me, but something kept going wrong. First, field service technicians seemed to frequently grab parts without filling out a transaction. That made our records go to pot. As a matter of fact, we had a complete physical inventory a couple of months ago, and it showed our records to be less than 30% accurate! I suspect our records are almost that bad again, and we don't have another physical inventory scheduled for another nine months."

"Second, with our records so bad, the field service technicians can never tell if we really have the parts or not. Several of them have started to take large quantities of critical parts and are keeping their own inventory. When it comes time to replace their own 'private stock,' they take a bunch more. That has made the demand on the central inventory appear very erratic. One day we have plenty, and the next day we're out! You can imagine how happy purchasing is when the first time they see a purchase order that is requesting an immediate urgent shipment. We've made a policy that the technicians are only supposed to have a few specifically authorized parts with them, but I'm sure many of the technicians are violating that policy big time."

QUENTIN BATES, FIELD SERVICE TECHNICIAN: "Something is drastically wrong with our inventory, and it's driving me and the other techs crazy. We're not supposed to keep much inventory with us, only a few commonly used parts. If we have a field problem requiring a part, we're supposed to be getting it from the central inventory. Problem is, much of the time it's not there. We have to take time to pressure purchasing for it, and then have to try to calm our customers while we wait for delivery. In the meantime, the customers' systems are often unusable, and they're losing business. It doesn't take too long before they're really mad at us. I guess the people at purchasing don't care, since we have to take all the heat. Lately, I've been taking and keeping a bunch of parts I'm not really supposed to have in my inventory, and I know the other field technicians do as well. That's saved us a few times, but the situation seems to be getting worse."

Now that Rosa had some real information as to the nature of the problems, she needed to start developing solutions—and it appeared that it was important to come up with good solutions fast! The first thing she tried to do was take a couple of part numbers at random and see if she could improve on the ordering approach.

The first number she selected was the A233 circuit board. The average weekly usage was 32 with a weekly standard deviation of 47. The lead time was given as one week. The board cost $18, and the cost to place an order was given as $16. The quantity ordered to fill the bin was usually 64. The second number was the P656 power supply. It cost $35, but since the supplier only required a fax to order the cost was only $2 per order. Even with the fax, the delivery lead time was two weeks. The average weekly demand for the power supply was 120 with a weekly standard deviation of 14. The company typically ordered 350 units at a time. Recently, the supplier for the circuit board hinted that it might be able to give Carl's a price break of $2 per board if Carl's would order 200 or more at a time.

Assignment

1. Using the data on the two part numbers given, provide a comprehensive evaluation of the ordering policies. Compare the present annual average cost with the cost of using a system such as EOQ, and discuss any other order policies as appropriate.
2. Should Carl's pursue the price break? Why or why not?
3. What do you think the sources of the other problems are? Be specific and analyze as completely as possible.
4. Develop a comprehensive plan to help Rosa get the inventory back under control.

CHAPTER THIRTEEN

PHYSICAL INVENTORY AND WAREHOUSE MANAGEMENT

13.1 INTRODUCTION

Because inventory is stored in warehouses, the physical management of inventory and warehousing are intimately connected. In some cases, inventory may be stored for an extended time. In other situations, inventory is turned over rapidly, and the warehouse functions as a distribution center.

This chapter will deal with the physical management of inventory in a warehouse, including basic approaches to warehouse layout, the activities involved in handling goods, and the controls necessary to work efficiently while maintaining a desired level of customer service. Inventory accuracy is the responsibility of warehousing, and methods of determining the accuracy of stocks will be covered, along with a description of how to conduct an annual physical audit. The cycle counting method of auditing inventory will show the advantages of timely correction of errors, along with improved error prevention. Bar coding, radio frequency identification (RFID), the Internet of Things (IoT), and the concept of wearable technology will be introduced as a means to improve the speed and accuracy of gathering information. In addition, an introduction to the concept of cross-docking will be discussed.

In a factory, stores, or stockrooms, perform the same functions as warehouses and contain raw materials, work-in-process inventory, finished goods, supplies, and possibly repair parts. Since they perform the same functions, stores and warehouses are treated alike in this chapter.

13.2 WAREHOUSING MANAGEMENT

As with other elements in a distribution system, the objective of a warehouse is to minimize cost and maximize customer service. To do this, efficient warehouse operations perform the following:

- Provide timely customer service.
- Keep track of items so they can be found readily and correctly.
- Minimize the total physical effort and thus the cost of moving goods into and out of storage.
- Provide communication links with customers.

The costs of operating a warehouse can be broken down into capital and operating costs. Capital costs are dependent on the type of material to be stored and include those of building space, information systems, racking or shelving, tanks or containers, and material handling equipment. The space needed depends on the peak quantities that must be stored, the methods of storage, and the need for ancillary space for aisles, docks, offices, and so on.

The major operating cost is labor, and the measure of labor productivity is the number of units, e.g., pallets, that an operator can move in a day. The labor cost depends on the type of material handling equipment used, the location and accessibility of stock, warehouse layout, stock location system, and the order picking system used.

Warehouse Activities

Operating a warehouse involves several processing activities, and the efficient operation of the warehouse depends upon how well these are performed. These activities are as follows:

1. **Receive goods.** The warehouse accepts goods from outside transportation or an attached factory and accepts responsibility for them. This means the warehouse must:
   a. Check the goods against an order and the bill of lading.
   b. Check the quantities.
   c. Check for damage and fill out damage reports if necessary.
   d. Inspect goods if required.
2. **Identify the goods.** Items are identified with the appropriate stock keeping unit (SKU) number (part number) and the quantity received is recorded.
3. **Dispatch goods to storage.** Goods are sorted and put away, recording the location.
4. **Hold goods.** Goods are kept in storage and under proper protection until needed, such as cold, heat, dust/dirt, and explosives.
5. **Pick goods.** Items required from stock must be selected from storage and brought to a consolidating area. This requires goods to be accessible and have accurate location records.
6. **Consolidate the shipment.** Goods making up a single order are brought together and checked for omissions or errors. Order records are updated if required.
7. **Dispatch the shipment.** Orders are packaged, shipping documents prepared, and goods loaded and secured on the right vehicle.
8. **Record the information.** A record must be maintained for each item in stock showing the quantity on hand, quantity received, quantity issued, and location in the warehouse. The system can be very simple, depending on a minimum of written information and human memory, or it may be a sophisticated computer-based system.

In various ways, all these activities take place in any warehouse. The complexity depends on the number of SKUs handled, the quantities of each SKU, and the number of orders received and filled. To maximize productivity and minimize cost, warehouse management must work with the following:

1. **Maximum use of space.** Usually the largest capital cost is for space. This means not only floor space but cubic space as well since goods are stored in the space above the floor as well as on it.
2. **Effective use of labor and equipment.** Material handling equipment represents the second-largest capital cost and labor the largest operating cost. There is a trade-off between the two in that labor costs can be reduced by using more material handling equipment. Warehouse management will need to:

   - Select the best mix of labor and equipment to maximize the overall productivity of the operation.
   - Provide ready access to all SKUs. The SKUs should be easy to identify and find. This requires a good stock identification system, stock location system, and layout.
   - Move goods efficiently. Most of the activity that goes on in a warehouse is material handling: the movement of goods into and out of stock locations.

Several factors influence effective use of warehouses. Some are as follows:

- Cube utilization and accessibility.
- Stock location.
- Order picking and assembly.
- Packaging.

With the exception of packaging, these are discussed in the following sections.

Cube Utilization and Accessibility

Goods are stored not just on the floor, but also in the cubic space of the warehouse. Although the size of a warehouse can be described as so many square feet, warehouse capacity depends on how high goods can be stored and the density of the storage.

Space is also required for aisles, receiving and shipping docks, offices, and order picking and assembly. In calculating the space needed for storage, some design figure for maximum inventory is needed. Suppose that a maximum of 90,000 cartons is to be inventoried and 30 cartons fit on a pallet. Space is needed for 3000 pallets. If pallets are stacked three high, 1000 pallet positions are required. A pallet is a platform usually measuring $48'' \times 40'' \times 4''$.

Pallet positions Suppose a section of a warehouse is as shown in Figure 13.1. Since the storage area is $48''$ deep, the $40''$ side is placed along the wall. The pallets cannot be placed tight against one another; a $2''$ clearance must be allowed between them so they can be moved. This then leaves room for $(120' \times 12'') \div 42'' = 34.3$, or 34, **pallet positions** along each side of the aisle. Since the pallets are stacked three high, there is room for 204 $(34 \times 3 \times 2)$ pallets.

Example Problem

A company wants to store an SKU consisting of 13,000 cartons on pallets, each containing 30 cartons. How many pallet positions are needed if the pallets are stored three high?

Answer

Number of pallets required = $13{,}000 \div 30 = 433.33 \rightarrow 434$ pallets
Number of pallet positions = $434 \div 3 = 144.67 \rightarrow 145$ pallet positions

Notice one pallet position will contain only two pallets.

Accessibility **Accessibility** means being able to get at the goods wanted with a minimum amount of work. For example, if no other goods had to be moved to reach an SKU, the SKU would be 100% accessible. As long as all pallets contain the same SKU, there is no problem with accessibility. The SKU can be reached without moving any other product. When several SKUs are stored in the area, each product should be accessible with a minimum of difficulty.

Cube utilization Suppose items are stacked along a wall, as shown in Figure 13.2. There will be excellent accessibility for all items except item 9, but cube utilization is not maximized. **Cube utilization** is the use of space horizontally and vertically. There is room for 30 pallets, but only 21 spaces are being used for a cube utilization of 70%: $(21 \div 30) \times 100\%$. Some method must be devised to increase cube utilization and

FIGURE 13.1 Cube utilization.

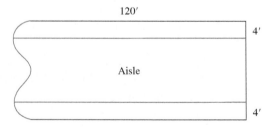

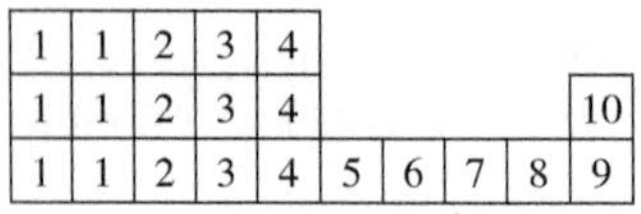

FIGURE 13.2 Cube utilization versus accessibility.

maintain accessibility. One way is to install tiers of racks so that lower pallets can be removed without disturbing the upper ones. This represents a trade-off between the capital cost of the racking and the savings in the operating cost of extra handling. Whether the additional cost is worthwhile will depend on the amount of handling and the savings involved.

Example Problem

A small warehouse stores five different SKUs in pallet loads. If pallets are stacked three high and there is to be 100% accessibility, how many pallet positions are needed? What is the cube utilization?

SKU A	4 pallets
SKU B	6 pallets
SKU C	14 pallets
SKU D	8 pallets
SKU E	5 pallets
Total	37 pallets

ANSWER

SKU	Pallet positions
A: 4 pallets	2
B: 6 pallets	2
C: 14 pallets	5
D: 8 pallets	3
E: 5 pallets	2
Total	14

In 14 pallet positions, there is room to store $14 \times 3 = 42$ pallets.

Number of pallets actually stored = 37

Cube utilization = $(37 \div 42) \times 100\% = 88\%$

Stock Location

Stock location, or warehouse layout, is concerned with the location of individual items in the warehouse. There is no single universal stock location system suitable for all occasions, but there are a number of basic systems that can be used. Which system, or mix of systems, is used depends on the type of goods stored, the type of storage facilities needed, the throughput or volume of items picked, and the size of orders. Whatever the system, management must maintain enough inventory of safety and working stock to provide the required level of customer service, keep track of items so they can be found easily, and reduce the total effort required to receive goods, store them, and retrieve them for shipment.

The following are some basic systems of locating stock:

- **Group functionally related items together.** Group together items similar in their use (functionally related). For example, put all hardware items in the same area of the warehouse. If functionally related items are ordered together, order picking is easier. Warehouse personnel become familiar with the locations of items.

- **Group fast-moving items together.** If fast-moving items are placed close to the receiving and shipping area, the work of moving them in and out of storage is reduced. Slower-moving items can be placed in more remote areas of the warehouse.

- **Group physically similar items together.** Physically similar items often require their own particular storage facilities and use similar handling equipment. Small packaged items may require shelving, whereas heavy items, such as tires or drums, require different facilities and handling equipment. Frozen foods need freezer storage space.

- **Locate working stock and reserve stock separately.** Relatively small quantities of **working stock**, stock from which withdrawals are made, can be located close to the consolidating and shipping area, whereas **reserve stock** used to replenish the working stock can be located more remotely. This allows order picking to occur in a compact area and replenishment of the working stock in bulk by pallet or container load. Warehouses with racking often use the floor level for order picking with reserve stock stored in the upper levels.

There are two basic systems for assigning specific locations to individual stock items: fixed location and random location. Either system may be used with any of the location systems cited in the preceding paragraphs.

Fixed location In a **fixed-location storage** system, an SKU is assigned a permanent location or locations, and no other items are stored there. This system makes it possible to store and retrieve items with a minimum of recordkeeping. In some small, manual systems, no records are kept at all. It is like always keeping cornflakes on the same shelf in the kitchen cupboard at home. Everything is nice and simple so things are readily found. However, fixed-location systems usually have poor cube utilization. If demand is uniform, presumably the average inventory is half the order quantity, and enough space has to be allocated for a full-order quantity. On the average, only 50% of the cube space is utilized. Fixed-location systems are often used in small warehouses where space is not at a premium, where throughput is small, and where there are few SKUs. Fixed location systems have particularly poor space utilization when product design changes are frequent and yet stock of the older design must be maintained for a period of time.

Random location (sometimes called floating location) In a **random-location storage** system (**floating inventory location system**), goods are stored wherever there is appropriate space for them. The same SKU may be stored in several locations at the same time and different locations at different times. The advantage to this system is improved cube utilization. However, it requires accurate and up-to-date information on item location and the availability of empty storage space so items can be put away and retrieved efficiently. If either part number or location information is incorrect, then someone looking for the material will have a major problem trying to locate the material, as it could be anywhere in the warehouse. Modern warehouses using random location systems are usually computer-based. The computer assigns available locations to incoming items, records what items are on hand and where they are located, and directs the order picker to the right location to find the item. Thus, cube utilization and warehouse efficiency are greatly improved.

Over time, other storage systems have evolved to address special cases or problems with the two basic location systems.

Zone random storage This is really a hybrid system of the two basic systems, combining advantages and some disadvantages of both. First, zones are established in the warehouse where closely related products are stored. Then within each zone, the material can be randomly located as in a floating location system. For example, if a zone is established for locating all fasteners and if a location or part number for one fastener is incorrectly placed in the record, then the entire warehouse does not need to be searched, only the zone for fasteners. Because of the floating storage within the zone, the system is fairly efficient with good cube utilization.

Point-of-use storage Sometimes, particularly in repetitive manufacturing and in a lean production environment, inventory is stored close to where it will be used. There are several advantages to **point-of-use storage**.

- Materials are readily accessible to users.
- Material handling is reduced or eliminated.
- Central storage costs are reduced.
- Materials are accessible at all times.

This method is excellent as long as inventory is kept low and operating personnel can keep control of inventory records. Sometimes "C" items are issued as floor stock, where manufacturing is issued a large quantity of the material, which is used as needed. Inventory records are adjusted when the stock is issued, not when it is used.

Central storage As opposed to point-of-use storage, **central storage** contains all inventory in one central location. There are several advantages:

- Ease of control.
- Inventory record accuracy is easier to maintain.
- Specialized storage can be used.
- Reduced safety stock, since users do not need to carry their own safety stock.

Cross-docking

The *APICS Dictionary*, 16th edition, defines **cross-docking** as: "The concept of packing products on incoming shipments so they can be easily sorted at intermediate warehouses or for outgoing shipments based on final destination. The items are carried from the incoming vehicle docking point to the outgoing vehicle docking point without being stored in inventory at the warehouse. Cross-docking reduces inventory investment and storage space requirements." There can be several types of cross-docking, including:

- Order makeup—implies an order can arrive sorted and/or identified by SKU but then the load needs to be broken up and repackaged by customer order.
- Hybrid—incoming material may be combined with additional material from the warehouse, and/or some of the incoming material may be separated for storage in the warehouse.
- "Hot" items—including items arriving late or on a backorder. Customer service might be helped by cross-docking such items.
- Transportation consolidation—including items to be transferred from truck to rail and back.

Categories of products that are well suited for cross-docking include:

- High value products.
- Promotional or seasonal in nature.
- Products with high volume demand.
- Products on backorder.

Order Picking and Assembly

Once an order is received, the items on the order must be retrieved from the warehouse, assembled, and prepared for shipment. All these activities involve labor and the movement of goods. The work should be organized to provide the level of customer service required and at least cost. There are several systems that can be used to organize the work, among which are the following:

1. **Area system.** The order picker circulates throughout the warehouse selecting the items on the order, much as a shopper would in a supermarket. The items are then taken to the shipping area for shipment. The order is self-consolidating in that when the order picker is finished, the order is complete. This system is generally used in small warehouses where goods are stored in fixed locations.

2. **Zone system.** The warehouse is divided into zones, and order pickers work only in their own area. An order is divided up by zone, and each order picker selects those items in their zone and sends them to the consolidating area, where the order is assembled for shipment. Each order is handled separately and leaves the zone before another is handled. If applied to the supermarket example above, one individual would go to say the fresh fruit department, another to the freezer section, and so on. They would then consolidate the items from each department at the checkout.

> **KITTING SAVES TIME WHEN IT IS NEEDED THE MOST**
>
> In 2014, Supply Logistics, a division of New York City Transit, which is the operator of the city's bus and subway systems, issued nearly 2.5 million line items from its network of 65 storerooms. This material is used in the maintenance and repair of over 5600 buses and 6500 subway cars. With the average bus or subway storeroom less than 5000 sq. ft., space is always at a premium. While Supply Logistics storerooms utilize a variety of space-saving strategies (mezzanines, vertical carousels, specialized racks for glass, pipe, hoses, etc.), labor is another limiting constraint. Too many people in a small space receiving and issuing material simultaneously can become counterproductive and could pose a serious safety hazard.
>
> One way that Supply Logistics has maximized space in the storerooms is by repackaging material into kits. The kitting operation began in 1989. In 2001 it reached the milestone of 1 million kits assembled and by 2014 it had assembled 3 million. In 2014 alone Supply Logistics assembled 216,000 kits.
>
> What is the impact of kitting to New York City Transit? The average kit contains 12 line items. If these kits had not been produced that would mean that Supply Logistics would have had to issue an additional 2.4 million lines from its storerooms, nearly doubling the current annual issues. This would have resulted in a much larger investment in storeroom space and equipment, as well as hiring additional staff. Kitting has saved time and money for Supply Logistics, New York City Transit, and the citizens of New York City.
>
> Submitted by:
>
> Gary A. Smith, CFPIM, CSCP
> Vice President, Division of Supply Logistics
> New York City Transit

Zones are usually established by grouping together related parts. Parts may be related because of the type of storage needed for them, e.g., freezer storage, or because they are often ordered together.

A variation of the zone system is to have the order move to the next zone rather than to the consolidating area. By the time it exits the last zone, it is assembled for shipment.

3. **Multi-order system.** This system is the same as the zone system except that rather than handling individual orders, a number of orders are gathered together and all the items divided by zone. The pickers then circulate through their area, collecting all the items required for that group of orders. The items are then sent to the consolidating area where they are sorted to individual orders for shipment.

The area system is simple to manage and control, but as the warehouse throughput and size increases, it becomes unwieldy. The zone systems break down the order-filling process into a series of smaller areas that can be better managed individually. The multi-order system is probably most suited to the situation in which there are many items or many small orders with few items.

Working stock and reserve stock In addition to the systems mentioned, reserve stock and working stock may be separated. This is appropriate when the pick unit for a customer's order may be a box or a case that is stored on pallets. A pallet can be moved into the working area by a lift truck and cartons or boxes picked from it. The working stock is located close to the shipping area so the work in picking is reduced. A separate workforce is used to replenish the working stock from the reserve stock.

13.3 PHYSICAL CONTROL AND SECURITY

Because inventory is tangible, items have a nasty habit of becoming lost, stray, or stolen, or of disappearing in the night. It is typically not that people are dishonest, rather that they are forgetful. What is needed is a system that makes it difficult for people to make mistakes or be dishonest. There are several elements that help:

- **A good-part numbering system.** Part numbering was discussed in Chapter 4. It must be clear and easy to use for the order pickers and material handlers.
- **A simple, well-documented transaction system.** When goods are received, issued, or moved in any way, a transaction occurs. There are four steps in any transaction: identify the item, verify the quantity, record the transaction, and physically execute the transaction.

1. **Identify the item.** Many errors occur because of incorrect identification. When receiving an item, the purchase order, part number, and quantity must be properly identified. When goods are stored, the location must be accurately specified. When issued, the quantity, location, and part number must be recorded.
2. **Verify quantity.** Quantity is verified by a physical count of the item by weighing or by measuring. Sometimes standard-sized containers are useful in counting.
3. **Record the transaction.** Before any transaction is physically carried out, all information about the transaction must be recorded.
4. **Physically execute the transaction.** Move the goods in, about, or out of the storage area.

- **Limited and controlled access.** Inventory must be kept in a safe, secure place with limited general access. It should be locked except during normal working hours. This is less to prevent theft than to ensure people do not take things without completing the transaction steps. If people can wander into the stores area at any time and take something, the transaction process integrity fails. A failed transaction process is not a trivial matter. If an item of inventory needed to fill demand or produce product for a customer order is not available when needed, the company pays a price in many possible ways, including expediting costs, premium shipping costs and possibly having a customer abandon the company as a supplier. If extra inventory is maintained "just in case the records are incorrect" then several costs increase including inventory holding costs, storage costs, increased handling, and sometime spoilage or obsolescence.

- **A well-trained workforce.** Not only should the stores staff be well trained in handling and storing material and in recording transactions, but other personnel who interact with stores must be trained to ensure transactions are recorded properly.

13.4 INVENTORY RECORD ACCURACY

The usefulness of inventory is directly related to its accuracy. Based on the inventory record, a company determines net requirements for an item, releases orders based on material availability, and performs inventory analysis.

These three pieces of information must be accurate: part description (part number), quantity, and location. Accurate inventory records enable firms to:

- **Operate an effective materials management system.** If inventory records are inaccurate, gross-to-net calculations will be in error.
- **Maintain satisfactory customer service.** If records show an item is in inventory when it is not, any order promising of that item will be in error.
- **Operate effectively and efficiently.** Planners can plan, confident that the parts will be available.
- **Analyze inventory.** Any analysis of inventory is only as good as the data it is based on.

Inaccurate inventory records will result in the following, with resulting costs to the company.

- Lost sales.
- Shortages and disrupted schedules.
- Excess inventory (of the wrong things).
- Low productivity.
- Poor delivery performance.
- Excessive expediting, since people will always be reacting to a bad situation rather than planning for the future.

Causes of Inventory Record Errors

Poor inventory record accuracy can be caused by many things, but they all result from poor recordkeeping systems and poorly trained personnel. Some examples of causes of inventory record error follow:

- **Unauthorized withdrawal of material.** Employees will often take items from inventory due to problems or shortages in their department. This usually solves a problem for the department needing the inventory, but the transactions are often not recorded or not recorded accurately and in a timely manner.
- **Unsecured stockroom.** Secure stockrooms prevent unauthorized withdrawal, which may or may not be legitimate.
- **Poorly trained personnel.** Most employees do not realize the consequences of not properly recording transactions or picking the wrong items.
- **Inaccurate transaction recording.** Errors can occur because of inaccurate piece counts, unrecorded transactions, delay in recording transactions, inaccurate material location, and incorrectly identified parts.
- **Poor transaction recording systems.** Most systems today are computer-based and can provide the means to record transactions properly. Errors, when they occur, are usually the fault of human input to the system. The documentation reporting system should be designed to reduce the likelihood of human error.
- **Lack of audit capability.** Some program of verifying the inventory counts and locations is necessary. The most popular one today is cycle counting, discussed in the next section.

Measuring Inventory Record Accuracy

Inventory accuracy ideally should be 100%. Banks and other financial institutions reach this level, but other companies can move toward this potential.

Figure 13.3 shows 10 inventory items, their physical count, and the quantity shown on their record. What is the inventory accuracy? The total of all items is the same, but only 2 of the 10 items are correct. Is the accuracy 100% or 20% or something else?

Tolerance To judge inventory accuracy, a tolerance level for each part must be specified. For some items, this may mean no variance; for others, it may be very difficult or costly to measure and control to 100% accuracy. An example of the latter might be nuts

Part Number	Inventory Record	Shelf Count
1	100	105
2	100	100
3	100	98
4	100	97
5	100	102
6	100	103
7	100	99
8	100	100
9	100	97
10	100	99
Total	1000	1000

FIGURE 13.3 Inventory record accuracy.

Part Number	Inventory Record	Shelf Count	Tolerance	Within Tolerance	Outside Tolerance
1	100	105	5%	X	
2	100	100	±0%	X	
3	100	98	±3%	X	
4	100	97	±2%		X
5	100	102	±2%	X	
6	100	103	±2%		X
7	100	99	±3%	X	
8	100	100	±0%	X	
9	100	97	±5%	X	
10	100	99	±5%	X	
Total	1000	1000			

FIGURE 13.4 Inventory accuracy with tolerance.

or bolts ordered and used in the thousands. For these reasons, tolerances are set for each item. **Tolerance** is the amount of permissible variation between an inventory record and a physical count.

Tolerances are set on individual items based on value, critical nature of the item, availability, lead time, ability to stop production, safety problems, or the difficulty of getting precise measurement.

Figure 13.4 shows the same data as the previous figure but includes tolerances. This information tells us exactly what inventory accuracy is.

Example Problem

Determine which of the following items are within tolerance. Item A has a tolerance of ±5%; item B, ±2%; item C, ±3%; and item D, ±0%.

Part Number	Shelf Count	Inventory Record	Tolerance
A	1500	1550	±5%
B	120	125	±2%
C	225	230	±3%
D	155	155	±0%

ANSWER

Item A. With a tolerance of ±5%, variance can be up to ±75 units.
Item A is within tolerance.

Item B. With a tolerance of ±2%, variance can be up to ±2 units.
Item B is outside tolerance.

Item C. With a tolerance of ±3%, variance can be up to ±7 units.
Item C is within tolerance.

Item D. With a tolerance of ±0%, variance can be up to ±0 units.
Item D is within tolerance.

Auditing Inventory Records

Errors occur, and they must be detected so inventory accuracy is maintained. There are two basic methods of checking the accuracy of inventory records: periodic (usually annual) counts of all items and cyclic (usually daily or weekly) counts of specified items. It is important to audit record accuracy, but it is more important to audit the record system to find the causes of inaccuracies and eliminate them. Cycle counting does this; periodic audits tend not to.

Periodic (annual) inventory The primary purpose of a **periodic inventory** (often referred to as **annual physical inventory**) is to satisfy the financial auditors that the inventory records represent the value of the inventory. To planners, the physical inventory represents an opportunity to correct any inaccuracies in the records. Whereas financial auditors are concerned with the total value of the inventory, planners are concerned with item detail.

The responsibility for taking the physical inventory usually rests with materials management and finance, who should ensure that a good plan exists and it is followed. George Plossl once said that taking a physical inventory was like painting; the results depend on good preparation.[1] There are three factors in good preparation: housekeeping, identification, and training.

Housekeeping Inventory must be sorted and the same parts collected together so they can easily be counted. Sometimes items can be pre-counted and put into sealed cartons.

Identification Parts must be clearly identified and tagged with part numbers. This can, and should, be done before the inventory is taken. Personnel who are familiar with parts identification should be involved and all questions resolved before the physical inventory starts.

Training Those who are going to do the inventory must be properly instructed and trained in taking inventory. Physical inventories are usually taken once a year, and the procedure is not always remembered from year to year.

Process Taking a physical inventory consists of four steps:

1. Count items and record the count on a ticket left on the item.
2. Verify this count by recounting or by sampling.
3. When the verification is finished, collect the tickets and list the items in each department.
4. Reconcile the inventory records for differences between the physical count and inventory dollars. Financially, this step is the job of accountants, but materials personnel are involved in adjusting item records to reflect what is actually on hand. If major discrepancies exist, they should be checked immediately.

---

[1]George W. Plossl, *Production and Inventory Control, Principles and Techniques*, 2nd ed., Appendix VI: Physical Inventory Techniques. Englewood Cliffs, NJ: Prentice Hall, 1985.

Taking a physical inventory is a time-honored practice in many companies mainly because it has been required for an accurate appraisal of inventory value for the annual financial statements. However, taking an annual physical inventory presents several problems. Usually, the factory has to be shut down, thus losing production time; labor and paperwork are expensive; the job is often done hurriedly and poorly since there is much pressure to get it done and the factory running again. In addition, the people doing the inventory are not used to the job or the inventory and are prone to making errors. As a result, more errors often are introduced into the records than are eliminated.

Another reason this process of inventory auditing can present a problem is that when it is done primarily to obtain an accurate monetary value of the inventory for financial reporting (as an asset on the balance sheet) then the focus on the need for accurate inventory records for production needs are downplayed or overlooked. A simple example can illustrate why a primary focus on just the financial needs are not suitable for the production system:

Part number	Book value ($)	Recorded quantity	Actual quantity (count)
123	$2	10	15
124	$5	4	0
125	$1	22	17
126	$3	8	13
Total value	$86	Total value	$86

It is clear that from a financial perspective the inventory records (in monetary terms) are perfect, while those same records for production requirements are 100% incorrect. It will be difficult to tell the person needing item 124 that the inventory is 100% accurate when they have a need for those four parts and they find they do not exist.

Because of these problems, the idea of cycle counting has developed.

Cycle counting Cycle counting is a system of counting inventory continually throughout the year. Physical inventory counts are scheduled so that each item is counted on a predetermined schedule. Depending on their importance, some items are counted frequently throughout the year whereas others are not. The idea is to count selected items each day.

The advantages to cycle counting are as follows:

- **Timely detection and correction of problems.** The purpose of the count is first to find the cause of error and then to correct the cause so the error is less likely to happen again.
- **Complete or partial reduction of lost production.** Cycle counting can usually be done without stopping production.
- **Use of personnel trained and dedicated to cycle counting.** This utilizes experienced inventory takers who will not make the errors personnel involved in the physical inventory do. Cycle counters are often also trained to identify problems and to correct them. Sometimes the people within production, who use the parts, are also the cycle counters.

It should be noted that while cycle counting is an important method to keep records accurate with minimal disruption of production, it is even more important to use the results of the cycle counts to discover the causes of the record inaccuracy. As causes of inaccuracy are discovered and process improvement in the transaction system address the causes, not only are the records more accurate but over time inaccuracies become very rare. In some organizations, cycle counters who also are responsible for finding causes of inaccuracy become highly valued personnel in the company, since the activity of evaluating processes to discover the cause of a problem makes them highly knowledgeable in both the company and in the approaches to process improvement. This activity is an important part of **data governance**, which is the set of policies, procedures, and standards for maintaining effective data and information systems. In summary it should be noted that many companies correctly consider that the primary function of cycle counting is to audit and correct the transaction processes. Correcting the count in the records, while important, is really the secondary, not primary, focus of the cycle count activity.

Classification	Number of Items	Count Frequency per Year	Number of Counts per Year	% of Total Counts	Counts per Day
A	1000	12	12,000	58.5	48
B	1500	4	6000	29.3	24
C	2500	1	2500	12.2	10
Total counts per year			20,500		
Workdays per year			250		
Counts per day			82		

FIGURE 13.5 Scheduling cycle counts.

In some organizations with highly effective cycle count programs, the financial auditors are encouraged to audit the records as kept through cycle counting rather than the annual physical inventory. If a high enough confidence level is reached by the financial auditors, they may decide to forgo the annual physical inventory (instead taking the monetary value of the existing inventory records), thereby saving the company substantial money and time.

Count frequency The basic idea is to count some items each day so all items are counted a predetermined number of times each year. The number of times an item is counted in a year is called its **count frequency**. For an item, the count frequency should increase as the value of the item and number of transactions (chance of error) increase. Several methods can be used to determine the frequency. Three common ones are the ABC method, zone method, and location audit system.

- **ABC method.** This is a popular method. Inventories are classified according to the ABC system (refer to Chapter 10). Some rule is established for count frequency. For example, A items might be counted weekly or monthly; B items, bimonthly or quarterly; and C items, biannually or once a year. On this basis, a count schedule can be established. Figure 13.5 shows an example of a cycle count scheduled using the ABC system. Note that there should be 82 items counted each day, of which 58.5% should be A items or 48 counts per day. This will result in 250 days × 48 A items per day or 12,000 counts per year.

Example Problem

A company has classified its inventory into ABC items. It has decided that A items are to be counted once a month; B items, four times a year; and C items, twice a year. There are 2000 A items, 3000 B items, and 5000 C items in inventory. Develop a schedule of the counts for each class of item.

ANSWER

Classification	Number of Items	Count Frequency per Year	Number of Counts per Year	% of Total Counts	Counts per Day
A	2000	12	24,000	52.2	96
B	3000	4	12,000	26.1	48
C	5000	2	10,000	21.7	40
Total counts per year			46,000		
Workdays per year			250		
Counts per day			184		

- **Zone method.** Items are grouped by zones to make counting more efficient. The system is used when a fixed-location system is used, or when work-in-process or transit inventory is being counted.

- **Location audit system.** In a floating-location system, goods can be stored anywhere, and the system records where they are. Because of human error, these locations may not be 100% correct. If material is mislocated, normal cycle counting may not find it. In using location audits, a predetermined number of stock locations are checked each period. The item numbers of the material in each bin are checked against inventory records to verify stock point locations.

A cycle counting program may include all these methods. The zone method is ideal for fast-moving items. If a floating-location system is used, a combination of ABC and location audit is appropriate.

When to count Cycle counts can be scheduled at regular intervals or on special occasions. Some selection criteria are as follows:

- **When an order is placed.** Items are counted just before an order is placed. This has the advantage of detecting errors before the order is placed and reducing the amount of counting work by counting at a time when stock is low. Items with high order frequency have more opportunities for transaction errors. This method counts fast moving items more often.
- **When an order is received.** Inventory is generally at its lowest level and easiest to count. Also, the effort of going to the stock location is already taking place (to stock the new order received).
- **When the inventory record reaches zero.** Again, this method has the advantage of reducing work.
- **When a specified number of transactions have occurred.** Errors occur most often when transactions occur. Fast-moving items have more transactions and are more prone to error.
- **When an error occurs.** A special count is appropriate when an obvious error is detected. This may be a negative balance on the stock record or when no items can be found although the record shows some in stock.

13.5 CONSIGNMENT INVENTORY AND VENDOR-MANAGED INVENTORY

A recent trend in many companies is to essentially outsource some of the inventory control and holding costs to suppliers. This type of inventory is called **consignment inventory**. The supplier typically places inventory in a specified customer location for the customer to use as needed. Under non-consignment conditions, when a customer orders material from a supplier, the customer is obligated to pay the supplier under predefined terms, for example, within 30 days. With consignment inventory, the customer is not obligated to pay for the material until after it is used or sold. This clearly has a possible cost reduction advantage for the customer, since they have no money invested in the inventory, but can cause some possible cash flow issues for the supplier. Having their inventory in a customer facility does, however, tend to create a stronger relationship between the supplier and the customer. Risk can also be reduced with strong consignment agreements, identifying such things as inventory levels and responsibility for lost or damaged inventory.

Another method being used in some companies is **vendor-managed inventory (VMI)**. VMI is actually an inventory management and planning system. Under a planning system not using VMI, the customer's planning system determines inventory need and the inventory is acquired through the purchasing process. With VMI, the supplier assumes the planning activities to determine what material is needed and when. Clearly, in order to do this, there has to be a strong relationship between the supplier and customer, since the way the supplier determines the customer material needs is by assuming the responsibility to have data on customer usage of the material. This implies that communication and planning systems need to be acquired, at a cost, but the improvements in customer service, uncertainty of supply, fewer forecasting requirements, and often a reduction in total inventory often makes up for the cost. The impact on storage should also be noted. An additional benefit could be a closer, stronger relationship between customer and supplier.

> ### INVENTORY ACCURACY: "A WAY OF LIFE"
>
> Inventory accuracy has become a way of life for Supply Logistics at New York City Transit. Each of the 65 storerooms is cycle counted quarterly while the main warehouses are cycle counted six times per year. Each cycle count consists of three activities: Stockkeeping Performance (SKP), Bin Stockkeeping Performance (Bin SKP), and Precision, Accuracy & Control Test (PACT). Cycle counts are done by a third-party contractor to avoid any conflicts.
>
> SKP is measured by choosing a number of items to test (fast, medium, and slow movers) and counting the selected inventory. For example, if 150 items are chosen, then all of the items are counted and the physical inventory is reconciled with the quantity on hand in the inventory management system. If two items fail, the SKP percentage is 98.7%(148/150).
>
> Bin SKP measures the accuracy at the location level. In the example above, if the 150 items were spread among 160 locations (bins), then the inventory in each is checked for accuracy. In the current example, if 7 locations failed to have the correct inventory, the Bin SKP percentage would be 95.6%(153/160).
>
> The final measure is the PACT. In this test, the auditors look for locations or items that may be suspect (open cartons or loose items). The items are counted and compared with inventory. For example, if 100 locations were tested and five locations were incorrect, the PACT would be 95%(152/100).
>
> It must be realized however, that cycle counting is a measurement of inventory accuracy, not the cause. The reason an inventory is accurate is that the procedures used are disciplined and balanced, focusing on the fundamentals of best practice inventory management. This is true for Supply Logistics as the cycle count results across all warehouses and storerooms averaged 99.6% for 2014.
>
> Submitted by:
>
> Gary A. Smith, CFPIM, CSCP
> Vice President, Division of Supply Logistics
> New York City Transit

13.6 TECHNOLOGY APPLICATIO$NS

Most imbalances in inventory records are caused by human error. Reading stock codes and entering count quantities can be a source of many errors in any transaction, including during the audit itself. **Bar codes** can reduce this error as they are machine-readable symbols and are widely used to gather information at all levels of retailing, distribution, and manufacturing. The error rate for this method is extremely low compared to human error, which is estimated to be as high as 3% for repetitive entries. Bar codes are standardized by industry and are usually printed on a paper label or tag. Typically, they only contain a unique identifier, such as part number, which can be referred to a database for further information, such as price or description, as required. The automotive industry requires labels designed to their specifications for layout and the type of code used. These labels will include, in addition to the product code, the manufacturer, package quantity, date of manufacture, and so forth.

Bar codes are read using a laser light, which picks up the reflection from the bars and spaces on the label and is usually read from a short distance, although range is improving with new designs. The use of bar codes improves the speed of data entry but, more importantly, improves the accuracy of the data retrieved.

Radio frequency identification (RFID) works in a way similar to bar codes, but rather than light, it uses reflected radio waves from a small device or tag to receive its information. Unlike bar codes, RFID devices do not require a line of sight between the label and the reader and can accurately identify products that are within containers or otherwise hidden from view. This feature makes RFID suited for use as a security device or under conditions where labels are hard to read or access. Railroads in North America use RFID tags on all their rolling stock and locomotives. Trains are required to work in extreme conditions of dirt and weather and the passive RFID tags located on each side of all vehicles allows the accurate transfer of information such as owner, car number, and type of equipment. RFID tags are more costly than printed bar codes, but the price is falling rapidly to only a few cents per tag, encouraging their use in wider applications. Major retailers such as Walmart see the value in this method of gathering information and are demanding its use in many of their products, further reducing costs throughout the supply chain.

Similar to bar codes, RFID tags provide identification numbers that can be used to reference databases, but often have further information about the tagged item. In the railroad system, a car's number can reference the car's contents, destination, and so forth, allowing delivery tracking and even safety/hazardous information access. Many organizations use RFID pass cards to access parking or secured areas. The user is identified by the RFID signal and is allowed access accordingly. Management can then easily exercise central control over these areas. When your season parking pass expires, access is automatically restricted.

Supply chains are driven by information, and the ability to retrieve timely, accurate, and comprehensive information can lead to cost reductions. In previous chapters lead time was shown to affect inventories and customer service levels. Bar code and RFID technologies can reduce lead time, allowing all members of a supply chain to reduce costs while at the same time drastically reducing the number of occurrences of out of stock.

A **warehouse management system (WMS)** is a computer application that manages all processes carried out by a warehouse, including receiving, picking, and shipping. The *APICS Dictionary* (16th edition) defines WMS as, "a computer application system designed to manage and optimize workflows and the storage of goods within a warehouse. Often interfaces with automated data capture and enterprise resources planning systems." Output generated from ERP is executed by the WMS, such as the management and optimization of storage locations, cross-docking, cycle counting, automatic replenishment, and the management of order picking and reverse logistics.

Wearable technologies are becoming more common, presenting an opportunity to increase visibility, efficiency, and effectiveness in warehouses as well as other facilities. These are devices that, through connections on Wi-Fi or the internet, can be used to rapidly and frequently more effectively enhance communication to and from personnel in the warehouse. In addition, some wearables can potentially be used to track activity levels, distances moved to execute a transaction, and even (through GPS monitors) the exact location of workers in the warehouse. Some examples of devices being used or with the potential to be used include:

- Smartphones
- Electronic tablets
- "Smart" glasses and watches
- Voice control headsets
- Fitness bands

In spite of the obvious advantages these technologies provide, there are increased risks that should be addressed prior to their adoption. In addition to the costs involved in acquiring and implementing the technologies, there are the normal security concerns that accompany any use of the internet and Wi-Fi applications, including malware, ransomware, computer viruses, and risks of "hacking" into the systems. There can also be some privacy concerns expressed by warehouse personnel.

The Internet of Things (IoT). The 16th edition of the *APICS Dictionary* defines IoT as: "An environment in which objects, animals or people are provided with unique identifiers and the ability to transfer data over a network without requiring human-to-human or human-to-computer interaction. This allows objects to be sensed and controlled remotely across existing network infrastructure, creating opportunities for more direct integration between the physical world and computer-based systems." This is a rapidly growing opportunity to rapidly and effectively link, through internet technologies, both computer systems and connected technologies. Several examples already exist in the daily lives of many people. For example, many people have systems in their homes where they can control things such as door locks, heating and cooling systems, and turning lights off and on. Similarly, the concepts can be applied in a warehouse setting to communicate for needs such as lift truck status and location systems, and also link into transportation systems to provide truck locations and expected delivery windows of time. Information from the production planning and control systems can also be linked directly into warehouse activity.

Inventory Traceability

It is often necessary or even required to be able to track material as it becomes a final product and in the customer's use. Sometimes it is discovered after a product has moved through the supply chain that a problem exists with some material. It might be that the material has a previously undetected quality problem or perhaps even might risk the health or well-being of the customer (examples are food products, pharmaceuticals, and automobiles). In such cases, it is important, for both cost and customer comfort, that the specific products are located and recalled or repaired as quickly and efficiently as possible. Batches of material are often given lot numbers, and the lot numbers can be included in the bar codes. Standards have been developed (for example, a **Global Trade Item Number (GTIN),**) for tracking material, and RFID can also be used in certain cases.

SUMMARY

Previous chapters in this book have used theory and common sense to effectively manage inventories. This chapter looks at the physical handling and storage of inventory. Warehousing is changing from a place to simply store material to a strategic activity affecting a major investment to companies and a significant factor in improving customer service. The basic warehouse activities of receiving, identifying, storing, retrieving, and eventually shipping goods are all still required, but technology and the need for speed in the supply chain are changing the way we do things. "A place for everything and everything in its place" has given way to floating locations and point-of-use storage. As movements are being sped up, there is an increasing need for control and security, and it is important that workers know and understand proper procedures and the consequences of errors. To provide high levels of customer service, inventory records must be accurate. Cycle counting improves record accuracy but, more importantly, finds errors in the inventory systems so that processes can be continually improved. Bar code and RFID technologies are widely used in any modern inventory and warehousing system, improving data accuracy and speeding the movement of product. Consignment inventory and VMI are changing the replenishment of inventory and can impact issues of inventory storage. Other approaches, such as cross-docking, using wearable technology, and using the Internet of Things continue to make the warehouse and inventory management an important part of global supply chains

KEY TERMS

Accessibility 318	Internet of Things (IoT) 331
Annual physical inventory 326	Pallet positions 318
Bar codes 330	Periodic inventory 326
Central storage 321	Point-of-use storage 320
Consignment Inventory 329	Radio frequency identification (RFID) 330
Count frequency 328	Random-location storage 320
Cross-docking 321	Reserve stock 320
Cube utilization 318	Tolerance 325
Cycle counting 327	Vendor-managed inventory (VMI) 329
Data governance 327	Warehouse management system (WMS) 331
Fixed-location storage 320	
Floating inventory location system 320	Wearable technologies 331
Global Trade Item Number (GTIN) 332	Working stock 320

QUESTIONS

1. What are four objectives of warehouse operation?
2. Describe the eight warehouse activities as they would apply to a supermarket. Include in your description where each activity takes place and who performs the activity.
3. What are cube utilization and accessibility?

4. Why is stock location important in a warehouse? Name and describe four basic systems of stock location and give examples of each system from a retail setting.
5. Describe fixed and random systems for assigning locations to SKUs.
6. Name and describe three order-picking systems.
7. What is the difference between working stock and reserve stock?
8. What are the four steps in any transaction?
9. What are some of the results of poor inventory accuracy?
10. Six causes of poor inventory accuracy are discussed in the text. Name and describe each.
11. How should inventory accuracy be measured? What is tolerance? Why is it necessary?
12. What is the basis for setting tolerance?
13. What are the four major purposes of auditing inventory accuracy?
14. In taking a physical inventory, what are the three factors in preparation? Why is good preparation essential?
15. What are the four steps in taking a physical inventory?
16. Describe cycle counting. On what basis can the count frequency be determined?
17. Why is cycle counting a better way to audit inventory records than an annual physical inventory?
18. When are some good times to count inventory?
19. What information is typically stored in a bar code or RFID tag?
20. Give three examples where RFID would be preferred over a bar code.
21. How do bar codes and RFID reduce costs in the supply chain?
22. What is consignment inventory?
23. What is VMI?
24. How does VMI differ from consignment inventory?
25. How does cross-docking work? Under what conditions might it be valuable and why?
26. What is the Internet of Things (IoT) and how will it impact warehousing?

PROBLEMS

13.1. A company wants to store an SKU consisting of 5000 cartons on pallets, each containing 30 cartons. They are to be stored three high in the warehouse. How many pallet positions are needed?

Answer. 56 pallet positions

13.2. A company has 6600 cartons to store on pallets. Each pallet takes 20 cartons, and the cartons are stored four high. How many pallet positions are needed?

13.3. A company has an area for storing pallets as shown in the following diagram. How many pallets measuring 48″ × 40″ can be stored four high if there is a 2″ space between the pallets?

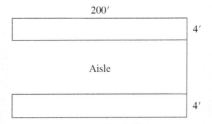

Answer. 456 pallets

13.4. A company has a warehouse with the dimensions shown in the following diagram. How many pallets measuring 48″ × 40″ can be stored three high if there is to be a 2″ space between the pallets?

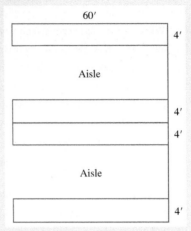

13.5. A company wishes to store the following SKUs so there is 100% accessibility. The items are stored on pallets that can be stacked three high.

a. How many pallet positions are needed?

b. What is the cube utilization?

c. If the company bought racking for storing the pallets, how many pallet positions are needed to give 100% accessibility?

SKU	Number of Pallets	Pallet Positions Required
A	13	
B	4	
C	10	
D	13	
E	14	
Total		

Answer. a. Pallet positions needed = 21

b. Cube utilization = 86%

c. 18 pallet positions

13.6. A company wants to store the following seven SKUs so there is 100% accessibility. Items are stored on pallets that are stored four high.

a. How many pallet positions are needed?

b. What is the cube utilization?

c. If the company bought racking for storing the pallets, how many pallet positions are needed to give 100% accessibility?

SKU	Number of Pallets	Pallet Positions Required
A	14	
B	18	
C	40	
D	32	
E	55	
F	23	
G	35	
Total		

13.7. a. Which of the following items are within tolerance?

b. What is the percent accuracy by item?

Part Number	Shelf Count	Inventory Record	Difference	% Difference	Tolerance	Within Tolerance?
A	650	635			±3%	
B	1205	1205			±0%	
C	1350	1500			±5%	
D	77	80			±5%	
E	38	40			±3%	
Total						

Answer. a. A, B, and D are within tolerance.

b. 60%

13.8. a. Which of the following items are within tolerance?

b. What is the percent accuracy by item?

Part Number	Shelf Count	Inventory Record	Difference	% Difference	Tolerance	Within Tolerance?
A	75	80			±3%	
B	120	120			±0%	
C	1435	1500			±5%	
D	75	76			±5%	
E	68	66			±2%	
Total						

13.9. A company does an ABC analysis of its inventory and calculates that out of 5000 items, 22% can be classified as A items, 33% as B items, and the remainder as C items. A decision is made that A items are to be cycle counted once a month, B items every three months, and C items twice a year. Calculate the total counts and the counts per day by classification. The company works five days per week and 50 weeks per year.

Classification	Number of Items	Count Frequency per Year	Number of Counts per Year	% of Total Counts	Counts per Day
A					
B					
C					
Total counts per year					
Workdays per year					
Total counts per day					

Answer. Counts per day A = 53; B = 26; C = 18

13.10. A company does an ABC analysis of its inventory and calculates that out of 10,000 items, 17% can be classified as A items, 30% as B items, and the remainder as C items. A decision is made that A items are to be cycle counted twice a month, B items every three months, and C items twice a year. Calculate the total counts and the counts per day by classification. There are 250 working days per year.

Classification	Number of Items	Count Frequency per Year	Number of Counts per Year	% of Total Counts	Counts per Day
A					
B					
C					
Total counts per year					
Workdays per year					
Total counts per day					

CASE STUDY 13.1

Lesscost Warehouse

Amy Gordon could not have been more pleased when she was first appointed as the new inventory management supervisor for the Lesscost regional warehouse. She had previously worked part time as a clerk in the local Lesscost Department Store while she finished her university degree. After she got the degree, she was named as the section head in charge of roughly one-fourth of the store. Now, a year later, she started to wonder about that old adage, "Be careful what you ask for—you just might get it."

Background

One constant problem Amy had complained about when she was head clerk was the difficulty she had with the warehouse replenishing supplies for her areas of responsibility. She was sure the problem was not hers. The store used point-of-sale terminals, in which the cash register doubled as a computer, instantly recognizing inventory movement. She also realized that shoplifting and other forms of loss were a constant problem in retail stores, so she instructed all her clerks to spot count inventory in their areas of responsibility whenever there was a lull in store traffic. The store computer had a built-in program to suggest replenishment orders when the stock reduced to a certain quantity. Amy had learned, of course, that these were only suggestions, since she knew that some items were "faddish" and would have to be ordered sooner or not reordered at all depending on how the fad was progressing. Some items were seasonal in nature, which needed to be accommodated, and she was also aware when an item would go on sale or have a special promotional campaign. These were announced well in advance during the monthly managerial meetings, and she had good estimates as to the projected impact on demand.

It was because she was so effective at managing the inventory in her area that she was so vocal about the problems at the warehouse. It seemed that almost everything she ordered for replenishment from the warehouse was a problem. Some items were late, occasionally by as much as six weeks. Other items were replenished in quantities far larger or smaller than what was ordered, even if they were occasionally delivered on time. It finally seemed to her that every warehouse delivery was a random event instead of the accurate filling of her orders. Her complaints to general management stemmed from the impact of the warehouse problems. Customers in her area were complaining more often and louder as stockouts of various items became a pattern. Several customers had vowed to never again shop at Lesscost because of their frustration.

In other cases, the quantity delivered was two to three times the amount she ordered. She would often have to hold special unannounced sales to avoid being burdened with the excessive inventory, especially since one of her performance metrics was inventory dollars. Of course, one of the major performance metrics was profitability, and both the stockouts and unannounced sales impacted that adversely. Finally, after one particularly frustrating day, she told the general manager, "Maybe you should put me in charge of the inventory over at the warehouse. I can control my own area here—I bet I could put that place back in shape pretty fast!" Two weeks later, she was notified she was promoted to inventory management supervisor for the warehouse.

The Current Situation

One of the first issues Amy faced was some not-so-subtle resentment from the warehouse general supervisor, Henry "Hank" Anderson. Hank had been a supervisor for over 10 years, having worked his way up from an entry-level handler position. The inventory supervisor position had been created specifically for Amy, as Hank had previously had responsibility for the inventory. Their mutual boss had explained to Hank that the reduction in overall responsibility was not a demotion, but that growth in the warehouse made splitting the responsibilities a necessity. Although Hank outwardly acknowledged the explanation, everyone knew that in reality he felt the change was a "slap in the face." As soon as Amy started the job she discovered that Hank had mentioned to several workers that he thought Amy was wrong for the job since she was young and had no previous warehouse experience. He strongly believed that the only way you could really understand how a warehouse worked was to start at the "bottom" and learn by experience.

Amy knew that the Hank situation was one she would have to work on, but in the meantime, she had to understand how things were run, and specifically why the warehouse was causing all the problems she experienced at the store. Her first stop was to talk to Kailin Cheng, who was responsible for processing orders from the store. Kailin explained the situation from her perspective.

"I realize how much it must have bothered you to see how your store requests were processed here, but it frustrates me too. I tried to group orders to prioritize due dates and still have a full truckload to send to the store, but I was constantly having problems thrown back at me. Sometimes I was told the warehouse couldn't find the inventory. Other times I was told that the quantity you ordered was less than a full box, and they couldn't (or wouldn't) split the box up, so they were sending the full box. Then they would find something they couldn't find when it was ordered a long time ago, so now that they found it they were sending it. That order would, of course, take up so much room in the truck that something else had to be left behind to be shipped later. Those problems, in combination with true inventory shortages from supplier-missed shipments, always seem to put us behind and we are never able to ship what we are supposed to. None of this seemed to bother Hank too much. Maybe you can do something to change the situation."

Amy's concern with what Kailin told her was increased when she asked Kailin if she knew the accuracy of their inventory records and was told that she wasn't sure, but the records were probably no more than 50% accurate. "How can that be?" Amy asked herself. She knew they had recently installed a new computer system to handle the inventory, they did cycle counting on a regular basis, and they used a "home base" storage system, where each SKU had its own designated space in the warehouse racks. She realized she needed to talk to one of the workers. She decided on Deshawn Carson, who had been with the company for about five years and had a reputation for being a dedicated and effective worker. Amy told Deshawn what she already knew and asked him if he could provide any additional information.

According to Deshawn, "What Kailin told you is true, but what she didn't tell you is that a lot of it is her fault. If she would only give us some advance warning about what she wants to send for the next shipment, we could probably do a better job of finding the material and staging it. What happens, though, is that she gives us this shipment list out of the blue and expects us to find it all and get it ready in very little time. For one thing, she doesn't understand that it's very impractical to break boxes apart in order to ship just the quantity she wants. We don't have a good way to package the partial box, and an open box increases the chance for the remaining goods to be damaged or get dirty. Even if we had a way to partially package, the time it would take would increase the chance we wouldn't make the shipment on time.

"Then there's the problem of finding material. When supplier shipments come in, they are often for more goods of a given SKU than we have room for on the rack. We put the rest in an overflow area, but it's really hard to keep track of. Even if we locate it in the system correctly, someone will soon move it to get to something behind it. That person will usually forget to record the move in the heat of getting a shipment ready. Since the cycle counts don't find it in the designated rack, the cycle counters adjust the count so the system doesn't even know it exists anymore. You might think we should expand the space in the rack to hold the maximum amount of each SKU, but we would need a warehouse at least double this size to do that—and there's no way management would approve that. I guess the only good thing about the situation is that when we do find some lost material that was requested earlier, we ship it to make up for not shipping it earlier."

Amy was beginning to feel a tightening in her stomach as she realized the extent of the problem here. She almost had to force herself to talk to Crista Chávez, who worked for the purchasing department and was responsible for warehouse ordering. Crista was also considered to be experienced, capable, and dedicated to doing a good job for the company. Crista added the following perspective:

"We have good suppliers, but they're not miracle workers. Since we beat them up so badly on price most of the time, I can understand why they're not interested in doing more than they already are. The problem is we can't seem to get our own house in order to give the suppliers a good idea of what we need, and when we really need it. To do that, we would need to know what the warehouse needs and when, and also the existing inventory of the item. We seem to have no idea what we need, and the inventory records are a joke. I spend most of my day changing order dates, order quantities, or expediting orders to fill a shortage, and often the shortage isn't really a shortage at all. Our only hope has been to order early and increase our order quantities to ensure we have enough safety stock to cover the inventory accuracy problems. I've complained to Hank several times, but all he says is that it's my job to pull the suppliers in line, that the problem is obviously theirs."

At least by this point, Amy had a better perspective about the problems. Unfortunately, it was now up to her to fix them. She wished she had never opened her mouth to complain about the problems. Too late for that—she now had to develop a strategy to deal with what she had been handed.

Assignment

1. Structure what you think the problems are. Be sure to separate the problems from the symptoms.
2. Assume Amy needs to build a data-based case to convince her boss and start to "win over" Hank. What data should she gather to help her build the case?
3. Develop a model of how you think the warehouse should work in this environment.
4. Develop a time-phased plan to move from the present situation to the model you developed in question 3.

CHAPTER FOURTEEN

PHYSICAL DISTRIBUTION

14.1 INTRODUCTION

In Chapter 1 it was pointed out that a supply chain is comprised of a series of suppliers and customers linked together by a physical distribution system. Usually, the supply chain consists of several companies linked in this way. This chapter will discuss the channels of physical distribution, which cover the physical movement of goods as well as the change of ownership that occurs throughout the supply chain. Physical distribution involves the movement of goods through the various transportation modes, the inventories that exist in transit and in distribution centers and production facilities, the physical handling of goods, and the need for protective packaging. Multiple warehouse decisions involve the costs associated with adding more warehouses and the effect on customer service.

Physical distribution is the movement of materials from the producer to the consumer, and is the responsibility of the distribution or logistics department. Figure 14.1 shows the relationship of the various functions in this type of system.

In Figure 14.1, the movement of materials is divided into two functions: physical supply and physical distribution. **Physical supply** is the movement and storage of goods (incoming items) from suppliers to manufacturing. Depending on the conditions of sale, the cost may be paid by either the supplier or the customer, but it is ultimately passed on to the customer. **Physical distribution**, on the other hand, is the movement and storage of finished goods from the end of production to the customer (outgoing items). The particular path in which the goods move, through distribution centers, wholesalers, and retailers, is called the channel of distribution.

Channels of Distribution

A channel of distribution is one or more companies or individuals who participate in the flow of goods and/or services from the producer to the final user or consumer. Sometimes a company delivers directly to its customers, but often it uses other companies or individuals to distribute some or all of its products to the final consumer. These companies or individuals are called intermediaries. Examples of intermediaries are wholesalers, agents, transportation companies, and warehouses.

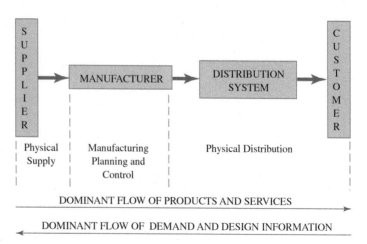

FIGURE 14.1 Supply chain (logistics system).

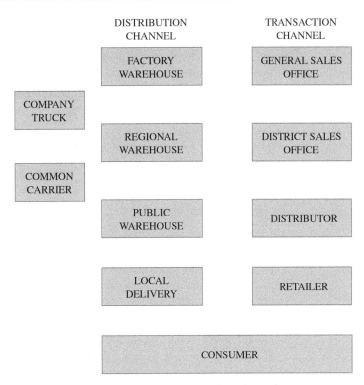

FIGURE 14.2 Separation of distribution and transaction channels.

There are two related channels involved: transaction channel and distribution channel. The **transaction channel** is concerned with the transfer of ownership and funds. Its function is to negotiate, contract, and sell. The **distribution channel** is concerned with the transfer or physical delivery of the goods or services. The same intermediary may perform both functions, but not necessarily.

Figure 14.2 shows an example of the separation of distribution and transaction channels. An example would be a company distributing a major appliance such as a refrigerator or stove. In such a process, the retailer usually carries only display models. When the customer orders an appliance, delivery is made from either the regional warehouse or the public warehouse.

Although it can be argued that one firm's physical supply is another firm's physical distribution, frequently there are important differences, particularly as they relate to the bulk and physical condition of raw materials and finished goods. The logistics problems that occur in moving and storing iron ore are quite different from those that occur with sheet steel. These differences influence the design of a logistics system and are important in deciding the location of distribution centers and factories. This text refers to both physical distribution and physical supply as physical distribution, but the differences for any particular company should be remembered.

Physical distribution is vital in everyday life. Usually, manufacturers, customers, and potential customers are widely dispersed geographically. If manufacturers serve only their local market, they restrict their potential for growth and profit. By extending its market, a firm can gain economies of scale in manufacturing, reduce the cost of purchases by volume discounts, and improve its profitability. However, to extend markets requires a larger and well-run distribution process. Manufacturing adds value to a product by taking the raw materials and creating something more useful. Bread is made from grain and is far more useful to humans than the grain itself. Distribution adds value by placing goods in markets where they are available to the consumer at the time the consumer wants them.

The specific way in which materials move depends upon many factors. For example:

- **The number and types of channels of distribution that the firm is using.** How many intermediaries the organization is going through to travel from producer to consumer, such as producer to wholesaler to retailer to consumer.

- **The types of markets served.** Market characteristics such as the geographic dispersion of the market, the number of customers, and the size of orders.
- **The characteristics of the product.** Factors such as weight, density, fragility, and perishability, and other special handling and storage considerations.
- **The type of transportation available to move the material.** Modes of transportation such as trains, ships, planes, and trucks.

All of these factors are closely related. For instance, florists selling a perishable product to a local market will sell directly and probably use their own trucks. However, a national canning company selling a nonperishable product to a national market through a distribution channel composed of wholesalers and retailers may use trucks and rail transport.

Reverse Logistics

As discussed in Chapter 2, reverse logistics is becoming increasingly important in the management of a supply chain because of the growth of online shopping and the desire to reuse and recycle materials once a customer is finished with the product. Companies must deal with the flow of goods coming back from the final customer or other companies in the distribution channel. This reverse supply chain may be caused by several factors, including the following:

- Quality issues in the goods resulting in customer returns, warranty replacements, and so forth.
- Financial pressures on distributors to reduce slow-moving or unwanted inventories.
- The return of reusable materials, such as packaging, after delivery.
- The return of components and materials for recycling or disposal, such as electronics.

The *APICS Dictionary*, 16th edition, defines **reverse logistics** as a "complete supply chain dedicated to the reverse flow of products and materials for the purpose of returns, repair, remanufacture, and/or recycling."

In some distribution channels reverse logistics can represent major costs, which are growing partly due to the increased use of the internet. Sales through the internet tend to be to a wider geographic area and may have a higher-than-normal frequency of returns. The total costs associated with reverse logistics are estimated to exceed $50 billion per year in the United States alone. The amount of goods returned in the publishing industry can be as high as 50% of the original shipments as magazines go out of date, or 90% in the automotive parts industry for the rebuilding of starter motors and alternators. In addition, companies are being forced to take responsibility for the return of packaging. There are two main categories of reverse logistics: asset recovery, which is the return of actual products, and **green reverse logistics**, which represents the responsibility of the supplier to dispose of packaging materials or environmentally sensitive materials such as heavy metals and other restricted materials.

The costs of green reverse logistics are reduced through the use of reusable packaging, such as bins or racks rather than corrugated containers, or an overall reduction in the amount of packaging. Environmentally sensitive materials can be sorted and either reused in manufacturing or disposed of in the most cost-effective method possible, hopefully avoiding landfills. Refillable beverage containers reduce the need for landfill space but impose a cost on the producer for sorting and handling.

Reducing costs associated with asset recovery involves coordinating the handling of the materials, perhaps with the outbound flow of new goods. Information on the returned materials is necessary to ensure proper reuse or disposal of the materials. If the return will generate a credit to the sender, this will also require information. An example of this working effectively occurs with the replacement of a starter motor in a car. The installer orders a starter motor and installs it. The old starter motor, which should be the same size and model as the replacement, is placed in the original carton and sent back to the supplier for a credit. The carton will have all the information describing the motor and can be used for identification as it travels back to the rebuilder. To reduce costs, distributors will coordinate the outbound shipments of new motors with the return of old motors for rebuilding.

Assets are returned for many reasons, which can include the following:

- Quality demands by final customers (both real and perceived).
- Damaged or defective products.
- Inventories that result from over-forecast demand.
- Seasonal inventories.
- Out-of-date inventories.
- Remanufacturing and refurbishment of products.

Returned goods can be:

- Returned to inventory.
- Refurbished for resale.
- Sold into alternate markets.
- Broken down into reusable components.
- Sorted to recover valuable materials (further reducing disposal costs).

If the distribution channel is very simple, as in the return of starter motors example, organizations can reduce reverse logistics costs by coordinating outbound and inbound shipments. Should the distribution channel, however, be very complex, third-party logistics companies (3PLs), discussed later in this chapter, can be used to centralize the handling and disposition of the returned goods. The 3PLs are also very good at providing the information to track the flow of the goods. Companies that take a strategic approach to reverse logistics can recover significant costs and provide better inventory levels throughout the supply chain. These companies will also build reputations as good corporate citizens.

14.2 PHYSICAL DISTRIBUTION

Physical distribution is responsible for delivering to the customer what is wanted on time and at minimum cost. The objective of distribution management is to design and operate a distribution channel that attains the required level of customer service and does so at least cost. To reach this objective, all activities involved in the movement and storage of goods must be organized into an integrated system.

Activities in Physical Distribution

A physical distribution system involves six interrelated activities that affect customer service and the cost of providing it:

1. **Transportation.** Transportation involves the various methods of moving goods outside the firm's buildings. For most firms, transportation is the single highest cost in distribution, usually accounting for 30% to 60% of distribution costs. Transportation only adds value to the product if the customer is willing to pay for the inventory to be located closer to them.

2. **Distribution inventory.** Distribution inventory includes all finished goods inventory at any point in the distribution channel. In cost terms, it is the second most important item in distribution, accounting for about 25% to 30% of the cost of distribution. Again, if the customer is willing to pay for it, inventories can create value by placing the product close to the customer.

3. **Warehouses** or **distribution centers (DC).** Warehouses are used to store inventory. The management of warehouses includes decisions on site selection, number of distribution centers in the system, layout, and methods of receiving, storing, and retrieving goods.

4. **Material handling.** Material handling is the movement and storage of goods inside the distribution center. The type of material handling equipment used affects the efficiency and cost of operating the distribution center. Material handling equipment represents a capital cost, and a trade-off exists between this capital cost and the operating costs of the distribution center. The labor associated with material handling represents a cost of carrying inventory.

5. **Protective packaging.** Goods moving in a distribution channel must be contained, protected, and identified. In addition, goods are moved and stored in packages and must fit into the dimension of the storage spaces and the transportation vehicles.
6. **Order processing and communication.** Order processing includes all activities needed to fill customer orders. Order processing represents a time element in delivery and is an important part of customer service. Many intermediaries are involved in the movement of goods, and good communication is essential to a successful distribution system.

Total Cost Concept

As mentioned, the objective of distribution management is to provide the required level of customer service at the **least total cost**. This does not mean that transportation costs or inventory costs or any one activity cost should be a minimum but that the total of all costs should be a minimum. What happens to one activity has an effect on other activities, total cost, and the service level. Management must treat the process as a whole and understand the relationships among the activities.

Example Problem

A company normally ships a product by rail. Transport by rail costs $200, and the transit time is 10 days. However, the goods can be moved by air at a cost of $1000 and will take one day to deliver. The cost of inventory in transit for a particular shipment is $100 per day. What are the costs involved in their decision?

ANSWER

	Rail	Air
Transportation Cost	$ 200	$1000
In-Transit Inventory Carrying Cost	1000	100
Total	$1200	$1100

There are two related principles illustrated here:

1. **Cost tradeoff.** The cost of transportation increased with the use of air transport, but the cost of carrying inventory decreased. There was a cost trade-off between the two.
2. **Total cost concept.** By considering all of the costs and not just any one cost, the total cost is reduced. Note also that even though no cost is attributed to it, customer service is improved by reducing the transit time. The total cost should also reflect the effect of the decision on other departments, such as production and marketing.

The preceding example does not mean that using faster transport always results in savings. For example, if the goods being moved are of low value and inventory carrying cost is only $10 per day, rail will be cheaper. In addition, other costs may have to be considered.

Most of the decisions in distribution, and indeed much of what is done in business and in our own lives, involve trade-offs and an appreciation of the total costs involved. In this section, the emphasis is on the costs and trade-offs incurred and on improvement in customer service. Generally, but not always, an increase in customer service requires an increase in cost, which is one of the major trade-offs.

Global Distribution

Global distribution is the movement of goods to and from locations around the world. Organizations are moving toward the global sourcing and selling of goods due to lower manufacturing costs in other nations and the ability of both foreign and domestic manufacturers to supply a global market. Global distribution of goods is similar to movement within one particular continent since information is needed to control inventories, customer needs must be satisfied, and carriers depend on communications to reach their destinations. Some differences, however, must be taken into consideration when dealing with organizations around the world: distance, language, culture, currency, and measurement.

The longer the distance that goods must travel, the greater the time to reach markets. Such distances may require conducting business across different time zones. As the goods cross borders, the languages spoken may differ for the manufacturer, warehouse agencies, and carriers. Cultural differences can include the methods of conducting business, the occurrence of religious holidays, and the local work ethic. Currency exchange and international fund transfers are becoming increasingly easy; however, fluctuations in currencies can change costs dramatically and must be taken into account when assessing the risks involved in deciding where to source goods. Measurement systems will differ around the world, and a good example of this is weight. A ton is 2000 pounds, a long ton is 2240 pounds, and a metric ton is 2205 pounds. The units of measure for weight can depend on the country that is handling the goods. Overall, global distribution can be very complicated for individuals and companies facing this challenge.

Fortunately, technologies and supply chain management processes in place now can alleviate some of these problems. Time is less of an issue. The internet allows companies to conduct business at all hours of the day, with fewer misinterpretations of order information such as sizes, quantities, and descriptions. International standards help solve many distribution issues. The International Organization for Standardization (ISO) (discussed in Chapter 17), for example, has established standards for ocean shipping containers that allow the seamless handling of goods from ship, to rail, to truck across virtually all nations. Some standards have a long history. As early as 1936, Incoterms were developed by the Paris-based International Chamber of Commerce to provide internationally accepted regulations for trade terms such as export packing costs, customs clearance, inland and ocean transportation costs, and damage insurance. Incoterms are explained later in this chapter.

Global distribution will continue to grow, become easier, and increasingly allow companies to manufacture goods at competitive rates and to sell their products in a global market. As global distribution becomes essential, people working in distribution will need to expand their knowledge of global-based supply systems and international business practices to remain effective.

3PLs: Third-Party Logistics Providers

Buyers and suppliers of goods often work with a third party who is in a position to offer physical distribution services at less cost than the buyer or seller would incur performing these services themselves. Beyond providing delivery services, a **third-party logistics (3PL)** provider may supply services such as warehousing, electronic data interchange (EDI), packaging, warehousing, freight forwarding, order processing, product tracking, and delivery. They can provide these services at an economical rate since they already have the infrastructure in place and combine one company's distribution needs with those of their other clients. FedEx is an example of this, providing customers over geographically dispersed markets with local warehousing, inventory management, labeling requirements, and more.

3PLs allow businesses to concentrate on their core competencies and leave the problems associated with delivering or servicing their products to someone else. They also react well to problems of scale where the volumes shipped or the number of customers either increase or decrease. 3PLs are especially effective in emerging markets where a supplier needs the capabilities of a local warehouse but does not have the resources or volumes to justify their own facility.

4PLs: Fourth-Party Logistics Providers

A supply chain partner similar to a 3PL is a **fourth-party logistics(4PL)** provider. The 4PL is a logistics specialist who manages the entire logistics operation for the customer. It includes those activities mentioned under 3PL, but also includes the possible subcontracting of other parties to perform some of the services, coordinating the logistics efforts of all parties. According to the *APICS Dictionary*, 16th edition, it differs from 3PL as it serves as "… an interface between the client and multiple logistics providers."

14.3 PHYSICAL DISTRIBUTION INTERFACES

By taking the goods produced by manufacturing and delivering them to the customer, physical distribution provides a bridge between marketing and production. As such, there are several important interfaces among physical distribution and production and marketing.

Marketing

Although physical distribution interacts with all departments in a business, its closest relationship is probably with marketing. The "marketing mix" is made up of product, promotion, price, and place, and the latter is created by physical distribution. Marketing is responsible for creating demand. This is accomplished by such methods as personal selling, advertising, sales promotion, merchandising, and pricing. Physical distribution is responsible for giving the customer possession of the goods and does so by operating distribution centers, transportation channels, inventories, and order processing. It has the responsibility of meeting the customer service levels established by marketing and the senior management of the firm.

Physical distribution contributes to creating demand. Prompt delivery, product availability, and accurate order filling are important competitive tools in promoting a firm's products. The distribution system is a cost, so its efficiency and effectiveness influence the company's ability to price competitively. All of these affect company profits.

Production

Physical supply establishes the flow of material into the production process. The component and raw material service level must usually be very high because the cost of interrupted production schedules caused by raw material shortages is usually enormous.

There are many factors involved in selecting a site for a factory, but an important one is the cost and availability of transportation for raw materials to the factory and the movement of finished goods to the marketplace. Sometimes the location of factories is decided largely by the sources and transportation links of raw materials. This is particularly true where the raw materials are bulky and of relatively low value compared to the finished product. The location of steel mills on the Great Lakes of the United States is a good example. The basic raw material, iron ore, is bulky, heavy, and of low unit value. Transportation costs must be kept low to make a steel mill profitable. Iron ore from mines in either northern Quebec, Canada, or the state of Minnesota is transported to the mills by cargo ship, which is typically the least costly mode of transportation. In other cases, the availability of low-cost transportation makes it possible to locate in areas remote from markets, but where labor is inexpensive.

Unless a firm is delivering finished goods directly to a customer, demand on the factory is created by the distribution center orders and not directly by the final customer. As noted in Chapter 12, this can have severe implications on the demand pattern at the factory. Although the demand from customers may be relatively uniform, the factory reacts to the demand from the distribution centers for replenishment stock. If the distribution centers are using an order point technique, the demand on the factory will not be uniform and will be dependent rather than independent. The distribution channel is the factory's customer, and the way that distribution interfaces with the factory will influence the efficiency of factory operations.

14.4 TRANSPORTATION

Transportation is an essential ingredient in the economic development of any area. It brings together raw materials for production of marketable commodities and distributes the products of industry to the marketplace. As such, it is a major contributor to the economic and social fabric of a society and aids economic development of regional areas.

The carriers of transportation can be divided into five basic modes:

1. Rail.
2. Road, including trucks, buses, and automobiles.
3. Air.
4. Water, including oceangoing, inland, and coastal ships.
5. Pipeline.

Each mode has different cost and service characteristics. These determine which method is appropriate for the types of goods to be moved. Certain types of traffic are simply more logically moved with one mode than they are with another. For example, trucks are best suited to moving small quantities to widely dispersed markets, but trains are best suited to moving large quantities of bulky cargo, such as grain.

Costs of Carriage

To provide transportation service, any carrier, whatever mode, must have certain basic physical elements. These elements are ways, terminals, and vehicles. Each results in a cost to the carrier and, depending on the mode and the carrier, may be either capital (fixed) or operating (variable) costs. **Fixed costs** are costs that do not change with the volume of goods carried. The purchase cost of a truck owned by the carrier is a fixed cost. No matter how much it is used, the cost of the vehicle does not change. However, many costs of operation, such as fuel, maintenance, and driver's wages, depend on the use made of the truck. These are **variable costs**.

Ways are the paths over which the carrier operates. They include the right-of-way (land area being used), plus any road, tracks, or other physical facilities needed on the right-of-way. The nature of the way and how it is paid for vary with the mode. Ways may be owned and operated by the government, by the carrier, or provided by nature.

Terminals are places where carriers load and unload goods to and from vehicles and make connections between local pickup and delivery service and line haul service. Other functions performed at terminals are: weighing; connecting with other routes and carriers; vehicle routing, dispatching, and maintenance; and administration and paperwork. The nature, size, and complexity of the terminal varies with the mode and size of the firm and the types of goods carried. Terminals are generally owned and operated by the carrier but, in some special circumstances, may be publicly owned and operated.

Vehicles of various types are used in all modes except pipelines. They serve as carrying and power units to move the goods over the ways. The carrier usually owns or leases the vehicles, although sometimes the shipper owns or leases them.

Besides ways, terminals, and vehicles, a carrier will have other costs such as maintenance, labor, fuel, and administration. These are generally part of operating costs and may be fixed or variable.

Rail

Railways provide their own ways, terminals, and vehicles, all of which represent a large capital investment. This means that most of the total cost of operating a railway is fixed. Thus, railways must have a high volume of traffic to absorb the fixed costs. They will not want to install and operate rail lines unless there is a large enough volume of traffic. Trains move goods by trainloads composed of perhaps a hundred cars, each with a carrying capacity of around 160,000 pounds.

Therefore, railways are best able to move large volumes of bulky goods over long distances. Their frequency of departure will be less than trucks, which can move when one truck is loaded. Rail speed is good over long distances, the service is generally reliable, and trains are flexible about the goods they can carry. Train service is cheaper than road service for large quantities of bulky commodities, such as coal, grain, potash, and containers moved over long distances.

Road

Trucks do not provide their own ways (roads and highways) but pay a fee to the government in the form of licenses, gasoline, and other taxes and tolls for the use of roads. Terminals are usually owned and operated by the carrier but may be either privately owned or owned by the government. Vehicles are owned, or leased, and operated by the carrier. If owned, they are a major capital expense. However, in comparison to other modes, the cost of a vehicle is small. This means that for road carriers, most of their costs are operating (variable) in nature.

Trucks can provide door-to-door service as long as there is a suitable surface on which to drive. In the United States and Canada, the road network is comprehensive. The unit of movement is a truckload, which can be up to about 100,000 pounds. These two factors—the excellent road system and the relatively small unit of movement—mean that trucks can provide fast flexible service almost anywhere in North America. Trucks are particularly suited to distribution of relatively small-volume goods to a dispersed market.

Air

Air transport does not have ways in the sense of fixed physical roadbeds, but it does require an airway system that includes air traffic control and navigation systems. These are usually provided by the government. Carriers pay a user charge that is a variable cost to them. Terminals include all of the airport facilities, most of which are provided by a local government. However, carriers are usually responsible for providing their own cargo terminals and maintenance facilities, either by owning or renting the space. The carrier provides the aircraft through either ownership or leasing. The aircraft are expensive and are the single most important cost element for the airline. Since operating costs are high, airlines' costs are mainly variable.

The main advantage of air transport is speed of service, especially over long distances. Some cargo travels in passenger aircraft, and thus many delivery schedules are tied to those of passenger service. The service destination is flexible provided there is a suitable landing strip. Transportation cost for air cargo is higher than for other modes. For these reasons, air transport is most often suitable for high-value, low-weight cargo or for emergency items.

Water

Waterways are provided by nature or by nature with the assistance of the government. The St. Lawrence Seaway system is an example of this. The carrier thus has no capital cost in providing the ways but may have to pay a fee for using the waterway.

Terminals may be provided by the government but are increasingly privately owned. In either case, the carrier will pay a fee to use them. Thus, terminals are mainly a variable cost. Vehicles (ships) are either owned or leased by the carrier and represent the major capital or fixed cost to the carrier.

The main advantage of water transport is cost. Operating costs are low, and since the ships have a relatively large capacity, the fixed costs can be absorbed over large volumes. Ships are slow and operate door to door only if the shipper and the consignee are on a waterway. Therefore, water transportation is most useful for moving low-value, bulky cargo over relatively long distances where waterways are available.

Pipelines

Pipelines are unique among the modes of transportation in that they move only gas, oil, and refined products on a widespread basis. As such, they are of little interest to most users of transportation. Capital costs for ways and pipelines are high and are borne by the carrier, but operating costs are very low.

Intermodal

When deliveries cannot be accomplished using a single method of transportation, multiple methods can be combined into a hybrid mode known as **intermodal transport**. Product travels part of the way on one mode, and is then transferred to another. Examples of this include:

- **Piggyback**, referring to placing a truck trailer or container on the back of a railcar.
- **Fishyback**, referring to truck trailers or containers that are transported partway by ship or barge.
- **Birdyback**, referring to cargo that is shipped using both road and air shipments.

Transportation Scheduling

Whichever mode of transportation is selected, the scheduling of shipments is a critical part of physical distribution. Shipments must be planned to optimize the transportation network, taking speed and cost into consideration, as well as the customer's needs. Routes must be selected, and may include the ability to perform **backhauling**, which refers to transporting a full or partial load on the return trip from the destination point to the point of origin. Since the bulk of a logistics budget may be used for transportation costs, constraints must be managed to ensure that goods are delivered on a timely and cost-effective basis. Some constraints are inherently based on the type of carrier selected, such as access to ports or railroads, air freight schedules, and so forth. Other constraints may occur unplanned, such as worker strikes, natural disasters that limit or prohibit accessibility and the movement of goods, and terrorism.

Transportation management systems (TMS) can be used to automate the transportation planning and scheduling process and deal with any bottlenecks that occur. These applications automate the planning and decision-making for activities such as determining the most efficient routes, carrier rate acceptance, dispatching, tracing shipments, fleet management, and generating required documents for domestic and import/export shipments. The TMS load planning/building process determines the appropriate mode of transportation and optimizes the shipment based on volume, density, and cost. For example, it can provide information required to make a decision on whether a shipment should be shipped LTL, or held for additional goods for a truckload shipment.

14.5 LEGAL TYPES OF CARRIAGE

Carriers are legally classified as public (for hire) or private (not for hire). In the latter, individuals or firms own or lease their vehicles and use them to move their own goods. Public transport, on the other hand, is in the business of hauling for others for pay. All modes of transport have public and for-hire carriers.

For-hire carriers are subject to economic regulation by federal, state, or municipal governments. Depending on the jurisdiction, economic regulation may be more or less severe, and in recent years, there has been a strong move by government to reduce regulations. Economic regulation has centered on three areas:

1. Regulation of rates.
2. Control of routes and service levels.
3. Control of market entry and exit.

Private carriers are not subject to economic regulation but, like public carriers, are regulated in such matters as public safety, license fees, and taxes.

For Hire

A for-hire carrier may carry goods for the public as a common carrier or under contract to a specified shipper.

Common carriers make a standing offer to serve the public. This means that whatever products they offer to carry will be carried for anyone wanting their service. With some minor exceptions, they can carry only those commodities they are licensed to carry. For instance, a household mover cannot carry gravel or fresh vegetables. Common carriers provide the following:

- Service available to the public.
- Service to designated points or in designated areas.
- Scheduled service.
- Service of a given class of movement or commodity.

Contract carriers haul only for those with whom they have a specific formal contract of service, not the general public. Contract carriers offer a service according to a contractual agreement signed with a specific shipper. The contract specifies the character of the service, performance, and charges.

Private

Private carriers own or lease their equipment and operate it themselves. This means investment in equipment, insurance, and maintenance expense. A company normally considers operating its own fleet only if the volume of transport is high enough to justify the capital expense.

Service Capability

Service capability depends on the availability of transportation service, which in turn depends on the control that the shipper has over the transportation agency. The shipper must go to the marketplace to hire a common carrier and is subject to the schedules and regulations of that carrier. Least control is exercised over common carriers. Shippers can exercise most control over their own vehicles and have the highest service capability with private carriage.

Other Transportation Agencies

There are several transportation agencies that use the various modes or combinations of the modes. Some of these are the post office, freight forwarders, couriers, and shippers. They all provide a transportation service, usually as a common carrier. They may own the vehicles, or they may contract with carriers to move their goods. Usually, they consolidate small shipments into large shipments to make economic loads.

14.6 TRANSPORTATION COST ELEMENTS

There are four basic cost elements in transportation. Knowledge of these costs enables a shipper to get a better price by selecting the right shipping mode. The four basic costs are as follows:

1. Line haul.
2. Pickup and delivery.
3. Terminal-handling.
4. Billing and collection.

Motor transport will be used as an example, but the principles are the same for all modes.

Goods move either directly from the shipper to the consignee or through a terminal. In the latter, they are picked up in some vehicle suitable for short-haul local travel. They are then delivered to a terminal where they are sorted according to destination and loaded onto highway vehicles for travel to a destination terminal. There they are again sorted, loaded on local delivery trucks, and taken to the consignee. Figure 14.3 shows this pattern schematically.

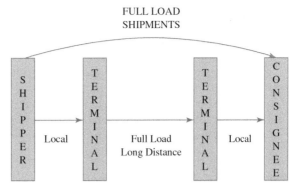

FIGURE 14.3 Shipping patterns.

Line Haul Costs

When goods are shipped, they are sent in a moving container that has a weight and volume capacity. The carrier, private or for hire, has basic costs to move this container, which exist whether the container is full or not. These are called **line haul costs**. For a truck, these include such items as the driver's wages and depreciation due to usage. These costs vary with the distance traveled, not the weight carried. The carrier has essentially the same basic costs whether the truck moves full or empty. If it is half full, the basic costs must be spread over only those goods in the truck.

Therefore, the **total line haul cost** varies directly with the cost per mile and the distance shipped, not on the weight shipped. For example, if for a given product the line haul cost is $3 per mile and the distance is 100 miles, the total line haul cost is $300. If the shipper sends 50,000 pounds, the total line haul cost is the same as if 10,000 pounds is sent.

Even though weight is not a factor in line haul costs, it can be used to designate the cost to the shipper. The unit of measure used for weight and mass may vary according to country, and includes ton, metric ton, pound, kilogram, and hundredweight (cwt.). The following example uses a calculation of **line haul cost per hundredweight**, which is a common designation for shipping commodities within North America.

$$\text{LHC/cwt.} = \frac{300}{500}$$
$$= \$0.60 \text{ per cwt. [for 50,000 lbs. (500 cwt.)]}$$
$$\text{LHC/cwt.} = \frac{300}{100}$$
$$= \$3 \text{ per cwt. [for 10,000 lbs. (100 cwt.)]}$$

Weight can also be used to compare line haul costs by determining the cost per a certain weight.

Example Problem

For a particular commodity, the line haul cost is $2.50 per mile. For a trip of 500 miles and a shipment of 6.5 tons, what is the cost of shipping per ton? If the shipment is increased to 10 tons, what is the saving in cost per ton?

ANSWER

$$\text{Total line haul cost} = \$2.5 \times 500 = \$1250$$
$$\text{Cost per ton} = \$1250 \div 6.5 = \$192.31$$

If shipping 10 tons:

$$\text{Cost per ton} = \$1250 \div 10 = \$125.00$$
$$\text{Savings per ton} = \$192.31 - \$125.00 = \$67.31$$

Volume

The carrier has two limitations or capacity restrictions on how much can be moved on any one trip: the weight limitation and the cubic volume limitation of the vehicle. With some commodities, their density is such that the volume limitation is reached before the weight limitation. If the shipper wants to ship more, a method of increasing the density of the goods must be found. This is one reason that some lightweight products are made so they nest (e.g., disposable cups) and bicycles and wheelbarrows are shipped in an unassembled state. This is not to frustrate the poor mortals who try to assemble them but to increase the density of the product so more weight can be shipped in a given vehicle. The same principle applies to goods stored in distribution centers. The more compact they are, the more can be stored in a given space. Therefore, if shippers want to reduce transportation cost, they should (a) increase the weight shipped and (b) maximize density.

Example Problem

A company ships barbecues fully assembled. The average line haul cost per shipment is $12.50 per mile, and the truck carries 100 assembled barbecues. The company decides to ship the barbecues unassembled and figures it can ship 500 barbecues in a truck. Calculate the line haul cost per barbecue assembled and unassembled. If the average trip is 300 miles, calculate the saving per barbecue.

Answer

Line haul cost assembled = $12.50 ÷ 100 = $0.125 per barbecue per mile
Line haul cost unassembled = $12.50 ÷ 500 = $0.025 per barbecue per mile

Physical Distribution

Saving per mile = $0.125 − 0.025 = $0.10
Trip saving = 300 × $0.10 = $30.00 per barbecue

Pickup and Delivery Costs

Pickup and delivery costs are similar to line haul costs except that the cost depends more on the time spent than on the distance traveled. The carrier will charge for each pickup and the weight picked up. If a shipper is making several shipments, it will be less expensive if they are consolidated and picked up on one trip.

Terminal-Handling Costs

Terminal-handling charges depend on the number of times a shipment must be loaded, handled, and unloaded. If full truckloads are shipped, the goods do not need to be handled in the terminal but can go directly to the consignee. If partial loads are shipped, they must be taken to the terminal, unloaded, sorted, and loaded onto a highway vehicle. At the destination, the goods must be unloaded, sorted, and loaded onto a local delivery vehicle.

The process of receiving includes not only the physical receipt of the material, but may also encompass inspection of the shipment to make sure it complies with the quantity on the purchase order, as well as any damage or other quality issues. Bulk packages may need to be broken down for storage or packaged for delivery.

Preparing an order for shipping from the warehouse typically involves picking the required quantities from warehouse storage, performing any necessary repackaging, labeling the package with any required delivery and safety information, and then loading it onto the carrier.

From the time a product is received until it is reshipped, an individual parcel may be handled multiple times. A shipper who has many customers, each ordering small quantities, will expect the terminal-handling costs to be high because there will be a handling charge for each package.

A basic rule for reducing terminal-handling costs is to reduce handling effort by consolidating shipments into fewer parcels. In addition, technology can help to lower the time it takes to perform these activities. Bar coding and RFID technology can simplify the process, as well as automated guided vehicle systems, automated storage retrieval systems (AS/RS), and robotics. Consideration must be given to the changes in packaging and installation of technology required to utilize the tools.

Billing and Collection Costs

Every time a shipment is made, paperwork must be done and an invoice made out. **Billing and collection costs** can be reduced by consolidating shipments and reducing the pickup frequency.

Total Transportation Costs

The total cost of transportation consists of line haul, pickup and delivery, terminal-handling, and billing and collection costs. To reduce shipping costs, the shipper needs to do the following:

- Decrease line haul costs per unit by increasing the weight shipped.
- Decrease pickup and delivery cost by reducing the number of pickups. This can be done by consolidating and increasing the weight per pickup.
- Decrease terminal-handling costs by decreasing the number of parcels by consolidating shipments.
- Decrease billing and collection costs by consolidating shipments.

For any given shipment, the line haul costs vary with the distance shipped. However, the pickup and delivery, terminal-handling, and billing costs are fixed. The total cost for any given shipment thus has a fixed cost and a variable cost associated with it. This relationship is shown in Figure 14.4. The carrier will consider this relationship and either charge a fixed cost plus so much per mile or offer a tapered rate. In the latter, the cost per mile for short distances far exceeds that for longer distances.

The rate charged by a carrier will also vary with the commodity shipped and will depend upon the following:

- **Value.** The more valuable the item, the greater the carrier's liability for damage will be.
- **Density.** The more dense the item, the greater the weight that can be carried in a given vehicle.
- **Perishability.** Perishable goods often require special equipment and methods of handling.
- **Packaging.** The method of packaging influences the risk of damage and breakage.
- **Hazardous materials.** Loads containing hazardous materials may require special considerations and/or handling.

In addition, carriers have two rate structures, one based on full loads called **truckload (TL)** or carload (CL) and one based on **less than truckload (LTL)** and less

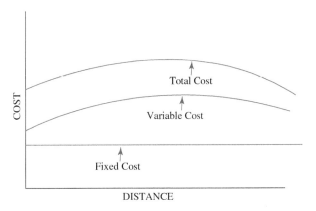

FIGURE 14.4 Distance versus cost of transportation.

than carload (LCL). For any given commodity, the LTL rates can be up to 100% higher than the TL rates. The basic reason for this differential lies in the extra pickup and delivery, terminal-handling, and billing and collection costs. Truckers, airlines, and water carriers accept less than full loads, but rail companies usually do not accept LCL shipments.

Transportation Terms

The transporters of goods have used terminology for many years to specify who pays the cost of transportation and who owns the goods in-transit. The four basic terms in common use are based on the term **free on board (FOB)**, which specifies who is responsible for the goods and pays the insurance to protect against risk for the goods.

The four FOB terms used allow astute buyers and sellers to take advantage of their own area of expertise to keep costs as low as possible. A small company with a low shipping volume often pays higher transportation fees than a large supplier. In the case of a small buyer, it is often better to get the shipper to pay for transportation and bill the buyer at a lower total cost. A company with their own fleet of trucks would want to take responsibility for the freight charges.

- **FOB origin, freight collect.** The seller makes the goods available at their dock and all risk and costs of transportation are assumed at this point by the buyer.
- **FOB origin, freight pre-paid.** The seller delivers the goods to the buyer but the buyer has the risk of the goods while in transit and pays the insurance.
- **FOB destination, freight pre-paid.** The seller is responsible for all costs of risk and transportation to the buyer's location.
- **FOB destination, freight collect.** The seller is responsible for the risk to the buyer's location but the buyer is responsible for the freight.

Incoterms

The term FOB indicates the point of transfer of ownership for North American shipments only, which does not necessarily conform to the international standards of **Incoterms**. International Commercial Terminology (Incoterms) was developed by the International Chamber of Commerce in 1936 to establish standards governing the responsibilities between buyers and sellers when transporting goods within and across international boundaries. The terms have been subsequently updated, and the definitions in this text are based on Incoterms 2010. In-transit goods represent a major investment to buyer and seller and run many risks of damage or loss. It is essential that a clear understanding exists as to who actually owns the goods at a given time or location in the supply chain and who will pay the expenses incurred to transport the goods. Another risk when transporting hazardous goods is the potential for accidents or spills and the potential environmental cleanup costs.

There are three responsibilities to be addressed:

- Cost of transportation of the goods.
- Ownership or insurance against risk of the goods.
- Preparation of customs documentation for the movement of the goods.

Incoterms are divided into two groups:

- Rules for any mode or modes of transport.
- Rules for sea and inland waterway transport.

Terms for any mode(s) of transport

- **EXW (named place): Ex Works.** The seller makes the goods available at their premises. The buyer is responsible for all charges to transport and insure the goods, including customs documentation. The buyer bears all risks for transporting the goods.
- **FCA (named carrier): Free carrier.** The seller is responsible for preparing the export documentation and delivering the goods to the transporter named after the Incoterm. The risk then passes to the buyer, who takes responsibility for all subsequent transportation costs.

- **CPT (named destination port): Carriage paid to.** The seller is responsible for export documentation and pays the freight to the named destination. However, the risk passes to the buyer when the goods have been delivered to the first carrier.
- **CIP (named destination): Carriage and insurance paid to.** The seller pays the costs of transportation to the destination and prepares the export documentation. Risk transfers to the buyer when the goods have been delivered to the first carrier.
- **DAT (named place): Delivered at terminal.** The seller prepares the export documentation and bears transportation costs to deliver the goods to the named terminal or port. Both parties must agree at which point within the terminal the risk is transferred from seller to buyer.
- **DAP (named place): Delivered at place.** The seller delivers the goods to the buyer, who assumes responsibility for the unloading at the named place. Both parties must agree at which point at the destination place the risk is transferred from seller to buyer.
- **DDP (named destination): Delivered duty paid.** The seller pays all the costs of shipping, export documentation, and any duties to the named destination, which is typically the buyer's location.

Terms for sea and inland waterway transport

- **FAS (named ship): Free alongside ship.** The seller is responsible for export documentation and the transportation of the goods to the point of ship loading at the port of export. Risk is passed to the buyer once the goods are delivered to the port of shipment. The buyer pays loading costs, freight, insurance, unloading costs, and transportation from the import port.
- **FOB (named ship): Free on board.** The seller is responsible for export documentation and delivery of the goods onto the named vessel or "past the ship's rail." FOB destination designates that the seller is responsible until the buyer takes possession. Note that this term is similar but not the same as the term *FOB* introduced earlier for North American freight.
- **CFR (named destination port): Cost and freight.** The seller is responsible for export documentation and pays for the freight to the port of destination. However, when the goods "pass the ship's rail" in the port of shipment, the buyer assumes the risk for the goods. This Incoterm is intended for use when shipping product that is not containerized.
- **CIF (named destination port): Cost, insurance, and freight.** The selling price includes the cost of all transportation costs and insurance to the named destination port. The seller is responsible for acquiring the insurance policy, but the responsibility of the goods ends when the goods are aboard the vessel.

14.7 WAREHOUSING

Chapter 13 discussed the management of warehouses. This section is concerned with the role of warehouses in a physical distribution system.

Warehouses include factory warehouses, regional warehouses, and local warehouses. They may be owned and operated by the supplier or intermediaries such as wholesalers, or they may be public warehouses. The latter offer a general service to their public, which includes providing storage space and warehouse services. Some warehouses specialize in the kinds of services they offer and the goods they store. A freezer storage is an example. The service functions that warehouses perform can be classified into two kinds:

1. The **general warehouse** is where goods are stored for long periods and where the prime purpose is to protect goods until they are needed. There is minimal handling, movement, and relationship to transportation. Furniture storage or depositories for documents are examples of this type of storage. It is also the type used for inventories accumulated in anticipation of seasonal sales.

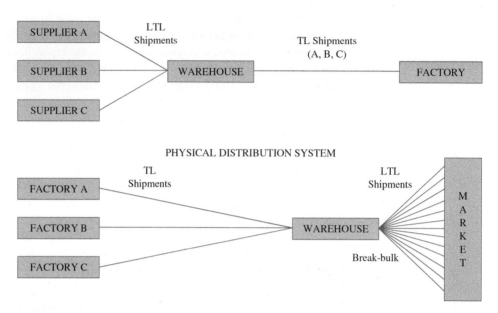

FIGURE 14.5 Transportation consolidation.

2. The **distribution warehouse** has a dynamic purpose of movement and mixing. Goods are received in large volume uniform lots, stored briefly, and then broken down into small individual orders of different items required by the customer in the marketplace. The emphasis is on movement and handling rather than on storage. This type of warehouse is widely used in distribution channels. The size of the warehouse is not so much its physical size as it is the throughput, or volume of traffic handled.

As discussed in the previous chapter, warehouses or distribution centers, are places where raw materials, semifinished, or finished goods are stored. They represent an interruption in the flow of material and thus add cost to the system. Items should be warehoused only if there is an offsetting benefit gained from storing them.

Role of Warehouses

Warehouses serve three important roles: transportation consolidation, product mixing, and service.

Transportation consolidation As shown in the preceding section, transportation costs can be reduced by using warehouses. This is accomplished by consolidating small (LTL) shipments into large (TL) shipments.

Consolidation can occur in both the supply and distribution systems. In physical supply, LTL shipments from several suppliers can be consolidated at a warehouse before being shipped as a TL to the factory. In physical distribution, TL shipments can be made to a distant warehouse and LTL shipments made to local users. Figure 14.5 shows the two situations graphically. Transportation consolidation in physical distribution is sometimes called **break-bulk**, which means the bulk (TL) shipments from factories to distribution centers are divided into small shipments going to local markets. **Cross-docking** (explained in more detail in Chapter 13) may also be used, meaning the items are moved from the incoming docking area directly to the outgoing carrier without being stored at the warehouse.

Product mixing Although transportation consolidation is concerned with reduction of transportation costs, product mixing deals with the grouping of different items into an order and the economies that warehouses can provide in doing this. When customers

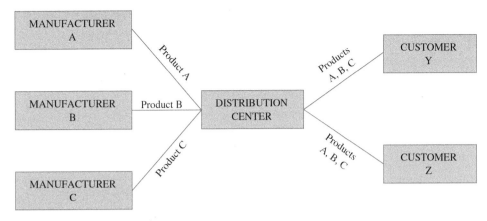

FIGURE 14.6 Product mixing.

place orders, they often want a mix of products that are produced in different locations. The same can be true for shipments being made from a distribution center to a retailer. Rather than ordering many items from multiple sources, fulfillment can be done through a distribution center that mixes the various products supplied from multiple sources.

Without a distribution center, customers and retailers would have to order from each source and pay for LTL transport from each source. Using a distribution center, orders can be placed and delivered from a central location. Figure 14.6 illustrates the concept.

Service Distribution centers improve customer service by positioning goods close to markets so the markets can be served more quickly.

Warehousing and Transportation Costs

Any distribution channel should try to provide the highest service level (the number of orders delivered in a specified time) at the lowest possible cost. The particular shipping pattern will depend largely upon the following:

- Number of customers.
- Geographic distribution of the customers.
- Customer order size.
- Number and location of plants and distribution centers.

Suppliers have little or no control over the first three but do have some control over the last. They can establish local distribution centers in their markets. With respect to transportation, it then becomes a question of the cost of serving customers directly from the central distribution center or from a regional distribution center. If truckload shipments are made, the cost is less from the central distribution center, but if LTL shipments are made, it may be cheaper to serve the customer from the local distribution center.

Example Problem

Suppose a company with a plant located in Toronto, Canada, is serving a market in the northeastern United States with many customers located in Boston. If the company ships direct to customers from the Toronto plant, most shipments will be less than truckload. However, if it locates a distribution center in Boston, it can ship truckloads to Boston and distribute by local cartage (LTL) to customers in that area. Whether this is economical or not depends on the total cost of shipping direct compared to shipping via the distribution center. Assume the following figures represent the average shipments to the Boston area:

Plant to customer LTL: $100/cwt.

Plant to distribution center TL: $50/cwt.

Inventory carrying cost (distribution center): $10/cwt.

Distribution center to customer LTL: $20/cwt.

Is it more economical to establish the distribution center in Boston? If the annual shipped volume is 10,000 cwt., what will be the annual saving?

ANSWER

Costs if a distribution center is used:

$$\begin{aligned}\text{TL Toronto to Boston} &= \$50 \text{ per cwt.} \\ \text{Distribution center costs} &= \$10 \text{ per cwt.} \\ \text{LTL in Boston area} &= \underline{\$20 \text{ per cwt.}} \\ \text{Total cost} &= \$80 \text{ per cwt.} \\ \text{Saving per cwt.} &= \$100 - \$80 = \$20 \\ \text{Annual saving} &= \$20 \times \$10,000 = \$200,000\end{aligned}$$

Market Boundaries

Continuing with the previous example problem, the company can now supply customers in other locations directly from the factory in Toronto or through the distribution center in Boston. The question is to decide which locations should be supplied from each source. The answer, of course, is the source that can service the location at least cost.

In order to determine least cost, the logistics costs from different locations need to be determined. **Landed cost**, sometimes referred to as total landed cost, laid down cost or landed duty price, is the delivered cost of a product to a particular geographic point. The delivered cost includes all logistics costs of moving the goods from A to B, including transportation, warehousing, handling, and any taxes and fees associated with logistics. In the previous example problem, the landed cost of delivering from Toronto would be the transportation cost per mile times the miles to a particular destination. The landed cost from Boston would include all costs of getting the goods to Boston, inventory costs in the Boston distribution center, and the transportation costs in getting to a particular destination.

$$\text{Landed cost} = P + TX + F$$

where

P = product costs
T = transportation costs per mile
X = distance
F = fees

The product cost includes all costs in getting the product to the supply location and storing it there. In the previous example, the product cost at Boston includes the TL cost of delivery to Boston and the inventory cost at Boston.

Example Problem

Syracuse is 300 miles from Toronto. The product cost for an item is $10 per cwt., and the transportation cost per mile per cwt. is $0.20. What is the landed cost per cwt.?

ANSWER

$$\begin{aligned}\text{Landed cost} &= \text{Product cost} + (\text{transportation cost per mile})(\text{distance}) \\ &= \$10 + (\$0.20 \times 300) = \$70 \text{ per cwt.}\end{aligned}$$

Market boundary The **market boundary** is the line between two or more supply sources where the landed cost is the same. Consider Figure 14.7. There are two sources of supply: A and B. The market boundary occurs at Y, where the landed cost from A is the same as from B.

In the example shown in Figure 14.7, the distance between A and B is 100 miles. If the distance from A to Y is X miles, then the distance from B to Y is $(100 - X)$ miles. Assume supply A is the factory and supply B is a distribution center. Assume the product cost at A is $100 and product cost from B is $100 plus TL transportation

Packages are arranged on it so that several packages may be moved at one time. Loaded with packages, it forms a cube that is a unit load.

Unitization can be successive. Shippers place their products into primary packages, the packages into shipping cartons, the cartons onto pallets, and the pallets into warehouses, trucks, or other vehicles.

To use the capacity of pallets, trucks or other vehicles, and warehouses, there should be some relationship between the dimensions of the product, the primary package, the shipping cartons, the pallet, the truck, and the warehouse space. The packages should be designed so space on the pallet is fully utilized and so the cartons interlock to form a stable load. Figure 14.9 shows two unit loads, each using the total space of the pallet. However, load B does not interlock and is not stable.

Pallets fit into trucks and railway cars. The standard dimensions are selected so pallets would fit into nominal 50′ railway cars and 40′ or 53′ truck trailers with a minimum of lost space. Figure 14.10 shows the layout in railcars and trailers.

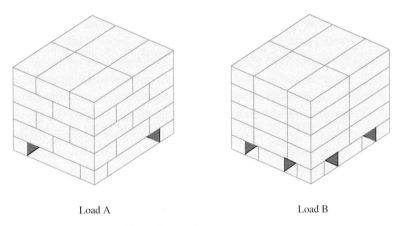

Load A　　　　　　　　　　Load B

FIGURE 14.9 Stable and unstable pallet loads.

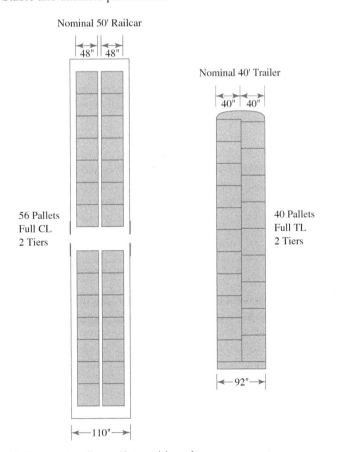

FIGURE 14.10 Railcar and trailer pallet position plan.

Thus, to get the highest cube utilization, consideration must be given to the dimensions of the product, the carton, the pallet, the vehicle, and the warehouse.

14.9 MATERIAL HANDLING

Material handling is the short-distance movement that takes place in or around a building such as a plant or distribution center. For a distribution center, this means the unloading and loading of transport vehicles and the dispatch and recall of goods to and from storage. In addition, the racking systems used in distribution centers are usually considered to be part of material handling.

Some objectives of material handling are as follows:

1. To increase cube utilization by using the height of the building and by reducing the need for aisle space as much as possible.
2. To improve operating efficiency by reducing handling. Increasing the load per move will result in fewer moves.
3. To improve the service level by increasing the speed of response to customer needs.

There are many types of material handling equipment. For convenience, they can be grouped into three categories: conveyors, industrial trucks, and cranes and hoists.

Conveyors are devices that move material (or people) horizontally or vertically between two fixed points. They are expensive, create a fixed route, and occupy space continuously. As a result, they are used only where there is sufficient throughput between fixed points to justify their cost.

Industrial trucks are vehicles powered by hand, electricity, or propane. Diesel and gasoline are not used indoors because they are noxious and lethal. Industrial trucks are more flexible than conveyors in that they can move anywhere there is a suitable surface and no obstructions. They do not occupy space continuously. For these reasons, they are the most often used form of material handling in distribution centers and in manufacturing.

Cranes and hoists can move materials vertically and horizontally to any point within their area of operation. They use overhead space and are used to move heavy or large items. Within their area of operation, they are very flexible.

As technology improves and becomes more affordable, other types of more automated material handling equipment is becoming more common, especially in larger warehouses. Those include:

Automated guided vehicle systems (AGVS) as defined in the *APICS Dictionary* (16th edition), is "A transportation network that automatically routes one or more material handling devices, such as carts or pallet trucks, and positions them at predetermined destinations without operator intervention."

Automated storage/retrieval systems (AS/RS) is defined in the *APICS Dictionary* (16th edition) as "A high-density rack inventory storage system that uses vehicles to automatically load and unload the racks."

14.10 MULTI-WAREHOUSE SYSTEMS

This section will look at the result of adding more distribution centers to the system. As might be expected, there is an effect on the cost of warehousing, material handling, inventories, packaging, and transportation. The objective will be to look at how all of these costs and the total cost behave, as well as what happens to the service level as more distribution centers are added to the system.

Transportation Costs

In the section on transportation, it was discussed that if shipments to customers are in less-than-full vehicle lots, the total transportation cost is reduced by establishing a distribution center in a market area. This is because more weight can be shipped for greater distances by truck or carload and the LTL shipments can be made over relatively short distances. Generally, then, as more distribution centers are added to a system, the following can be expected:

- The cost of TL shipments increases.
- The cost of LTL shipments decreases.
- The total cost of transportation decreases.

The major savings are made with the addition of the first distribution centers. Eventually, as more distribution centers are added, the marginal savings decrease.

Inventory Carrying Cost

The average inventory carried depends on the order quantity and the safety stock. The average order quantity inventory in the system should remain the same since it depends on demand, the cost of ordering, and the cost of carrying inventory.

The total safety stock will be affected by the number of warehouses in the system. Safety stock is carried to protect against fluctuations in demand during the lead time and depends, in part, on the number of units sold. In Chapter 12, it was shown that the standard deviation varies as the square root of the ratio of the forecast and lead time intervals. Similarly, for the same SKU, the standard deviation varies approximately as the square root of the ratio of the different annual demands. Suppose that the average demand is 1000 units and, for a service level of 90%, the safety stock is 100 units. If the 1000 units is divided between two distribution centers each having a demand of 500 units, the safety stock in each is:

$$SS = 100\sqrt{\frac{500}{1000}} = 71 \text{ units (in each warehouse)}$$

With two distribution centers and the same total sales, the total safety stock increases to 142 from 100. Thus, with a constant sales volume, as the number of distribution centers increases, the demand on each decreases but there is an increase in the total safety stock in all distribution centers.

Warehousing Costs

The fixed costs associated with distribution centers are space and material handling. The space needed depends on the amount of inventory carried. As noted previously, as more distribution centers are added to the channel, more inventory has to be carried, which requires more space.

In addition, there will be some duplication of non-storage space such as restrooms and offices. So, as the number of distribution centers increases, there will be a gradual increase in distribution center space costs.

Operating costs also increase as the number of distribution centers increases. Operating costs depend largely on the number of units handled. If there is no increase in sales, the total number of units handled remains the same, as does the cost of handling. However, the non-direct supervision and clerical costs increase.

Material Handling Costs

Material handling costs depend upon the number of units handled. If the sales volume remains constant, the number of units handled should also remain constant. There will be little change in material handling costs as long as the firm can ship unit loads to the distribution center. However, if the number of distribution centers increases to the point that some non-unitized loads are shipped, material handling costs increase.

Packaging Costs

Per-unit packaging costs will remain the same, but since there will be more inventory in the multi-warehouse system, total packaging costs will rise with inventory.

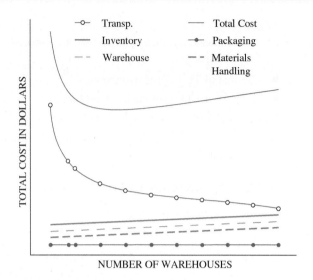

FIGURE 14.11 Total system cost.

Total Cost

With the assumption that total sales remain the same, Figure 14.11 shows graphically how the costs of transportation, warehousing, material handling, inventory, and packaging behave as distribution centers are added to the system. Up to a point, total costs decrease and then start to increase. Similar to the determination of an economic order quantity (EOQ), as discussed in Chapter 11, it is the objective of logistics to determine this least-cost point.

System Service Capability

The service capability of the system must also be evaluated. One way of assessing this is by estimating the percentage of the market served within a given period. Figure 14.12 represents such an estimate.

As expected, the service level increases as the number of distribution centers increases. It increases rapidly from one to two distribution centers and much less rapidly as the number is further increased. The first distribution center is built to serve the best market, the next to serve the second-best market, and so on.

Assume that a study has been made of a system of 1 to 10 distribution centers and the costs are as shown in Figure 14.13.

A three-distribution-center system would provide the least total cost. Figure 14.13 shows that by moving from 3 to 10 distribution centers, the one-day service level increases by 8%. Management must decide which distribution system to select. The decision must be based on adequate analysis of the choices available and a comparison of the increase in costs and service level.

Number of Warehouses	Percentage of Market Reached in 1 Day
1	30
2	70
3	87
10	95

FIGURE 14.12 Estimate of market reached versus number of warehouses.

Cost ($1000)	Number of Locations			
	1	2	3	10
Transportation	$8000	$6000	$5000	$4500
Warehousing	500	600	700	900
Materials Handling	1000	1000	1100	1400
Inventory	400	425	460	700
Packaging	100	100	100	100
Total Cost	$10,000	$8125	$7360	$7600

FIGURE 14.13 Cost versus number of warehouses.

14.11 IMPACT OF TECHNOLOGY IN DISTRIBUTION

As is the case in warehouse management (and introduced in Chapter 13) rapid growth in technology applications is also impacting physical distribution, including transportation. Tracking goods in the distribution network can become easier, more effective, more efficient, faster, and more accurate. Information about the condition (maintenance issues, for example) and location of fleets can also be more effective and efficiently done with rapid information linkages available through technology applications.

Chapter 13 introduced the concept of the Internet of Things (IoT) and its potential application in a warehouse setting. The connectivity implied by the IoT concept can also provide great benefits in the transportation network as well. Unfortunately, the same risks mentioned in Chapter 13 related to reliance on the internet also exist in its use for transportation. Those risks include unintentionally downloading malware, ransomware, computer viruses, and also the risks of someone "hacking" into the systems. Some of the potential applications include:

- Effectively connecting a truck driver with the dispatcher for real-time information can reduce delays and costs. There is likely to be better overall visibility into current transportation activities.
- Sensors on trucks can detect truck issues such as fuel efficiencies and possible maintenance issues.
- Linkages into fleet management systems can recommend optimal routes, speeds, loading and unloading hours, and alternative routes in case of construction problems, commuting congestions, weather problems, or other issues such as congestion at target loading docks.
- Data gathered from IoT linked trucks can reduce the time necessary to fill out logbooks, comply with regulations, etc.
- Sensors on pallets and other item in a truckload can provide linkages into the truck drivers and fleet managers. This can potentially assist with damage control, capacity management, and total visibility.
- Real-time data gathering provides for monitoring, information analysis, anti-counterfeiting of products, and automation and optimization of many supply chain and logistic actions.

SUMMARY

Physical distribution is the movement of goods into and out of a company and is changing in both the global economy and in the individual delivery of goods to internet customers. The majority of goods are still shipped in large economical units through distribution channels. Third- and fourth-party logistics providers are applying their area of specialization to help companies drive down costs while improving customer service. Sound packaging design is critical to a company not only because it affects the efficient handling of product all the way through the supply chain but also because it can have a high environmental cost. Reverse logistics along with a change in the distribution channel to internet customers is another reason for the growth of 3PLs. Warehouses strategically placed through the supply chain can help provide improved customer service and reduce costs.

The global sourcing and distribution of goods across international boundaries has increased the need for accurate, comprehensive documentation to determine things such as who pays the freight, who prepares the documentation, who owns the goods, and who may be responsible for any environmental problems. Incoterms, standard terms accepted around the world, address most of these concerns and are a requirement of any freight movement. Scheduling of shipments is a critical part of physical distribution in order to optimize transportation methods while taking speed, cost, and customers' needs into consideration.

KEY TERMS

Automated guided vehicle systems (AGVS) 362
Automated storage/retrieval systems (AS/RS) 362
Backhauling 349
Billing and collection costs 353
Birdyback 349
Break-bulk 356
Common carrier 350
Contract carrier 350
Conveyor 362
Cost tradeoff 344
Cranes and hoists 362
Cross-docking 356
Density 353
Distribution center (DC) 343
Distribution channel 341
Distribution inventory 343
Distribution warehouse 356
Fishyback 349
Fixed costs 347
Fourth-party logistics (4PL) 345
Free on board (FOB) 354
General warehouse 355
Green reverse logistics 342
Hazardous material 353
Incoterms 354
Industrial trucks 362

Intermodal transport 349
Landed cost 358
Least total cost 344
Less than truckload (LTL) 353
Line haul cost per hundredweight 351
Line haul costs 351
Market boundary 358
Material handling 343
Order processing and communication 344
Packaging 353
Pallet 360
Perishability 353
Physical distribution 340
Physical supply 340
Pickup and delivery costs 352
Piggyback 349
Private carrier 350
Protective packaging 344
Reverse logistics 342
Terminal-handling charges 352
Terminals 347
Third-party logistics (3PL) 345
Total cost concept 344
Total line haul cost 351
Transaction channel 341
Transportation 343
Transportation management system (TMS) 349

Truckload (TL) 353
Unitization 360
Unit loads 360
Value 353

Variable cost 347
Vehicle 347
Warehouse 343
Ways 347

QUESTIONS

1. Name and describe the two functions in the flow of materials from supplier to consumer. What are the differences between physical supply and physical distribution?
2. What is the primary function of the transaction channel and the distribution channel?
3. The particular way that goods move depends in part on four factors. What are they?
4. Describe the reverse logistics system of a beverage company, including the movement of goods and the flow of payment. What steps does the company take to reduce their costs?
5. Why are the total costs associated with reverse logistics increasing? Will this trend continue into the future?
6. What are the objectives of a physical distribution system?
7. Name and describe each of the six activities in a physical distribution system.
8. What are the cost trade-off and total cost concepts? Why are they important?
9. Describe the relationship between marketing and physical distribution. How does physical distribution contribute to creating demand?
10. Why is the demand placed on a central distribution center or a factory by distribution centers considered dependent?
11. What are the five basic modes of transportation?
12. What are the three physical elements in a transportation system? For each of the five modes, describe who provides them and how they are funded.
13. Describe why train service is cheaper than road transport for large quantities of bulky commodities moving over long distances.
14. Why can trucks provide a fast, flexible service for the distribution of small volumes of goods to a dispersed market?
15. What are the major characteristics of water and air transport?
16. What are the major legal types of carriage? What are the three areas of economic regulation? To which legal type of carriage do they apply?
17. Compare common and contract carriage. How do they differ from private carriage? Which will give the highest level of service?
18. On what do total line haul costs and line haul costs per hundredweight depend? What two ways can shippers reduce line haul costs?
19. Describe how a shipper can reduce the following:
    a. Pickup and delivery costs.
    b. Terminal-handling costs.
    c. Billing and collection costs.
20. The rates charged by a shipper vary with the commodity shipped. Name and describe five factors that affect the rates.
21. Why are LTL rates more expensive than TL rates?
22. Name and describe the two basic types of warehouses.
23. Name and describe the three important roles warehouses serve.
24. Name four factors that affect shipping patterns. Which can a supplier control?
25. What is the landed cost? What is a market boundary? Why are landed costs important in determining market boundaries?
26. As more distribution centers are added to a system, what happens to the cost of truckload, less than truckload, and total transportation costs?
27. What are the three roles of packaging in a distribution system? Describe why each is important.
28. What is unitization? Why is it important in physical distribution? Why is it successive?

29. What are three prime objectives of material handling? Describe the characteristics of conveyors, industrial trucks, cranes, and hoists.

30. As more warehouses are added to the system, what would we expect to happen to the following?
    a. Transportation costs.
    b. Inventory costs.
    c. Material handling costs.
    d. Packaging costs.
    e. Total costs.
    f. System service capability.

31. How does the use of the transportation term *FOB* for domestic shipments in North America differ from the use of the term in international shipments?

32. Which North American shipping term has the minimum risk for:
    a. the buyer?
    b. the seller?

33. A buyer in England would like to import a Harley-Davidson motorcycle directly from a dealer in the United States. Which Incoterm would minimize the buyer's efforts? Would arranging the shipment themselves save money for the buyer?

34. What is meant by the term *green reverse logistics*?

35. Describe how a 3PL would benefit a company that sells their product to individuals through the internet.

36. Describe how reverse logistics would apply to your school bookstore.

PROBLEMS

14.1. A company normally ships to a customer by rail at a cost of $400 per load. The transit time is 15 days. The goods can be shipped by truck for $700 per load and a transit time of four days. If transit inventory cost is $36 per day, what does it cost to ship each way?

Answer. Rail, $940; truck, $844

14.2. A company manufactures component parts for machine tools in North America and ships them to Southeast Asia for assembly and sale in the local market. The components are shipped by sea, transit time averages six weeks, and the shipping costs $2700 per shipment. The company is considering moving the parts by air at an estimated cost of $7500; the shipment taking two days to get there. If inventory carrying cost for the shipment in transit costs $100 per day, should they ship by air?

Why should the fact that forecasts are more accurate for nearer periods of time be considered? What activities are affected by the shorter lead time?

14.3. For a given commodity, the line haul cost is $13 per mile. For a trip of 200 miles and a shipment of 300 lbs., what is the cost per pound? If the shipment is increased to 400 lbs., what is the saving in cost per pound?

Answer. $2.17 per lb.

14.4. A company ships a particular product to a market located 1600 miles from the plant at a cost of $4.50 per mile. Normally it ships 350 units at a time. What is the line haul cost per unit?

14.5. In problem 14.4, if the company can ship the units unassembled, it can ship 550 units in a truck. What is the line haul cost per unit now?

14.6. A company processes feathers and ships them loose in a covered truck. The line haul cost for an average shipment is $600, and the truck carries 2000 pounds of feathers. A bright new graduate has just been hired and has suggested that they should bale the feathers into 500-pound bales. This would make them easier to handle and also allow them to be compressed into about one-tenth of the space they now occupy. How many pounds of feathers can the truck now carry? What is the present line haul cost per pound? What will it be if the proposal is adopted?

14.7. A company in Calgary serves a market in the northwestern United States. Now it ships LTL at an average cost of $30 per unit. If the company establishes a distribution center in the market, it estimates the TL cost will be $15 per unit, inventory carrying costs will be $5 per unit, and the local LTL cost will average $7 per unit. If the company forecasts annual demand at 200,000 units, how much will they save annually?

Answer. Annual saving = $600,000

14.8. A company ships LTL to customers in a market in the Midwest at an average cost of $40 per cwt. It proposes establishing a distribution center in this market. TL shipment costs to the DC would be $20 per cwt., the estimated inventory carrying costs are $5 per cwt., and the local cartage (LTL) cost is estimated at $10 per cwt. If the annual shipped volume is 100,000 cwt., what will the annual savings be by establishing the distribution center?

14.9. A company has a central supply facility and a distribution center located 400 miles away. The central supply product cost is $75, TL transportation rates from central supply to the DC are $60 per unit, and handling costs at the DC are $4 per unit. Calculate the market boundary location and the landed cost at the market boundary. LTL rates are $2 per unit per mile.

Answer. Market boundary is 216 miles from central supply. Landed cost = $507

14.10. Suppose the company in problem 14.8 had another market area located between the parent plant and the proposed distribution center. The LTL costs from the plant to that market are $35 per cwt. The company estimates that LTL shipments from the distribution center will cost $7 per cwt. Should it supply this market from the distribution center or central supply?

14.11. A company can ship LTL direct to customers in city A or use a public warehouse located in city B. It has determined the following data.
Cost per pound for shipping LTL to city A is $0.70 + $0.30 per mile.
Cost per pound for shipping TL to warehouse is $0.40 + $0.15 per mile.
Warehouse handling costs are $0.30 per pound.
Distances: Plant to city A = 120 miles
Plant to city B = 145 miles
From city B to city A = 40 miles

a. What is the total cost per pound to ship from the plant direct to customers in city A?

b. What is the total cost per pound to ship via the warehouse in city B?

c. In this problem, the cost per pound has a fixed and a variable component. Why?

CASE STUDY 14.1

Metal Specialties, Inc.

Metal Specialties is a wholesaler of specialty metals such as stainless steel and tool steel. The company purchases its stainless steel from a mill located some 200 miles away. At present the company operates its own truck. However, the truck is in need of repairs estimated at $20,000. Annual operating costs are $30,000 and the line haul costs are $2.20 per mile. Janet Jones (JJ), the traffic manager, wants to reduce the cost of bringing in the stainless steel, and because of the impending repair expense, she feels now is a good time to look at alternatives. She has solicited a number of proposals and has narrowed her choices down to a motor carrier and a rail carrier.

Heavy Metal Transport (HMT), a contract motor carrier, has an excellent reputation for service and reliability. It has submitted an incremental rate, $4.00/cwt. for shipments weighing less than 150 cwt., $3.80 for shipments between 150 and 200 cwt., $3.60 for shipments between 200 and 250 cwt., and $3.40 for shipments over 250 cwt. up to a maximum of 400 cwt.

Midland Continental Railway has submitted a piggyback rate of $3.25 per cwt. with a minimum load of 200 cwt. The piggyback rate includes pickup by truck at the steel mill, line haul by trailer on flat car, and delivery by truck to Metal Specialties' warehouse. They are considered to be a reliable carrier as well.

The finance department estimates that Metal Specialties' annual inventory carrying cost is 20%, the cost of inventory in transit is 10%, and the cost of capital is 8%. The cost of placing an order for stainless steel is estimated to be $40 per order. Stainless steel presently costs $300 per cwt.

Assignment

1. JJ has to make a decision soon. Given the information provided, what would you advise her to do?

CASE STUDY 14.2

Rictok Fabrication Products Delivery Problem

The management team of Rictok Fabrication Products Company (RFPC) was deeply involved in a discussion about a recent request from one of their major long-time customers. Although the customer was located some 2300 miles from the RFPC location, the relations between the two companies had been very strong for many years. Recently, however, the customer had approached RFPC with what appeared to be a simple request. The customer was starting a pilot project to determine if moving to a leaner production approach would work for them. The usual approach for one of the RFPC components that would have been impacted by the project was to have RFPC produce and load a full truckload of the parts and ship them directly to the customer warehouse. The customer would assume ownership once the parts arrived. That full truckload represented approximately five to six months' worth of average production, meaning that RFPC would send a truck roughly twice a year. They would normally plan to complete and ship another load when the customer still has from four to six weeks' worth of stock remaining in their warehouse. What the customer was asking for is to have the delivery be made about every two weeks and set to arrive when their existing inventory represented only about one weeks' worth of production. The wanted to significantly reduce their inventory holding cost and the cost they had been incurring related to damage by material handling and obsolescence (the design of the parts sometimes needed to be modified per customer request).

The supply chain manager of RFPC started the meeting with an overall statement "This customer has been and is likely to be a good, steady customer and since their purchases from us represent over 60% of our annual revenue, I think we need to take their request and a requirement and not just a simple request. Even though they are currently only asking about this one component, since this is a pilot project of theirs, I would assume that if it is successful, they will be expanding it to most of the other components we make for them. I talked with their supply chain manager and they realize that this may have an impact on our cost structure. They are willing to negotiate sharing that cost, but we have to be ready to give them details of what we can do and what it will mean for delivery and cost. One final point—whatever solution we develop, we need to keep in mind that we cannot reduce our production lot size. The setup time for the equipment we use for that component is large, and even once we get the setup completed it will often take several days for the output to meet the tight production quality specifications. What that means is that once we have the setup complete and running correctly, we need to make a large production run to absorb all that setup cost and equipment downtime while the setup is being done. Right now we are looking at a minimum production lot run of at least nine months' worth of demand to justify the setup. Also keep in mind that while we own enough warehouse space we have to contract with shippers for delivery."

Prior to the meeting the supply chain manager had the transportation and warehouse specialists develop some alternative approaches with both qualitative issues and some cost estimates. He now presented these alternatives, in no particular order of attractiveness:

Continue delivery as is, but stop requiring the customer from accepting ownership until they use the inventory in their production (consignment inventory).

Keep the inventory at the RFPC warehouse, and then deliver it every two weeks as requested by the customer.

Deliver full truckloads of inventory to the city where the customer facility is, but instead of delivering it all to the customer rent space in a public warehouse and then use short distance delivery to the customer every two weeks as requested.

Deliver by rail to the city where the customer facility is (again to a public warehouse), then use short distance delivery to the customer every two weeks as requested.

To help analyze the options, the supply chain manger handed out a sheet of data and issues that could be used for the analysis:

- Distance from RFPC to customer facility: 2300 miles

- Pallets: each pallet is four feet wide, four feet deep. There are two components in each box, ten boxes on each tier of the pallet, and five tiers per pallet, making the height 4.1 feet. Completely loaded, the pallet weighs 450 pounds.

- Each component has a book cost (cost of goods sold) of $23.00. Inventory holding cost for RFPC averages about 17% per year.

- A full semitruck trailer is 53 feet long, 13.5 feet high, and 8.5 feet wide. The truck then can hold a maximum of 52 pallets as a full load.

- The customer uses an average of 220 components per week under normal circumstances.

- The analysis did not obtain a specific cost for an LTL (less than truckload) shipment since a determination has not yet been made as to how much would be shipped by partial load. A two-week shipment would be roughly two pallets. The basic cost was estimated to be about 50% higher than the full truckload, not including extras. The extras included extra mileage (LTL shipping typically uses a hub and spoke approach, meaning the shipment would be offloaded to the shipper's hub, after which it would be reloaded to another truck destined for the customer's area). Other extras would be extra transit time (while the load waits at the shipper's hub), consolidation services (the shipper charges for scheduling the delivery), and a greater risk of damage from the increased material handling as the shipment is offloaded, stored, and reloaded at the shopper's hub. If this option is taken, they would need to go to a broker to obtain a concrete estimate of cost.

- Full truckload rates average around $3.00 per mile

- Rail rates would be roughly $0.07 per mile, but there are additional costs of delivery to the rail facility, additional time to handle the shipment from one rail car to another at a switching area, pickup charges at the destination rail facility, and delivery to whatever warehouse is used.

- Rental public warehouse space, for the area surrounding the customer facility has been about $1.20 per square foot of space per month, including both the basic space usage and the operating expenses for the warehouse.

- The product history has shown that a design change is required for the part approximately every 1.2 years. When required and existing inventory of the old style needs to be reworked at an average cost of 40% of the component value. Also, about once every two design changes the production tooling needs to be reworked at an average cost of $600 and usually taking at least a week of work. In the past the customer has notified RFPC of the new design requirement with about 10 weeks advanced notice prior to an expected delivery of the new design.

- Short-term pickup and delivery to and from a warehouse has not yet been determined. They would first need to establish warehouse locations and distances, then contact short-term haulers for estimates.

Assignment

1. Use what data and information you have to try to come up with what looks like a reasonable approach to the problem. In each approach you consider, list the additional data and information you would need to make your decision stronger. Also describe how you would use the information. Keep in mind that some information surrounding qualitative issues may need to be estimated (increased time and damage using LTL delivery, for example). You may also wish to estimate numerical issues (such as the cost of short distance delivery from a local warehouse) to help make the decision more solid. You may wish to try more than one estimate for such information to determine the sensitivity of the decision based on that data point. Also, for each approach try to estimate the qualitative issue that should be considered for the approach.

2. Are there additional approaches not mentioned that could be used? What would you need to know to analyze those approaches?

CHAPTER FIFTEEN

PRODUCTS AND PROCESSES

INTRODUCTION

The effect and the efficiency of operations management, lean manufacturing, and total quality management all depend on the way products are designed and the processes selected to produce them. The way products are designed determines the processes that are available to make them. The product design and the process determine the quality and cost of the product. Quality and cost determine the profitability of the company. This chapter studies the relationship between product design and process design and the costs associated with different types of processes. Finally, the chapter looks at the improvement of existing processes.

NEED FOR NEW PRODUCTS

Products, like people, have a limited life span. A product passes through several stages, known as the **product life cycle**, beginning with its introduction and ending with its disappearance from the marketplace. Figure 15.1 gives a simplified view of the profit and volume relationships in each phase of the cycle. No time scale is implied. The life cycle may take weeks, months or years to complete depending on the products and the market. **Life cycle analysis** is used to not only determine how much to produce and keep in inventory, but is also beneficial as a forecasting technique for estimating demand for new products. Additional information on life cycle is covered in the Fundamentals of Supply Chain Management chapter.

Introduction phase This phase is the most expensive and risky stage. To get customer acceptance of the product, the firm will usually spend heavily on advertising and sales promotion, hoping these costs will be recovered in future sales. If the introduction fails, or the demand forecast is inaccurate, the firm loses money, a fact that underlines the importance of thoroughly researching a new product before introducing it. In addition, design changes may be incurred as the market for the product develops.

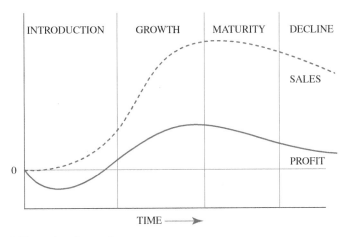

FIGURE 15.1 Life cycle of a product.

Growth phase In this phase, sales of a successful product increase at a rapid rate. The design of the product has stabilized and production increases, resulting in a decrease of the unit cost of the product. The increased sales volume and the lower unit cost cause profits to increase rapidly. Customer response time is critical, so it is vital to forecast as accurately as possible in order to have product available. However, the success of the product usually attracts the attention of competitors. Their entrance into the market forces prices down, possibly reducing the firm's sales and profit margin.

Maturity or saturation phase Nearly everyone interested in the product has sampled or owns the product and sales begin to level off. The market is saturated, price competition is often severe and profits start to decline. Depending on the product, this phase may last a very long time, or cause the company to begin considering replacing it with a newer, improved product.

Decline phase Sales drop as customers begin to lose interest in the product or to buy improved versions from the company or its competitors. Again, forecasting may prove difficult. As profits decline still further, companies must determine whether or not to continue producing the item. They may look for ways to maintain profitability, using one of the three following decisions:

- Introduce new similar products that replace the existing product.
- Improve the existing product by adding new features and functions.
- Improve the methods of production in order to increase remaining profit margins. This may also include moving production offshore or outsourcing.

Depending on the firm's resources, it may do these things through its own research and development, by copying competitors' products, or by relying on customers or suppliers to provide the necessary research and development. The upcoming discussion focuses on the firm that does its own research, development, and engineering.

PRODUCT DEVELOPMENT PRINCIPLES

A few organizations supply a single product, but most supply a range of similar or related products. There are two conflicting factors to be considered in establishing the range of products to supply.

- If the product line is too narrow, the customer base may be too small or customers may be lost.
- If the product line is too wide, the customer base may be expanded, but operating costs will increase because of the lack of specialization.

Sales organizations are responsible for increasing sales and revenue, and typically want to offer product variety to consumers. Often this means the organization must offer a variety of products, many of which sell in small volumes.

Operations, on the other hand, would like to produce as few products as possible and make them in long runs. In this way it can reduce the number of setups (and cost) and probably reduce run costs by using specialized equipment or labor. It would fulfill its mandate to produce at the lowest cost.

Somehow the needs of the market and the economics of production must be balanced. Usually this balance can be obtained with good programs of:

- Product simplification.
- Product standardization.
- Product specialization.

Simplification

Simplification is the process of making something easier to do or produce. It seeks to cut out waste by getting rid of needless product varieties, sizes, and types. The emphasis is not in cutting out products simply to reduce variety but to remove unnecessary products and variations.

As well as reducing the variety of parts, product design can often be simplified to reduce operations and material costs. For example, the use of a snap-on plastic cap instead of a screw cap reduces the cost of both materials and labor.

Standardization

In product design, a **standard** is a carefully established specification covering the product's material, configuration, measurements, and so on. Thus, all products made to a given specification will be alike and interchangeable. Light bulbs are a good example of standardization: the sockets and wattage are standardized and the light bulbs are interchangeable.

A range of standard specifications can be established so it covers most uses for the item. Men's shirts are made in a range of standard collar sizes and sleeve lengths so nearly everyone can be fitted. Most shirt manufacturers also use the same standards so the consumer can get the same size shirt from any manufacturer and expect it to fit.

Because product **standardization** allows parts to be interchangeable, as long as the range of standard specifications has been well chosen, a smaller variety of components is needed. Using the example of light bulbs, the wattages are standardized, such as 40, 60, and 100 watts. This allows users to pick wattages that satisfy their needs and manufacturers to reduce the number of different bulbs, thus reducing inventory.

Another aspect of standardization is the way parts fit together. If the designs of assemblies are standardized so various models or products are assembled in the same way, then mass production is possible. The automotive industry designs automobiles so many different models can be assembled on the same assembly line. For example, several different engines can be mounted in a chassis because the engines are mounted in the same way and designed so they will all fit into the engine compartment.

Modularization **Modularization** uses standardized parts for flexibility and variety. Standardization does not necessarily reduce the range of choice for the customer. By standardizing on component parts, a manufacturer can make a variety of finished goods, one of which will probably satisfy the customer. Automobile manufacturing is a prime example of this. Cars are usually made from a few standard components and a series of standard options so the consumer has a selection from which to choose. For example, the Mazda Miata contains 80% parts standard to other Mazda cars, which enables Mazda to produce the car quickly and at low cost, thus making a profit even though sales are comparatively small. Dell provides a series of modularization choices for consumers who want to custom configure their own laptop, including size of hard drives, processors, screen size, and memory.

Modularization also permits the practice of postponement, as introduced in Chapter 1, where the components to manufacture the final product for the customer are made ahead of time. The final product is then configured to customer specifications once the order is placed. This reduces the lead time to the customer without the need for final product inventory.

Specialization

Specialization is concentration of effort in a particular area or occupation. Electricians, doctors, and lawyers specialize in their chosen fields. In product specialization, a firm may produce and market only one or a limited range of similar products. This leads to process and labor specialization, which increases productivity and decreases costs.

With a limited range of products, productivity can be increased and costs reduced by:

- Allowing the development of machinery and equipment specially designed to make the limited range of products quickly and cheaply.
- Reducing the number of setups because of fewer task changes.
- Allowing labor to develop speed and dexterity because of fewer task changes.

Specialization is sometimes called focus and can be based either on product and market or on process.

> ### A SPECIAL CASE: PROCESS INDUSTRIES
>
> Many products are not produced by discrete production, and are typically classified as process industries. Some of the products in this category include chemicals (including gasoline), paper, glass, and some food products. Some of these are produced in a process flow production method but then packaged in a more discrete production mode (many food products, for example). While the fundamental concepts of inventory, capacity, scheduling, and so forth are used in process industries, the specific application of these can often be different than for discrete production.
>
> As examples, consider that many of the process industries produce price sensitive commodity products (even though they can be packaged differently). Often process operations are designed to run essentially on a continuous basis, implying major business activities tend to focus on raw material storage, finished goods storage, selling the output of the capacity, and on packaging and transportation scheduling. This is because the fundamental production typically uses specialized equipment with a narrow product definition, which in some cases is difficult to shut down and restart. The end of the process for many products tends to then diverge as the output may be packaged in several ways and sent to several customers all over the world. An additional issue that is common for process companies is the issue of lot tracing for the material produced.
>
> Specialization has the disadvantage of inflexibility. Often it is difficult to use highly specialized labor and equipment for tasks other than those for which they were trained or built.
>
> In summary, the three concepts of simplification, standardization, and specialization are different but interrelated. Simplification is the elimination of the superfluous and is the first step toward standardization. Standardization is establishing a range of standards and standard components that will meet most needs. Finally, specialization would not be possible without standardization. Specialization is concentration in a particular area and therefore implies repetition, which cannot be arrived at without standard products or procedures.
>
> A program of product simplification, standardization, and specialization allows a firm to concentrate on the things it does best, provides the customers with what they want, and allows operations to perform with a high level of productivity. Reducing part variety will create savings in raw material, work-in-process, and finished goods inventory. It will also allow longer production runs, improve quality because there are fewer parts, and improve opportunities for automation and mechanization. Such a program contributes significantly to reducing cost.

Product focus **Product focus** can be based on characteristics such as customer grouping (serving similar customers), demand characteristics (volume), or degree of customization. For example, one company may specialize in a limited range of high-volume products, whereas another may specialize in providing a wider range of low-volume products with a high level of customization.

Process focus **Process focus** is based on the similarity of process. For example, automobile manufacturers specialize in assembling automobiles. Other factories and companies supply the assemblers with components and the assembler specializes in assembly operations.

Focused factory Currently there is a trend toward more specialization in manufacturing whereby a factory specializes in a narrow product mix for a niche market. Generally, a **focused factory** is thought to produce more effectively and economically than more complex factories, the reason being that repetition and concentration in one area allow the workforce and management to gain the advantages of specialization. The focused factory may be a "factory within a factory," an area in an existing factory set aside to specialize in a narrow product mix.

PRODUCT SPECIFICATION AND DESIGN

Product design is responsible for producing a set of specifications that manufacturing can use to make the product. Products should be designed to be:

- Functional.
- Capable of low-cost processing.
- Environmentally sensitive.

Functional The product will be designed to perform as specified in the marketplace. The marketing department, through the inclusion of knowledge of the marketplace as well as customer requests, produces a market specification laying down the expected performance, sales volume, selling price, and appearance values of the product. Product design engineers design the product to meet the market specifications. Engineers establish the dimensions, configurations, and specifications so the item, if properly manufactured, will perform as expected in the marketplace.

Low-cost processing The product must be designed so it can be made at least cost. The product designer specifies materials, tolerances, basic shapes, methods of assembling parts, and so on and, through these specifications, sets the minimum product cost. Usually, many different designs will satisfy functional and appearance specifications. The job then is to pick the design that will also minimize manufacturing cost.

Poor design can add cost to processing in several ways:

- The product and its components may not be designed to be made using the most economical methods impossible.
- Parts may be designed so excessive material has to be removed.
- Parts may be designed so operations are difficult or time-consuming.
- Lack of standardized components may mean batches of work have to be small. Using standard parts across a range of products reduces the number of parts in inventory, tooling, and operator training and permits the use of special-purpose machinery. All this reduces product cost.
- Finally, product design can influence indirect costs such as production planning, purchasing, inventory management, and inspection. For example, one design may call for twenty different nonstandard parts, whereas another uses fifteen standard parts. The effort required to plan and control the flow of materials and the operations will be greater, and more costly, in the first case than the second, due to additional risks of shortages and dealing with many parts and suppliers.

Environmental or "green" sensitivity When designing products, there are several environmental issues that should be included or considered. One of these issues concerns the materials and processes used. Similar to reverse logistics, consideration should be given as to whether materials or packaging can be:

- Reduced to efficiently use resources.
- Designed to reduce consumption of energy during the manufacturing process.
- Easily separated for reuse.
- Recycled.

If not, the company should question if they can make alternative choices that could be less harmful to the environment. Many of these environmental concerns come under the overall term **sustainability**. As discussed in Chapter 2, sustainability includes concepts for being a responsible community "citizen" and being ethical in the approach to doing business. More information about sustainability is provided in the chapters on Fundamentals of Supply Chain Management and Purchasing.

Simultaneous Engineering

To design products for low-cost manufacture requires close coordination between product design and process design, which is called **simultaneous** or **concurrent engineering**. If the two groups can work together, they have a better chance of designing a product that will function well in the marketplace and can be manufactured at least cost. This relationship between product design and process design can determine the success or failure of a product. If a product cannot be produced at a cost that will allow a profit to be made, then it is a failure for the firm.

The traditional approach to product and process design has been a little like a relay race. When the product design was finished, the work would move to process design and that department would figure out how to make it. This system has proved time consuming and expensive and leads to less efficient outcomes. Figure 15.2 shows, with some humor, what can happen without strong communication and interaction between all parties in the product development cycle.

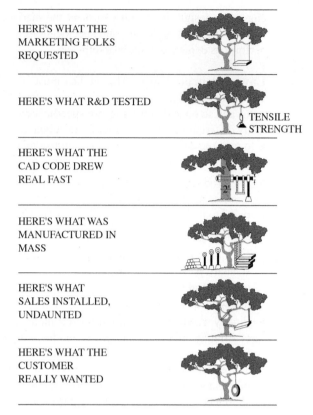

HERE'S WHAT THE MARKETING FOLKS REQUESTED

HERE'S WHAT R&D TESTED — TENSILE STRENGTH

HERE'S WHAT THE CAD CODE DREW REAL FAST

HERE'S WHAT WAS MANUFACTURED IN MASS

HERE'S WHAT SALES INSTALLED, UNDAUNTED

HERE'S WHAT THE CUSTOMER REALLY WANTED

FIGURE 15.2 Communication is essential.

Today, many organizations concurrently develop the design for the product and the processes used to make it. Often a team is made up of people from product design, process design, quality assurance, production planning and inventory control, manufacturing, purchasing, marketing, field service, and others who contribute to, or are affected by, the delivery and use of the product to the customer. These groups work together to develop the product design so it meets the needs of the customer and can be made, delivered to the customer, and serviced cost-effectively.

There are several advantages to this approach:

- **Time to market is reduced.** The organization that gets its products to market before the competition gains a strong competitive advantage.
- **Cost is reduced.** Involving all stakeholders early in the process means less need for costly product or design changes later.
- **Better quality.** Because the product is designed for ease of production and ease of implementing quality during manufacturing, the number of rejects will be reduced. Because quality is improved, the need for after-sales service and warranty costs are reduced.
- **Lower total system cost.** Because all groups affected by the product design are consulted, all concerns are addressed in advance. For example, field service might need a product that is designed so it is easy to service in the field, thus reducing servicing costs.

Supply Chain Collaboration

As part of simultaneous or concurrent engineering, collaboration with suppliers and customers across the product supply chain can be instrumental in designing a high quality, low cost, and successful product or service. This can be done informally through conversations with all entities, or more formally through the use of voice of the customer (VOC), or utilizing quality function deployment (QFD). This will be discussed further in Chapter 17.

In addition to the results already mentioned for simultaneous engineering, customer satisfaction is improved by aligning products with customer needs, and the service or product features is perceived as value-added.

PROCESS DESIGN

Operations management is responsible for producing the products and services the customer wants, when wanted, with the required quality, at a desired cost, and with high effectiveness and productivity. Processes are the means by which operations management reaches these objectives.

A **process** is a method of doing something, generally involving a number of steps or operations. Process design is the developing and designing of the steps.

Every activity involves a process of some type. Going to the bank to deposit or withdraw money, preparing a meal, or going on a trip involves a process or series of processes. Sometimes consumers are personally involved in the process. Most have waited at a checkout counter in a store and wondered why management has not devised a better process for serving customers.

Nesting   Another way of looking at the hierarchy of processes is the concept of **nesting**. Several small processes are linked to form a larger process. Consider Figure 15.3. Level zero shows a series of steps, each of which may have its own series of steps. One of the operations on level zero is expanded into its component parts and shown on level one. The nesting can continue to further levels of detail.

Mass customization   Recent changes in process flexibility and agile manufacturing have allowed for the development of a concept called **mass customization**. If the operation is designed to be flexible and efficient enough, it will allow the production of customized products (specific to customer demand) at virtually the same cost and speed as mass-produced product. Customization in general requires the ability to quickly redesign and produce a product or service based on customer need. A simple example of mass customization would be the ability to provide an exact, unique shade of paint to a customer. Companies such as Ross Controls have a division called Ross/Flex which works directly with customers to design custom valves by using existing standard products. This eliminates the need for design lead time, life-cycle testing, and custom packaging.

The objective of mass customization is to provide value to the customer by creating opportunities for unlimited variety, rather than forcing them to compromise and purchase something "close enough." This requires a commitment from not only operations

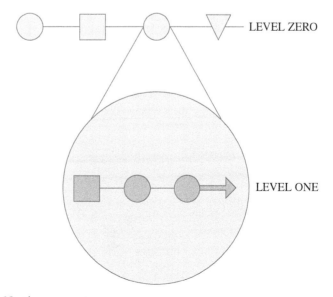

FIGURE 15.3 Nesting concept.

management, marketing, research and development and accounting, in order to support the changes in process, cost drivers, relationships with suppliers.

In some cases, the customization can occur at the final stage, referred to as **postponement**, discussed in Chapter 1. Whatever type of customization is incorporated, the key is to design a product and process that combines flexibility, agility, and knowledge of customer needs.

FACTORS INFLUENCING PROCESS DESIGN

Six basic factors must be considered when designing a process.

Product design and quality level   The product's design determines the basic processes needed to convert the raw materials and components into the finished product. For example, if a steak is to be barbecued, then the process must include a barbecue operation. The process designer can usually select from a variety of different machines or operations to do the job. The type of machine or operation selected will depend upon the quantity to be produced, the available equipment, and the quality level needed.

The desired quality level affects the process design because the process must be capable of achieving that quality level and doing it repeatedly. If the process cannot do that, operations will not be able to produce what is wanted except with expensive inspection and rework. The process designer must be aware of the capabilities of machines and processes and select those that will meet the quality level at least cost.

Demand patterns and flexibility needed   If there is variation in demand for a product, the process must be flexible enough to respond to these changes quickly. For example, if a full-service restaurant sells a variety of foods, the process must be flexible enough to switch from broiling hamburgers to making pizzas. Conversely, if a pizza parlor sells nothing but pizzas, the process need not be designed to cook any other type of food. Flexible processes require flexible equipment and personnel capable of doing a number of different jobs. During the pandemic of 2020, many manufacturing companies quickly adapted their processes and workforce to be able to meet the sudden global need for personal protective equipment. For example, a company that normally makes paper and receipts and labels was able to turn their operations into making disposable facemasks.

Quantity/capacity considerations   Product design, the quantity to produce, and process design are closely related. Both product and process design depend on the quantity needed. For example, if only one of an item is to be made, the design and the process used will be different than if the volume is 100,000 units. The quantity needed and the process design determine the capacity needed. Figure 15.4 shows this relationship. Note that all three are directly connected to the customer.

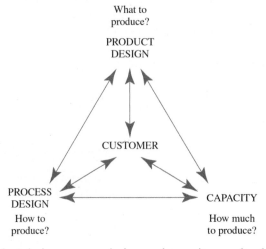

FIGURE 15.4 Product design, process design, and capacity are closely related.

Customer involvement Chapter 1 discussed five manufacturing strategies—engineer-to-order, make-to-order, configure-to-order, assemble-to-order, and make-to-stock—and the extent of customer involvement in each. Process design will depend on which strategy is chosen.

Environmental concerns Just as in product design, the process design should have minimal impact on the environment, if at all possible. In addition, concern should be taken to minimize the amount of energy utilized. This consideration also directly impacts the cost of running the process. Painting often involves the use of solvents and other chemicals that are harmful to the environment and workers. Water-based paints have been developed that reduce emissions and cleanup. The paint and the equipment used to apply it are often expensive but the company saves in cleanup and worker protection costs.

Make or buy decision A manufacturer has the alternative of making parts in-house or of buying them from an outside supplier. Few companies make everything or buy everything they need. Indeed, on the average, North American manufacturers purchase more than 50% of the cost of goods manufactured. A decision has to be made about which items to make and which to buy. Although cost is the main determinant, other factors such as the following are usually considered:

Reasons to Make In-house

- Produces for less cost than a supplier.
- Utilizes existing equipment to fullest extent.
- Keeps confidential, proprietary knowledge within control of the firm.
- Maintains quality.
- Maintains workforce.

Reasons to Buy

- Requires less capital investment.
- Uses specialized expertise of suppliers.
- Allows the firm to concentrate on its own area of specialization.
- Provides known and competitive prices.
- Accommodates large changes in volume.

The decision to make or buy is clear for many items such as nuts and bolts, motors, or components that the firm does not normally manufacture. For other items that are in the firm's specialty area, a specific decision will have to be made.

As supply chains are becoming more tightly linked and sources of supply are becoming more global, the issue to buy (outsource) or make (insource) becomes more complex. For example, exchange rates, transit inventory levels, impact of transit time on lead time, tariffs, and government controls are all becoming issues that need to be considered in making the decision to make or buy.

PROCESSING EQUIPMENT

Processing equipment can be classified in several ways. This discussion will focus on classification by the degree of specialization of machinery and equipment.

General-purpose machinery **General-purpose machinery** can be used for a variety of operations or can do work on a variety of products within its machine classification. For example, a home sewing machine can sew a variety of materials, stitches, and patterns within its basic capability. Different auxiliary tools can be used to create other stitches or for particular sewing operations.

Special-purpose machinery **Special-purpose machinery** is designed to perform specific operations on one work piece or a small number of similar work pieces.

For example, a sewing machine built or equipped to sew shirt collars would be a special-purpose machine capable of sewing collars on any size shirt of any color but not capable of performing other sewing operations unless it was modified extensively.

General-purpose machinery is generally less costly than special-purpose machinery. However, its run time can be slower, and because its operations often require more human input, the quality level tends to be more variable than when using special-purpose machinery. One exception is robotic machines, which automate the human aspect and tend to produce high quality on repetitive processes, but the cost is much higher. Special-purpose machinery is less flexible but parts can generally be produced with it much quicker than with general-purpose machinery.

PROCESS SYSTEMS

Depending on the product design, volume, and available equipment, the process engineer must design the system to make the product. As mentioned in Chapter 6, based on material flow, processes can be organized in three ways:

- Flow.
- Intermittent.
- Project.

The process used will depend on the demand for the item, range of products, and the ease or difficulty of moving material. All three can be used to make discrete units such as automobiles or textbooks, or to make nondiscrete process products such as gasoline, paint, or fertilizer.

Flow Processes

Work centers needed to make the product, or family of similar products, are grouped together in one department and are laid out in the sequence needed to make the product. Examples are assembly lines, cafeterias, oil refineries, and steel rolling mills. In **flow processing**, work flows from one work center to another at a nearly constant rate and with no delays. **Continuous flow** processing is used to manufacture products such as liquids, basic metals, and commodities such as petroleum products. There is some method of moving goods between work centers. If the units are discrete, such as automobiles, flow manufacturing is called **repetitive manufacturing**. The typical flow pattern is shown in Figure 15.5. Flow process layout is sometimes called **product layout** because the system is set up for a limited range of similar products.

Flow systems produce only a limited range of similar products. For example, an assembly line that produces a certain type of refrigerator cannot be used to assemble washing machines. The operations and materials used to make one are different and in a different sequence than those used for making the other. Demand for the family of products has to be large enough to justify setting up the line economically. If sufficient demand exists, flow systems are extremely efficient, for several reasons:

- Work centers are designed to produce a limited range of similar products, so machinery and tooling can be specialized.
- Because material flows from one work center to the next, there is very little buildup of work-in-process inventory.
- Because of the flow system and the low work-in-process inventory, lead times are short.
- In most cases, flow systems substitute capital for labor and standardize what labor there is into routine tasks.

FIGURE 15.5 Material flow: Flow process.

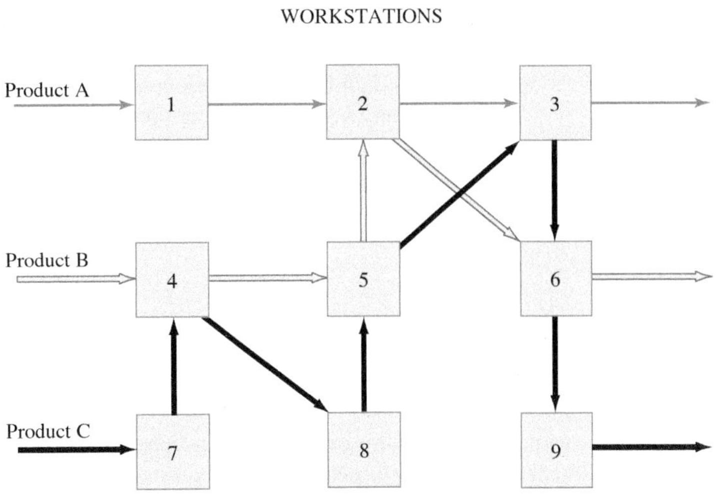

WORKSTATIONS

[Figure: material flow diagram with workstations 1-9 arranged in a 3x3 grid, showing Product A, Product B, and Product C flowing through various workstations]

FIGURE 15.6 Material flow: Intermittent process.

Intermittent Processes

In **intermittent production**, goods are not made continuously as in a flow system but are made at intervals in lots or batches. Work centers must be capable of processing many different parts. Thus, it is necessary to use general-purpose work centers and machinery that can perform a variety of tasks.

General-purpose work centers do not produce goods as quickly as special-purpose work centers used in flow manufacturing. Usually, work centers are organized into departments based on similar types of skills or equipment. For example, all welding and fabrication operations are located in one department, machine tools in another, and assembly in yet another department. Work moves only to those work centers needed to make the product and skips the rest. This results in the jumbled flow pattern shown in Figure 15.6. This is referred to as a **process layout**, or **functional layout**, in which similar functions or equipment are grouped together, and product flows from work center to work center.

Intermittent processes are flexible. They can change from one part or task to another more quickly than can flow processes. This is because they use general-purpose machinery and skilled flexible labor that can perform the variety of operations needed.

Control of work flow is managed through individual work orders for each lot or batch being made. Because of this and the jumbled pattern of work flow, manufacturing planning and control of the shop floor are critical. Often, many work orders exist, each of which can be processed in different ways.

Provided the volume of work exists to justify it, flow manufacturing is less costly than intermittent production. There are several reasons for this:

- Setup costs are low. Once the line is established, changeovers are needed infrequently to run another product.
- Since work centers are designed for specific products, run costs are low.
- Because products move continuously from one work center to the next, work-in-process inventory will be low.
- Costs associated with controlling production are low because work flows through the process in a fixed sequence.

But the volume of specific parts must be enough to use the capacity of the line and justify the capital cost.

Project Processes

Project manufacturing is mostly used for large, complex projects such as ships, bridges or buildings, and is typically a one-time endeavor, used to create a unique product or service. The product may remain in one location for its full assembly period known as a

384 CHAPTER FIFTEEN

fixed-position layout, as with a ship, or it may move from location to location after considerable work and time are spent on it. Project manufacturing avoids the very high costs of moving the product from one work center to another.

There are many variations and combinations of these three basic types of processes. Companies try to find the best combination to make their particular products. In any one company it is not unusual to see examples of all three being used.

PROCESS COSTING

There are two common methods for determining product costs. **Job costing** is used when multiple products are produced within a period, and costs such as labor, material, and overhead are allocated to each product. The other method is **process costing**, which is most commonly used in industries that manufacture product in a continuous process, such as paper, petroleum, or concrete. In these process industries, it is impossible to allocate specific costs for a period to a specific lot, as the flow is continuous. Materials, labor, and overhead consumed during a particular period are accumulated, and then allocated to departments or operations, rather than a specific product. Overall product cost can be determined by summing the accumulated costs, and then dividing that by the volume of products produced during the time period.

SELECTING THE PROCESS

Generally, the larger the volume (quantity) to be produced, the greater the opportunity to use special-purpose processes. The more special purpose an operation, the faster it will produce. Often the capital cost for such machinery or for special tools or fixtures is high. Capital costs are known as **fixed costs** and the production, or run, costs are known as **variable costs**.

Fixed costs These are costs that do not vary with the volume being produced. Purchase costs of machinery and tools and setup costs are considered fixed costs. No matter what volume is produced, these costs remain the same. Suppose it costs $200 to set up a process; this cost will not change no matter how much is produced.

Variable costs These are costs that vary with the quantity produced. Direct labor (labor used directly in the making of the product) and direct material (material used directly in the product) are the major variable costs. If the runtime for a product is 12 minutes per unit, the labor cost $10 per hour, and the material cost $5 per unit, then:

$$\text{Variable cost} = \frac{12}{60} \times \$10.00 + \$5.00 = \$7.00 \text{ per unit}$$

Let:
$$FC = \text{Fixed cost}$$
$$VC = \text{Variable cost per unit}$$
$$x = \text{Number of units to be produced}$$
$$TC = \text{Total cost}$$
$$UC = \text{Unit(average) cost per unit}$$

Total cost = Fixed cost + (Variable cost per unit)(number of units produced)

Then:
$$TC = FC + VCx$$
$$\text{Unit cost} = \frac{\text{Total cost}}{\text{number of units produced}} = \frac{TC}{x}$$

Example Problem

A process designer has a choice of two methods for making an item. Method A has a fixed cost of $2000 for tooling and jigs and a variable cost of $3 per unit. Method B requires a special machine costing $20,000 and the variable costs are $1 per piece. Let x be the number of units produced.

	Method A	Method B
Fixed cost	$2000.00	$20,000.00
Variable unit cost	$3.00	$1.00
Total cost	$2000.00 + 3x$	$20,000.00 + 1x$
Unit (average) cost	$\dfrac{2000.00 + 3x}{x}$	$\dfrac{2000.00 + 1x}{x}$

Table 15.1 shows what happens to the total cost as quantities produced are increased. The total cost data in this table is shown graphically in Figure 15.7. From Table 15.1 and Figure 15.7, we can see that initially the total cost and unit cost of method A are less than method B. This is because the fixed cost for method B is greater and has to be absorbed over a small number of units. Although the total cost for both methods increases as more units are produced, the total cost for method A increases faster until, at some quantity between 8000 and 10,000 units, the total cost for method B becomes less than for method A.

Similarly, the unit cost for both methods decreases as more units are produced. However, the unit cost for method B decreases at a faster rate until, at some quantity between 8000 and 10,000 units, unit costs for both methods are equal.

Cost Equalization Point

Knowing the quantity beyond which the cost of using method B becomes less than for method A enables the decision of which process to use to minimize the total cost (and the unit cost).

This quantity is called the **cost equalization point (CEP)** and is the volume for which the total cost (and unit cost) of using one method is the same as another. For the example, the total cost calculations are as follows:

$$TC_A = TC_B$$
$$FC_A + VC_A x = FC_B + VC_B x$$
$$\$2000 + \$3x = \$20,000 + \$1x$$
$$3x - 1x = 20,000 - 2000$$
$$2x = 18,000$$
$$x = 9000 \text{ units}$$

Volume	Total Cost (Dollars)		Unit Cost (Dollars)	
(Units)	Method A	Method B	Method A	Method B
2000	$8000	$22,000	$4.00	$11.00
4000	14,000	24,000	3.5	6
6000	20,000	26,000	3.33	4.33
8000	26,000	28,000	3.25	3.5
10,000	32,000	30,000	3.2	3
12,000	38,000	32,000	3.17	2.67
14,000	44,000	34,000	3.14	2.43
16,000	50,000	36,000	3.13	2.25

TABLE 15.1 Total and average cost versus quantity produced.

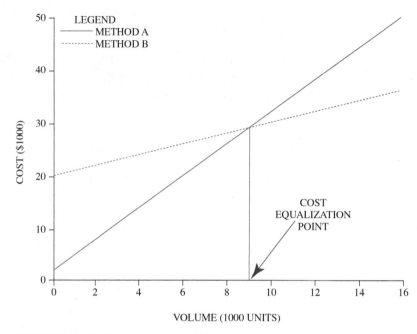

FIGURE 15.7 Total cost versus quantity produced.

The CEP is 9000 units. At this quantity, the total cost of using method A will be the same as for method B.

$$TC_A = FC_A + VC_A x - 2000 + (3 \times 9000) = \$29,000$$
$$TC_B = FC_B + VC_B x = 20,000 + (1 \times 9000) = \$29,000$$

We can get the same results by using unit costs instead of total costs.
From the preceding calculations it is clear that:

- If the volume (quantity to produce) is less than the CEP, the method with the lower fixed cost will cost less.
- If the volume is greater than the CEP, the method with the greater fixed cost will cost less.

For example, if the volume were 5000 units, method A would cost less, and if the volume were 10,000 units, method B would cost less.

Variable costs, mostly direct labor and material, can be reduced by substituting machinery and equipment (capital) for direct labor. This increases the fixed costs and decreases the variable costs. But to justify this economically, the volume must be high enough to reduce the total or unit cost of production.

Increasing Volume

The obvious way to increase volume is to increase sales. However, a finished product is usually made up of several purchased or manufactured components. If the volume of these components can be increased, then the unit cost of the components, and the final product, will be reduced.

The volume of components can be increased without increasing sales by a program of simplification and standardization, discussed previously in this chapter.

If a subassembly or component part can be standardized for use in more than one final product, then the volume of the subassembly or part is increased without an increase in the total sales volume. Thus, more specialized and faster-running processes can be justified and the cost of operations reduced.

Standardization of parts is a major characteristic of modern mass production. At the turn of the twentieth century, Henry Ford revolutionized manufacturing by standardizing the finished product—one model of car. The joke often heard then was that you could have any color you wanted as long as it was black. Today a vast range of models are made, but if each model was exploded into its subassemblies and component parts, one would find specific components common to a great number of models. In this way, modern manufacturers can provide the consumer with a wide choice of finished products made from standard parts and components.

CONTINUOUS PROCESS IMPROVEMENT

People have always been concerned with how best to do a job and the time it should take to do it. Process improvement is concerned with improving the effective use of human and other resources. *Continuous* implies an ongoing activity; *improvement* implies an increase in the productivity or value of quality or condition. Hence, the name continuous process improvement.

Continuous process improvement (CPI) consists of a logical set of steps and techniques used to analyze processes and to improve them.

Improving productivity Productivity can be improved by spending money (capital) on better and faster machines and equipment. However, with any given amount of capital, a method must be designed to use the machinery and equipment most productively. A work center might consist of highly sophisticated machinery and equipment worth $1 million or more. Its productivity and return on investment depend on how the equipment is used and how the operator manages it.

Continuous process improvement is a low-cost method of designing or improving work methods to maximize productivity. The aim is to increase productivity by better use of existing resources. Continuous process improvement is concerned with removing work content, not with spending money on better and faster machines.

Peter Drucker has said, "Efficiency is doing things right; effectiveness is doing the right things." Continuous process improvement aims to do the right things and to do them efficiently.

People involvement Today management recognizes the need to maximize the potential of flexible, motivated workers. People are capable of thinking, learning, problem-solving, and contributing to productivity. With existing processes and equipment, people are the primary source of improvement because they are the experts in the things they do.

Process improvement is not solely the responsibility of industrial engineers. Everyone in the workforce must be given the opportunity to improve the processes they work with. Techniques that help to analyze and improve work are not complicated and can be learned. Indeed, the idea of continuous improvement is based on the participation of operators and improvement in methods requiring relatively little capital.

Workers have two jobs:

- Their "as defined" job.
- To improve their "as defined" job.

Teams One of the features of CPI is team involvement. A team is a group of people working together to achieve common goals and objectives. The members of the team should be all those who are involved with the process. Teams are successful because of the emphasis placed on people. Not all problems can be solved by teams, nor are all people suited to teams. However, they are often effective, as problems often cross functional lines and thus multifunctional teams are common.

Continuous process improvement can still be effectively carried out by the individuals.

The Six Steps in Continuous Process Improvement

The general method used to solve many kinds of problems with CPI includes six steps:

1. Select the process to be studied.
2. Record the existing method to collect the necessary data in a useful form.
3. Analyze the recorded data to generate alternative improved methods.
4. Develop the best method of doing the work by evaluating the alternatives.
5. Implement the method as standard practice by training the operator.
6. Maintain the new method.

Select the Process

The first step is to decide what to study. This depends on the ability to recognize situations that have good potential for improvement. Observation of existing methods comes first.

Observe The important feature in observation is a questioning attitude. Questions such as why, when, and how must be asked whenever something is observed. This attitude needs development because people tend to assume that the familiar method is the only one. Often heard is, "We have always done it this way!" However, "this way" is not necessarily the only, most productive, or most effective way.

Any situation can be improved but some have better potential than others. Indicators in manufacturing that show areas most needing improvement include the following:

- High scrap, reprocessing, rework, and repair costs.
- Backtracking of material flow caused by poor plant layout.
- Bottlenecks.
- Excessive overtime.
- Excessive manual handling of materials, both from workplace to workplace and at the workplace.
- Use of environmentally hazardous materials.
- Employee grievances without true assignable causes.

Select The purpose of continuous process improvement is to improve productivity to reduce operating, product, or service costs. In selecting jobs or operations for method improvement, there are two major considerations: economic and human.

Economic considerations The cost of the improvement must be justified. The cost of doing the study and implementing the improvement must be recovered from the savings in a reasonable time. One to two years is a commonly used period.

The job size must justify the study. Almost anything can be improved, but the improvement must be worthwhile. Suppose a process improvement saves 1 hour on a job taking 5 hours, performed once a month or 12 times a year. The reduction in time is 20% and the total time saved in a year is 12 hours. Another process improvement saves 1 minute on a job taking 10 minutes, performed 200 times per week. The time saved in this case is only 10% but will be 173 ($1 \times 200 \times 52 \div 60 = 173.3$) hours per year, a much higher rate of return on the investment made in the study.

The human factor The human factor governs the success of continuous improvement. The resistance to change, by both management and worker, must always be remembered. Working situations characterized by high fatigue, accident hazards, absenteeism, and dirty and unpleasant conditions should be identified and improved. Sometimes it is difficult to give specific economic justification for such improvements, but the intangible benefits are extensive and should weigh heavily in selecting studies.

Pareto diagrams **Pareto analysis** can be used to select problems with the greatest economic impact. The theory of Pareto analysis is the same as that used in the ABC analysis discussed in Chapter 10. This theory says that a few items (usually about 20%) account for most of the cost or problems. It separates the "vital few" from the "trivial many." Examples of the "vital few" are as follows:

- A few processes account for the bulk of scrap.
- A few suppliers account for most rejected parts.
- A few problems account for most process downtime.

The steps in performing a Pareto analysis are as follows:

1. Determine the method of classifying the data: by problem, cause, nonconformity, and so forth.
2. Select the unit of measure. This is usually dollars but may be the frequency of occurrence.

3. Collect data for an appropriate time interval, usually long enough to include all likely conditions.
4. Summarize the data by ranking the items in descending order according to the selected unit of measure.
5. Calculate the total cost.
6. Calculate the percentage for each item.
7. Construct a bar graph showing the percentage for each item and a line graph of the cumulative percentage.

Example Problem

A product has failed in the field a number of times. Data is collected according to the type of failure with the following results: type A—11, type B—8, type C—5, type D—60, type E—100, type F—4, other—12. Construct a table summarizing the data in descending order of magnitude. From this table, construct a Pareto diagram.

ANSWER

Type of Failure	Number of Failures	Percent	Cumulative Percentage
E	100	50.0	50.0
D	60	30.0	80.0
A	11	5.5	85.5
B	8	4.0	89.5
C	5	2.5	92.0
F	4	2.0	94.0
O (Other)	12	6.0	100.0
Total	200	100.0	

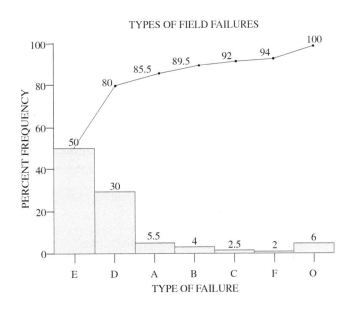

Note that Pareto analysis does not report what the problems are, only where they seem to occur. In the previous example, further investigation into the causes of failure types E and D will give the best return for the effort spent. It is important to select the categories carefully. For example, in the previous problem, if the location of failures were recorded rather than the type of failure, the results would be quite different and perhaps not significant.

Cause-and-effect diagram Sometimes called a fishbone or **Ishikawa diagram**, the **cause-and-effect diagram** is a very useful tool for identifying root causes. Figure 15.8 shows such a diagram.

The **fishbone diagram** is best used by a group or team working together. It can be constructed by discussion and brainstorming. The steps in developing a fishbone diagram are as follows:

1. Identify the problem to be studied and state it in a few words. For example, the reject rate on machine A is 20%.
2. Generate some ideas about the main causes of the problem. Usually all probable root causes can be classified into six categories.
   - **Materials.** For example, from consistent to inconsistent raw materials.
   - **Machines.** For example, a well-maintained machine versus a poorly maintained one.
   - **People.** For example, a poorly trained operator instead of a well-trained one.
   - **Methods.** For example, changing the speed on a machine.
   - **Measurement.** For example, measuring parts with an inaccurate gauge.
   - **Environment.** For example, increased dust or humidity.
3. Brainstorm all possible causes for each of the main causes.
4. Once all the causes have been listed, try to identify the most likely root causes and work on these.

Record

The next step is to record all the facts relating to the existing process. To be able to understand what to record, it is necessary to define the process being studied. Recording defines the process. The following must be determined to properly define the process.

- **The process boundaries.** All processes, big or small, begin and end somewhere. Starting and ending points form the boundaries of the process. For example, the starting point in the process of getting to work in the morning might be getting out of bed. The ending point might be arrival at the desk or classroom.

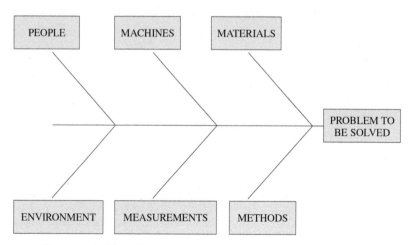

FIGURE 15.8 Cause-and-effect diagram.

- **Process flow.** This is a description of what happens between the starting and ending points. Usually this is a listing of the steps taken between the start and finish of the process. There are several recording techniques to help perform this step. Some of these will be discussed later in this chapter.
- **Process inputs and outputs.** All processes change something. The things that are changed are called inputs and they may be physical, such as raw materials, or informational, such as data. Outputs are the result of what goes on in the process. For example, raw materials are converted into something more useful or data is manipulated to produce reports.
- **Components.** Components are the resources used in changing inputs to outputs. They are composed of people, methods, and equipment. Unlike process inputs, components do not become part of the output but are part of the process. For example, in producing a report, the graphics program, computer, and printer are all components.
- **Customer.** Processes exist to serve customers and customers ultimately define what a process is supposed to do. If customer needs are not considered, there is a risk of improving things that do not matter to the users of the output.
- **Suppliers.** Suppliers are those who provide the inputs. They may be internal to the organization or external.
- **Business environment.** The process is controlled or regulated by external and internal factors. The external factors are beyond the firm's control and include customers' acceptance of the process output, competitors, and government regulation. Internal factors are within the organization and can be controlled.

Figure 15.9 shows a schematic of the process.

The next step is to record all facts relating to the existing method. A record is necessary because it is difficult to record and maintain a large mass of detail by memory for the duration of the analysis. Recording helps the team consider all elements of the problem in a logical sequence and makes sure all the steps in the process are considered. The record of the present method also provides the basis for both the critical examination and the development of an improved method.

Classes of activity Before discussing some of the charts used, the kinds of activities recorded will be discussed. All activity can generally be classified into one of six types. As a method of shorthand, there are six universally used symbols for these activities. The activities and symbols are shown in Figure 15.10.

Following are descriptions of some of the various charting techniques.

Operations process charts **Operations process charts** record in sequence only the main operation and inspections. They are useful for preliminary investigation and give a bird's-eye view of the process. Figure 15.11 shows such a chart.

The description, and sometimes the times, for each operation is also shown. An operations process chart would be used to record product movement.

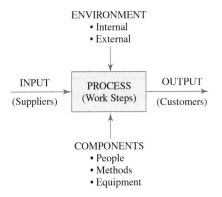

FIGURE 15.9 Schematic of a process.

392 CHAPTER FIFTEEN

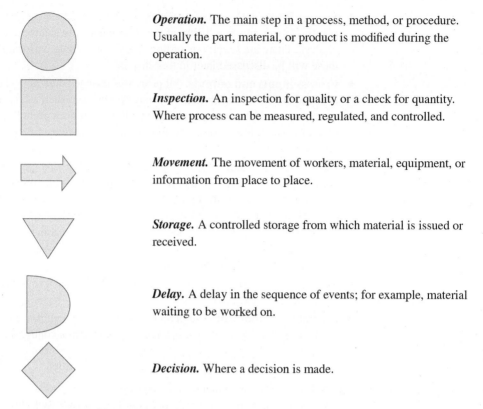

FIGURE 15.10 Classes of activity.

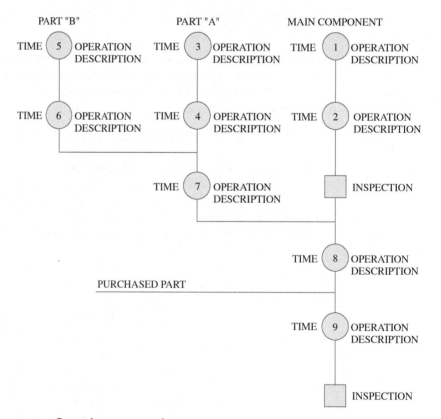

FIGURE 15.11 Operations process chart.

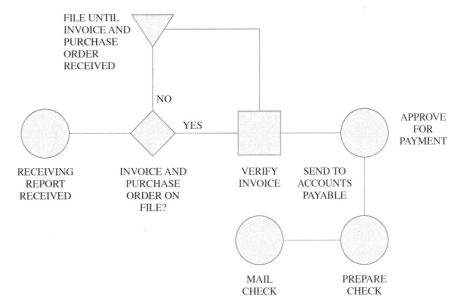

FIGURE 15.12 Process flow diagram.

Process flow diagram   A **process flow diagram** shows graphically and sequentially the various steps, events, and operations that make up a process. It provides a picture, in the form of a diagram, of what actually happens when a product is made or a service performed. In addition to the six symbols shown in Figure 15.10, others may be used to show such things as rework and documentation. Figure 15.12 shows an example of a process flow diagram. In this example, the process starts when the goods are received and ends when a check is sent to the supplier.

Analyze

Examination and analysis are the key steps in continuous process improvement. Although all the other steps are significant, they either lead up to, or result from, the critical analysis. This step involves analyzing every aspect of the present method and evaluating all proposed possible methods.

Find the root cause   Frequently it is difficult to separate symptoms from the root causes of problems. Often the only things observed are symptoms and it is difficult to trace back to the root cause. To find root causes requires a questioning attitude. For the analyst, "why" is the most important word. Every aspect of the existing method should be questioned.

A rule of thumb common to many problem-solving methods says it is necessary to ask (and answer) the "why" question up to five times before one reaches the root cause of a problem. As each "why" is answered, the question is asked of "why" that was the correct answer.

Three approaches can help in finding the root causes of problems:

- **A questioning attitude.** This implies an open mind, examining the facts as they are, not as they seem, avoiding preconceived ideas, and avoiding hasty judgments.
- **Examining the total process to define what is accomplished, how, and why.** The answers to these questions will determine the effectiveness of the total process. The results may show that the process is not even needed.
- **Examining the parts of the process.** Activities can be divided into two major categories: those in which something is happening to the product (worked on, inspected, or moved) and those in which nothing constructive is happening to the product (delay or storage, for example). In the first category, value is added only when the part is being worked on. Setup, put away, and move, while necessary, add cost to the product but

do not increase its value, so must be minimized. Value will be added when the product is being worked on but, again, the goal is to maximize the productivity of these operations.

- **Analysis of the relationship between production rate, item throughput, and process inventory.** For processes that have the rate of material input equal to the rate of output, there is a basic relationship between the amount of inventory in the process and the time it takes for a single item to be processed through the process (throughput time). If the processing rate of the process is called R, the inventory is called I, and the throughput time is called T, then the relationship is given as

$$I = RT$$

Example Problem

A process has the ability to produce two items every minute. The process currently has 100 of the items being processed. If another item were introduced into the process, how long would it take until the new item is produced?

Answer

The production rate is two items per minute, and the inventory is 100 units. The solution would then be as follows:

$$I = RT$$
$$100 \text{ units} = (2 \text{ units/minute}) \times T$$
$$T = 50 \text{ minutes}$$

This relationship is often called **Little's Law**, which can be applied to any process that includes time, inventory, and throughput. For example, to estimate the time a customer waits in line for a service could be computed as

$$I(\# \text{ of customers}) = R(\text{minutes of service}) \times$$
$$T(\text{minutes a customer will wait to be served})$$
$$8 \text{ customers} = 0.5 \text{ people per minute} \times T$$
$$T = 16 \text{ minutes}$$

Develop

When developing possible solutions, there are four approaches to take to help develop a better method.

- *Eliminate* all unnecessary work. Question why the work is being done in the first place and if it can be eliminated.
- *Combine* operations wherever possible. Thus, material handling will be reduced, space saved, and the throughput time reduced. This is a major thrust of lean manufacturing.
- *Rearrange the sequence* of operations for more effective results. This is an extension of the previous approach. If sequences are changed, then they can possibly be combined.
- *Simplify* wherever possible by making the necessary operations less complex. If the questioning attitude is used, then complexity should be reduced. Usually the best solutions are the simple ones.

Principles of motion economy There are several principles of motion economy, including the following:

1. Locate materials, tools, and workplace within normal working areas and pre-position tools and materials.
2. Locate the work done most frequently in the normal working areas and everything else within the maximum grasp areas.

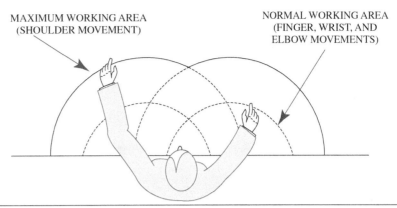

FIGURE 15.13 Working areas.

3. Arrange work so motions of hands, arms, legs, and so on are balanced by being made simultaneously, in opposite directions, and over symmetrical paths. Both hands should be working together and should start and finish at the same time. The end of one cycle should be located near the start of the next cycle.

4. Conditions contributing to operator fatigue must be reduced to a minimum. Provide good lighting, keep tools and materials within maximum working areas (see Figure 15.13), provide for alternate sitting and standing at work, and design workplaces of proper height to eliminate stooping.

Human and environmental factors   In addition to the principles of motion economy, other important matters that influence human and environmental conditions must be considered. These include safety, comfort, cleanliness, and personal care, so provision must be made for lighting, ventilation and heating, noise reduction, seating, and stimulation.

Of these, stimulation may be the least obvious. In highly repetitive work, workers may become bored and dissatisfied, which may lead to emotional problems. A pleasant environment created by attractive color schemes in plants or offices, location of windows, and music during working hours can do much to reduce stress and absenteeism.

Methods improvement is based on the concepts of scientific management. It concentrates on the task and ways of removing waste from tasks. It gives little consideration to the human being and higher-level needs, such as self-esteem and self-fulfillment, and work can become repetitive and boring. **Job design** is an attempt to provide more satisfying meaningful jobs and to use the worker's mental and interpersonal skills. These improvements include the following:

Job enlargement expands a worker's job by clustering similar or related tasks into one job. For example, a job might be expanded to include a sequence of activities instead of only one activity. This is called *horizontal enlargement.*

Job enrichment adds more meaningful, satisfying, and fulfilling tasks. The job not only includes production operations but also many setup, scheduling, maintenance, and control responsibilities.

Job rotation trains workers to do several jobs so they can be moved from one job to another. This is called **cross-training**.

All these factors help to create a more motivated and flexible workforce. In modern manufacturing, where quick response to customers' needs is essential, these characteristics can mean the difference between business success and failure.

Employee empowerment and **self-directed work teams** assume the workers have more knowledge and responsibility for understanding and performing the work. Empowering the workers to take more responsibility can lead to self-directed work teams, where there is typically no supervisor. The workers themselves understand what is to be done and manage their own activities to accomplish the required work efficiently and effectively with little or no supervision.

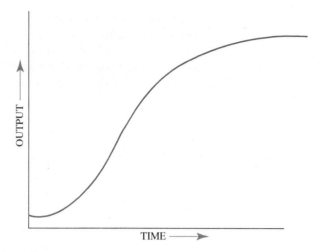

FIGURE 15.14 Learning curve.

Implement

Once the analyst has completed planning the improvement, the plans must be put into action by implementing the new procedure.

In planning the implementation process, consideration must be given to the best time to implement, the method of implementation, and the people involved. The analyst needs to be sure that equipment, tooling, information, and the people are all available. At the time of implementation, a dry run will show whether all equipment and tooling are working properly.

Training the operator is the most important part of the implementation process. If the operator has been involved in designing the change, this should not be difficult. The worker will be familiar and comfortable with the change and will hopefully feel some sense of ownership in the new process.

Learning curve Over time, as the operator does the tasks repetitively, speed will increase and errors decrease. This process is known as the learning curve and is illustrated in Figure 15.14. Note there is no time scale shown. Depending on the task, a worker may progress through the learning curve in a few minutes or, for high-skill jobs, several months or years.

Maintain

Maintaining is a follow-up activity that has two parts. The first is to be sure that the new method is being done as it should be. This is most critical for the first few days, and close supervision may be necessary. The second is to evaluate the change to be sure that the planned benefits are accomplished. If not, the method must be changed.

SUMMARY

Products and the processes used to make them are continually being redesigned to create products more appealing to customers, to improve productivity, or to make the products and their associated processes friendlier to the environment. Producers follow well-established principles of product development. The powerful principles of simplification, standardization, and specialization can help to improve productivity and make products more reliable. With information easy to share in today's environment, product design and process design can work simultaneously to bring better quality products to market faster and with reduced costs. As the design of the product is being established, the process is also being designed based on: the quality level desired, the ability of the process to react quickly to changes in customer demand (flexibility), the overall volume of demand, and how much the customer wants to be involved in the production of the product. Depending on the volume, the decision may be made to buy the product rather than make it, or to use one process rather than another. This is best determined with the cost equalization point (CEP), which determines a volume below which the low fixed cost, high variable cost alternative is used, and above which a company may invest in the high fixed cost and low variable cost alternative.

Continuous process improvement (CPI) can be applied to all processes to make them more cost effective and competitive. It is done on an ongoing basis, not just when new products are introduced. CPI uses the traditional tools of the scientific methods in six steps: select, record, analyze, develop, implement, and maintain. All the steps have one thing in common and that is the involvement of people.

With improvements to product and process, design goods tend to flow quickly and smoothly, resulting in lower costs and improved profits. Continuous improvement is a major factor in lean production which is discussed in the next chapter.

KEY TERMS

Cause-and-effect diagram 390	Modularization 375
Concurrent engineering 377	Nesting 379
Continuous flow 382	Operations process chart 391
Continuous process improvement (CPI) 387	Pareto analysis 388
Cost equalization point (CEP) 385	Postponement 380
Cross-training 395	Process 379
Employee empowerment 395	Process costing 384
Fishbone diagram 390	Process flow diagram 393
Fixed cost 384	Process focus 376
Fixed-position layout 384	Process layout 383
Flow processing 382	Product focus 376
Focused factory 376	Product layout 382
Functional layout 383	Product life cycle 373
General-purpose machinery 381	Project manufacturing 383
Intermittent production 383	Repetitive manufacturing 382
Ishikawa diagram 390	Self-directed work team 395
Job costing 384	Simplification 374
Job design 395	Simultaneous engineering 377
Job enlargement 395	Specialization 375
Job enrichment 395	Special-purpose machinery 381
Job rotation 395	Standard 375
Life cycle analysis 373	Standardization 375
Little's Law 394	Sustainability 377
Mass customization 379	Variable cost 384

QUESTIONS

1. What are the phases of a product life cycle?
2. Describe simplification, standardization, and specialization. Why are they important and why are they interrelated?
3. What are the advantages and disadvantages of standardization?
4. What are the advantages and disadvantages of specialization?
5. What are product focus and process focus based on? What is a focused factory?

6. What is the advantage to modular design?
7. How are production costs affected by:
   a. Standard sizes?
   b. Universal fit parts?
8. What are the three criteria for designing a product?
9. Why is product design important to operations costs?
10. Why is product design important to quality?
11. What is simultaneous engineering, and what are some of its advantages?
12. What is a process? What is process nesting?
13. What is the objective of mass customization?
14. What are the five basic factors that must be considered when designing a process?
15. Why are product design, quantity to produce, and process design intimately related?
16. Give four reasons why companies will make a product in-house and why they will outsource.
17. Describe general-purpose and special-purpose machinery. Compare each for flexibility of use, operator involvement, run time per piece, setup time, quality, capital cost, and application.
18. What is flow processing, and what are its advantages and disadvantages?
19. What is intermittent processing and when is it used? Contrast it with flow processing.
20. When is project processing used?
21. Define fixed and variable costs and give examples of each in manufacturing. What is total cost, and what is the equation for it?
22. What is the cost equalization point?
23. How can the variable cost be reduced? What does this do to the fixed costs, and what is needed to economically justify this course of action?
24. What are the six steps in continuous process improvement?
25. Name and describe the two considerations in selecting a job to be studied.
26. What is a Pareto diagram, and why is it useful?
27. What is a cause-and-effect diagram, and why is it useful?
28. Why is it necessary to record?
29. Describe each of the following as it relates to recording:
    a. Process boundaries.
    b. Process flow.
    c. Process inputs and outputs.
    d. Process components.
    e. Suppliers.
    f. Environment.
30. What are the six symbols used in method analysis? Are there other symbols that can be used?
31. Describe each of the following:
    a. Operations process chart.
    b. Process flow diagram.
32. What is the purpose of the analysis step in continuous process improvement? What is the basic question?
33. What are the four approaches that should be taken when developing a better method?
34. What are the four principles of motion economy?
35. Describe job design.
36. What is the learning curve?
37. What might be some of the advantages and disadvantages of self-directed work teams?
38. How might the learning curve impact standard production times used for planning? How would you possibly deal with this impact?
39. How might an application of Little's Law prove useful?
40. Select a product with which you are familiar. How do you think it might be redesigned to make it easier to manufacture and possibly more useful to the user?

PROBLEMS

15.1. Given the following fixed and variable costs and the volumes, calculate the total and unit costs.

Fixed cost	Variable cost	Volume (units)	Total cost	Unit cost
$200.00	$8.00	100		
$200.00	$7.00	1000		
$50.00	$15.00	50		
$800.00	$2.00	2000		
$500.00	$20.00	500		

15.2. A process costs $200 to set up. The run time is 5 minutes per piece and the run cost is $36 per hour. Determine the following:

a. The fixed cost.

b. The variable cost.

c. The total cost and unit cost for a lot of 600.

d. The total cost and unit cost for a lot of 1200.

Answer. a. Fixed cost = $200.00

b. Variable cost = $3.00

c. Total cost = $2000.00

Unit cost = $3.33

d. Total cost = $3800.00

Unit cost = $3.17

15.3. A manufacturer has a choice of purchasing and installing a heat-treating oven or having the heat treating done by an outside supplier. The manufacturer has developed the following cost estimates:

	Heat Treat In-house	Purchase Services
Fixed cost	$29,000.00	$0.00
Variable cost	$10.00	$18.00

a. What is the cost equalization point?

b. Should the company have the heat treating done by an outside supplier if the annual volume is 3000 units? 5000 units?

c. What would be the unit (average) cost for the selected process for each of the volumes in b above?

Answer. a. CEP = 3625 units

b. Volume 3000 units purchase services; Volume 5000 units heat treat in-house

c. Unit cost for 3000 units is $18.00; for 5000 units is $15.80

15.4. Bananas are on sale at the Cross Towne store for 89¢ per pound. They normally sell for $1.03 per pound at your corner store. If round-trip bus fare costs $4.20 to the Cross Towne store, is it worth going? What is the cost equalization point? Discuss other ways of taking advantage of this bargain.

400 CHAPTER FIFTEEN

15.5. Given the following costs, which process should be used for an order of 400 pieces of a given part? What will be the unit cost for the process selected?

	Buy	Process A	Process B
Setup		$40.00	$135.00
Tooling		$15.00	$20.00
Labor/unit		$4.30	$4.10
Material/unit		$2.00	$2.00
Purchase cost	$7.20		

15.6. The Light Company is planning on producing a new type of light shade, the parts for which may be made or bought. If purchased, they will cost $2 per unit. Making the parts on a semiautomatic machine will involve a $5000 fixed cost for tooling and $1.30 per unit variable cost. The alternative is to make the parts on an automatic machine. The tooling costs are $15,000, but the variable cost is reduced to 60¢ per unit.

a. Calculate the cost equalization point between buying and the semiautomatic machine.

b. Calculate the cost equalization point between the semiautomatic and automatic machines.

c. Which method should be used for expected sales of the following?

   i. 5000 units

   ii. 6000 units

   iii. 8000 units

   iv. 10,000 units

   v. 25,000 units

d. What is the unit cost for the selected process for each of the sales in c above?

Answer.   i. Unit cost = $2.00

   ii. Unit cost = $2.00

   iii. Unit cost = $1.925

   iv. Unit cost = $1.80

   v. Unit cost = $1.20

15.7. A major mail order house collected data on the reasons for return shipments over a three-month period with the following results: wrong selection 62,000; wrong size 50,000; order canceled 15,000; wrong address 3000; other 15,000. Construct a Pareto diagram.

Reason	Number	Percent	Cumulative Percent
Total			

15.8. A firm experienced abnormal scrap and collected data to see which parts were causing the problem, with the following results: part A—$6540, part B—$11,100, part C—$920, part D—$1440, and part E—$710. Complete the following table, listing the errors in descending order of importance. Construct a Pareto diagram.

Part	Number	Percent	Cumulative Percent
Total			

15.9. A process has a production rate of 30 units per hour. If it takes 40 minutes for a single unit to be completed in the process, what is the inventory of units in the process?

15.10. Draw an operations process chart for the assembly of a ballpoint pen. The pen is made from three subassemblies (see figure):

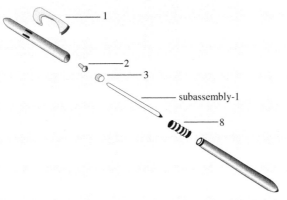

1. Upper barrel
2. Cartridge
3. Lower barrel

Operation	1	Attach clip to upper barrel
	2	Insert button into upper barrel
	3	Insert rotor into upper barrel
Subassembly-1	4	Press ball into tip of cartridge
	5	Insert tip into cartridge
	6	Fill cartridge with ink
Inspection	1	Test cartridge
Operation	7	Insert cartridge into upper barrel
	8	Slip spring over cartridge
	9	Screw lower barrel and upper barrel together
Inspection	2	Test operation of pen
	3	Final inspection

15.11. Draw a process flow diagram for your activities from the time you wake up until you arrive at work. Can you think of ways to improve the process?

15.12. A process can produce a final product at the rate of two products per minute. If an individual product takes 30 minutes to pass through the process, what is the inventory of products in the process at any time?

15.13. A university registration process can register a person every 4 minutes. If there are an average of 12 people in line for the registration, how long will each person have to wait in line?

CASE STUDY 15.1

Cheryl Franklin, Production Manager

Cheryl Franklin had been production manager at Cooper Toy Company for only a week when she started to wonder if her promotion had really been a good thing. Her boss was concerned that production was falling behind on one of their newest toys. The company had spent a lot to advertise the toy and the demand that had been created was not being met by production.

Cheryl decided to try to investigate to discover the problem and try to get the production volume back on track. Unfortunately, she found that almost everyone involved was blaming someone else.

The toy in question was one of those plastic water squirters popular at beaches and swimming pools. Water is held in a reservoir and then air is pumped into the squirter, allowing a fairly large stream of water to be shot about 10 meters.

Cheryl knew that the slowest part of the production line that she was using to produce the toy should be able to produce one every 3 minutes, and she also knew that the demand averaged only about one toy every 5 minutes. When she started her investigation at the assembly area, however, she found that assembly was averaging only one toy every 10 minutes. The assembly supervisor had a good explanation, however. He said that they didn't have enough inventory to work on. They were getting the water tanks being released from inspection on average only one every 8 minutes, so they were assembling the toys as fast as they could.

When Cheryl went to the inspection area, she discovered that there were lots of tanks waiting there to be inspected. The inspector assigned to the tanks appeared to be working hard, but obviously that was the problem.

Cheryl talked to the quality manager and got quite a bit more information. To quote the quality manager, "Look, we need to put those tanks through some important testing. If a child pumps too much air into the tank they are designed to let some of the air out through a relief valve. Unfortunately, a lot of the tanks have improperly glued seams. If one of those bad tanks gets on a toy then the tank is liable to rupture before the relief value activates. That rupture could send pieces of plastic flying. The last thing we need is to have some child get injured with flying plastic. We find that we have to inspect every tank since the reject rate on the gluing of the seams has been running close to 20 percent!"

Cheryl continued her investigation and discovered the following:

- The supervisor of the production area where the glue was applied claims the machine used to spread the glue will often clog, and even when it works, the glue will often be applied unevenly.

- The maintenance people in charge of the glue machine claim the machine doesn't work well because the glue fixture is not correct for the application and the glue itself is too thick for the type of application.

- The engineers in charge of the design and installation of the glue fixture claim that the toy was rushed into production so quickly that they did not have time to design a proper fixture so they were forced to use an old one that was originally designed for a different application.
- The purchasing agent responsible for buying the glue claims that they were not given detailed specifications for the glue and therefore relied on the recommendation from the supplier as to which glue to use.
- The marketing people claimed that the toy market is so competitive that they need to get a toy to market quickly, especially given that the life cycle of the average toy is very short.

Cheryl needs your help. Specifically, develop answers to the following:

Assignment

1. What are the underlying problems here? Try to be as specific as possible.
2. What is the best solution for these problems? Again, try to be specific.
3. Suggest how continuous process improvement techniques could be applied to the problems.

CHAPTER SIXTEEN

LEAN PRODUCTION

INTRODUCTION

In the past several years manufacturing has become much more competitive and the global economy is now a forceful reality. Producers in many countries can produce goods of consistently superior quality and deliver them to international markets at a competitive price and schedule. They have responded to changing market needs and often have detected those needs before the consumer. Because of such competition, some countries have lost their previous edge in the manufacture of such goods as radios, televisions, cameras, and ships.

How have some of these companies been able to do this? It is not necessarily only because of their culture, geography, government assistance, new equipment, or cheap labor, but because many of them practice lean production. **Lean production** is a philosophy that relates to the way a manufacturing company organizes and operates its business. It includes both the approach to organizing the business and the practice of **just in time (JIT)** production. These are not magic formulas or a set of new techniques that suddenly makes a manufacturer more productive. Rather, they are the very skillful and coordinated application of existing industrial and manufacturing engineering principles.

Another philosophical approach to production, developed originally by Eliyahu M. Goldratt in his book *The Goal*, was adding yet another perspective to those assumptions about how companies approach production. This approach is called the Theory of Constraints (TOC), which was introduced and discussed in Chapter 6, including the scheduling approach often called drum-buffer-rope.

This chapter will introduce and discuss the concepts of lean production and JIT. It will also elaborate more about TOC and discuss some of the environments where each of these concepts can best be applied.

LEAN PRODUCTION

Lean production is a concept that has evolved from JIT concepts over the past several years. JIT was brought to global attention in the late 1970s and early 1980s based on its successful development and implementation in the Toyota production system. There were many less-than-successful implementations of JIT worldwide during the 1980s, in spite of the well-documented advantages and benefits resulting from JIT. There is strong evidence that many manufacturers, eager to take advantage of the documented advantages, attempted to implement JIT without first understanding the fully integrated approach and impacts of such a highly integrated system. This condition led to many disappointing implementations or outright failures of JIT implementations during that period. Part of the misunderstanding can be attributed to confusion between the higher-level JIT "philosophy" (the fundamental changes in business practices) and the more specific JIT production, which in its simplest form can be interpreted as delivering material to a production process "just in time" for its need.

As is often the case when fundamental concepts are not fully understood, the JIT concepts were viewed by many companies as invalid or inappropriate for their particular environment. In the meantime, the development of highly integrated production systems was rapidly evolving. Manufacturing resource planning (MRP II) was recognized to be an effective engine to drive an integrated enterprise-wide information system that is today called **enterprise resource planning (ERP)**. Purchasing and logistics activities were similarly being integrated with fundamental internal materials management principles into an enterprise-wide approach, today called supply chain management. Similarly, the fundamental concepts of the higher-level JIT approaches evolved to an enterprise-wide perspective called lean production, and the JIT term has come to be reserved for the specific concept of JIT production, which often implies a pull production system.

Lean production, on the other hand, implies understanding and correctly implementing the major enterprise-wide changes required to truly eliminate or significantly reduce waste in the system. It is the system-wide philosophical approach used to integrate the system toward an ultimate goal of maximized customer service with minimal system waste.

Lean production can be defined in many ways, but the most popular is the *elimination of all waste and continuous improvement of productivity*. Waste means anything other than the minimum amount of equipment, parts, space, material, production time, and workers' time absolutely necessary to add value to the product. This means the ideal approach is that there should be no surplus, there should be no safety stocks, and lead times should be minimal: "If you can't use it now, don't make it now." From an organizational view, waste refers to *anything* that is unnecessary and adds no value in the customer's eyes and for which they are unwilling to pay.

The long-term result of eliminating waste is a cost-efficient, quality-oriented, fast-response organization that is responsive to customer needs. Such an organization has the potential to acquire a huge competitive advantage in the marketplace. While the fundamental objective focused on waste reduction, the waste reduction approaches that were developed also led to rapid responses to market demand through reduction of lead times, increased levels of quality, and lower production costs.

A principal result of the lean production approach is that removal or reduction of excessive (waste) inventory or capacity between activities in a process, regardless of the reason that inventory or capacity existed, serves to force a tighter coupling of the activities in the system. In other words, the organization must be managed as a system instead of a set of relatively disjointed activities. This is an important point, in that organizations can sometimes produce failed implementations as they may focus on only certain aspects of lean production rather than recognizing that it is really an integrated system.

Adding Value

What constitutes value to the user? It is having the right parts and quantities at the right time and place. It is having a product or service that does what the customer wants, does it well and consistently, and is available when the customer wants it. Value satisfies the actual and perceived needs of the customer and does it at a price the customer can afford and considers reasonable. Another word for this is **quality**. Quality is meeting and exceeding customers' expectations.

Value starts in the marketplace when marketing must decide what the customer wants. Design engineering must design the product so it will provide the required value to the customer. Manufacturing engineering must first design a process to make the product and then build the product according to certain specifications. The loop is complete when the product is delivered to the customer, as shown in Figure 16.1. If any part of the chain does not add value for the customer, there is waste.

FIGURE 16.1 Product development cycle.

Adding value to a product does not mean adding cost. Users are not concerned with the manufacturer's cost but only with the price they must pay and the value they receive. Many activities increase cost without adding value and, as much as possible, these activities should be eliminated.

WASTE

Anything in the product development cycle that does not add value to the product is **waste**. This section will look at the causes of waste in each element of the product cycle.

Waste Caused by Poor Product Specification and Design

The creation of waste tends to start with the policies set by management in responding to the needs of the marketplace. Management is responsible for establishing policies for the market segments the company wishes to serve and for deciding how broad or specialized the product line is to be. These policies affect the costs of manufacturing. For example, if the range and variety of product are large, production runs will be short and then machines must be changed over frequently. There will be little opportunity to use specialized machinery and fixtures. On the other hand, a company with a limited product range can probably produce goods on an assembly line basis and take advantage of special-purpose machinery. In addition, the greater the diversity of products, the more complex the manufacturing process becomes, and the more difficult it is to plan and control.

Component standardization As stated in Chapter 15, companies can specialize in the products they make and still offer customers a wide range of options. If companies standardize on the component parts used in the different models they make, they can supply customers with a variety of models and options made from standard components. Parts standardization has many advantages in manufacturing. It creates larger quantities of specific components that allow longer production runs. This, in turn, makes it more economical to use more specialized machinery, fixtures, and assembly methods. Standardization reduces the planning and control effort needed, the number of items required, and the inventory that has to be carried.

The ideal product is one that meets or exceeds customer expectations, makes the best use of material, and can be manufactured with a minimum of waste (at least cost). As well as satisfying the customer, the product's design determines both the basic manufacturing processes that have to be used and the cost and quality of the product. The product should be designed so it can be made by the most productive process with the smallest number of operations, motions, and parts and includes all of the features that are important to the customer. Chapter 15 discussed the principles of product design in more detail.

Waste Caused in Manufacturing

Manufacturing takes the design and specifications of the product and, using the manufacturing resources, converts them into useful products. First, however, manufacturing engineering must design a process capable of making the product. They do so by selecting the manufacturing steps, machinery, and equipment and by designing the plant layout and work methods. Manufacturing must then plan and control the operation to produce the goods. This involves manufacturing planning and control, quality management, maintenance, and labor relations.

Shigeo Shingo identified seven important sources of waste in manufacturing. The first four relate to the design of the manufacturing system and the last three to the operation and management of the system:

1. **Processing.** The best process is one that has the capability to consistently make the product with an absolute minimum of scrap, in the quantities needed, and with the least cost added. Waste, or cost, is added to the process if the wrong type or size of machine is used, if the process is not being operated correctly, or if the wrong tools

and fixtures are used. Overprocessing is a redundant effort that adds no value to the product or service.

2. **Motion.** Waste is added if the methods of performing tasks by the operators cause wasted movement, time, or effort. Activities that do not add value to the product should be eliminated. Searching for tools, walking, or unnecessary motions are all examples of waste of motion.

3. **Transportation.** Moving and storing components adds cost but not value. For example, goods received may be stored and then issued to production. This requires labor to put away, find, and deliver to production. Records must be kept and a storage system maintained. Poorly planned layouts may make it necessary to move products over long distances, thus increasing the movement cost and possibly storage and recordkeeping costs. Any movement that does not directly support immediate use is considered waste.

4. **Defects.** Defects interrupt the smooth flow of work. If the scrap is not identified, the next workstation receiving it will waste time trying to use the defective parts or waiting for good material. Schedules must be adjusted. If the next step is the customer, then the cost will be even higher. Sorting out defects or the repair or reworking of products or services are also waste.

5. **Waiting time.** There are two kinds of waiting time: that of the operator and that of material. If the operator has no productive work to do or there are delays in getting material or instructions, there will be waste. Ideally, material passes from one work center to the next and is processed without waiting in queue. Waste occurs when activities are not fully synchronized.

6. **Overproduction.** Overproduction is producing products beyond those needed for immediate use. When this occurs, raw materials and labor are consumed for parts not needed, resulting in unnecessary inventories. Considering the costs of carrying inventory, this can be very expensive. Overproduction causes extra handling of material, extra planning and control effort, and quality problems. Because of the extra inventory and work-in-process, overproduction adds confusion, tends to bury problems in inventory, and often leads to producing components that are not needed instead of those that are. Overproduction is not necessary as long as market demand is met. Machines and operators do not always need to be fully utilized.

7. **Inventory.** As discussed in Chapter 10, inventory costs money to carry, and excess inventory adds extra cost to the product. However, there are other costs in carrying excess inventory. Any supply of inventory, from raw materials to finished goods, in excess of immediate requirements is waste.

To remain competitive, a manufacturing organization must produce better products at lower cost while responding quickly to the marketplace. The role inventory plays in each of these steps is discussed below.

A better product suggests one that has features and quality superior to others. The ability to take advantage of product improvement opportunities depends on the speed with which engineering changes and improvements can be implemented. If there are large quantities of inventory to work through the system, it takes longer and is more costly prior to engineering changes reaching the marketplace.

Lower inventories also improve quality. Suppose that a component is made in batches of 1000 and that a defect occurs on the first operation. Eventually, the defect will be caught, very often after several more operations have been completed. Thus, all 1000 pieces have to be inspected. Because much time has elapsed since the first operation when the defect occurred, it is also difficult to pinpoint the cause of the problem. If the batch size had been 100 instead of 1000, it would have moved through the system more quickly and been detected earlier, and there would only be 100 to inspect.

Companies can offer better prices if their costs are low. Lower inventories reduce cost. Also, if work-in-process inventory is reduced, less space is needed in manufacturing, resulting in cost savings.

Responsiveness to the marketplace depends on being able to provide shorter lead times and better due date performance. In Chapter 6 it was observed that manufacturing lead time depends on queue, and queue depends on the number and the batch size of the orders in process. If batch size is reduced, the queue and lead time will be reduced. Chapter 9 noted that forecasts were more accurate for nearer periods of time. Reducing lead time improves forecast accuracy and provides better order promising and due-date performance.

Poka-Yoke (Mistake-Proof)

Poka-yoke was introduced by Shigeo Shingo of Japan. It implies the concept of removing faults at the first instance and making a process or product "foolproof." Shingo argued that statistical quality control does not prevent defects. He differentiated between errors and defects, stating that errors will always be made, but defects can be prevented. Errors are primarily related to the process while defects primarily are associated with the product. Process errors are not always directly related to the creation of a defect, but usually imply the need for focused inspection at least until process improvement can be made. Corrective action should take place immediately after a mistake is made, which implies 100% inspection as soon as an action occurs. This inspection can take one of three forms: successive check, self-check, and source inspection. Successive check inspection is done by the next person in the process who passes the information back to the worker who just performed the operation who can then make any needed repair. *Self-check* is done by the worker and can be used on all items where a sensory perception is sufficient. Scratches and paint blobs are examples of these. Source inspection also is done by the individual worker who, instead of checking for defects, checks for errors that will cause defects.

Poka-yoke tries to change either the process or its resources, thus eliminating the need to rely on human experience and knowledge. Examples include the following:

- Use color-coded parts.
- Put a template over an assembled component to show operators where specific parts go.
- Use counters to tell an operator how many operations have been performed.
- Have one prong larger than the other so the electric plug will fit only one way, such as found in electrical wiring.

THE LEAN PRODUCTION ENVIRONMENT

Many elements are characteristic of a lean production environment. They may not all exist in a particular manufacturing situation, but in general they provide some principles to help in the development of a lean production system. These can be grouped under the following headings:

- Flow manufacturing.
- Process flexibility.
- Quality management.
- Total productive maintenance.
- Uninterrupted flow.
- Continuous process improvement.
- Supplier partnerships.
- Total employee involvement.

Flow Manufacturing

The lean production concept was developed by companies such as Toyota and some major appliance and consumer electronics manufacturers. These companies manufacture goods in a repetitive manufacturing environment.

WORKSTATIONS

FIGURE 16.2 Flow manufacturing.

Repetitive manufacturing is the production of discrete units on a flow basis. In this type of system, the workstations required to make the product, or family of products, are located close together and in the sequence needed to make the product. Work flows from one workstation to the next at a relatively constant rate and often with some material handling system to move the product. Figure 16.2 shows a schematic of **flow manufacturing**.

These systems are discussed in Chapter 15. They are suitable for a limited range of similar products such as automobiles, televisions, or microwave ovens. Because work centers are arranged in the sequence needed to make the product, the system is not suitable for making a variety of different products. Therefore, the demand for the family of products must be large enough to justify economically setting up the line. Flow systems are usually very cost effective.

Work cells Many companies do not have a product line that lends itself to flow manufacturing. For example, many companies do not have sufficient volume of specific products to justify setting up a line. Companies with this kind of product line usually organize their production on a functional basis by grouping together similar or identical operations. Lathes will be placed together, as will milling machines, drills, and welding equipment. Figure 16.3 shows a schematic of this kind of layout, including routing for a hypothetical product (saw, lathe, grinder, lathe, drill). Product moves from one workstation to the other in lots or batches. This type of production produces long queues, high work-in-process inventory, long lead times, and considerable material handling.

Usually, this kind of layout can be improved. It depends on the ability to detect product flows. This can be done by grouping products together into product families. Products are in the same family if they use common work flow or routing, materials, tooling, setup procedures, and cycle times. Workstations can then be set up in miniature flow lines or **work cells**. For example, suppose the product flow shown in Figure 16.3 represents the flow for a family of products. The work centers required to make this family can be laid out according to the steps to make that family. Figure 16.4 shows a schematic of such a layout.

Parts can now pass one by one, or in very small lots, from one workstation to the next. This has several benefits:

- Queue and lead times going through the cell are reduced drastically.
- Production activity control and scheduling are simplified. The cell has only one work center to control as opposed to five in a conventional system.

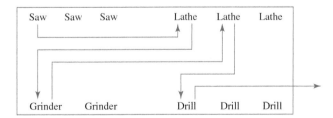

FIGURE 16.3 Functional layout.

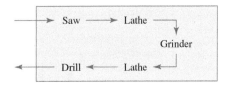

FIGURE 16.4 Work cell layout.

- Floor space needed is reduced.
- Feedback to preceding operations is immediate. If there is a quality problem, it will be found out immediately.

Work cells permit high-variety, low-volume manufacturing to be repetitive. For work cells to be really effective, product design and process design must work together so parts are designed for manufacture in work cells. Component standardization becomes even more important. Work cells are sometimes called **cellular manufacturing**.

Process Flexibility

Process flexibility is desirable so the company can react swiftly to changes in the volume and mix of their products. To achieve this, operators and machinery must be flexible, and the process must be configured to be changed over quickly from one product to another.

Machine flexibility To achieve **machine flexibility**, it often makes more sense to have two relatively inexpensive general-purpose machines than one large, expensive special-purpose piece of equipment. Smaller general-purpose machines can be adapted to particular jobs with appropriate tooling. Having two instead of one makes it easier to dedicate one to a work cell. Ideally, the machinery should be low-cost and movable. Figure 16.5 illustrates the concept.

Quick changeover **Quick changeover** requires short setup times, and improves responsiveness to customer changes. Short setup times also have the following advantages:

- **Reduced order quantity.** The lot size is often dependent on the setup cost. If the setup time can be reduced, the lot size can be reduced. For example, if the order quantity is 100 units and the setup can be reduced to 25% of its former value, the order quantity decreases to 50. Inventory is cut in half, and queue and lead times are reduced. Reductions in setup of even greater magnitude are possible. The general opinion is that setup can be cut 50% simply by organizing the work and having the right tools and fixtures available when needed. For example, in one instance, a changeover on a die press was videotaped. The operator doing the changeover was not in view for more than 50% of the time, as he was away from the machine getting tools, dies, and so on. One system for setup reduction, called the "four-step method," claims that reductions of 90% can be achieved without major capital expense. This is accomplished by organizing the preparation, streamlining the setup, and eliminating adjustments.
- **Reduced queue and manufacturing lead time.** Manufacturing lead time depends mostly on the queue. In turn, queue depends on the order quantity and scheduling. Reducing setup time reduces the order quantity and queue and lead times.

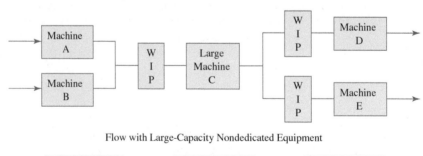

FIGURE 16.5 Large versatile equipment versus small dedicated equipment.

- **Reduced work-in-process (WIP) inventory.** WIP inventory depends on the number of orders in process and their size. If the order quantity is reduced, the WIP is reduced. This frees up more floor space, allowing work centers to be moved closer together, thus reducing handling costs and promoting the creation of work cells.
- **Improved quality.** When order quantities are small, defects have less time to be buried. Because they are more quickly and easily exposed, their cause will more likely be detected and corrected.
- **Improved process and material flow.** Inventory acts as a buffer, burying problems in processes and in scheduling. Reducing inventory reduces this buffer and exposes problems in the production process and in the materials control system. This gives an opportunity to correct the problems and improve the process.

A lean principle that applies to quick changeover is **single-minute exchange of die,** or **SMED**. SMED is a methodology that identifies which setup activities can be done while the machine is running (**external setup time**) and which can only be done while the machine is not operating (**internal setup time**). Once defined, a process is put in place to convert internal steps to external steps when possible, organize all tools required to support the activities, streamline all steps, and constantly evaluate the process for opportunities to improve.

Operator flexibility Flexible machinery and flexible processes need flexible people to operate them. To achieve **operator flexibility**, people should not only be trained in their own jobs but should also be cross-trained in other skills and in problem-solving techniques. Only with well-trained people can the benefits of process flexibility be realized.

Standardized work Standardized work is the process of documenting and standardizing tasks through the value stream, including processes, best practices, and standards. These then become the baseline for continuous improvement activities. As the standard is improved, it becomes the new baseline. Fundamentals for standardized work include utilizing best practices, and determining the who, what, where, when, why, and how of a process. Standardization is not the goal, but is the foundation used to drive continuous improvement, and can be applied at all levels and in all areas of a business.

Quality Management

Quality management is the subject of Chapter 17. This section will focus on quality from the point of view of manufacturing and lean production.

Quality is important for two reasons. If quality is not present in what is supplied to the customer and the product is defective, the customer will be dissatisfied. If a process produces scrap, it creates disrupted schedules that delay supplying the customer, increases inventory or causes shortages, wastes time and effort on work centers, and increases the cost of the product.

Who is the user? Ultimately, it is the company's customer, but the user is also the next operation in the process. Quality at any one work center should meet or exceed the expectations (needs) of the next step in the process. This is important in maintaining the uninterrupted flow of material. If defects occur at one work center and are not detected until subsequent operations, time will be wasted, and the quantity needed will not be supplied.

For manufacturing, quality does not mean inspecting the product to segregate good from bad parts. Manufacturing must ensure that the process is capable of producing the required quality consistently and with as close to zero defects as possible. Manufacturing must do all it can to improve the process to achieve this and then monitor the process to make sure it remains in control. Daily monitoring can best be done by the operator. If defects are discovered, the process should be stopped, and the cause of the defects corrected.

The benefits of a good-quality program are less scrap, less rework, less inventory (inventory just in case there is a problem), better on-time production, timely deliveries, and more satisfied customers.

As the concepts of quality management evolved, several of the concepts were structured, formalized, and enhanced to eventually become six sigma quality. Several of the key aspects of the quality approach are discussed in Chapter 17.

Quality at the source **Quality at the source** means doing it right the first time and, if something does go wrong, stopping the process and fixing it. People become their own inspectors, personally responsible for the quality of what they produce.

Total Productive Maintenance

Traditional maintenance might be called breakdown maintenance, meaning maintenance is done only when a machine breaks down. The motto of breakdown maintenance is "If it ain't broke, don't fix it." Unfortunately, breakdowns occur only when a machine is in operation, resulting in disrupted schedules, excess inventory, and delayed deliveries. In addition, lack of proper maintenance results in wear and poor performance. For example, if a car is not properly maintained, it will break down, not start, or perform poorly on the road.

For a process to continue to produce the required quality, machinery must be maintained in excellent condition. This can best be achieved through a program of **preventive maintenance**. This is important for more reasons than quality. Low work-in-process inventories mean there is little buffer available. If a machine breaks down, it will quickly affect other work centers. Preventive maintenance starts with daily inspections, lubrication, and cleanup. Since operators usually understand how their equipment should "feel" better than anyone else, it makes more sense to have them handle this type of regular maintenance.

Total productive maintenance (TPM) takes the ideas of preventive maintenance one step further. According to the *APICS Dictionary*, 16th edition, total productive maintenance is "preventive maintenance plus continuing efforts to adapt, modify, and refine equipment to increase flexibility, reduce material handling, and promote continuous flow." As such, it emphasizes the lean production principle of eliminating waste. Machine operators are typically a major part of a total productive maintenance program. Not only can they be involved with small daily maintenance responsibilities and lower-skilled routine tasks, but they can also assist skilled repair technicians who work on the equipment should a breakdown occur. They can also help the skilled technicians on start-up and shutdown tasks.

An additional advantage of operator involvement is the reduction in time and work requirements on the skilled technicians, allowing those skilled technicians more time to focus on equipment overhaul and processing improvements.

Uninterrupted Flow

Ideally, material should flow smoothly from one operation to the next with no delays. This is most likely to occur in repetitive manufacturing, where the product line is limited in variety. However, the concept should be the goal in any manufacturing environment. Several conditions are needed to achieve uninterrupted flow of materials: uniform plant loading, a pull system, valid schedules, and linearity.

Uniform plant loading **Uniform plant loading** means that the work done at each workstation is distributed to be as close to equal as possible. In repetitive manufacturing, this is called **line balancing**, which means that the time taken to perform tasks at each workstation on the line is the same or very nearly so. The result will be no bottlenecks and no buildup of work-in-process inventory.

Valid schedules There should be a well-planned valid schedule. The schedule sets the flow of materials coming into the factory and the flow of work through manufacturing. To maintain an even flow, the schedule must be level. In other words, the same amount should be produced each day. **Rate-based scheduling** can be used, which bases the production on a rate for a given period, whether it be hour, day, or week. Furthermore, the mix of

Week	On Hand	1	2	3
Economy	0	1500		
Standard	600		1500	300
Deluxe	800			1200
Total	1400	1500	1500	1500

FIGURE 16.6 Master production schedule.

Week	On Hand	1	2	3
Economy	250	500	500	500
Standard	300	600	600	600
Deluxe	200	400	400	400
Total	750	1500	1500	1500

FIGURE 16.7 MPS leveled by week.

products should be the same each day. For example, suppose a company makes a line of dog clippers composed of three models: economy, standard, and deluxe. The demand for each is 500, 600, and 400 per week, respectively, and the capacity of the assembly line is 1500 per week. The company can develop the schedule shown in Figure 16.6. This will satisfy demand and will be level based on capacity. However, inventory will build up and, if there is no safety stock, a variation in demand will create a shortage. For example, if there is a surge in demand for the deluxe model in week 1, none may be available for sale in week 2.

An alternate schedule is shown in Figure 16.7. With this schedule, inventory is reduced, and the ability to respond to changes in model demand increases. The number of setups increases, but this is not a problem if setup times are small. The idea can be carried further by making some of each model each day. Now it would mean producing 100, 120, and 80 of each model each day. If the line has complete flexibility, these can be produced in the following mixed sequence of 15. This is repeated 20 times during the day for a total output of 300:

Sequence: ESD, ESD, ESD, ESD, SES
E: Economy
S: Standard
D: Deluxe

The company makes some of everything each day in the proportions to meet demand. Inventories are at a minimum. If demand shifts between models, the assembly line can respond daily. This is called **mixed-model scheduling**. The schedule is leveled, not only for capacity, but also for material.

No matter what type of scheduling is used in a lean environment, the critical principle is that it be leveled, or smoothed, as much as possible, to facilitate the use of kanbans, and to eliminate waste. Another term for this is **heijunka**, which is a strategy for redistributing production volume and mix in order to minimize extremes and meet varying customer demand. The *APICS Dictionary*, 16th edition, defines it as "an approach to level production throughout the supply chain to match the planned rate of end product sales."

Linearity The emphasis in lean production is on achieving the plan—no more, no less. This concept is called **linearity** and is usually reached by scheduling to less than full capacity. If an assembly line can produce 100 units per hour, it can be scheduled for perhaps 700 units for an eight-hour shift. If there are problems during the shift, there is extra time so the 700-unit schedule can be maintained. If there is time left over after the 700 units are produced, it can be spent on jobs such as cleanup, lubricating machinery, getting ready for the next shift, quality improvements, total productive maintenance activities, process improvements, or solving problems.

Continuous Process Improvement

This topic is an element in both lean production and quality management and is discussed in Chapters 15 and Chapter 17. Elimination of waste depends on improving processes continuously. Thus, continuous improvement is a major feature of lean production.

Supplier Partnerships

If good schedules are to be maintained and the company is to develop a lean environment, it is vital to have good, reliable suppliers. They establish the flow of materials into the factory.

Partnering Partnering implies a long-term commitment between two or more organizations to achieve specific goals. Lean production places much emphasis not only on supplier performance but also on supplier relations. Suppliers are looked on as co-producers, not as adversaries. The relationship with them should be one of mutual trust and cooperation.

There are three key factors in partnering:

- **Long-term commitment.** This is necessary to achieve the benefits of partnering. It takes time to solve problems, improve processes, and build the relationship.
- **Trust.** Trust is needed to eliminate an adversarial relationship. Both partners must be willing to share information and form a strong working relationship. Open and frequent communications are necessary. In many cases the parties have access to each other's production plans and technical information.
- **Shared vision.** All partners must understand the need to satisfy the customer. Goals and objectives should be shared so that there is a common direction.

If properly done, partnering is a win–win situation. The benefits to the buyer include the following:

- The ability to supply the quality needed 100% of the time so there will be no need for inbound inspection. This implies that the supplier will have, or develop, an excellent process quality improvement program.
- The ability to make frequent deliveries on a JIT basis, which implies that the supplier will apply lean principles to become a supply chain partner.
- The ability to work with the supplier to improve performance, quality, and cost. For a supplier to become a valuable supply chain supplier, a long-term relationship must be established. Suppliers need to have that assurance so they can plan their capacity and make the necessary commitment to a single customer.

In return, the supplier has the following benefits:

- A greater share of the business with long-term security.
- Ability to plan more effectively.
- A competitive advantage as a lean supplier.

Supplier selection Chapter 8 noted that the factors to be considered when selecting suppliers were technical ability, manufacturing capability, reliability, after-sales service, supplier location, and lean capabilities. In a partnership there are other considerations based on the partnership relationship. They include the following:

1. The supplier has a stable management system and is sincere in implementing the partnership agreement.
2. There is no danger of the supplier breaching the organization's proprietary information.
3. The supplier has an effective quality system.
4. The supplier shares the vision of customer satisfaction and delighting the customer.

Supplier certification Once the supplier is selected, the next step is a certification process that begins after the supplier has started to ship the product. Organizations can set up their own criteria for **supplier certification** or can use one such as what has been developed by the American Society for Quality, whose criterion emphasizes the absence of defects both in product and nonproduct categories (e.g., billing errors) and use of a good documentation system, such as the ISO 9000:2015 system, which is discussed in Chapter 17.

Total Employee Involvement

A successful lean production environment can be achieved only with the cooperation and involvement of everyone in the organization. The ideas of elimination of waste and continuous improvement that are central to the lean production philosophy can be accomplished only through people cooperating.

Instead of being receivers of orders, operators must take responsibility for improving processes, controlling equipment, correcting deviations, and becoming vehicles for continuous improvement. Their jobs include not only direct labor but also a variety of traditionally indirect jobs such as preventive maintenance, some setup, data recording, and problem-solving. As discussed previously in this chapter, employees must be flexible in the tasks they do. Just as machines must be flexible and capable of quick changeover, so must the people who run them.

The role of management must change. Traditionally, management has been responsible for planning, organizing, and supervising operations. In a lean environment many of their traditional duties are done by line workers. More emphasis is placed on the leadership role, meaning managers and supervisors must become coaches and trainers, develop the capability of employees, and provide coordination and leadership for improvements.

Traditionally, specialized staff have been responsible for such things as quality control, maintenance, and recordkeeping. Under lean production, line workers do many of these duties. Staff responsibilities then become those of training and assisting line workers to do the staff duties assigned to them.

MANUFACTURING PLANNING AND CONTROL IN A LEAN PRODUCTION ENVIRONMENT

The philosophy and techniques of lean production discussed in this chapter are related to how processes and methods of manufacture are designed. The major responsibility for designing processes and methods lies with manufacturing and industrial engineering. Manufacturing planning and control are responsible for managing the flow of material and work through the manufacturing process, not designing the process. However, manufacturing planning and control are governed by, and must work with, the manufacturing environment, whatever it is. Figure 16.8 shows the relationship.

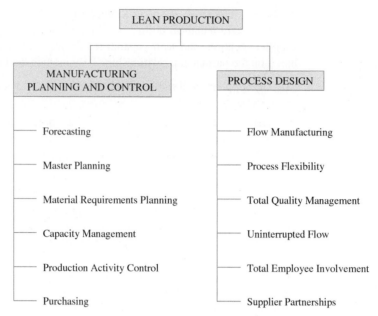

FIGURE 16.8 Lean production.

No matter what planning and control system is used, these basic questions have to be answered:

1. What are we going to make?
2. How much are we going to make?
3. When do we need to make it?
4. What do we need to make it?
5. What do we have?
6. What must we get?

The logic of these questions always applies, whether cooking a meal or making a jet aircraft. Systems for planning and control vary based on the industry and product. The manufacturing planning and control system discussed in this text has proved effective in any manufacturing environment. The complexity of the manufacturing process, the number of finished items and parts, the levels in the bill of material, and the lead times have made the planning and control problems either simple or complex. If anything can be done to simplify these factors, the planning and control system will be simpler. In general, lean production simplifies these factors, thus making the planning and control problems easier to identify and resolve.

The sections that follow will look at how the various parts of the manufacturing planning and control system relate to a lean production environment. In general, lean production does not make the manufacturing planning and control system obsolete but, in some ways, does change the focus. Lean production is not primarily a planning and control system. It is a philosophy and a set of techniques for designing and operating a manufacturing plant. Planning and control are still needed in a lean production environment.

Forecasting

The major effect that lean production has on forecasting is shortened lead time. This does not affect forecasting for business planning or production planning, but it does for master production scheduling. If lead times are short enough that production rates can be matched to sales rates, forecasting for the master production schedule (MPS) becomes less important.

Sales and Operations Planning/Production Planning

Long-term planning must extend far enough into the future to allow for any changes in the production flow and required resources, long lead time purchases, and the development of any new long-term relationships with suppliers. Lean production has the potential for

Level Capacity Schedule

Week	1	2	3
Model A	900		
Model B	300	900	
Model C		300	1200
Total	1200	1200	1200

Level Material and Capacity Schedule

Week	1	2	3
Model A	300	300	300
Model B	400	400	400
Model C	500	500	500
Total	1200	1200	1200

FIGURE 16.9 MPS leveled by capacity and by material.

reducing long lead times, but more importantly, it provides an environment in which the supplier and buyer can work together to plan the flow of material.

Master Production Scheduling

Several scheduling factors are influenced by lean production:

- Master scheduling tries to level capacity, while lean production tries to level the schedule based on capacity and material flow. Figure 16.9 illustrates the difference.
- The shorter lead times reduce time fences and make the MPS more responsive to customer demand. The ideal lead time is so short that the company can respond to actual sales, not to forecast. Whether the company builds to a seasonal demand or to satisfy promotion, a forecast is still necessary. Planning horizons can also be reduced.
- Lean production requires a stable schedule to operate. This principle is supported by using time fences. These are established based on lead times and the commitment of materials and resources. If lead times can be reduced through lean production, the time fences can be reduced.
- Traditionally, weekly time buckets have been used for planning. Because of reduced lead times and schedule stability, it is possible to use daily time buckets or a **bucketless system** in a lean environment, due to the uninterrupted flow.

Material Requirements Planning

Material requirements planning (MRP) plans the material flow based on the bill of material, lead times, and available inventory. Lean practices may modify this approach in several ways:

- The MRP time buckets were originally one week. As lead times are reduced and the flow of material improved, these can be reduced to daily buckets.
- The MRP netting logic is based on generating order quantities calculated using the planned order releases of the parent, the inventory on hand, and any order quantity logic used. In a pure lean production environment, there is very little or no inventory on hand, and the order quantity logic is to make exactly what is needed, or lot-for-lot.

If the lead times are short enough, component production occurs in the same time bucket as the gross requirement, and no offsetting is required.
- Bills of material can frequently be flattened in a lean environment. With the use of work cells and the elimination of many inventory transactions, some levels in bills of material become unnecessary.

Both MRP and lean production are based on establishing a material flow. In a repetitive manufacturing environment, this is set by the model mix and the flow rate. The product to be made is decided by the need of the following workstation, which is ultimately the assembly line.

However, many production situations do not lend themselves to level scheduling and the pull system and therefore must rely more heavily on MRP logic. Some examples are as follows:

- The demand pattern is unstable.
- Custom engineering is required.
- Quality is unpredictable.
- Volumes are low and occur infrequently.

Capacity Management

Capacity planning's function is to determine the need for labor and equipment to meet the priority plans. Leveling schedules should make the job easier, as will the lean production emphasis on cutting out waste and problems that cause ineffective use of capacity. Capacity control focuses on adjusting capacity daily to meet demand. Capacity across work centers is synchronized so that work flows smoothly. Leveling should make this task easier, but so will the lean production emphasis on cutting out waste and problems that cause ineffective use of capacity. Linearity, the practice of scheduling extra capacity, will improve the ability to meet priority schedules.

Capacity management in some ways becomes more important in a lean production environment. Since a lean production environment tends to have very low inventory, it becomes critical to have the right capacity at the right time to make the inventory according to customer demand. This also relies on the importance of total productive maintenance, since that approach provides assurance that the process capacity will likely be ready to use when needed.

Even if an MRP system is not used to launch production orders, the logic in a good MRP system can effectively be used to help plan production capacity in a lean production environment.

Inventory Management

Because lean production reduces the inventory in the system, in some respects this should make inventory management easier. However, if order quantities are reduced and annual demand remains the same, more work orders and more paperwork must be tracked, and more transactions recorded. The challenge then is to reduce the number of transactions that have to be recorded. One way to do this is to **backflush** inventory, also called **post-deduct inventory transaction processing**.

Material flows from raw material to work-in-process (WIP) to finished goods. In a post-deduct system, raw materials are recorded into WIP. When work is completed and the raw material becomes finished goods, the WIP inventory is relieved by multiplying the number of units completed by the number of parts in the bill of material. In some cases, primarily for those products with short throughput times, material is not electronically issued into WIP, meaning it is removed directly from raw material once work is completed. The system works if the bills of material are accurate and if the manufacturing lead times are short. Backflushing can also be done at intermediate work stations rather than waiting until finished goods for all components, to gain greater visibility of WIP.

One benefit of backflushing is the reduction in the number of transactions, which reduces waste, a major target of lean production. This replacement of detailed WIP accounting requires a very high level of bill of material data accuracy. A possible disadvantage with backflushing as the only accounting approach for materials is the inability to capture possible problems with a component, such as scrapping the component for quality problems, damage to the component, or just losing it in the inventory system. When such problems occur, those losses should be reported to the transaction system to prevent possible future shortages of components.

Little's Law, introduced in Chapter 15, which shows the relationship between rate of the process, R, the inventory, I, and the throughput time, T, as $I = RT$, demonstrates the impact of reducing inventory in a lean environment. By removing the need for buffer inventory through quality improvements, reduced setup times and cycle inventory, total inventory is reduced. Assuming a relatively constant value-added production rate, the reduction in inventory would also reduce throughput time. That reduction in time would imply the ability to respond quicker to demand, meaning better information about the demand due to a reduction in forecast error. The improved information continues to allow for more inventory reduction, and the continuous improvement cycle continues.

The Push System

Traditional manufacturing and inventory control uses a **push system**, meaning production is performed based on an advance schedule with dates and quantities. In a distribution system, a push system replenishes warehouses based on decisions made at the central facility. MRP is often called a push system, meaning that the material needs are calculated ahead of time (planned order releases) and, assuming there are no significant changes to the plans, pushed out to the production system as a production order. The trigger for the entire plan is the projection of the final product need, as represented by the MPS. Part of the difficulty with MRP is that often the plans are not effective because of problems or changes, including:

- Changes in customer requirements, both in timing and quantity.
- Supplier delivery problems, including timing, quantity, and quality.
- Inaccurate databases that can make the plans invalid.
- Production problems, including the following:
    - Absenteeism in the workforce.
    - Productivity and/or efficiency problems.
    - Machine downtime.
    - Quality problems.
    - Poor communication.

These problems generally promote an environment that, despite the best-laid plans, can allow for ineffective execution and a growth in inventory levels.

The Pull System

The **pull system** is an alternative to the push system, and is often identified as the primary aspect of lean production. Instead of making product based on a predetermined schedule, the pull concept produces only what is needed, when it is needed, to satisfy customer order demand. The same concept is used throughout the production facility, each work center not producing product until the subsequent work center has a demand. In distribution, replenishments would not be made until the warehouse had a demand for the item. Essentially, this system is much the same as the basic reorder point system used for independent inventory.

Reorder points normally do not work well in a dependent inventory environment due to a significant violation of the assumption of relatively constant demand that allows a reorder point to work well in some independent inventory environments. A simple example may help illustrate the problem.

Suppose the product is a specific model of bicycle. The bicycles are made in batches, which is a typical mode of production for an assemble-to-order environment. The batch size is 200 bicycles.

A dependent inventory item that is one level lower on the bill of materials is the bicycle seat. Suppose it has a lot size of 300, a two-week lead time, and a reorder point of 80.

Example 1—In this case suppose there is an inventory of 290 seats. A new batch of bicycles has just been ordered, requiring the use of 200 of the seats in a very short time. The balance left is 90 seats, 10 above the reorder point. The seat is not reordered since the reorder point has not been reached. The 90 will stay in inventory until the next order for the bicycles is generated, which may be a significant time. When that order does come to build another 200 bicycles, only 90 can be built because that is the number of seats in inventory. Another lot of 300 must be immediately ordered, but it will be two weeks before they are available.

Example 2—Now assume that there are 270 seats in inventory. The order for 200 bicycles comes in, 200 seats are used, and the seat reorder point is reached, causing an immediate reorder. Two weeks later the 300 seats arrive and are added to the 70 left in stock. There are now 370 seats that will stay in inventory (costing a lot of money) until the next time the bicycles are made, which may be a very long time.

As the example illustrates, the lot-sizing problem with dependent inventory often results in either a crisis shortage or a replenishment of stock well before it is actually needed. This example shows that the critical conditions causing the problem are the large lot sizes and the long lead times, both of which are major targets of lean production waste reduction.

First, look again at the standard EOQ model that helps determine the most economical lot size. It is, of course, the basic trade-off of inventory holding cost and order cost, as described in detail in Chapter 10 and shown in Figure 16.10.

A fundamental assumption of this model is that the two major costs involved are known and relatively fixed. While this is essentially true with holding cost, it is not true with order cost. If the order cost is equipment setup, then a major lean production effort is to reduce this setup cost. If it is a purchased item, the major effort is to work with suppliers to reduce the cost and time of purchase order and delivery. With these efforts, the order cost curve is driven downward and to the left, as shown in Figure 16.11.

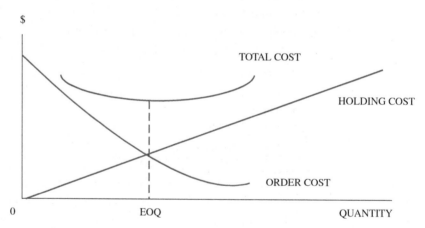

FIGURE 16.10 Balancing total cost.

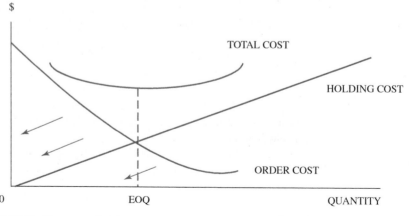

FIGURE 16.11 Downward order cost.

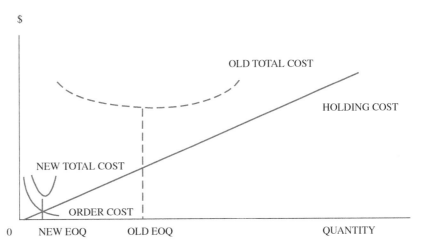

FIGURE 16.12 New order cost curve.

When these actions are taken, a new total cost curve based on the new order cost curve is generated, resulting in a significantly smaller EOQ, as Figure 16.12 illustrates.

This implies that the economic order quantities and the reorder points are very small, meaning they will be ordered frequently but in very small batches. Since the actions are also taken at the final product level, there will be, in the preceding bicycle example, frequent lots of a very small quantity of bicycles built requiring small lots of seats frequently reordered.

Reviewing the scenario illustrates the impact. The lot size for the bicycles is very small, as is the lot size for the seats. Lead time to replenish the seats, also a target for lean production improvement, has shrunken as well. Suppose the bicycle lot size is now 7 and the seat lot size is 10. The reorder point for the seats is now zero. If one lot size of bicycles is built, the seat reorder point will not be reached, and there will not be enough seats to make another lot of bicycles. With such a small seat lot size, however, it is affordable to keep two, three, or even more lots on hand (the number being dependent on the new replenishment lead time). Therefore, the next lot of bicycles can be produced with the second lot of seats while the first lot of seats is being replenished.

A lean production approach that emphasizes this reduction in batch size is **one-piece flow**. This philosophy builds on the principle that the optimal batch size is one at a time. This reduction in batch size is reliant on eliminating the constraints that require larger batch sizes.

The downside of the change While the average inventory is clearly lower in the small lot size scenario, there is a cost involved beyond the one-time cost to reduce the order cost and the lead time. Given that the overall customer demand has not diminished, batches will need to be built much more frequently since each batch is smaller in size. Each time the inventory of a given batch gets close to the reorder point, there is a risk of stockout if the demand during the replenishment lead time exceeds expectations. In order to maintain service levels, such as discussed in Chapter 12, a small safety stock may be required to protect against the additional risks of stockouts. Figure 16.13 illustrates the condition.

The Kanban System

With shortened lead times a constant goal in lean production, something is needed to generate the reorder point signal without having to rely on a formal, structured system that could take time to react. The developers of the JIT production concepts utilized a simple card system called **kanban**, which roughly translated from Japanese means *card* or *sign*.

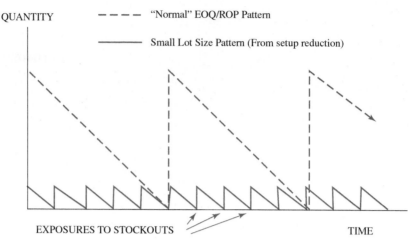

FIGURE 16.13 Exposure to stockouts.

The system works very simply. The kanban signal identifies the material to which it is attached. The information on the kanban will often include the following:

- Component part number and identification.
- Storage location.
- Container size (if the material is stored in a container).
- Work center (or supplier) of origin.

How it works The following figures illustrate the use of what is often called a two-card kanban system. The two types of cards are a production card (authorizing production of whatever part number is identified on the card in the quantity specified) and a withdrawal card (authorizing the movement of the identified material).

At the start of the process there is no movement, since all the cards are attached to full containers, as shown in Figure 16.14. It is only when a card is unattached that activity is allowed. In this way, the number of cards will clearly limit the inventory authorized to be at any location.

At some point, a downstream process needs some of the parts produced by work center 2 (in its "Finished Production" stock). It takes a container of the material, leaving the work center 2 production card with the center. This illustrates two additional rules of the system—all material movement is in full containers (recall that the container lot size is supposed to be very small) and kanban cards are associated with a work center. This initial movement is illustrated in Figure 16.15.

The unattached production card is the signal to start the work center 2 production to replace the container that was taken. To do that work they need raw material, which is in the containers in front of the work center with the move cards attached. When that material is used to replace the work center 2 finished material, the raw material container is now empty and the associated move card is unattached, as shown in the Figure 16.16.

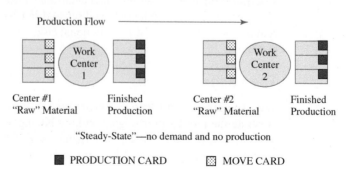

FIGURE 16.14 Kanban steady state.

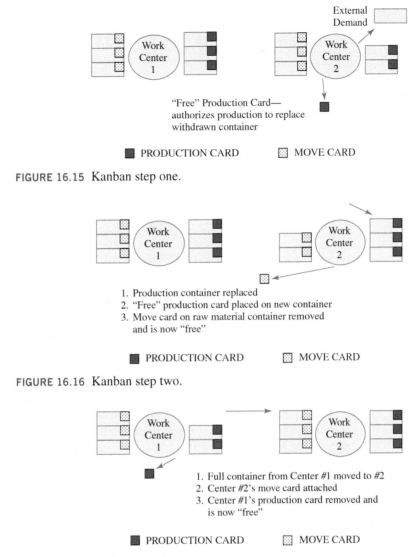

FIGURE 16.15 Kanban step one.

FIGURE 16.16 Kanban step two.

FIGURE 16.17 Kanban step three.

The unattached move card authorized movement of material to replace the material that was used. That material is found in the finished goods section of work center 1. The operator (or material handler) will now move the material and place the move card on the container as proof of the authorization to move the material. Before doing so, however, they must remove the production card that had first authorized its production. That represents another critical rule for kanban: every container with material must have one, but only one, card attached. Therefore, when the move card is attached the production card must be removed. That is illustrated in Figure 16.17.

Now, of course, there is an unattached production card for work center 1, allowing it to produce, using some of the raw material for work center 1 and freeing a move card for that material, as shown in Figure 16.18.

This process continues upstream even to the suppliers, who can also receive the kanban move cards as a signal for their next shipment to the facility.

Notice that there are no schedules with this system. Production and movement of material are only authorized purely as a reaction to the utilization of material for production downstream. The production of the final product may be the customer taking material. In some facilities there is a final assembly schedule for customer orders. In those facilities that may be the only formal schedule used.

Also note that the cards only circulate within and between work centers, as shown in Figure 16.19.

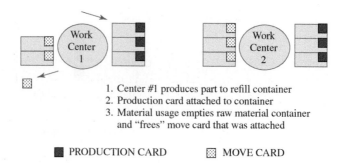

1. Center #1 produces part to refill container
2. Production card attached to container
3. Material usage empties raw material container and "frees" move card that was attached

■ PRODUCTION CARD   ▦ MOVE CARD

FIGURE 16.18 Kanban step four.

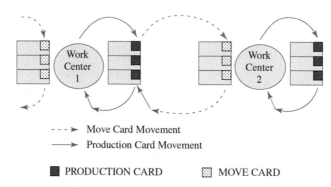

- - -▶ Move Card Movement
⎯⎯▶ Production Card Movement

■ PRODUCTION CARD   ▦ MOVE CARD

FIGURE 16.19 Kanban circulation.

The question is often raised about what the quantity or batch size of the kanban should be. There are several rules of thumb around the logic. If containers are being used, it should be a container's worth. If there is no container, and it needs to be more than one unit (which would eliminate all wasted inventory), then a standard quantity should be set that keeps the work center running while another kanban is being produced; in other words, an amount that covers process lead time. Many believe that a kanban should not be more than 10% of a day's requirements. However, the more important rule is that it be a disciplined quantity that keeps the flow moving at a level pace.

Once the amount of the kanban has been determined, the calculation of the number of kanban cards can be done. Again, there are several ways of computing this. The following is one that is simple and easy to use, and shows the similarity to reorder point.

# of kanban cards = (Demand During Lead Time)/kanban quantity

Kanban "rules" Even though there are no formal schedules in a kanban system, there is a fairly important set of rules or set of constraints under which the kanban system can most effectively operate. Those rules are summarized here:

- Every container with parts shall have one, but only one, kanban.
- There will be no partial containers stored. Every container will be filled, empty, or in the process of being filled or emptied. This rule makes inventory accounting easy. Only containers need to be counted, instead of individual parts. The number of containers is then multiplied by the number of parts in a container.
- There will be no production or movement without an authorization in the form of an unattached kanban card.

Card alternatives Since the development and successful implementation of kanban systems in many facilities, many alternatives have been designed and implemented. Some of the alternative methods include the following:

- Single card systems. The single card is the production card, with the empty container serving as the move signal.
- Color coding of containers.

- Designated storage spaces.
- Computer systems, often with bar coding serving as the signal generator.
- Workflow signals to suppliers to release product against a purchase order.

It should be noted that the method used is not important. What is important is that there is a clear reactive signal to generate activity that everyone clearly understands.

Using the Kanban System for Process Improvement

Because the kanban system allows for a controlled inventory of relatively small containers, there is a great opportunity for using the system to promote continual process improvement. Specifically, whenever the process is working smoothly for an extended period of time, there is a possibility that there actually is too much inventory in the system. The analogy that is often used is a river. If the water level is high enough, it will cover all the rocks in the river and appear to be running smoothly without any obstructions. The water in the analogy is inventory, and the rocks are process problems, including quality problems, worker skills, equipment breakdown, and so forth.

The approach is to gradually remove the water until the first "rock" is exposed, thereby establishing a priority of the most important obstacle to work on. It would be dangerous, of course, to remove too much "water" at a time, because the obstacles may stop the flow altogether. This is where the small lot size of kanban is a benefit. Removal of one kanban card will remove one container, and since the containers are small, so too will be the impact of the removal. The important aspect of this is that some process problem will ultimately emerge, signaling the next target for lean production process improvement efforts.

This is not an easy approach to implement. What is implied is that every time a process is working smoothly, there may be too much inventory and what is needed is to remove inventory until it "hurts." That is certainly not a natural action for most people, and the performance evaluation system needs to be altered to reflect this type of activity.

Some Additional Lean Production Tools and Concepts

Value stream mapping Value stream mapping is a tool to map and understand the flow of materials from supplier to customer, focusing on not only understanding the current state of process and flow but also specifying the value-added and nonvalue-added time of all process steps. This includes all activities, even the in-process storage of inventory and key metrics for the process. This visible understanding of the current flow of activities allows for developing a future state map that will significantly reduce waste, decrease flow time, and make the process flow more efficiently and effectively. Strategies can then be developed for specific actions that will lead the operation toward the future state.

Kaizen Kaizen is a philosophy of continual improvement that emphasizes employee participation. The **kaizen event** usually focuses on a fairly small part of the overall process to improve that part of the value stream. It is a structured approach to understand and redesign the process to meet specific process goals that are often part of the overall implementation of a lean production system. Kaizen events are generally considered to be a major part of the overall approach to continuous improvement for an operation. Since kaizen events are generally designed to be accomplished in short periods of time (one to two weeks), the term **kaizen blitz** is frequently used. An example of a goal where a kaizen event might be used would be to reduce the changeover time on a machine from 16 to less than 8 hours. Another example might be applying continuous improvement steps to be able to process customer orders within 4 hours of receipt.

Takt time **Takt time** is thought of as the heartbeat of the overall process. It is usually defined to be the rate of production that is synchronized with the rate of customer demand. If the production process is well synchronized with demand, the implication is that customer demand will be met with little or no excess inventory or other forms of waste. Takt time is a common metric used to help design a future state process during value stream mapping. Line balancing can be achieved by aligning cycle times and process times to takt time.

A formula for determining takt time is:

$$\frac{\text{Time available during production period}}{\text{Unit demand during production period}} = \frac{\text{Time}}{\text{Volume}}$$

A BRIEF DESCRIPTION OF AN ACTUAL APPLICATION

A few years ago, one of the authors had an opportunity with a small group of people to visit a medium sized manufacturing facility in the southeast United States that had successfully implemented many of the concepts described conceptually in this chapter. This description summarizes that visit.

The visit started with the general manager inviting us to a small conference room where he had a brief presentation about the company, the products, their markets, and so forth. At the conclusion of that presentation, he invited us to visit the production floor. Once he opened the door to the production areas, he ushered us in and then immediately fell to the back of our group. Since he did not direct us or tell us anything, we naturally started walking further into the production areas. Whenever we came close to one of the work cells, we noticed that one of the workers would leave their workstation and approach us, generally welcoming us to their work cell and asking if they could show us what was happening in the cell. They not only showed us the production process but proudly showed a large bulletin board where they had carefully kept track of key measurements and accomplishments (production, quality, productivity, etc.). Once they finished showing us the cell and had successfully answered any questions we had, they thanked us for visiting and they returned to their work as we moved on to the next work cell. Our experience in each cell was close to identical. During this entire time the general manager said not one word unless one or more of the workers greeted him, after which he would then return the greeting.

When we had finished the tour and returned to the conference room, the first thing we asked the general manager was why he had been so quiet during the production tour. He stated "the production floor is not my world—it belongs to the production workers. My world consists of running the business, while theirs is to make the product we need in the most effective and efficient way they can. My main role in production is primarily to provide the information and resources to allow the production workers to effectively and efficiently do their "job." When we asked more about that, he gave us additional information on how he was running the overall business. Examples include:

- **The way engineers and quality professionals work with the cell teams.** If the workers in the cells feel they need help with a problem, they can request professional help from the engineering department. If the workers don't like or don't agree with what the professional is telling them, the workers can "fire" the professional in that they can ask them to leave and request someone else. Of course, the professionals can also make suggestions and recommendations for improvement at any time, but the suggestion has to be approved by the workers before it can be implemented. If an accepted recommendation is implemented, the impacts (production or quality changes, for example) are then included into the planning system and the measurement system. The major exception to this approach is for customer-requested or safety related product or process changes.

- **The role of the supervisors in the teams.** Supervisors and workers function as a team to understand problems, capabilities, planned maintenance schedules, possible employee absenteeism, etc. They then work with the production planning personnel to develop projected production needs for each cell over a given unit of time. Supervisors then are responsible to make sure workers have the materials they need when they need them and also work to get resources as necessary (engineering help, for example). Supervisors don't really dictate which workers work on which machine for which product at which time. Those decisions are made by the workers as a team. Supervisors have a "vote" but do not intervene unless the workers need them to. Essentially each cell contains a worker team which is self-directed.

- **The role of human resources in staffing the teams.** The workers and their supervisor have ultimate control for many human resource decisions. They can ask to have a worker removed, for example, if they feel that worker is incapable of fitting in to the production needs or the "culture" of the work cell. The same is true for hiring a new worker. The workers in a cell will interview proposed new workers and have the final decision if they want the proposed new worker into the cell or not. Once again, the supervisor is involved but is only one vote. Workers are very sensitive that they maintain a positive cell "culture." If a worker is removed from one team for lack of "cultural fit," they are welcome to join a different work team, but they must be approved by that team.

Ultimately the general manager gets the summary reports from the production floor and as long as overall production and quality goals are met, he sees no reason to even consider any intervention. The workers were extremely proud of their work in the cells and also their approach to continuous improvement and the reputation of the company with customers. They truly felt "ownership" of their cells and what they produce.

Example

AVAILABLE WORK TIME

Shift 1: 6 am to 2:30 pm with 20 minute lunch and 10 minute break

$$8.5 \text{ hours} \times 60 \text{ minutes} = 510 \text{ minutes}$$
$$510 \text{ minutes} - 20 \text{ minute lunch} = 490 \text{ minutes}$$
$$490 \text{ minutes} - 10 \text{ minute break} = 480 \text{ minutes}$$
$$480 \text{ minutes} \times 60 \text{ seconds} = 28,800 \text{ seconds}$$

CUSTOMER DEMAND

$$230,400 \text{ units annually}$$
$$240 \text{ working days per year}$$
$$230,400 \text{ units} / 240 \text{ days} = 960 \text{ units per day}$$
$$960 \text{ units per day} / 2 \text{ shifts} = 480 \text{ units per shift}$$

TAKT TIME

$$\frac{28,800 \text{ seconds per shift}}{480 \text{ units per shift}} = \frac{60 \text{ seconds}}{\text{takt time}}$$

5S While originally developed from a series of Japanese words, a rough English translation is sort, straighten, shine, standardize, and sustain. In general, it is a structured housekeeping approach to organizing the operation for more effectiveness and less waste, which is the overall goal of a lean production system.

- **Sort.** Determine things that are needed and those not needed in the workplace. Remove those that are not needed.
- **Straighten.** Put the necessary things in good order so they may be readily available when needed.
- **Shine.** Clean the area, and keep it clean.
- **Standardize.** Maintain the order and cleanliness that have been developed.
- **Sustain.** Train and develop attitudes to keep the orderliness as an expected and ongoing part of the organization culture.

Some organizations have added a sixth S to the principle, and include safety, as good housekeeping can be beneficial to providing a safe work environment.

Visual management Similar to the use of kanbans as a visual signal, **visual management** allows people to see what is happening and rapidly respond to issues as they occur. Visual systems eliminate waste as the status of the process is very apparent, and no time or energy is wasted trying to determine it. Many tools can be used to provide visual management, including charts, colors, lights, sound, or empty spaces. Display boards or charts posted where they are visible can easily display key information. One tool, an **andon**, is a signaling device within a process. In some cases, it works similar to a traffic signal, using green, yellow, and red, to alert personnel of what action is required. Examples of everyday andons include displays of customer wait times, the fuel warning light in an automobile, or the walk signal at a crosswalk.

Lean accounting Accounting in a lean environment is much different than a traditional cost accounting structure. Typical cost accounting gathers detailed information from each work center, applies a certain level of overhead to each work center, and determines what costs should be applied to each unit of work produced. The time associated with this task is nonvalue-added, and contradicts the idea of eliminating wasteful activities, including the energy given to identifying and analyzing variances. In addition, traditional standard costing rewards production for overproducing due to favorable overhead absorption. In contrast, lean accounting measures the costs and savings associated with the elimination of waste and continuous improvement. For example, value stream accounting focuses on actual, real costs within the value stream and compares them with the revenues and profitability generated by the value stream during the time period. Although it is time and transaction heavy, and not necessarily a long-term solution, **activity-based costing (ABC**

		Current State Before Lean	Lean Step One	Lean Step Two	Future State
Operations	Sales per person	$220,833	$220,833	$220,838	$225,033
	Inventory turns	6.5	10	15	20
	Average cost per unit	31.32	30.19	29.75	25
	First time through	82%	84%	88%	95%
Capacity	Productive	58%	62%	65%	75%
	Non productive	36%	30%	30%	10%
	Available	6%	8%	5%	15%
Financial	Revenue monthly	$ 7,062,000	$ 7,062,000	$ 7,062,000	$ 8,686,000
	Material cost	$ 2,164,184	$ 2,164,184	$ 2,109,327	$ 2,552,839
	Labor cost	$ 2,483,416	$ 2,483,416	$ 2,483,416	$ 2,657,500
	Value stream gross profit	$ 2,414,400	$ 2,414,400	$ 2,469,257	$ 3,475,661

FIGURE 16.20 Box score.

can be used to identify those areas that are generating the most cost and need improvement. Another tool used in lean accounting is referred to as **box scores**, similar to balanced scorecards mentioned in Chapter 2. Box scores show a 3-dimensional view of value stream performance, operational performance measurements, financial performance, and how value stream capacity is being used, comparing productive time to nonproductive time. They can be modified to reflect specific information needed for decision-making. An example of a box score is shown in Figure 16.20.

Finance and accounting functions can be of great support to lean initiatives by providing financial measures that benefit the processes that are transforming. People begin to notice when the lean processes begin to generate additional revenue due to the cost savings inherent in lean principles. Following is an example of the areas where lean accounting can make an impact on an organization by providing beneficial cost information:

- Visual Management.
    - Work cell performance measurements aligned with lean initiatives.
    - Reduction in number of transactions.
- Value Stream Management.
    - Value stream measurements.
    - Financial benefits of lean change.
    - Value stream costing.
    - Lean decision-making.
    - Life cycle costing.
- Continuous improvement.
    - Value stream cost and capacity.
    - Costs of various product features and characteristics.
    - Target costing.
    - Capital project justification.

COMPARING ERP, KANBAN, AND THEORY OF CONSTRAINTS

As the various systems are developed and promoted, proponents of each claim their favorite to be "the answer" to maximizing service to the customer and minimizing inventory and other process costs. There are, however, certain production and business environments in which one may provide better benefits and another may not work as well. This brings to light the importance of understanding the business environment and developing a specific production strategy to best meet the needs of customers in that environment.

Enterprise Resource Planning (ERP)

Advantages ERP is by its very nature a forward-looking system. Among its tools, it uses the master schedule to project when products need to be made and projects the need for component production or purchase as well as capacity at all levels based on a defined set of parameters. Because of the forward-looking nature of the system, ERP can be very effective in an environment with a great deal of variability and uncertainty. For example, it can handle variability of demand as well as, or better than, most other systems and tends to be quite effective in dealing with product design changes and process changes.

Disadvantages ERP has one major disadvantage—it is highly data dependent. It is not only critical to have a lot of data, but the data also needs to be both accurate and timely on an ongoing basis. The data burden on the infrastructure can be high and costly, as well as the cost associated with the hardware, implementation, and training required for a successful ERP process. However, this disadvantage is typically far outweighed by the advantages gained from an integrated system, especially when the production system is characterized by the need for frequent change and/or uncertainty.

Lean Production (Kanban)

Advantages Lean production and kanban are almost the opposite of ERP from the standpoint of the type of preferred environment. In the lean approach, most of the implementation recommendations fall under the categories of eliminating or reducing the process uncertainties and making production more stable and predictable. This makes sense when it is realized that it is a very reactive system. Very little actual production is planned ahead. Instead, kanbans cause replacement of material used in a totally reactive mode. Changes in product design or demand patterns can cause real problems in a kanban system since the system tends to not look ahead. For these reasons, kanbans work best in a highly stable and predictable environment.

Disadvantages Kanbans can quickly fail in a highly volatile environment because of the reactive nature of the system. Volatility in customer demand, processing problems, and extensive changes in product designs make it very difficult for a kanban system to work effectively.

Kanbans do not plan production ahead of time but merely provide a signal to replace what has been used. As part of a lean production system, there is also an implication that little excess inventory exists. If there is a change in product design, the kanban has little advance notice of the change, meaning some inventory of the old design may be left over and demand for the new design can exist before any has been produced in the system. In addition, since the system relies on a fairly small amount of in-process inventory, spikes in demand or even large trends in demand can often run the system out of inventory with little chance to react. While the pull approach of a kanban system may be inappropriate, in a volatile environment, the facility can still obtain some benefit by utilizing other aspects of lean production, such as the elimination of waste.

Theory of Constraints (Drum-Buffer-Rope)

Advantages Since the TOC system has as its basis the identification and effective management of a constraint, it assumes that such a constraint can be identified and will be a constraint long enough to be managed effectively. In operations where that is the case, the concepts of TOC can bring great benefits to managing the process, as has been proven in many operations.

Disadvantages TOC may not be as effective where the constraint cannot be easily identified or managed. If, for example, the product mix for the operation changes in such a way that the constraint may change many times a day, then it may be virtually impossible to effectively manage a constraint, even if it can be identified. Much like lean and kanban, TOC probably works more effectively in a more stable demand environment.

Hybrid Systems

As might be expected, some operations have managed to take the best features of each system and combine them into hybrid systems.

Kanban and MRP The combination of these two systems is quite common. An ERP system is used for advanced planning, including long lead time purchased materials, adding or changing resources (capacity), and implementing product design changes. Once ERP has the materials and resources scheduled, however, kanban is used as an execution system, bringing with it the characteristics of rapid response to customer orders and reduced inventory levels throughout the process. It should be emphasized that ERP is primarily a planning system and kanban is primarily an execution system. As such they are totally compatible if the ERP system is used to make effective plans in the context of the production plan and master schedule, while using kanbans to execute those plans as customer demand is recognized. This approach is fairly commonly used in many environments especially if there is a limited amount of volatility or uncertainty. While a pull system generates the actual production of products, the ERP system is very important for advanced planning aspects such as capacity (extremely important in lean inventory conditions), scheduling maintenance, scheduling long lead-time supplier components, and planning product design changes. The pull system alone can seldom deal effectively with those outputs from an effective ERP system.

Lean production and theory of constraints Lean production often promotes process improvements throughout the system, but as TOC teaches, there is almost always one process that constrains the overall throughput. This implies that the TOC approach can help prioritize the areas of improvement, and lean production principles can help with this improvement. In this sense, the TOC concepts can be thought of as an effective approach to implementing the concepts of lean production without trying to change everything at once.

Conwip The **CONWIP** (which stands for **constant work-in-process**) approach is similar to kanban in that it is a pull type system that is used to limit the material in a processing system, but the application is slightly different. CONWIP operates almost as if the entire system is operated as one large kanban-controlled process rather than as a series of linked operations. As is the case with kanban, when one item is removed from the finished goods buffer after the last process, a signal is generated, but instead of that signal going to the most immediate upstream process, the signal goes all the way back to the release of raw material into the system. Once the raw material is released into the system, it is processed at each operation as soon as possible until it reaches the finished goods buffer. Essentially, the entire system is operated as if it were one large kanban cell. Note that this implies if the system is in an idle state that there are no in-process buffers as there would be in a kanban system.

Overall, this approach leads to a potential for inventory reduction. Clearly, the finished goods buffer has to be large enough to protect customers from disruptions in the system as well as to accommodate processing lead time, but the in-process inventory can be significantly smaller. If one process in the system has a problem, the usage of finished goods will continue to generate raw material releases, but it will be processed all the way to the problem process. Then, as soon as the process is back on line, it can continue to work through the material build up. In kanban, by contrast, if a downstream process has problems, then it will release no replacement signals and the entire upstream system will stop.

A pull system with "spike" control This hybrid system works primarily with an environment in which the overall demand tends to be fairly stable, but there are occasional spikes in demand. Examples of upward demand spikes may be the result of a marketing promotion campaign or a large product purchase from a specific customer or set of customers. Downward spikes can also occur, such as the loss of a specific customer order or the aggressive marketing moves from a competitor.

In order to handle the spike, the cause needs to first be understood. It may be unwise, for example, to include the spike information in determining the average demand and variability of demand. But, for example, if the loss of a customer represents the loss of that customer permanently rather than just for an order, that information should be used to calculate the new demand statistics. The same is true about competitive moves. That information should be available as an outcome of sales and operations planning (and, of course, the development of a master schedule).

Once a determination is made as to the expectation of what the average demand and variability is to be, then the number of kanban containers can be calculated and the pull system be used in the typical reactive way. The master schedule and MRP can be used to determine the possible existence of possible spikes such as a large one-time order or a marketing sales event. Knowing how large the spike is and the timing for that demand provides for a special anticipation order to be generated from the MRP system to deal with the coming demand spike. This special production order can be either a typical dispatch order above normal kanban use, or could be the insertion of special one-use-only kanbans in addition to the regular kanbans. Anytime the order is for a special demand situation, the spike should NOT be used to calculate summary statistics for kanban as they would skew the results and produce excess inventory for normal times.

Note that a variation of this system can be used for products with cyclical demand (such as seasonal products). In those cases, MRP can be used to determine both the timing for a cyclical demand change and the average level of demand expected during the cycle. MRP output can then be used to determine the number of kanbans in the system.

Dealing with special replenishment problems It is possible that a facility may have demand patterns that allow for using a "pure" pull system internally, but perhaps has special circumstances that prevent including one or more of the suppliers (either internal or external) in the pull system environment. There are several possible examples, including suppliers with very long lead times and suppliers or transportation modes that require very large quantities for an order. In such an environment it may be helpful to allow MRP to calculate the unique orders from the "problem" suppliers, yet still allow the pure pull system to operate the rest of the production system.

Dealing with an environment with frequent design changes If the production environment has stability in both average demand and variability of demand, then using a pull system is very attractive. A major characteristic of a pure pull system is, however, that it is totally reactive. If a component is to be subjected to a design change, how does a company deal with the change in the reactive system? Specifically, how is production started for the new design and how can the obsolescence cost for the old design be minimized? When will the new design be ready for production, can we the old design inventory be consumed, and when should inventory of the old design be expected to run out?

The forward-looking capability of MRP should be able to answer these questions. Specifically, the MRP system can be used to:

- Generate new kanban cards in the right quantity for the new design and determine the timing for converting to the new design. Then, based on the timing from MRP the new design kanban signals can be inserted into the production system. The goal is to have the new design inventory ready to use according to the phase-out plan for the existing design inventory.
- Prevent obsolete material. The plan generated from MRP can also be used to generate how the existing inventory can be phased out. A simple approach, for example, is to include an additional notation of some sort on the kanban signals indicating that even though the kanban has been freed from a now-empty container, it is NOT to be refilled.

The overall conclusion from all these examples should be that both "pure" MRP and pull systems have advantages and disadvantages, and the advantage that one has often represents a disadvantage for the other. Understanding each system and knowing the weaknesses and strengths of each can, however, allow modification to be used to develop unique approaches to unique production environments. The choice is not necessarily using one or the other, but by understanding completely how the systems work one can use parts of each system to create an effective approach to planning and execution of production.

EXPANDING LEAN PRINCIPLES TO WAREHOUSING

Many of the lean concepts can be applied to other environments, including service industries and warehouses. The approaches, as in the case of a manufacturing setting, tend to reduce inventory, increase productivity, load level, and improve process flows and sometimes utilize digital kanbans. Examples of waste reduction generally fall under the categories of material flow in the supply chain and the use of buffer (safety) stock. Specifically, lean principles in a warehouse focus on:

- Transportation—this includes restocking of material when not stored in logical locations. It also includes ineffective parking and routing of material handling vehicles.
- Waiting—this includes product unavailability as well as trucks waiting to be unloaded or loaded in a parking lot due to poor scheduling techniques.
- Motion—material is stored in poor or inconvenient locations or material handling equipment being inappropriately or improperly placed.
- Overproduction—if orders are picked too early then temporary storage is required. Sometimes more material is picked than necessary which then implies they may need to be restocked.
- Quality—this includes issues such as picking the incorrect item, the incorrect quantity, or damage from handling.
- Overprocessing—this includes a requirement for need to enter certain information more than once, unnecessary inspection checks, and overpacking.

SUMMARY

The lean production philosophy and techniques that seek the elimination of waste and continuous improvement were developed for repetitive manufacturing and are perhaps most applicable there. However, many basic concepts are appropriate to any form of manufacturing organization. They are now also being adapted and adopted by service-oriented operations as well.

Waste can be caused by poor product design or the manufacturing process itself. Seven types of waste include processing, motion, transportation, defects, waiting time, overproduction, and inventory. Poka-yoke techniques can be applied to eliminate the ability to produce defects.

Major concepts within lean production include flow manufacturing, process flexibility, quality management, total productive maintenance, uninterrupted flow, continuous process improvement, supplier partnerships, and total employee involvement. Push systems are replaced by pull systems, and the work authorized by the use of kanbans, or signals, for replenishment of inventory.

Other tools utilized within lean production include value stream mapping, kaizen, takt time, 5S housekeeping, and visual management. Accounting must also change in a lean environment, and be focused on measuring the costs and savings associated with the elimination of waste and continuous improvement.

The lean production environment requires a planning and control system. The enterprise resource planning (ERP) system is complementary to lean production and JIT. The way in which some functions are used changes to reflect the differences in the manufacturing environment, but, in general, the tools are advantageous by utilizing the elements in each that work best in that environment.

KEY TERMS

Activity-based costing (ABC Costing) 427
Andon 427
Backflush 418
Box score 428
Bucketless system 417
Cellular manufacturing 410
Constant work-in-process (CONWIP) 430
Enterprise resource planning (ERP) 405
External setup time 411
5Ss 427
Flow manufacturing 409
Heijunka 413
Internal setup time 411
Just in time (JIT) 404
Kaizen 425
Kaizen blitz 425
Kaizen event 425
Kanban 421
Lean production 404
Linearity 414
Line balancing 412
Little's Law 419
Machine flexibility 410

Mixed-model scheduling 413
One-piece flow 421
Operator flexibility 411
Partnering 414
Poka-yoke 408
Post-deduct inventory transaction processing 418
Preventive maintenance 412
Process flexibility 410
Pull system 419
Push system 419
Quality 405
Quality at the source 412
Quick changeover 410
Rate-based scheduling 412
Single-minute exchange of die (SMED) 411
Standardized work 411
Supplier certification 415
Takt time 426
Total productive maintenance (TPM) 412
Uniform plant loading 412
Value stream mapping 425
Visual management 427
Waste 406
Work cell 409

QUESTIONS

1. What is the definition of waste as it is used in this text?
2. What is value to the user? How is it related to quality?
3. What are the elements in a product development cycle? For what is each responsible?
4. Why is product specialization important? Who is responsible for setting the level of product specialization?
5. What is meant by component standardization? Why is it important in eliminating waste?
6. Why is product design important to manufacturing? How can the design add waste in manufacturing?
7. After a product is designed, what is the responsibility of manufacturing engineering?
8. Explain why each of the following are sources of waste:
    a. Processing
    b. Motion
    c. Transportation
    d. Defects
    e. Waiting time
    f. Overproduction
    g. Inventory
9. Explain how inventory affects product improvement, quality, prices, and the ability to respond quickly to the marketplace.

10. What is repetitive manufacturing? What are its advantages? What are its limitations?
11. What is a work cell? How does it operate? What conditions are necessary to establish one? What are its advantages?
12. Why is process flexibility desirable? What two conditions are required?
13. Name and describe five advantages of short setup time.
14. What are the two reasons why quality is important?
15. What is quality for manufacturing? How is it obtained?
16. Why is preventive maintenance important?
17. Describe the differences between preventive maintenance and total productive maintenance.
18. What are the four conditions needed for uninterrupted flow? Describe each.
19. What is the difference between leveling based on capacity and leveling based on material flow?
20. Why would a lean production manufacturer schedule 7 hours of work in an 8-hour shift?
21. Why are supplier relations particularly important in a lean production environment?
22. Why is employee involvement important in a lean production environment?
23. What are the differences in a master production schedule in a lean production environment?
24. What effect does a lean production environment have on MRP?
25. Describe the backflushing or post-deduct system of inventory recordkeeping.
26. Why is there sometimes difficulty with the MRP system?
27. What is the major difference between the MRP push system and the pull system?
28. What is the kanban system? How does it work?
29. What is the difference between production cards and move cards?
30. Where does an MRP system work best?
31. Where does a kanban system work best?
32. Where does a drum-buffer-rope system work best?
33. Can MRP and kanban be used at the same time and if so, how?
34. Provide an example of value stream mapping.
35. How does takt time work?
36. Give an example of a visual management system.
37. What are the 5Ss of housekeeping? What is the sixth?
38. How can a lean system with pull production control manage changes in product designs?
39. How can a lean system with pull production control erratic or seasonal demand in an efficient manner?
40. Will moving to a lean production system alter the approach to sales and operations planning? If so, how and if not, why not?
41. Can you effectively operate a lean production system without using pull production? Why or why not?

PROBLEMS

16.1. A company carries 10 items in stock, each with an economic order quantity of $30,000. Through a program of component standardization, the 10 items are reduced to 5. The total annual demand is the same, but the annual demand for each item is twice what it was before. In Chapter 11, we learned that the economic order quantity varies as the square root of the annual demand. Since the annual demand for each item is now doubled, calculate:

   a. The new EOQ.

   b. The total average inventory before standardization.

   c. The total average inventory after standardization.

16.2. In problem 16.1, if the annual carrying cost is 18% per unit, what will be the annual savings in carrying cost?

16.3. A company has an annual demand for a product of 2400 units, a carrying cost of $20 per unit per year, and a setup cost of $100. Through a program of setup reduction, the setup cost is reduced to $20. Run costs are $3 per unit. Calculate:

   a. The EOQ before setup reduction.

   b. The EOQ after setup reduction.

   c. The total and unit cost before and after setup reduction.

16.4. A company produces a line of golf putters composed of three models. The demand for model A is 500 per month, for B is 400, and for C is 300. What would be the mixed-model sequence if some of each were made each day?

   Answer. ABC, ABC, ABC, ABA

16.5. The following MPS summary schedule is leveled for capacity. Using the following table, level the schedule for material as well.

Week	1	2	3	4	5
Model A			800	1600	1600
Model B	600	1600	800		
Model C	1000				
Total					

Answer.

Week	1	2	3	4	5
Model A	800	800	800	800	800
Model B	600	600	600	600	600
Model C	200	200	200	200	200
Total					

16.6. Bill builds benches in a small shop and he plans to operate five 8-hour days per week. Each bench has two ends and a top. It takes 5 minutes to cut and sand each end and 2 minutes to make each top. Assembly requires 8 minutes per bench, and painting requires 5 minutes per bench. Bill has one employee who makes the tops and ends. Bill will do the final assembly and painting. He also plans 1 hour per day for setup and teardown for each department.

   a. How many benches should he plan to build per week?

   b. How many hours per week are required for each person?

   c. Suggest ways to increase the number of benches produced per week.

16.7. Parent W requires one of component B and two of component C. Both B and C are run on work center 10. Setup time for B is 2 hours, and run time is 0.12 hour per piece. For component C, setup time is 2 hours, and run time is 0.18 hour per piece. If the rated capacity of the work center is 80 hours, how many Ws should be produced in a week?

   Answer. 166 Ws

16.8. Parent S requires three of component T and two of component U. Both T and U are run on work center 30. Setup time for T is 7 hours, and run time is 0.1 hour per piece. For component U, setup time is 8 hours, and run time is 0.2 hour per piece. If the rated capacity of the work center is 120 hours, how many Ts and Us should be produced in a week?

   Answer. 450 Ts and 300 Us a week

CASE STUDY 16.1

Maxnef Manufacturing

When Joe Vollbrach, vice president of operations for Maxnef Manufacturing, was given the CEO's directive to investigate the lean production concepts and to implement them if appropriate, he was slightly apprehensive. Everyone knew, he thought, that ERP was the best way to run a manufacturing operation, and they had been pretty successful with their ERP system. Once he read a couple of books and a magazine article or two about lean production, however, he thought maybe there was something to it and it sure seemed simple enough. Dozens of companies had reported great reductions in inventory cost and other forms of waste, and with Maxnef Manufacturing having only five to six inventory turns per year, the prospect of significant inventory reductions was very appealing.

Encouraged with the success stories and very mindful of the CEO directive, Joe wasted no time. He put out the directive to all his people to implement lean production the way it was working in the book examples he had read. A few months later, however, he was beginning to wonder about the truth of the success stories in those books. The following are some of the examples of the complaints he was getting and the problems he was facing.

JAMARI, THE PURCHASING MANAGER: "Joe, this JIT and lean production are a disaster for us. It's not only costing us a lot more money, but the suppliers are getting really angry with us. Since our raw material inventory had typically been high, you said we should order smaller quantities and have it delivered just in time for its need. Sure, that cut down on the raw material inventory, but that cost savings has been more than made up for with all the increased cost. First, purchase orders are not cost free, and we're making a lot more of them. That's also taking up a lot more of our buyer's time. Then there's the transportation cost. Since most of the trucking companies charge a lot more for less-than-full truckloads, our costs are going sky-high with more frequent deliveries of smaller loads. Combine that with expediting costs, and it gets really bad. Our schedules are changing even more frequently, and without the raw material the production people are often asking for next day delivery of material they need for a schedule change. We're flying more parts in, and you know how much that costs!

"That's not all. Our suppliers are really wondering if we know how to run our business. We're changing the schedule to them much more frequently, and the only way they can hope to meet our needs is to keep a lot of our inventory in their finished goods. That's costing them a lot of money in inventory holding costs as well as administrative costs to manage the inventory and to process all our requests. They not only have more requests from us, but it seems like everything is a rush order. They're pressing us hard for price increases to cover their increased cost to keep us as a customer. I've held most of them off for a while but not much longer. Unfortunately, I agree with them, so it's hard for me to make any kind of logical argument to counter their requests for price increases."

OSCAR, SUPERVISOR OF SHIPPING/RECEIVING: "If you're going to keep this up, I have to ask for two more truck bays and about four more receiving clerks. There's a lot more trucks making a lot more deliveries. We can't schedule perfectly when a truck will show up, so many times during the day there are several trucks waiting for an open truck bay. Production people are pushing us because they see a certain truck and they know a critical part is on it, but we have to wait because the trucks currently in the bays being unloaded also have critical parts. It seems like everything is critical these days. A couple more bays with enough people to staff them will help, although it's not the only answer. I figure about a $3 million renovation to the receiving area and an increase in our budget of about $500,000 ought to be about enough to keep us afloat, at least most of the time. Oh, I forgot—better make that $600,000 more. I also need another records clerk to keep up with the big increase in paperwork from all those deliveries."

Joe was just finished taking a couple of aspirin to deal with that headache, when Sofia, a production supervisor, caught up with him:

> "Joe, I don't know where you came up with these silly ideas, but you're killing us here. This so-called kanban system seems just plainly no good. It reminds me of the reorder points we used before the MRP system, and we all know how much better MRP made things work. I feel like I just slipped 35 years back in time. We all know our customers tend to be a fickle bunch who are always changing their minds on orders. We used to have visibility with the change when we used MRP, but now all we have is what the final assembly area is working on. That not only causes us big, inefficient downtimes for setup changes but also doesn't give us much time to get the materials we need to make the different parts. When we do figure out what we need, most of the time our suppliers don't have any in raw material stock. Now we have to waste more time and energy to try to rush the buyers. What makes it even worse is your directive to eliminate finished goods. We not only don't have material to make what we need half the time, but we're not allowed to make what we can. It would seem that would make sense, so that when a customer does request a certain model we have in stock, at least that order wouldn't end up being a crisis.
>
> "The crisis scenario I just described is what happens when everything else works. When we have some other problems, such as a piece of equipment going down, the real disaster hits. What little order we can force goes out the window and everyone starts really getting uptight. And don't talk about preventive maintenance, either. We need every piece of equipment running as much as we can to even come close to making the orders we need.
>
> "By the way, some of our best people are threatening to quit. Some are being paid on a piece rate, and the lack of material combined with the directive to avoid finished goods has cut seriously into their paychecks. Even the ones on straight hourly pay are unhappy. All we supervisors are being evaluated on labor productivity and efficiency, and our numbers are looking really bad. Naturally, we put the heat on the workers to do better, but I'm not too sure they can do much about it most of the time.
>
> "And another thing while I have your attention—in some of our models we have a lot of engineering design changes. The ERP system used to give us some warning, but now we have none. Suddenly one of the engineers will show up and tell us to use a different part. When we check with purchasing, often they've only just gotten the notification themselves and have only just started to work with the suppliers for the part. Not only do we have to obsolete all the old parts, but we also have to wait for the new ones. The salespeople are even starting to get angry with us.
>
> "While I'm on a roll here, you've got to do something with those quality control people. We could deal with some scrap or rejected parts when we had plenty of inventory, but now we need every part in the place. Tell them to quit rejecting parts and let us use them. At least we'll have a better chance of shipping the order, even if it isn't perfect."

Almost as soon as Sofia left, Valorie (the sales manager) came in. Joe braced himself for more of a headache even before she spoke. The expression on her face foreshadowed what she had on her mind:

> "Joe, my job is to make sales and keep the customers happy. The sales have been going fine, but our delivery stinks. Our on-time delivery record has fallen from 95 to less than 50% in the past 6 months. Some of our customers are threatening to leave us, and some are pushing for price cuts. According to them, our delivery record is so poor they feel compelled to keep more of our inventory in their raw material stores to account for their lack of faith in our delivery promises. They say that since it's our fault they have increased raw material inventory costs, we should compensate them with a price cut. It's pretty hard to argue with their logic.

I'm sure we would do the same with our suppliers if they treated us the way we've been treating our customers lately.

"We all know our customers have to sometimes change orders to reflect what their own customers want. Now, however, the changes are becoming more frequent and radical. It seems since we are so poor in delivery, they order more in advance from us to buffer the time for our late deliveries. Ordering further into the future gives them a lot less certainty of what they really need, so naturally they have to change once they do know. I may have to get another order-entry clerk to deal with all the changes, and if I do, you better believe I'll let everyone know the extra expense is your fault!

"The bottom line here is simple, Joe. You've got to get after your people to improve the delivery drastically or we may be in big trouble, and soon."

After Valorie left Joe's secretary came in with a rush memo from the CEO:

"I just got the preliminary financial report for the last quarter, and for the first time in over 5 years we show a loss—and it's a big one. The details show sales goals have basically been met. The loss comes from a very large increase in expenses in virtually every area in operations except a modest decrease in inventory cost. I'm calling an emergency staff meeting for two o'clock today. Please be ready to explain the situation completely."

As Joe shut his door to insulate himself from the complaints and to prepare for the meeting, he began to wonder if some of the people who claimed that lean production was a culturally based system and impossible to implement outside of Japan were correct. He knew most of the problems were related to his attempt to implement the new system. What, he wondered, went wrong, and what should he do about it? These were certainly two critical questions that would be major parts of the two o'clock meeting, and he needed good and complete answers. The CEO was reasonable and could deal with the fact that mistakes might have been made, but she would expect a detailed analysis and complete action plan to get things back on track. The aspirin was definitely wearing off, and it was less than an hour ago that he took them!

Assignment

Prepare a complete and comprehensive report for Joe to use for his two o'clock meeting. This should include both the analysis of what went wrong and why, as well as a comprehensive and time-phased plan to implement lean production the correct way. If you do not think lean production is appropriate, explain why in detail and develop a comprehensive alternative plan.

CASE STUDY 16.2

Catskill Metal Products

Several years ago, Catskill Metal Products successfully implemented a lean production environment using kanban signals to run their production lines. They had worked not only with their internal environment to make kanban pull production work, but also worked with suppliers to almost seamlessly have them deliver components from kanban signals rather than from formal purchase orders. They first established long-term relationship agreements with the suppliers to achieve that.

For several years their system had worked quite well, but recently there have been problems and it appears that the disruptions have become increasingly bothersome. The general manager of Catskill Metal Products (Betsy) felt the issue had gotten to the point that the management team needed to discuss the situation and generate a plan to correct what issues seemed to be causing the disruptions.

The production manager (Josh), started off the discussion—"What seems to be happening is that there is a growth in design changes in our components. Since our production system is based on kanban, which is totally reactive to need, it does not seem to handle the changes at all. A reactive signal gets transmitted to the upstream work center when a part is use, but suddenly we are supposed to use a different designed part with a different part number. Since it has never been used before, we have none in the system and it really screws us up."

Betsy then looked at Elana, the engineering manager and said "Elana, what's going on? Why are you changing the designs frequently during the last several months?" Elana responded, "We feel badly about it, but we have no choice. The changes apparently are being driven by the customers." Ravi, the sales and marketing manager, supported the point. "The customers are basically demanding such changes from us or they may drop us as a supplier. Apparently, there are some new competitors in the market that are trying to gain market share by giving the customers lots of flexibility in designs, and are doing so by still being price competitive. I have some information that they are losing money doing what they are doing, but consider it a cost of growing their market. I suspect that if they were to drive us out of the market, we would see them start to raise prices to a profitable level. We need to ask ourselves what we need to do to fight off these attacks and keep the customers happy, or are we willing to lose several of them?"

Pheng, the purchasing manager, then added, "The problem is also causing lots of complaints from our suppliers. We spent a lot of time and energy to develop the suppliers to essentially be like a seamless part of the overall process for making our parts. These design changes hit them perhaps even harder than they hit our own production processes. Over the last few years, the product stability has been so good that we have great stability without the use of individual purchase orders. Suddenly the suppliers are demanding lead times to be able to make design changes and also demanding a return to negotiations for price consideration since they have a lot more work and pressure on them to supply us. This will lead to pressure to raise our own product price just when we have all this competitive pressure on us from these new companies." Josh then added to that: It's not just purchase component price pressure we have to deal with. This growing need for design changes to be flexible means that the entire kanban pull system may need to be scrapped. Going back to our old system of production orders and work centers will cost us lots of money for inventory and disruptions, and I also know it would be very stressful for our workforce."

Fortunately, Catskill Metal Products had recently hired a new management trainee (Maria) who had just graduated from a well-known business school where she had specialized in supply chain management and was very familiar with the concepts developed in "The Introduction to Materials Management" (okay, so it is shameless self-promotion). Betsy had already asked Maria to sit in on the meeting. Betsy then turned to Maria and asked "Do you understand the problem, and if so, do you think you could come up with a plan that might be able to deal with the problem without reverting back to the way we did business 10 or more years ago?" Luckily, Maria confidently replied that she was sure she could recommend an approach that would satisfy customers while minimizing disruptions in the current production process. "Great," said Betsy. "Please do so and be ready to present the approach to our next meeting on this in two days. Time is of the essence."

Assignment

Since you, too, have become familiar with the content in "The Introduction to Materials Management," put yourself in the role of Maria. Try to come up with a good solution recommendation that will minimize disruption while still trying to satisfy the changing customer demand patterns as much as possible. In addition to the plan, try to see if you can come up with a brief set of key implementation steps in the order that they should be accomplished.

CHAPTER SEVENTEEN

TOTAL QUALITY MANAGEMENT

INTRODUCTION

The quality of a product or service is sometimes difficult to define since it is often biased by people's perceptions and measured against their own experiences. This chapter will cover definitions of quality and how these can be used in the design of products and services to meet customer expectations. Constant improvement in quality is the responsibility of all members of an organization, and this chapter will introduce some management practices such as quality function deployment, ISO standards, and six sigma, which help to guide an organization in this improvement. Also included is a brief introduction to process variation and its measurement, along with inspection and benchmarking.

WHAT IS QUALITY?

Everyone knows, or thinks they know, what quality is. But often it means different things to different people. When asked to define quality, people's responses are influenced by personal opinion and perception. Answers are often general or vague, for example, "It's the best there is," "Something that lasts a long time and gives good service," and "Something that's perfect."

While the definitions of **quality** vary, the one to be discussed in this chapter contains the most commonly accepted ideas in business today: *Quality means user satisfaction: that goods or services satisfy the needs and expectations of the user.*

To achieve quality according to this definition, companies must consider quality and product planning, product design, manufacturing, and final use of the product.

Quality and product planning Product planning involves decisions about the products and services that a firm will market. A product or service is a combination of tangible and intangible characteristics that a company hopes the customer will accept and be willing to pay a price for. Product planning must decide the market segment to be served, the level of performance expected, and the price to be charged, and it must estimate the expected sales volume. The basic quality level of a product is thus specified by senior management according to its understanding of the wants and needs of the market segment.

Quality and product design A firm's studies of the marketplace should yield a general specification of the product, outlining the expected performance, appearance, price, and volume. **Form-fit-function** refers to designing a product that meets or exceeds what the customer expects. Product designers must then build into the product the quality level described in the general specification. They determine the materials to be used, dimensions, tolerances, product capability, and service requirements. If product designers do not do this properly, the product or service will be unsuccessful in the marketplace because it may not adequately satisfy the needs and expectations of the user.

Quality and manufacturing At the least, manufacturing is responsible for meeting the minimum specifications of the product design. **Tolerances** establish the acceptable limits and are usually expressed as the amount of allowable variation around the desired amount or nominal. For example, the length of a piece of lumber may be expressed as $7'6 \pm 8''.⅛''$. This means that the longest acceptable piece would be $7'6\ ⅛''$ and the shortest acceptable piece would be 7'5 ⅞" long. If an item is within tolerance, then the product should perform adequately. If it is not, it is unacceptable. However, the closer an item is to the nominal or target value, the better it will perform and the less chance there is of creating defects.

Quality in manufacturing means that, at a minimum, all production must be within specification limits, and the less variation from the nominal value, the better the quality. Manufacturing must strive to produce excellent, not merely adequate, products. Every product or service produced will have some form of tolerance expressed. For example, the weight of bars of soap, the frequency response of compact disks, or the time spent waiting in line will all have a plus and (in most cases) a minus tolerance.

Quality and use To the user, quality depends on an expectation of how the product should perform. This is sometimes expressed as **fitness for use**. Customers do not care *why* a product is defective, but they care *if* it is defective.

The customer may need some introduction to the proper use of a product or feature, especially with new products. A good example of this is a feature required on some new cars that turns off the daytime running light when its adjacent turn signal is activated. This is a safety feature that improves the visibility of the signal but customers, unaware of that design feature, thought their vehicles had a defect in the lighting. It was recommended that the manufacturer improve the wording in the owner's manual and introduce this improvement to the customer when purchasing the vehicle. If the product has been well conceived, well designed (meets customer needs), well made, well priced, and well serviced, then the quality is satisfactory. If the product exceeds the customer's expectations, that is superb quality.

Figure 17.1 shows the loop formed by product policy, product design, operations, and the user. Quality must be added by each link.

Quality has a number of dimensions, among which are the following:

- **Performance.** Performance is the primary operating characteristic that the customer wants from a product, such as the power of an engine. Fitness for use performance implies that the product or service is ready for the customer's use at the time of sale, and that the product does what it is supposed to do. Three dimensions to performance are important: reliability, durability, and maintainability.

    1. **Reliability** means consistency of performance. It is measured by the length of time a product can be used before it fails.
    2. **Durability** refers to the ability of a product to continue to function even when subjected to hard wear and frequent use.
    3. **Maintainability** refers to being able to return a product to operating condition after it has failed.

- **Features** and **options.** Features and options include secondary characteristics or extras, such as wireless capabilities on an appliance or entertainment unit, or the addition of extras in the purchase of an automobile.

- **Conformance.** Meeting established standards or specifications. This is manufacturing's responsibility.

FIGURE 17.1 Product development cycle.

- **Warranty.** An organization's public promise to back up its products with a guarantee of customer satisfaction.
- **Service.** An intangible generally made up of a number of things such as the customer service staff's availability, speed of service, courtesy, and competence. It may also include after sales service.
- **Aesthetics.** The product is pleasing to the senses; for example, the exterior finish or the appearance of a product.
- **Perceived quality.** Customers perceive quality not only in the product itself but also in satisfaction based on the complete experience with an organization. Many intangibles, such as a company's reputation or past performance, influence perceived quality.
- **Price.** Customers pay for value in what they buy. Value is the sum of the benefits the customer receives and can be more than the product itself. All the dimensions of quality listed are elements of value.

These dimensions are not necessarily interrelated. A product can be superb in one or a few dimensions and average or poor in others.

TOTAL QUALITY MANAGEMENT

Total quality management (TQM) is an approach to improving both customer satisfaction and the way organizations do business. TQM brings together all of the quality and customer-related process improvement ideas. It is people oriented. According to the *APICS Dictionary*, 16th edition, TQM "is based on the participation of all members of an organization in improving processes, goods, services, and the culture in which they work."

The objective of TQM is to provide a quality product to customers at a fair price. By increasing quality and decreasing cost, profit and growth will increase, which in turn, will increase job security and employment.

TQM is both a philosophy and a set of guiding principles that lead to a continuously improving organization.

Basic concepts   There are six basic concepts in TQM:

1. **A committed and involved management.** Since TQM is an organization-wide effort, management must direct and participate in the quality improvement program. TQM is a continuous process that must become part of the organization's culture. This requires senior management's absolute commitment, as they are directly responsible for the quality improvement.
2. **Focus on the customer.** This means listening to the customer so goods and services meet customer needs at a low cost while including value-adding features. It means improving design and processes to increase value from the customer's point of view and to reduce defects and cost. TQM embraces the idea that quality is defined by the customers' requirements.
3. **Total employee involvement.** TQM is the responsibility of everyone in the organization, and involves the total workforce. It means training and empowering all personnel in the techniques of product and process improvement and creating a new culture.
4. **Continuous process improvement.** Processes can and must be improved to reduce cost and increase quality. (This topic was discussed in Chapter 15.)
5. **Supplier partnering.** A partnership rather than an adversarial relationship with suppliers must be established. As discussed in Chapters 8 and 16, this should lead to mutual benefit to both parties.
6. **Performance measures.** Improvement is not possible unless there is some way to measure the results, including fact-based decision-making.

These basic concepts will be discussed in more detail in the following sections.

Management Commitment

Quality is a long-term investment. If senior management is not committed and involved, then TQM will fail. These managers must start the process and should be the first to be educated in the TQM philosophy and concepts. The chief executive officer and senior management should form a quality council or team whose purpose is to establish a clear vision of what is to be done, develop a set of long-term goals, and direct the program.

The quality team must establish core values that help define the culture of the organization. Core values include such principles as customer-driven quality, continuous improvement, employee participation, and quick response. The team must also establish quality statements that include a vision, mission, and quality policy statement. The vision statement describes what the organization is striving to become 5 to 10 years in the future. The mission statement describes the function of the organization: who they are, who their customers are, what the organization does, and how it does it. The quality policy statement is a guide for all inside and outside the organization about how products and services will be provided. The three statements must be compatible in that the quality policy statement must fulfill the mission statement, which in turn should support the vision statement. Finally, the quality team must establish a strategic plan that expresses the TQM goals and objectives of the organization and how it plans to achieve them.

Customer Focus

Total quality management implies an organization that is dedicated to delighting the customer by meeting or exceeding customer expectations. It means not only understanding present customer needs but also anticipating customers' future needs.

A customer is considered to be a person or organization who receives products or services. There are two types of customers: external and internal. *External customers* exist outside the organization and purchase goods or services from the organization. *Internal customers* are persons or departments who receive the output from another person or department in an organization. Each person or operation in a process is considered a customer of the preceding operation. If an organization is dedicated to delighting the customer, internal suppliers must be dedicated to delighting internal customers.

Customers have six requirements of their suppliers:

1. High quality level.
2. High flexibility to change such things as volume, specifications, and delivery.
3. High service level.
4. Accurate lead times.
5. Low variability in meeting targets.
6. Low cost.

Customers expect improvement in all requirements. These requirements are not necessarily in conflict. Low cost and high flexibility, for example, do not have to be trade-offs if the process is designed to provide them.

Employee Involvement

TQM is organization-wide and is everyone's responsibility. In a TQM environment, people come to work not only to do their jobs but also to work at improving their jobs. Motivating employees to improve their jobs needs employee commitment to the organization. To encourage this commitment through TQM requires the following:

1. **Training.** People should be trained in their own job skills and, where possible, cross-trained in other related jobs. They should also be trained to use the tools of continuous improvement, problem-solving, and statistical process control. Training provides the tools for continual people-driven improvement.

2. **Organization.** The organization must be designed to put people in close contact with their suppliers and customers, internal or external. One way is to organize into customer-, product-, or service-focused cells or teams.

3. **Local ownership.** People should be owners of the processes they work with. This results in a commitment to make their processes better and to push for continuous improvement. They should be empowered, which goes beyond providing a suggestion box for employee input.

Empowerment means giving people the authority to make decisions and take action in their work areas without getting prior approval. For example, a customer service representative can respond to a customer's complaint on the spot rather than getting approval or passing the complaint on to a supervisor. Giving people the authority to make decisions motivates them to take ownership of their jobs to accomplish the goals and objectives of the organization. The personnel involved in a particular process are usually the best source of information for how to improve that process and eliminate waste or unnecessary steps.

Teams A team is a group of people working together to achieve common goals or objectives. Good teams can move beyond the contribution of individual members so that the sum of their total effort is greater than their individual efforts. Working in a team requires skill and training, and to work in teams is part of TQM.

Communication Communication plays an important role in disseminating the TQM concepts across all levels of the organization. Communicating the goals, purposes, and benefits of TQM is critical for fostering continuous improvement, maintaining morale, and motivating employees to be involved in the total quality management effort.

Continuous Process Improvement

This topic, an element in both lean production and TQM, was discussed in detail in Chapter 15. Quality can and must be managed, and requires continuous process improvement. If a product is excellent in one dimension, such as performance, then improved quality in another dimension should be sought. A company should never be completely satisfied, as there are always methods that can be improved. This involves working smarter, not harder, by examining the source of problems and removing them.

In addition to the steps previously outlined in this text, many companies use the **Deming circle**, also known as the **Plan-do-check-act cycle (PDCA)** to drive continuous improvement. These steps are:

1. **Plan.** Create a plan for the change, define the steps needed and predict the results.
2. **Do.** Carry out the plan, often in a test environment.
3. **Check.** Examine the results and verify that the process was improved. If not, try again.
4. **Act.** Implement the changes that were verified in the previous step.

Supplier Partnerships

Supplier partnerships are very important in both lean production and total quality management. It is in the company's best interest that suppliers are provided with clear requirements and paid fairly and on time, in order to provide quality goods and services. This applies to both external and internal suppliers within the organization. This topic was discussed in Chapters 8 and 16.

Performance Measures

"What gets measured gets done" is a quote from Tom Peters, which illustrates management's need for quantifiable measures. To determine an organization's performance, its progress must be measured. An organization must continually collect and analyze data in order to:

- Discover which process(es) need(s) improvement.
- Evaluate alternative processes.
- Compare actual performance with targets so corrective action can be taken.

- Evaluate employee performance.
- Show trends.
- Improve decision-making.
- Achieve consensus.

It is not a question of whether performance measures are necessary but of selecting appropriate measures. There is no point in measuring something that does not give valid and useful feedback on the process being measured. There are many basic characteristics that can be used to measure the performance of a particular process or activity, such as the following:

- **Quantity.** How many units a process produces in a period of time. Time standards measure this dimension.
- **Cost.** The amount of resources needed to produce a given output.
- **Time/delivery.** Measurements of the ability to deliver a service or product on time.
- **Quality.** There are three dimensions to quality measurements:
  1. **Function.** Does the product perform as specified?
  2. **Aesthetics.** Does the product or service appeal to customers? For example, the percentage of people who like certain features of a product.
  3. **Accuracy.** This measures the number of nonconforming items produced. For example, the number of defects or rejects, or the ability to meet specifications.

Performance measures should be simple, easy for users to understand, relevant to the user, visible to the user, preferably developed by the user, designed to promote improvement, and few in number.

Measurement is needed for all types of processes. Some of the areas and possible measurements are as follows:

- **Customer.** Number of complaints, on-time delivery, dealer or customer satisfaction, customer response time, percentage of past due orders, perfect order fulfillment (right product, right customer, right time, right place, right condition, right quantity, right cost).
- **Scheduling.** Number of changes within the demand and planning time fences, percentage of orders completed on time, percentage past due MPS orders.
- **Material planning.** Percentage of orders released with no shortages, percentage of on-time order release, number of action messages, actual versus planned lead times.
- **Capacity planning.** Percentage of work centers overloaded or underloaded, percentage of overtime to available work time, utilization of work centers, work center downtime percentage.
- **Inventory control.** Percentage increase in inventory turns, percentage reduction in WIP inventory, percentage reduction in carrying costs, ratio of obsolete inventory to total inventory.
- **Production.** Inventory turns, scrap or rework, process yield, cost per unit, time to perform operations, performance to schedule, percentage of queue time in lead time.
- **Suppliers.** Billing accuracy, percentage of past due orders, perfect order fulfillment.
- **Sales.** Sales expense to revenue, new customers, gained or lost accounts, sales per square foot of facility.
- **Operating costs.** Ratio of value-added cost to total product cost, percentage reduction in setup costs, percentage reduction in logistics costs, percentage reduction in material handling costs.
- **Data accuracy.** Bill of material accuracy, routing accuracy, inventory record accuracy, work center record accuracy.

In addition to these measurements, organizations use **key performance indicators (KPIs)** to highlight those specific measures that are linked to specific strategic objectives. These measures, first mentioned in Chapter 2, are both financial and nonfinancial, and typically address key areas such as profitability, customer satisfaction, and other supply chain activities. These can be tracked using a **balanced scorecard**, which the

APICS Dictionary, 16th edition, defines as "a list of financial and operational measurements used to evaluate organizational or supply chain performance. The dimensions of the balanced score card might include customer perspective, business process perspective, financial perspective, and innovation and learning perspectives. It formally connects overall objectives, strategies, and measurements. Each dimension has goals and measurements."

QUALITY COST CONCEPTS

Quality costs fall into two broad categories: the cost of failure to control quality and the cost of controlling quality.

Costs of Failure

The costs of failing to control quality are the costs of producing material that does not meet specifications. They can be broken down into the following:

- **Internal failure costs** are the costs of correcting problems that occur while the goods are still in the production facility. Such costs are scrap, rework, and spoilage. These costs would disappear if no defects existed in the product before shipment.
- **External failure costs** are the costs of correcting problems after goods or services have been delivered to the customer. They include warranty costs, field servicing of customer goods, replacement goods, and all the other costs associated with trying to satisfy customer complaints. External failure costs can be the costliest of all if the customer loses interest in a company's product. These costs would also disappear if there were no defects.

Costs of Controlling Quality

The costs of controlling quality can be broken down into the following:

- **Prevention costs** are the costs of avoiding poor quality by doing the job right in the first place. They include training, statistical process control, machine maintenance, design improvements, and quality planning costs.
- **Appraisal costs** are the costs associated with checking and auditing quality in the organization. They include product inspection, quality audits, testing, and calibration.

Investment in prevention will improve productivity by reducing the cost of failure and appraisal. Figure 17.2 shows the typical pattern of quality costs before and after a quality

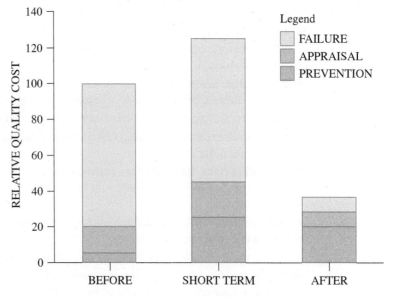

FIGURE 17.2 Impact of quality improvement on quality-related costs. Note that the increased effort put into prevention results in a reduction in both failure costs and appraisal costs.

improvement program. Investing in prevention will increase total costs in the short run, but in the long run, prevention will eliminate the causes of failure and reduce total quality costs.

VARIATION AS A WAY OF LIFE

Variability exists in everything—people, machines, and nature. People do not perform the same task in exactly the same way each time, nor can machines be relied upon to perform exactly the same way each time. In nature no two leaves are alike. It is really a question of how much variability there is.

Suppose that a lathe made 100 shafts with a nominal diameter of 1 inch. If those shafts are measured the diameters would tend to cluster about 1 inch, but some shafts would be smaller and some bigger. If the number of shafts of each diameter were plotted, a distribution, or **histogram**, such as shown in Figure 17.3 would be the result.

Random variation In nature or any manufacturing process, one can expect to find a certain amount of unexplained **random variation** that is inherent in the process and occurs by chance. This variation comes from everything influencing the process but is usually separated into the following six categories:

1. **People.** Poorly trained operators tend to be more inconsistent compared to well-trained operators.
2. **Machine.** Well-maintained machines tend to give more consistent output than a poorly maintained machine.
3. **Material.** Consistent raw materials give better results than poor quality, inconsistent, or ungraded materials.
4. **Method.** Changes in the method of doing a job will alter the quality.
5. **Environment.** Changes in temperature, humidity, dust, and so on can affect some processes.
6. **Measurement.** Measuring tools that may be in error can cause incorrect adjustments and poor process performance.

Dividing all possible variations into these six smaller categories makes it easier to identify the source of variation occurring in a process. If a connection can be found between variation in the product and variation in one of its six sources, then improvements in quality are possible.

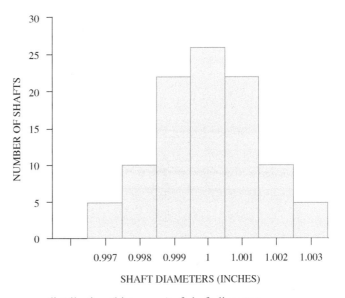

FIGURE 17.3 Frequency distribution (histogram) of shaft diameters.

There is no way to alter random variation except to change the process. If the process produces too many defects, then it must be changed.

Assignable variation Chance is not the only cause of variation. A tool may shift, a gauge may move, a machine may wear down, or an operator may make a mistake. There is a specific reason for these causes of variation, which is called **assignable variation**.

Statistical control As long as only random variation exists, the system is said to be in **statistical control**. If there is an assignable cause for variation, the process is not in control and is unlikely to consistently produce a good product. As will be shown later in this chapter, the objective of statistical process control is to detect the presence of assignable causes of variation. Statistical process control has two objectives:

- To help select processes capable of producing the required quality with minimum defects.
- To monitor a process to be sure it continues to produce at the required quality and no assignable cause for variation exists.

Patterns of Variability

The output of every process has a unique pattern that can be described by its **shape**, center, and spread. To determine the distribution or pattern of variability, a statistical process known as **sampling** is used to distribute values of a large amount of data using a relatively small number of observations, or samples of a given size.

Shape Suppose, instead of measuring the diameters of 100 shafts, one measured the diameters of 10,000 shafts. If the distribution of the diameters of the 10,000 shafts were plotted, the results in Figure 17.3 would be smoothed out and there would be a curve as shown in Figure 17.4. As discussed in Chapter 12, this bell-shaped curve is called **normal distribution** and is encountered in manufacturing processes that are running under controlled conditions. This curve exists in virtually all natural processes, from the length of grass on a lawn, to the heights of people, to student grades. Data tends to be distributed equally around a central value, with no bias to the left or right.

Center Figure 17.4 shows that the normal distribution has most results clustered near one central point, with progressively fewer results occurring away from this center. The **center** of the distribution, or **mean**, can be calculated as follows:

Let:

$$\Sigma x = \text{sum of all observation}$$
$$n = \text{number of observations}$$
$$\mu = \text{arithmetic mean (average or center)}$$

Then:

$$\mu = \frac{\Sigma x}{n}$$

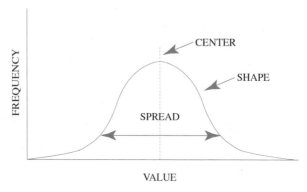

FIGURE 17.4 Normal distribution.

This example is using:

$$\Sigma x = 10{,}000$$
$$n = 10{,}000$$
$$\mu = \frac{10{,}000}{10{,}000} = 1 \text{ inch}$$

Spread To evaluate a process, one must know not only what the center of the distribution is but also something about the **spread** or variation. In statistical process control there are two methods of measuring this variation: the range and the standard deviation.

- **Range.** The range is simply the difference between the largest and smallest values in the sample. In the example shown in Figure 17.3, the largest value was 1.003 inches and the smallest was 0.997 inch. The range would be 1.003 inch − 0.997 inch = 0.006 inch.
- **Standard deviation.** The standard deviation (represented by the Greek letter σ or sigma) may be thought of as the "average spread" around the center. A distribution with a high standard deviation is wider than one with a low standard deviation. Higher-quality products have little variation (a low standard deviation). The following are noted in the measurement of the standard deviation discussed in Chapter 12 in the section on determining safety stock:

$$\mu \pm 1\sigma = 68.3\% \text{ of observations}$$
$$\mu \pm 2\sigma = 95.4\% \text{ of observations}$$
$$\mu \pm 3\sigma = 99.7\% \text{ of observations}$$

where

μ = mean or average
σ = standard deviation

The standard deviation can be used to estimate the amount of variation (quality) in a product.

Suppose in the example of the shafts that

$$\mu = 1.000 \text{ inch}$$
$$\sigma = 0.0016 \text{ inch}$$

Applying standard deviation to the previous example:

68.3% of the shafts will have a diameter of 1 inch ± 0.0016 inch (1σ)
95.4% of the shafts will have a diameter of 1 inch ± 0.0032 inch (2σ)
99.7% of the shafts will have a diameter of 1 inch ± 0.0048 inch (3σ)

In cases where the population of items is very large, it is very difficult to actually measure the entire population. In those cases, a large number of samples from the population is often taken. Then the mean of the sample means is calculated by adding all the sample means and dividing by the number of samples. Depending on the sample size (larger sample sizes tend to be better), the mean of the sample means can be used to approximate the population mean, and the standard deviation of the sample means will equal the population sample mean divided by the square root of the sample size.

PROCESS CAPABILITY

Tolerances are the limits of deviation from perfection and are established by the product design engineers to meet a particular design function. For example, a shaft might be specified as having a diameter of 1 inch ± 0.005 inch. Thus, any shaft having a diameter between 0.995 and 1.005 inches would be within tolerance. In statistical process control 0.995 inch is called the **lower specification limit (LSL)** and is the minimum acceptable level of output. Similarly, 1.005 inches is called the **upper specification limit (USL)** and is

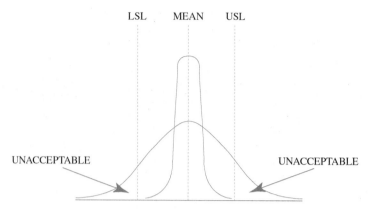

FIGURE 17.5 Effect of process spreads.

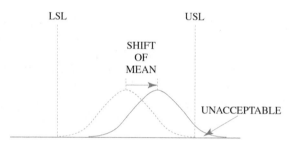

FIGURE 17.6 Effect of shift in mean.

defined as the maximum acceptable level of output. Both the USL and the LSL are related to the product specification usually determined by the customer and are independent of any process. The distance between the upper and lower specifications is often called the *specification doorway*. The doorway is similar to a tunnel with a vehicle (the process) passing through it. Parts of the vehicle that are wider than the doorway represent defective product or scrap. In this example it is not possible to change the width of the tunnel. Specifications are set by designers and ultimately by the customer and are normally not challenged. It is up to the producer to keep the spread of the process output well within these limits. Figure 17.5 illustrates this. One process, having a narrow spread (low sigma), will produce product within the specification limits. The other process, having a wide spread (high sigma), will produce defects. The first process is said to be capable, the second is not.

Besides spread, there is another way a process can produce defects. If there is a shift in the mean (average), defects will be produced. Figure 17.6 illustrates the concept.

In summary:

- The capability of the process is a measure of the process spread compared to the specification limits.
- A process must be selected that can meet the specifications.
- Processes can produce defects in two ways: by having too big a spread (sigma) or by a shift in the mean (average).

Process Capability Index, C_p

The **process capability index (C_p)** combines the process spread and the tolerance into one index making it easier for operators and managers to quickly determine the capability of a process. It assumes that the process is centered between the upper and lower specification limits—that there has been no shift of the mean. As well, the index assumes that processes are 6σ wide, representing 99.7% of the output of a normal process. If the 6σ process spread is smaller than the specification doorway then the process is said to be capable.

$$C_p = \frac{\text{USL} - \text{LSL}}{6\sigma}$$

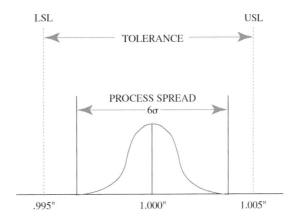

FIGURE 17.7 C_p index greater than 1.

In the example used under process spread, the tolerance was 1.000 inch ± 0.005 inch, and the standard deviation of the process (σ) was 0.0016 inch.

$$C_p = \frac{1.005 - 0.995}{6 \times 0.0016} = 1.04$$

In this example, the C_p is slightly greater than 1.00 so the process is said to be capable. If the capability index is greater than 1.00, the process is capable of producing 99.7% of parts within tolerance and is said to be capable. If C_p is less than 1.00, however, the process is said to be not capable. Because processes tend to shift back and forward, a C_p of 1.33 has become a standard of process capability. Some organizations use a higher value such as 2 for their quality standards. The higher the capability index, the fewer the rejects and the greater the quality. Figure 17.7 shows the concept of the capability index for the above example.

Example Problem

The specification for the weight of a chemical in a compound is 10 ± 0.05 grams. If the standard deviation of the weighing scales is 0.02, is the process considered capable?

ANSWER

$$C_p = \frac{10.05 - 9.95}{6 \times 0.02} = 0.83$$

The C_p is less than 1, and the process is not considered capable.

The process capability index indicates whether process variation is satisfactory, but it does not measure whether the process is centered properly. Thus, it does not protect against out-of-specification product resulting from poor centering. In some cases, this is important to know.

C_{pk} Index

This index measures the effect of both center and variation at the same time. The philosophy of the **C_{pk} index** is that if the process distribution is well within specification on the worst-case side, and is capable according to the C_p, then it is sure to be acceptable for the other specification limit. Figure 17.8 illustrates the concept.

The C_{pk} index is the lesser of

$$\frac{USL - Mean}{3\sigma} \quad \text{or} \quad \frac{Mean - LSL}{3\sigma}$$

The greater the C_{pk}, the further the 3σ limit is from the specification limits and the fewer rejects there will be.

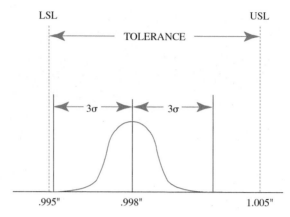

FIGURE 17.8 C_{pk} index.

Interpretation of the C_p index is as follows.

C_{pk} Value	Evaluation
Less than +1	Unacceptable process. Part of process distribution is out of specification.
+1 to +1.33	Marginal process. Process distribution barely within specification.
Greater than 1.33	Acceptable process. Process distribution is well within specification.

Example Problem

A company produces shafts on a lathe with a nominal diameter of 1 inch and a tolerance of ±0.005 inch. The process has a standard deviation of 0.001 inch. For each of the following cases, calculate the C_{pk} and evaluate the process capability.

a. A sample has an average diameter of 0.997 inch.

b. A sample has an average diameter of 0.998 inch.

c. A sample has an average diameter of 1.001 inches.

ANSWER

a. $C_{pk} = \dfrac{1.005 - 0.997}{3 \times 0.001} = 2.67$ or $= \dfrac{0.997 - 0.995}{3 \times 0.001} = 0.667$

$C_{pk} = 0.667$ which is less than 1. Process is not capable.

b. $C_{pk} = \dfrac{1.005 - 0.998}{3 \times 0.001} = 2.33$ or $= \dfrac{0.998 - 0.995}{3 \times 0.001} = 1.00$

C_{pk} is 1. Process is marginally capable.

c. $C_{pk} = \dfrac{1.005 - 1.001}{3 \times 0.001} = 1.33$ or $= \dfrac{1.001 - 0.995}{3 \times 0.001} = 2$

C_{pk} is 1.33. Process is capable.

Note that the lower C_{pk} is always used in each of the above calculations.

Example Problem

A sample has been taken from a recent shipment of laptop hinge pins and the following readings of pin lengths have been recorded:

  2.01 2.02 1.99 2.01 2.01 2.00 2.03 2.02 2.01 2.00

a. A specification of 2.00" ± 0.05" is required for the pins. Sketch the data and enter the specification limits.

ANSWER

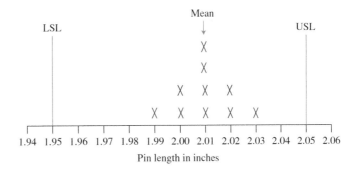

Pin length in inches

b. Calculate the standard deviation of the data, the C_p, and the C_{pk}

ANSWER

Sigma for the data = 0.01095; average = 2.01

$$C_p = \frac{2.05 - 1.95}{6 \times 0.01095} \qquad C_{pk} = \frac{2.01 - 1.95}{3 \times 0.01095} \qquad C_{pk} = \frac{2.05 - 2.01}{3 \times 0.01095}$$

= 1.52 Capable   = 1.83 Capable   = 1.21 Marginal

The data shows that the pins are slightly oversized since the average is above the nominal specification of 2.00". With a C_{pk} of 1.21, the process is marginally capable but would improve if the average was moved to 2.00. The supplier should adjust their equipment to reduce the length of the pins by 0.01 inch. Making the suggested improvement to the average, even though no parts out of specification were found is an example of continuous process improvement.

PROCESS CONTROL

Process control attempts to prevent the production of defects by showing that when the standard deviation increases or when there is a significant change in the average, there is an assignable cause for variation.

Since variation exists in all processes, the process must be so designed that the spread will be small enough to produce a minimum number of defects. Variation will follow a stable pattern as long as the random causes of variation remain the same and there is no assignable cause of variation. Once a stable process is established, the limits of the resulting pattern of variation can be determined and will serve as a guide for future production. When variation exceeds these limits, it shows a high probability that there is an assignable cause that needs to be corrected.

A process can produce defects if the spread is too great or if the center or the average is not correct. Some method is needed to measure these two characteristics continually in order to compare what is happening to the product specification. This is done using the $\bar{X}$ (X bar) and R control chart.

Control Charts

Run charts Suppose a manufacturer was filling bottles and wanted to check the process to be sure the proper amount of liquid was going into each. Samples are taken every half hour and measured. The average of the samples is then plotted on a chart as shown in Figure 17.9. This is called a **run chart**. While it gives a visual description of what is happening with the process, it does not distinguish between random variation and assignable cause variation.

$\bar{X}$ (X bar) and R chart A control chart for averages and ranges ($\bar{X}$ and R **charts**) track the two critical characteristics of a frequency distribution—the center and the spread.

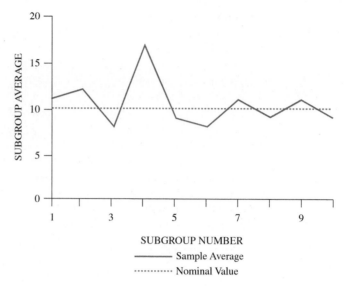

FIGURE 17.9 Run chart.

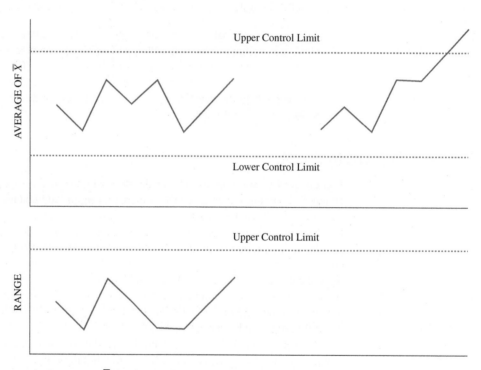

FIGURE 17.10 An $\overline{X}$ and R control chart.

Small samples (typically three to nine pieces) are taken on a regular basis over time and the sample averages and range plotted. The range is used rather than the standard deviation because it is easier to calculate. Samples are used in control charts rather than individual observations because average values will indicate a change in variation much faster. Figure 17.10 shows an example of an $\overline{X}$ and R chart.

Control Limits

Figure 17.10 also shows two dotted lines, the **upper control limit (UCL)** and the **lower control limit (LCL)**. These limits are statistically determined lines on a control chart that are established to help in assessing the significance of the variation produced by the process. If a value occurs outside of this limit, the process is deemed to be out of control. They should not be confused with specification limits, which are permissible limits of each unit of product, meaning the product is within the required tolerances. Specification limits are generally established by the customer and have nothing to do with the process. Showing specification limits on control charts is not recommended.

Control limits are set so that there is a 99.7% probability that if the process is in control, the sample value will fall within the control limits. When this situation occurs, the process is considered to be in statistical control and there is no assignable cause of variation. The process is stable and predictable. This is shown on the left portion of the chart in Figure 17.10. All the points lie within the limits and the process is in statistical control. Only random variation is affecting the process. However, when assignable causes of variation exist, the variation will be excessive and the process is said to be out of control. This is shown on the right portion of the chart, which indicates something has caused the process to go out of control.

As explained, two types of changes can occur in a process:

- **A shift in the mean or average.** This might be caused by a worn tool or a guide that has moved. This will show up on the $\bar{X}$ portion of the chart.
- **A change in the spread of the distribution.** If the range increases but the sample averages remain the same, this kind of problem might be expected. It might be caused by a gauge or tool becoming loose or by some part of the machine becoming worn. This will show up on the R portion of the chart.

The $\bar{X}$ and R chart is used to measure variables, such as the diameter of a shaft, which can be measured on a continual scale. Operators should construct and monitor their own control charts as a way of taking ownership of their jobs. They are the most likely to know what changes are affecting their process. In practice, operators promptly interpret changes in the mean or spread and can quickly take corrective action with the process, often before defects are produced.

Control Charts for Attributes

An **attribute** refers to quality characteristics that either conform to specification or do not, for example, visual inspection for color, missing parts, and scratches. A go/no-go gauge is a good example. Either the part is within tolerance or it is not. These characteristics cannot be measured, but they can be counted. Attributes are usually plotted using a proportion defective or *p* **chart**.

Other Quality Control Tools

In addition to some of the previously discussed tools such as histograms, run charts, and X bar and R charts, there are five other simple tools that are commonly used. These include the following:

1. **Pareto diagrams.** These diagrams are merely histograms that are reorganized in such a way as to show the highest bar first and all others in descending order from high to low. This approach allows one to easily focus on the most important issues and was discussed in Chapter 15 under continuous process improvement.
2. **Check sheets.** These represent a very simple method to collect data. Once an issue for monitoring has been determined (for example, customer complaints about some product or service), the sources of the complaints are listed as they occur. Whenever a complaint reason is repeated, a check is put beside the reason. Over time the number of checks should show clearly the major source of complaint, thereby allowing action to focus on that major source.
3. **Process flow diagrams.** These diagrams, discussed in Chapter 15, show in detail the steps required to produce a product or service. Once the specific tasks are identified, data can be collected about these tasks to determine bottlenecks or other types of problems that can be corrected to improve the process.

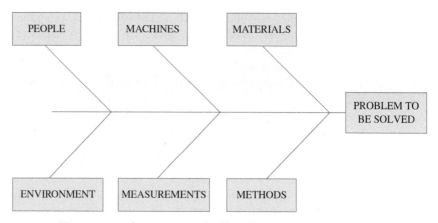

FIGURE 17.11 The structure for a cause-and-effect diagram.

4. **Scatter charts.** These show the relationship between two variables. For example, a bank may want to know whether there is a relationship between how long people have to wait in line and how satisfied they are with their service by the bank. For each customer, they would obtain a measure of the wait time and their satisfaction. They could then plot each person's score with the wait time on the horizontal axis and the satisfaction index on the vertical axis. Plotting many customer scores would then provide an indication of whether there is a relationship and also the possible strength of that relationship.

5. **Cause-and-effect diagrams (fishbone diagrams).** These diagrams, discussed in Chapter 15, are used to plot all the potential causes for an identified problem (effect). As specific problems are branched out from the major effect area, the result appears something like a fishbone. The potential causes can then be researched to find the root cause and correct it. Figure 17.11 shows the structure.

SAMPLE INSPECTION

Statistical process control monitors the process and detects when the process goes out of control, thus minimizing the production of defective parts. Traditional inspection methods inspect a batch of parts after completion and, on the basis of the inspection, accepts or rejects the batch. There are two inspection procedures: 100% and acceptance sampling.

A 100% inspection means testing every unit in the lot. This is appropriate when the cost of inspection is less than the cost of any loss resulting from failure of the parts, for example, when inspection is very easy to do or the inspection is part of the process. A light bulb manufacturer could easily build a tester to see if every light bulb works, or a baker could visually inspect all products prior to packaging them. In cases in which the cost of failure is exceptionally high, 100% inspection is vital. Products for the medical and aerospace industry may be checked many times because of the importance of product performance or the high cost of failure.

Acceptance sampling consists of taking a sample of a batch of product and using it to estimate the overall quality of the batch. Based on the results of the inspection, a decision is made to reject or accept the entire batch. Sampling inspection is necessary under some conditions.

Reasons for sampling inspection There are four reasons for using sample inspection:

1. **Testing the product is destructive.** The ultimate pull strength of a rope or the sweetness of an apple can be decided only by destroying the product.

2. **There is not enough time to give 100% inspection to a batch of product.** For many companies, inspecting every unit in a batch would cause shipments to be delayed. Experience may have proven that sampling is enough to ensure that the entire lot is of acceptable quality.

3. **It is too expensive to test all of the batch.** Market sampling is an example of this, as are surveys of public attitude.

4. **Human error is estimated to be as high as 20% when performing long-term repetitive testing.** There are good reasons to have a representative sample taken of a batch rather than to hazard this high an error.

Conditions necessary for sampling inspection The use of statistical sampling depends on the following conditions:

- **All items must be produced under similar or identical conditions.** Sampling the incoming produce to a food-processing plant would require separate samples for separate farmers or separate fields.
- **A random sample of the lot must be taken.** A random sample implies that every item in the lot has an equal chance of being selected.
- **The lot to be sampled should be a homogeneous mixture.** This means that defects will occur in any part of the batch. (The apples on the top should be the same quality as the apples on the bottom.)
- **The batches to be inspected should be large.** Sampling is rarely performed on small lots and is much more accurate in very large lot sizes.

Sampling Plans

Sampling plans are designed to provide some assurance of the quality of goods while taking costs into consideration. Lots are defined as good if they contain no more than a specified level of defects, called the **acceptable quality level (AQL)**. A plan is designed to have a minimum allowable number (or %) of defects in the sample in order to accept the lot. Above this level of defects, the lot will be rejected. There is a chance that a good batch will be rejected or a bad batch will be accepted.

Selecting a particular plan depends on three factors:

Consumer's risk The probability of accepting a bad lot is called the **consumer's risk**. Since sampling inspection does not produce results with 100% accuracy, there is always a risk that a lot containing more than the desired number of defects will be accepted. The consumer will want to be sure that the sampling plan has a low probability of accepting bad lots.

Producer's risk The probability of rejecting a good lot is called the **producer's risk**. Since sampling involves probabilities, there is a chance that a batch of good products will be rejected. The producer will want to ensure that the sampling plan has a low probability of rejecting good lots.

Cost Inspection costs money. The objective is to balance the consumer's risk and the producer's risk against the cost of the sampling plan. The larger the sample, the smaller the producer's and consumer's risks and the more the likelihood that good batches will not be rejected and poor batches will not be accepted. However, the larger the sample size, the greater the inspection cost. Thus, there is a balance between the producer's and consumer's risks and the cost of inspection.

Example For simplicity, a single sampling plan will be considered (there are others). The plan will specify the sample size (n)—the number of randomly selected items to be taken—from a given size of production lot (N). These will be inspected for some known characteristic, and the plan will set a maximum allowable number of defective products in the sample (c). If more than this number of defectives is found in the sample, the entire lot is rejected. If the allowable number of defects, or fewer, are found in the sample, the lot is accepted. Figure 17.12 illustrates an example of a single sampling plan.

The larger the sample, the smaller the producer's and consumer's risk and the more the likelihood that good batches will not be rejected and poor batches accepted. However, the larger the sample size, the greater the inspection cost. Thus, there is a trade-off between the producer's and consumer's risks and the cost of inspection.

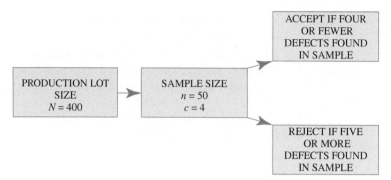

FIGURE 17.12 Sampling plan.

ISO 9000

The International Organization for Standardization (ISO) was established in 1947 as a nongovernmental organization based in Geneva, Switzerland. The acronym (ISO) is derived from the Greek *isos* meaning equality, and it is used in many languages. They provide a series of quality standards, the most recent version being ISO 9000:2015. The standards are intended to prevent nonconformities during all stages of business functions, such as purchasing, invoicing, quality, and design.

ISO 9000 describes the fundamental concepts and vocabulary of a universal **quality management system (QMS)**. It is based on seven quality management principles that are closely aligned with TQM:

1. **Customer focus.** Understanding current and future customer needs, and striving to exceed customer expectations.
2. **Leadership.** Establishing direction, unity of purpose, and encouraging people to work toward a common set of objectives.
3. **Engagement of people.** Ensuring that all employees at all levels are empowered and competent to fully use their abilities for the organization's benefit.
4. **Process approach.** Recognizing that all work is done through processes, and managed and controlled accordingly as a system.
5. **Improvement.** As a permanent organizational objective, recognizing and acting on the fact that ongoing improvement is possible and necessary.
6. **Evidence-based decision-making.** Acknowledging that sound decisions must be based on analysis of factual data and information.
7. **Relationship management.** Managing all relationships with suppliers, partners and all interested parties in order to achieve success.

Besides being a requirement for doing business in Europe, customers throughout the world have come to expect a quality standard and to demand ISO certification of their suppliers. **ISO 9001** is the set of standards or criteria for a quality management system, applying the principles of ISO 9000. Over a million companies in more than 170 countries have achieved ISO 9001 certification. Conformance to ISO standards is voluntary, and many organizations use the principles as a means of continuous improvement without actually going through the certification process.

A common misconception about ISO 9000 is that it applies only to the manufacturing sector. ISO's intent, however, is that the standard be applied to any organization seeking the following:

- Sustained success through implementation of a quality management system.
- Confidence in their ability to consistently provide products or services which conform to requirements.
- Confidence in their supply chain to meet requirements.
- Improvement in communication through a common understanding of quality management vocabulary.

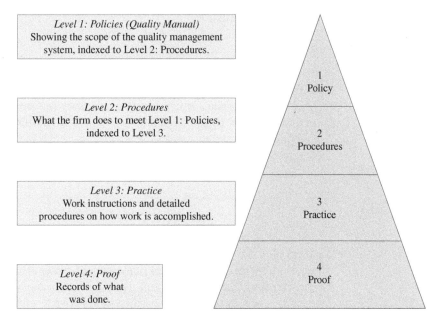

FIGURE 17.13 Documentation pyramid example.

- Ability to perform conformity assessments.
- Provide training, assessment or advice in quality management.
- Develop related quality standards.

Documentation

Documentation is a critical part of ensuring that standards and processes are being followed as part of a quality management system. An example of a documentation structure that can be easily used is shown in Figure 17.13. Documentation can be grouped into four levels: policies (quality manuals), procedures, practice, and proof. Policies define what management wants to accomplish. Procedures are established and closely linked to the policies to ensure that policies are incorporated in the activities of the organization. Practice is the demonstration of procedures, and documentation is the proof that the procedures have been carried out. An example of a policy in a retail environment may be the statement "Customers will be satisfied with the quality of our products." The resulting procedure could then be "A full refund will be issued, provided proof of purchase, and the approval of the department supervisor." The practice would involve giving the department supervisor the authority to issue the refund, and the proof would be documented by the proper filing of receipts and returns.

ISO 26000

Another standard provided by the ISO is **ISO 26000**, which offers organizations guidance for sustainability and social responsibility. It is not a management system, and is not intended for certification, but is provided for companies to use to integrate social responsibilities into its activities and decisions.

ISO 26000 is comprised of seven principles:

1. **Organization governance** and how it applies to social responsibility, the decision-making process and structures.
2. **Human rights**, including due diligence, risk situations, avoidance of complicity, resolving grievances, discrimination, civil and political rights, and fundamental principles and rights at work.
3. **Labor practices**, specifically around employment relationships, work conditions and social protection, social dialogue, health and safety at work, and human development and training in the workplace.

4. **The environment** and the issues of prevention of pollution, sustainable resource use, climate change, and restoration of natural habitats.

5. **Fair operating practices**, such as anticorruption, responsible political involvement, fair competition, social responsibility in the value chain, and respect for property rights.

6. **Consumer issues**, including fair marketing and contractual practices, consumer health and safety, sustainable consumption, customer service, complaint and dispute resolution, consumer data protection and privacy, and consumer education and awareness.

7. **Community involvement and development** and how it affects education and culture, employment creation and skills development, technology development, wealth and income creation, health, and social investment.

ISO 14001

A third series of standards by ISO is **ISO 14001**, which provides a structure and systems to help companies minimize harmful effects on the environment. ISO 14001 is the framework for an **environmental management system (EMS)**. These systems enable an organization to identify and control the impact of its activities, products, and services on the environment throughout the product or service life cycle, as well as improve its environmental performance and contribute to sustainability. There are currently over 300,000 organizations certified with ISO 14001 in 171 countries.

ISO standards cover all aspects of managing a business, including operations, administration, sales, and technical support. They put in place a process approach for all the day-to-day activities of generating products or services. Customers and suppliers of an ISO-certified organization can be assured that consistent management processes are in place, providing stability and continuous improvement in products and services.

BENCHMARKING

Benchmarking is a systematic method by which organizations can compare their performance in a particular process to that of a "best-in-class" organization, finding out how that organization achieves those performance levels and applying them to their own organization. Continuous improvement, as discussed in Chapter 15, seeks to make improvement by looking inward and analyzing current practices. Benchmarking looks outward to see what competitors and excellent performers outside the industry are doing.

Following are the steps in benchmarking:

1. **Select the process to benchmark.** This is much the same as the first step in the continuous improvement process.

2. **Identify an organization that is best in class in performing the process you want to study.** This may not always be a company in the same industry. The classic example is Xerox using L. L. Bean, an on-line and mail order sales organization, as a benchmark when studying its own order entry system.

3. **Study the benchmarked organization.** Information may be available internally, may be in the public domain, or may require some original research. Original research includes questionnaires, site visits, and focus groups. Questionnaires are useful when information is gathered from many sources. Site visits involve meeting with the best-in-class organization. Many companies select workers to be on a benchmark team that meets with counterparts in the other organization. Focus groups are panels that may be composed of groups such as customers, suppliers, or benchmark partners, brought together to discuss the process.

4. **Analyze the data.** What are the differences between your process and the benchmark organization? There are two aspects to this. One is comparing the processes and the other is measuring the performance of those processes according to some standard. The measurement of performance requires some unit of measure, referred to as metrics. Typical performance measures are quality, service response time, cost per order, and so forth.

SIX SIGMA

Modern equipment is becoming increasingly complex, often involving thousands of components that must work reliably to ensure that the entire system does not fail. Bill Smith, a reliability engineer at Motorola Corporation, found that to function properly, the company's complicated systems required individual component failure rates approaching zero. These extremely low rates were beyond the ability of inspectors to measure, and Smith worked with others to develop the six sigma breakthrough strategy aimed at defect rates of 3.4 parts per million:

- **Scope.** Systematic reduction of process variability.
- **Quality definition.** Defects per million possibilities.
- **Purpose.** Improve profits by reducing process variation.
- **Measurement.** Defects per million possibilities.
- **Focus.** Locating and eliminating sources of process error.

Six sigma, however, is about more than just measurements of parts production. It encourages companies to focus on improving all business processes. Process improvements result in reduced waste, costs, and lost opportunities, which all lead to higher profits for the producer and benefits to the customer. The six sigma methodology must be initiated by upper management since it sets the business goals of the company and guides the actions of all employees. Middle management is tasked with translating the business goals into process goals and measures. Six sigma is a highly focused system of problem-solving with two main elements: projects and project managers. Figure 17.14 illustrates the eight essential phases and responsibilities of six sigma projects, which may be expressed as **DMAIC (define-measure-analyze-improve-control)**.

When the project has been selected:

1. Select the appropriate metrics or key performance output variables.
2. Determine how the metrics will be tracked over time.
3. Determine the current project baseline performance.
4. Determine the input variables that drive the key performance output variables.
5. Determine what changes need to be made to the input variables to positively affect the key performance output variables.
6. Make the changes.
7. Determine if the changes positively affect the key performance output variables.
8. If the changes positively affect the key performance output variables, establish controls of the input variables at the new levels. If they do not, return to step 5.

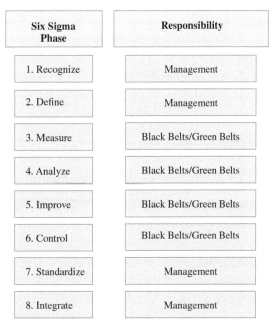

FIGURE 17.14 Six sigma responsibilities.

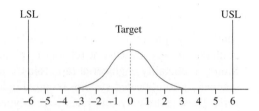

FIGURE 17.15 Six sigma occurs when the process doorway is twice the process spread: $C_p = \dfrac{\text{USL} - \text{LSL}}{6\sigma} > 2$.

From a technical perspective, six sigma is achieved when the process capability, C_p, is equal to 2 or greater. That is, the specification doorway is twice the width of the six sigma process spread, and the long-term failure rate would be 3.4 parts per million. With good controls in place, the minor shifts in the process will be detected prior to any defective parts being produced. Figure 17.15 shows this concept graphically. Note that the process is well centered.

The experience and abilities of six sigma certified individuals are designated by the following terms:

- White belts have achieved the basic level of certification that deals with fundamental six sigma concepts. They support change management in an organization and assist with local problem-solving teams.
- Yellow belts know the specifics of six sigma, as well as how and where to apply the principles, supporting and providing problem-solving expertise to project teams.
- Green belts understand advanced analysis and can resolve quality-related problems. They typically lead projects and assist black belts with data collection and analysis.
- Black belts are experts and agents of change, and can provide training to others as well as lead larger projects.
- Master black belts have achieved the highest level, and are responsible for shaping strategy, developing key metrics, and coaching black and green belts.

Many organizations, especially those involved in large projects, undertake a six sigma certification program for their employees. The level achieved may dictate the size of project the individual must manage. For example:

- Green belts may be required to complete a cost-saving project of approximately $10,000.
- Black belts may be expected to implement projects with savings in excess of $100,000.
- Master black belts usually have completed large-scale projects with savings of $1,000,000. Only a small portion of a company's managers will attain master black belt status, and they are tasked with the training and guidance of other trainees.

The six sigma method is an extension to other business processes of basic statistical process control and can be compared with other quality initiatives, such as continuous process improvement or ISO 9000. It encourages companies to take a customer focus and improve business processes through a series of well-defined steps and responsibilities. Process improvement of any kind will lead to reduced waste, decreased costs, and improved opportunities. It is the customer who ultimately enjoys lower costs and enhanced quality.

QUALITY FUNCTION DEPLOYMENT

Quality function deployment or **QFD** is a decision-making technique used in the development of new products or the improvement of existing products, which helps ensure that the wants, needs, and expectations of the customer are reflected in a company's designs. Organizations that ignore the relationship between what is provided in their products and what a customer wants will not remain competitive. QFD was originally developed in the 1960s by Dr. Yoji Akao and Dr. Shigeru Mizuno and has been adopted by notable manufacturers, including GM, Ford, Daimler Chrysler, IBM, Raytheon, Boeing, Lockheed Martin, and their suppliers. As defined in the *APICS Dictionary*, 16th edition, QFD is "a methodology designed to ensure that all the major requirements of the customer are identified and subsequently met or exceeded through the resulting product design process and the design and operation of the supporting production management system."

QFD begins by gathering the needs of the customer using various survey methods or by comparing a company's own products against the competition's. These requirements are referred to as the **voice of the customer (VOC)**, which must be translated into engineering specifications through a series of well-defined steps. A **house of quality (HOQ)**, as shown in Figure 17.16, is used to sort all the data into a structured process that takes the customer requirements, prioritizes them, and sets the engineering target values for the new design.

The house of quality in Figure 17.16 is a simple example used for the design of an insulated travel mug. On the left-hand side is a list of all the features that the customer feels are important in the design of the travel mug. On the right-hand side is an evaluation of how the sample mug holds up against the competition, listed from low to high and showing where the sample and the competition rank in each of the identified customer

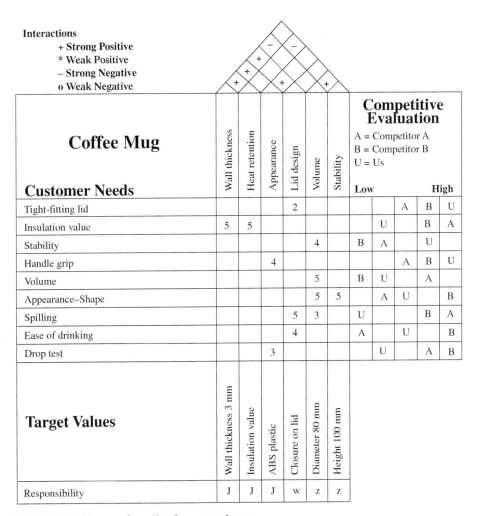

FIGURE 17.16 House of quality for a travel mug.

needs. In this example it seems the mug has a well-fitting lid and customers like the handle design. However, the mug spills easily, is not very durable, does not hold enough liquid, and does not insulate well. These are considered the "wants" of the customer. The "hows" of meeting these wants are listed near the top of the house with such features as the thickness of the material used and the volume and stability of the mug. The "roof" of the house is used to show the interaction of the features, and in this case only strong negatives and positives are shown. A strong positive, such as the thickness of the wall and the heat retention, shows that they are closely related to each other and that an increase in the wall thickness leads to an increase in the insulation value. Increasing the wall thickness, however, has a negative effect on the volume. In the center of the house, priorities are set by management showing where improvements should be made and their relative importance. Technical specifications to meet the desired features and priorities are entered at the bottom of the house along with a responsibility code showing which department or person will work on the design of the final product. In this case the design of the new product will call for a short (100 mm) and stout (80 mm) mug made of ABS plastic. The handle will also be of ABS plastic, which can be molded from the same plastic as the mug. The new design must ensure that any of the negative interactions such as wall thickness and volume are considered.

When QFD is properly administered and good group decision-making is used, it will ensure that design targets and features will reflect the needs of the customers and will avoid adding costs and features that are not required.

THE RELATIONSHIP OF LEAN PRODUCTION, TQM, AND ERP

Although the purposes of ERP, lean production, and TQM are different, there is a close relationship among them. Lean production is a philosophy that seeks to eliminate waste and focuses on decreasing nonvalue-added activities by improving processes and reducing lead time. TQM places emphasis on customer satisfaction and focusing the whole company to that end. While lean production seems to be inward looking (the elimination of waste in the organization and lead time reduction) and TQM outward looking (customer satisfaction), they both have many of the same concepts. Both emphasize management commitment, continuous process improvement, employee involvement, and supplier partnerships. Performance measurement is necessarily a part of process improvement in both lean production and TQM. Lean production places emphasis on quality as a means of reducing waste and thus embraces the ideas of TQM. TQM is directed to satisfying the customer, which is also an objective of lean production.

Lean production and TQM are mutually reinforcing. They should be considered two sides of the same coin—providing customers what they want at the lowest possible cost.

ERP is primarily concerned with managing the flow of materials into, through, and out of an organization. Its objectives are to maximize the use of the organization's resources and provide the required level of customer service. It is a planning and execution process that must work with existing processes, be they good or bad. Lean production is directed toward process improvement and lead time reduction, and TQM is directed toward customer satisfaction. Thus, both lean production and TQM are part of the environment in which ERP must work. Improved processes, better quality, employee involvement, and supplier partnerships can only improve the effectiveness of ERP. Figure 17.17 illustrates the relationship graphically.

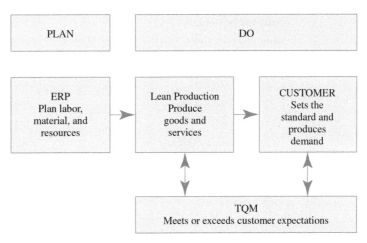

FIGURE 17.17 ERP, lean production, and TQM. *Source:* Adapted from *Basics of Supply Chain Management,* APICS—The Ed. Soc. For Res. Mgt., Falls Church, VA, 1997. Reprinted with permission.

SUMMARY

Quality of both products and services is ultimately established by the customer and the customer is continuously setting higher and higher standards. A company's commitment to quality must start with top management to ensure the entire company works toward improvements. A commitment to quality determines everything from the design policies of products and services, to the design of the production processes, to the quality of suppliers, and finally to the end use by the consumer. Worker involvement is essential since they have direct knowledge of the processes used and are the first to be aware of any changes.

The technical aspects of quality focus on reducing variation and the factors that cause variation of both common cause and assignable cause. Statistical tools are readily available to study variation and calculators and computers take a lot of the drudgery out of the calculations. Where possible, it is important to control the process rather than the product itself. Sampling plans test products after they are produced and do not directly improve the quality of products. But sampling is still necessary when the costs of failure are high in industries such as pharmaceuticals and aerospace.

Management can follow many accepted programs such as ISO standards, six sigma, or QFD to drive quality improvements. High quality is a necessary part of lean production as companies work with lower inventories and faster product flows, which has been the topic of all previous chapters in this book.

KEY TERMS

Acceptable quality level (AQL) 457
Acceptance sampling 456
Aesthetics 442
Appraisal costs 446
Assignable variation 448
Attribute 455
Balanced scorecard 445
Benchmarking 460
Cause-and-effect diagram 456
Center 448
Check sheet 455

Conformance 441
Consumer's risk 457
Control limit 455
C_{pk} index 451
Deming circle 444
DMAIC (define-measure-analyze-improve-control) 461
Durability 441
Empowerment 444
Environmental management system (EMS) 460

External failure costs 446
Feature 441
Fishbone diagrams 456
Fitness for use 441
Form-fit-function 440
Histogram 447
House of quality (HOQ) 463
Internal failure costs 446
ISO 14001 460
ISO 26000 459
ISO 9000 458
ISO 9001 458
Key performance indicator (KPI) 445
Lower control limit (LCL) 454
Lower specification limit (LSL) 449
Maintainability 441
Mean 448
Normal distribution 448
Option 441
Pareto diagram 455
p chart 455
Perceived quality 442
Performance 441
Plan-do-check-act cycle (PDCA) 444
Prevention costs 446
Price 442

Process capability index 450
Process flow diagram 455
Producer's risk 457
Quality 440
Quality function deployment (QFD) 462
Quality management system (QMS) 458
Random variation 447
Range 449
R chart 453
Reliability 441
Run chart 453
Sampling 448
Scatter chart 456
Service 442
Shape 448
Six sigma 461
Spread 449
Standard deviation 449
Statistical control 448
Tolerance 440
Total quality management (TQM) 442
Upper control limit (UCL) 454
Upper specification limit (USL) 449
Voice of the customer (VOC) 463
Warranty 442
$\bar{X}$ chart 453

QUESTIONS

1. What is the definition of quality?
2. In which four areas must quality be considered? How do they interrelate?
3. Name and describe the eight dimensions to quality.
4. What is total quality management, and what are its objectives?
5. What are the six basic concepts of TQM?
6. What does *customer focus* mean? Who is the customer?
7. What is empowerment, and why is it important in TQM?
8. What are the three key factors in employee involvement?
9. What is the purpose of performance measurement?
10. Quality costs fall under two categories. Name each category and think of an example to illustrate the category.
11. What is chance variation? What are its causes? How can it be altered?
12. What is assignable variation?
13. What is a normal distribution? Why is it important in quality control?
14. What is the arithmetic mean or average?
15. What is meant by *process spread*? What are the two measures of it?
16. What percentage of the observations will fall within 1, 2, and 3 standard deviations of the mean?

17. In which two ways can a process create defects?
18. What is tolerance, and how does it relate to the USL and LSL?
19. What is the purpose of the process capability index C_p and the C_{pk}? How do they differ?
20. What is the purpose of process control? What kind of variation does it try to detect?
21. What is a run chart?
22. What is an $\bar{X}$ and R chart?
23. What are upper and lower control limits?
24. What is the difference between variables and attributes?
25. When is it appropriate to use 100% inspection?
26. What is acceptance sampling? When is it appropriate to use?
27. What are the consumer's risk and the producer's risk?
28. What is benchmarking, and how is it different from continuous improvement?
29. What is the first step in quality function deployment? What is the output from QFD?
30. List three similarities and three differences between lean production and TQM.
31. What does the acronym VOC stand for?
32. What technique is used to organize the VOC?

PROBLEMS

17.1. The specification for the length of a shaft is 12 inches ± 0.001 inch. If the process standard deviation is 0.00033, approximately what percentage of the shafts will be within tolerance?

Answer. There are approximately 99.7% of shafts within tolerance.

17.2. In problem 17.1, if the tolerance changes to ± 0.0004 inch, approximately what percentage of the shafts will be within tolerance?

17.3. The specification for the thickness of a piece of steel is 0.5 inch ± 0.06 inch. The standard deviation of the band saw is 0.017. Using C_p, calculate whether the process is capable or not.

Answer. $C_p = 1.18$. Process is marginally capable.

17.4. The specification for the weight of a chemical in a compound is ± 0.06. If the standard deviation of the weighing scales is 0.02, is the process considered capable?

17.5. The specification for the diameter of a hole is 0.80 inch ± 0.020 inch. The standard deviation of the drill press is 0.007 inch. Using C_p, calculate whether the process is capable or not.

17.6. If, in problem 17.5, the process is improved so the standard deviation is 0.0035, is the process capable now?

Answer. Process is capable.

17.7. In problem 17.6, what is the C_{pk} when

a. The process is centered on 0.75? Is the process capable?

b. The process is centered on 0.74? Is the process capable?

Answer. a. $C_{pk} = 1.43$. Process centered on 0.75. The process is capable.

b. $C_{pk} = 0.24$. Process centered on 0.74. The process is not capable.

17.8. A company fills plastic bottles with 10 ounces of shampoo. The tolerance is ± 0.1 ounce. The process has a standard deviation of 0.02 ounce. For the following situations, calculate the C_{pk} and evaluate the process capability.

a. A sample has an average of quantity of 9.95 ounces.

b. A sample has an average of quantity of 9.98 ounces.

c. A sample has an average of quantity of 10.04 ounces.

17.9. Create a house of quality for a product familiar to you and complete the HOQ form. Pick at least three different brands of the product to make the competitive comparison. List at least three customer needs inherent in the products and three features that deliver these needs. Identify the interactions among the features and establish target values for your new and improved product.

Interactions
+ **Strong Positive**
* Weak Positive
− **Strong Negative**
o Weak Negative

Customer Needs

Competitive Evaluation
A = Competitor A
B = Competitor B
U = Us

Low   High

Target Values

17.10. A pallet manufacturer has measured the output of a board cutting operation. The boards have a specification of 24 inches ± 0.5 inch. A sample of the boards is shown below. Calculate the capability of the board cutting operation. What suggestions would you give to the operator?

24.0 24.1 24.2 23.7 24.3 24.0 24.1
23.8 24.4 23.9 24.4 24.2 23.9 24.2

CASE STUDY 17.1

Accent Oak Furniture Company

Accent Oak Furniture Company has been in business for more than 30 years, serving customers in Chicago and the surrounding area. The business consists of three different divisions: a custom furniture division that manufactures such crafted items as dining room suites, a kitchen cabinet division, and a banister division that manufactures and installs quality oak railings.

Each division reports to a vice president who in turn reports to Frank Johnson, the president and founder of the company. Total sales for all three divisions are projected to reach $10 million for the current fiscal year.

Frank has been quite pleased with the performance of the custom furniture division and the kitchen cabinet division, which together account for more than 85% of the sales revenue. He does, however, have some concerns about the banister division based on last year's profit performance and the first week of August's sales report for the installation crews as indicated in the Installation Department Monthly Report (see Figure 17.18).

It is now the second week of August and the new installation market is just starting up. Banisters are installed in new homes in the fall when the homes are nearly finished and ready for sale. There is some demand for banisters throughout the year for renovations. However, this business is concentrated in the fall and quickly dies off before Christmas. Any problems with the banisters should be resolved before the peak season begins.

On Monday Frank met with Jamari Smythe, the vice president of the division, and voiced his concerns. Frank suggested that they meet again on Friday to discuss some

| Installation Department Monthly Report—August 8. ||||||
Job#	Crew#	Budget $	Actual $	Comments
7156	1	1,100	1,127	Waited 1/2 hr for customer
7157	2	985	1,154	Fit problem with spindles
7158	5	1,200	996	
7160	4	1,500	1,854	Recalled for loose spindle
7163	2	850	865	Two split spindles
7166	5	1,200	1,385	Fit problem with spindles
7167	1	1,450	1,620	Customer changed design refit
7168	4	1,800	2,254	Spindle shims needed
7169	5	1,100	1,080	
7171	2	980	1,200	Handrail rough finish
7172	4	1,560	1,860	Loose spindles
7174	1	1,200	1,650	Not to drawings
7175	2	975	1,320	Handrail cracked
7177	4	1,400	1,875	Fit problem spindles
7179	3	2,250	3,200	Recalled for loose bannister
7181	5	1,900	2,520	Fit problem spindles
7182	3	1,800	2,260	Fit problem spindles
7184	3	1,750	1,780	Customer changed design
TOTAL		$25,000	$30,000	

FIGURE 17.18 Installation Department Monthly Report

further actions to get back on track. Jamari felt the concerns were a bit premature but agreed grudgingly. Jamari felt that some of his key people should attend, so Lui Chung, the sales manager, Judy Harburg, who was in charge of the five two-person installation crews, and Julia Guzman, the manufacturing supervisor, were invited to attend.

During the meeting, Jamari asked each one of his people to describe their individual concerns about the past week's installation report.

Julia began by giving a brief outline of the manufacturing process that she was responsible for in the plant.

Handrail Area

1. The oak was purchased in 16-foot lengths and inspected for any flaws or excessive knots to ensure it was #1 grade.
2. Any rejects were sent to the cabinet division to be used for interior shelves or to make spindles.
3. The approved boards were run through a multi head mill that planed the bottom side and shaped the top and sides of the handrail.
4. They were inspected again for rough grain and any knots that could cause quality problems. Any rejects were then cut into shorter usable lengths for shorter handrails or spindles.
5. The handrails were then sent to the drilling machine where the holes were automatically spaced and drilled by the operator on an industrial quality drill press.
6. Occasionally the operator inspects the dimensions of the holes using a dual purpose go/no-go gauge. On one end of the gauge is a metal rod that is inserted in the hole to measure the depth. The rod has a hatch marked area that has PASS stamped into the metal. If the top of the hole is in the PASS area, then the hole dimension is considered okay. The other end of the gauge is long and round and has three steps cut into the circumference. This makes the rod get bigger around at about $\frac{1}{4}$ from the end and $\frac{1}{2}$ from the end. If the first part goes into the hole but not the middle, the hole is undersized. If the second step but not the third enters the hole, the hole is of the correct diameter. The third step indicates oversize.
7. The handrails are then sent for sanding and varnishing. A final inspection is performed for appearance and finish. The handrail sections are then sorted by length for full size and shorts.

Spindle Area

1. Oak is purchased, inspected, and cut to length. Approximately 5% also comes from the handrail area.
2. Sections are placed on a planer and run through twice to give the square end dimensions. Pieces are then placed into a pattern lathe, which provides the product with the spindle design and tapered top end. The operator also lightly sands the spindle while it is on the lathe.
3. An occasional inspection is performed using a go/no-go gauge. The inspection procedure requires the operator to fit over the end of a spindle a gauge, which is an aluminum bar with a specified hole size. The spindle will pass inspection if $\frac{1}{4}$ to $\frac{1}{2}$ inch of spindle protrudes through the gauge. The gauge is actually marked with red paint outside the limits to assist the operator when inspecting.
4. The operator sets the pre-angled cutter to maintain the desired size of the spindle end.
5. The approved spindles are then sent to final sanding and varnishing and final inspection.

Julia felt that her people did the best job possible given the wood and machinery available.

Lui Chung, the sales manager, felt that his people were certainly doing their jobs since sales had increased at a rate of 15% over the last three years. One major concern that Brian did have was the increased number of complaints from his major customer, Lincoln Homes, for the quality of the installations. They have already told Lui they will take their business elsewhere if the quality does not improve immediately.

Judy was the most vocal of the group, since her area was always getting the flack. She felt her crews were being pushed as hard as possible but just couldn't keep on plan. Accent Oak banisters are designed to be easily installed in the field even for custom work. However, crews seem to have to refit every other piece and spend time chasing defects. Judy recently started having her crew keep track of reasons why they were over- or underbudget on each job.

Frank angrily told everyone that he didn't care whose fault the problems were, he just wanted them fixed immediately and then left the room.

Jamari and his group had just recently completed a quality management course, and all agreed that they might as well see if it could help them solve the problem and get Frank settled down.

Inspection Report—Spindle

The following figure shows the specified depth and allowable tolerance for the pilot study on spindle fit using the knife-edge gauge designed for this purpose (see Figure 17.19).

The following 50 readings have been taken and recorded as of August 14 (see Figures 17.20 and 17.21).

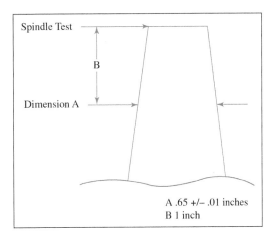

FIGURE 17.19 Allowabel Tolerance for Spindle Fit

Spindle Widths—Inches									
.60	.63	.60	.62	.60	.59	.61	.67	.57	.61
.62	.59	.61	.64	.67	.66	.69	.63	.69	.68
.61	.67	.68	.67	.60	.61	.68	.60	.62	.60
.66	.60	.63	.62	.68	.67	.62	.70	.67	.68
.58	.68	.67	.69	.58	.69	.65	.68	.59	.64

FIGURE 17.20 Spindle Widths in Inches

Hole Diameters—Inches									
.57	.56	.58	.59	.57	.58	.56	.58	.56	.58
.58	.59	.60	.55	.61	.60	.58	.57	.58	.60
.56	.61	.57	.59	.58	.57	.55	.57	.59	.57
.60	.58	.56	.60	.56	.62	.59	.58	.62	.59
.57	.62	.59	.61	.63	.59	.64	.60	.61	.63

FIGURE 17.21 Hole Diameters in Inches

Assignment

1. Analyze the customer complaints using Pareto analysis to identify the largest cause of complaints.
2. Construct histograms of the spindle and hole data to see if there is an assignable cause to the problem.
3. Complete the following three office memos to show how the Julia Guzman could implement your recommendations.

INTEROFFICE MEMO
TO: Lead Hand—Handrail   DATE: August 17
FROM: Julia Guzman
SUBJECT: Customer Complaints—Banisters

INTEROFFICE MEMO
TO: Lead Hand—Spindle Department   DATE: August 17
FROM: Julia Guzman
SUBJECT: Customer Complaints—Banisters

INTEROFFICE MEMO
TO: Jamari Smythe   DATE: August 17
FROM: Julia Guzman
SUBJECT: Customer Complaints—Banisters

READINGS

CHAPTER 1

Buffa, E. S., and R. K. Sarin. *Modern Production/Operations Management*, 8th ed., Chap. 2. India: Wiley, 2007.

Hill, T. E. *Manufacturing Strategy*, 3rd ed. New York, NY: McGraw-Hill, 2000.

Reid, R. D., and N. R. Sanders. *Operations Management, an Integrated Approach*, 7th ed. New York, NY: Wiley, 2019.

CHAPTER 2

Jacobs, F. R., W. L. Berry, D. C. Whybark, and T. E. Vollmann. *Manufacturing Planning and Control Systems for Supply Chain Management*, The CPIM Reference, 2nd ed. New York, NY: McGraw-Hill, 2018.

Nidumolu, R., C. K. Prahalad, and M. R. Rangaswami. "Why Sustainability Is Now the Key Driver of Innovation." *Harvard Business Review*, September 2009.

Wallace, T. F., and R. A. Stahl. *Sales and Operations Planning: The Executive's Guide*. Cincinnati, OH: T. F. Wallace & Company, 2014.

CHAPTER 3

Jacobs, F. R., W. L. Berry, D. C. Whybark, and T. E. Vollmann. *Manufacturing Planning and Control Systems for Supply Chain Management*, The CPIM Reference, 2nd ed. New York, NY: McGraw-Hill, 2018.

Schonberger, R. J., and E. M. Knod. *Operations Management: Meeting Customer Demands*, 7th ed. New York, NY: McGraw-Hill/Irwin, 2001.

CHAPTER 4

Buffa, E. S., and R. K. Sarin. *Modern Production Operations Management*, 8th ed., Chap. 6. New York, NY: Wiley, 2007.

Garwood, D. *Bills of Material for a Lean Enterprise*. Marietta, GA: Dogwood Publishing, 2004.

Jacobs, F. R., W. L. Berry, D. C. Whybark, and T. E. Vollmann. *Manufacturing Planning and Control Systems for Supply Chain Management*, The CPIM Reference, 2nd ed. New York, NY: McGraw-Hill, 2018.

Ptak, C., and C. Smith. *Orlicky's Material Requirements Planning*, 3rd ed. New York, NY: McGraw-Hill, 2011.

CHAPTER 5

Blackstone, J. H., Jr. *Capacity Management*. Cincinnati, Mason, OH: Cengage Learning, 2008.

Jacobs, F. R., W. L. Berry, D. C. Whybark, and T. E. Vollmann. *Manufacturing Planning and Control Systems for Supply Chain Management*, The CPIM Reference, 2nd ed. New York, NY: McGraw-Hill, 2018.

CHAPTER 6

Goldratt, E. M., and J. Cox. *The Goal, A Process of Ongoing Improvement*, 3rd rev ed. Great Barrington, MA: North River Press, 2012.

Haksever, C., and B. Render. *Service and Operations Management*. Hackensack, NJ: World Scientific Publishing Co., 2018.

Heizer, J., B. Render, and C. Munson. *Operations Management: Sustainability and Supply Chain Management*, 13th ed. Harlow, United Kingdom: Pearson Education, 2019.

Jacobs, F. R., W. L. Berry, D. C. Whybark, and T. E. Vollmann. *Manufacturing Planning and Control Systems for Supply Chain Management*, The CPIM Reference, 2nd ed. New York, NY: McGraw-Hill, 2018.

Pinedo, M. L. *Planning and Scheduling in Manufacturing and Services*, 2nd ed. New York, NY: Springer Publishing, 2008.

Project Management Institute. *A Guide to the Project Management Body of Knowledge*, 6th ed. Newtown Square, PA: PMI, 2017.

CHAPTER 7

Bowersox, D. J., D. J. Closs, M. B. Cooper, and J. C. Bowersox *Supply Chain Logistics Management*, 5th ed. New York, NY: McGraw-Hill Education, 2020.

Chang, Y., E. Iakovou, and W. Shi. *Blockchain in Global Supply Chains and Cross Border Trade: A Critical Synthesis of the State-of-the-Art, Challenges, and Opportunities*. International Journal of Production Research, Vol. 58, No. 7, 2019.

Cohen, S., and J. Roussel. *Strategic Supply Chain Management*, 2nd ed. New York, NY: McGraw-Hill, 2013.

Crandall, R. E., W. R. Crandall, and C. C. Chen *Principles of Supply Chain Management*, 2nd ed. Boca Raton, FL: CRC Press, 2015.

Heizer, J., B. Render, and C. Munson. *Operations Management: Sustainability and Supply Chain Management*, 13th ed. Harlow, United Kingdom: Pearson Education, 2019.

Jayaram, A. *Lean Six Sigma Approach for Global Supply Chain Management Using Industry 4.0 and IoT*. 2nd International Conference on Contemporary Computing and Informatics, 2016.

Kshetri, N., *Blockchain's Role in Meeting Key Supply Chain Management Objectives*. International Journal of Information Management, Vol. 39, 2018.

Mollenkopf, D., H. Stolze, W. L. Tate, and M. Ueltschy. *Green, Lean, and Global Supply Chains*. International Journal of Physical Distribution & Logistics Management, Vol. 40, No. ½, 2010.

Monzka, R. M., R. B. Handfield, L. C. Giunipero, and J. L. Patterson *Purchasing and Supply Chain Management*, 7th ed. Boston, MA: Cengage Learning, 2021.

O'Rourke, D., *The Science of Sustainable Supply Chains*, Science, Vol. 344, No. 6186, 2014.

Ritchie, B., and C. Brindley. *Supply Chain Risk Management and Performance*. International Journal of Operations and Production Management, Vol. 27, No. 3, 2007.

Schlegel, G. L., and R. J. Trent. *Supply Chain Risk Management*. Boca Raton, FL: CRC Press, 2015.

CHAPTER 8

Chopra, S., and P. Meindl, *Supply Chain Management: Strategy, Planning and Operation*, 6th ed. Englewood Cliffs, NJ: Prentice Hall, 2014.

Johnson, P. F., M. Leenders, and A. Flynn. *Purchasing and Supply Management*, 14th ed. Whitby, ON: McGraw-Hill Ryerson, 2010.

Monzka, R. M., R. B. Handfield, L. C. Giunipero, and J. L. Patterson. *Purchasing and Supply Chain Management*, 7th ed. Boston, MA: Cengage Learning, 2021.

CHAPTER 9

Buffa, E. S., and R. K. Sarin. *Modern Production Operations Management*, 8th ed., Chap. 4. New York, NY: Wiley, 2007.

Cecere, L. M., and C. W. Chase Jr. *Bricks Matter: The Role of Supply Chains in Building Market-Driven Differentiation*. Hoboken, NJ: John Wiley and Sons, Inc., 2013.

Jacobs, F. R., W. L. Berry, D. C. Whybark, and T. E. Vollmann. *Manufacturing Planning and Control Systems for Supply Chain Management*, The CPIM Reference, 2nd ed. New York, NY: McGraw-Hill, 2018.

Schonberger, R. J., and E. M. Knod. *Operations Management: Meeting Customer Demands*, 7th ed. New York, NY: McGraw-Hill/Irwin, 2001.

CHAPTER 10

Buffa, E. S., and R. K. Sarin. *Modern Production Operations Management*, 8th ed., Chap. 5. New York, NY: Wiley, 2007.

Jacobs, F. R., W. L. Berry, D. C. Whybark, and T. E. Vollmann. *Manufacturing Planning and Control Systems for Supply Chain Management*, The CPIM Reference, 2nd ed. New York, NY: McGraw-Hill, 2018.

Schonberger, R. J., and E. M. Knod. *Operations Management: Meeting Customer Demands*, 7th ed. New York, NY: McGraw-Hill/Irwin, 2001.

CHAPTERS 11 AND 12

Buffa, E. S., and R. K. Sarin. *Modern Production Operations Management*, 8th ed., Chap. 5. India: Wiley, 2007.

Reid, R. D., and N. R. Sanders. *Operations Management, an Integrated Approach*, 7th ed. New York: Wiley, 2019.

Summers, D. C. S. *Quality Management*, 2nd ed. Upper Saddle River, NJ: Pearson, 2008.

CHAPTERS 13 AND 14

Abushaikha, I., L. Salhieh, and N. Towers. *Improving Distribution and Business Performance Through Lean Warehousing*. International Journal of Retail & Distribution Management, Vol. 46, No. 8, 2018.

Ballou, R. H. *Business Logistics/Supply Chain Management*, 5th ed. Englewood Cliffs, NJ: Prentice Hall, 2003.

Benrqya, Y., M. Z. Babai, D. Estampe, and B. Vallespir. *Cross-Docking or Traditional Warehousing: What Is the Right Distribution Strategy for Your Product?* International Journal of Physical Distribution and Logistics Management, Vol. 50, No. 2, 2020.

Bowersox, D. J., D. J. Closs, M. B. Cooper, and J. C. Bowersox. *Supply Chain Logistics Management*, 5th ed. New York, NY: McGraw-Hill Education, 2020.

Coyle, J. J., E. J. Bardi, R. Novack, and B. Gibson. *Transportation: A Supply Chain Perspective*, 7th ed. Toronto, ON: Nelson, 2010.

Kong, X. T. R., H. Luo, G. Huang, and X. Yang. *Industrial Wearable System: The Human-Centric Empowering Technology in Industry 4.0*. Journal of Intelligent Manufacturing, 2019.

Kulwiec, R. *Crossdocking as a Supply Chain Strategy*. Target, Vol. 20, No. 3, 2004.

Murphy, P. R. Jr., and D. Wood. *Contemporary Logistics*, 11th ed. New Jersey, NJ: Prentice Hall, 2014.

CHAPTER 15

Garrison, R. H., E. W. Noreen, and P. C. Brewer. *Managerial Accounting*, 14th ed. New York, NY: McGraw-Hill/Irwin, 2012.

Schonberger, R. J., and E. M. Knod. *Operations Management: Meeting Customer Demands*, 7th ed. New York, NY: McGraw-Hill/Irwin, 2001.

Starr, M. K. *Production and Operations Management*. Boston, MA: Cengage Learning, 2014.

CHAPTER 16

Dennis, P. *Lean Production Simplified*, 3rd ed. Boca Raton, FL: CRC Press, 2018.

Jacobs, F. R., W. L. Berry, D. C. Whybark, and T. E. Vollmann. *Manufacturing Planning and Control Systems for Supply Chain Management*, The CPIM Reference, 2nd ed. New York, NY: McGraw-Hill, 2018.

Jasti, N. V. K., and A. Sharma. *Lean Manufacturing Implementation Using Value Stream Mapping as a Tool*, International. Journal of Lean Six Sigma, Vol. 3, No. 1, 2014.

Schonberger, R. J. *World Class Manufacturing: The Lessons of Simplicity Applied*. New York, NY: Free Press, 2008.

Suzaki, K. *The New Manufacturing Challenge: Techniques for Continuous Improvement.* New York, NY: Free Press, 2012.

Womack, J. P., and D. T. Jones. *Lean Thinking: Banish Waste and Create Wealth in Your Corporation,* 2nd ed. New York, NY: Simon & Schuster, 2003.

CHAPTER 17

Besterfield, D. H., C. Besterfield-Michna, G. H. Besterfield, and M. Besterfield-Sacre. *Total Quality Management*, 3rd ed. Englewood Cliffs, NJ: Prentice Hall, 2005.

Defeo, J. *Juran's Quality Handbook,* 7th ed. New York, NY: McGraw-Hill Professional, 2016.

Reid, R. D., and N. R. Sanders. *Operations Management, an Integrated Approach*, 7th ed. New York, NY: Wiley, 2019.

Schonberger, R. J., and E. M. Knod. *Operations Management: Meeting Customer Demands*, 7th ed. New York, NY: McGraw-Hill/Irwin, 2001.

Summers, D. C. S. *Quality Management*, 2nd ed. Upper Saddle River, NJ: Pearson, 2008.

INDEX

A
ABC classification, 258, 260, 328
ABC Costing (activity-based costing), 427–428
ABC inventory control, 258–261
Acceptable quality level (AQL), 457
Acceptance sampling, 456
Accessibility, 318–319
Action bucket, 86, 91–92
Activity-based costing (ABC Costing), 427–428
Actual cost, 258
Advanced planning and scheduling (APS), 26–27, 126
Aesthetics, 194, 442, 445
After-sales service, 198, 378, 415, 442
Aggregate inventory management, 245–246
Aggregate production plan, 20
Agility, 174–175, 380
AGVS (automated guided vehicle systems), 353, 362
AI (artificial intelligence), 185
Airways, 348
A items, 261, 269, 328
Allocation, 83
Andon, 427
Annual physical inventory, 326–328
Anticipation inventories, 247, 249, 279
Anti-corruption principles, 19, 182, 460
Appraisal costs, 446
APS (advanced planning and scheduling), 26–27, 126
AQL (acceptable quality level), 457
Artificial intelligence (AI), 185
AS/RS (automated storage/retrieval systems), 353, 362
Assemble-to-order (ATO), 3–4, 34, 51–52, 54, 381, 419
Asset, 173, 176, 245, 253–257, 327, 342–343
Assignable variation, 448, 453, 455
ATO (assemble-to-order), 3–4, 34, 51–52, 54, 381, 419
ATP (available-to-promise), 54–55, 57
Attribute, 195, 197, 455
Auditing, 316, 324, 326–329, 446
Automated guided vehicle systems (AGVS), 353, 362
Automated storage/retrieval systems (AS/RS), 353, 362
Available time, 118–119
Available-to-promise (ATP), 54–55
Average, 29, 78, 218–219, 221–227, 229–231, 250, 256, 271, 288–298, 363, 431, 449–450, 453–455

Average cost, 258
Average demand, 29, 218, 221, 224–227, 229, 250, 288–294, 431

B
B2B (business-to-business) commerce, 185
Backflush, 418–419
Backhauling, 349
Backlog, 23, 27, 31, 34, 152–153
Backorders, 31, 252, 297, 321
Back scheduling, 123–124, 141
Backward scheduling, 123–124, 140–141
Balanced scorecard, 19, 428, 445–446
Balance sheet, 245, 253–254, 256, 327
Bar codes, 316, 330–332
Bell curve, 292
Benchmarking, 460
Bias, 227–231, 290, 448
Big data, 185
Billing and collection costs, 353–354
Bill of material, 10, 73–91, 125, 137, 416–419, 445
 accuracy, 445
 indented bill, 77–78, 137
 lean production, 416–419
 multilevel bills, 75–78, 137
 multiple bill, 75–76, 88–91
 parent–component relationship, 74–75
 pegging report, 79
 phantom, 137
 planning bills, 78–79
 product tree, 71, 74, 79–81, 86–88, 90
 single-level bill, 74–77, 137
 summarized, 78
 uses for, 79–80
 where-used list, 78–79
Birdyback, 349
B items, 261, 328
Blanket purchase order, 203
Blockchain, 185–186
Bottlenecks, 35, 49, 114, 145–148, 250, 252, 349, 388, 412, 455
 managing, 147
 principles, 146
 scheduling, 145–147
 throughput, 145–146
Bottom-up replanning, 93
Box scores, 428
Brand, 195–196
Break-bulk, 356
Break-even point, 201
Bribery and corruption, 182, 206
Bucketless system, 86, 417
Buffer, 10, 147, 149, 177–178, 247–248, 411, 419, 430, 432
Bullwhip effect, 175, 178, 182

Business plan, 17, 20–23, 26–27, 53, 216
Business-to-business (B2B) commerce, 185
By-product, 180

C
Calculated capacity, 119
Capacity, 10–11, 16–17, 21–22, 26–30, 35–36, 45, 48–49, 51, 54, 58, 72, 86, 113–114, 116–123, 125–126, 139, 141, 144–147, 152, 154, 203–204, 214, 250, 380, 413, 417–418, 430
Capacity available, 113, 117–120, 123, 125, 141
Capacity control, 114, 418
Capacity cushion, 120
Capacity management, 21–22, 86, 113–126, 141, 145, 249, 418, 445
 available time, 118–119
 back scheduling, 123–124, 141
 balancing, 125–126
 calculated capacity, 119
 capacity available, 113, 117–120, 123, 125, 141
 capacity control, 114, 418
 capacity planning, 114–115, 126, 418, 445
 capacity required, 22, 113, 116–117, 120–123, 145, 249
 capacity requirements planning (CRP), 86, 115–117, 126
 demonstrated capacity, 119–120
 efficiency, 118–119
 lead time, 115–116, 123
 lean production environments, 126, 418
 levels of capacity, 118
 load, 113–115, 119–123, 125–126
 measuring capacity, 117–120
 move time, 116, 123–124
 open order, 115–116, 123
 planned order releases, 116, 121
 queue time, 116
 rated capacity, 119
 resource planning, 114–115, 126
 rough-cut capacity planning, 49–50, 114, 126
 scheduling orders, 123–125
 services, 122
 shop calendar, 116–117
 standard time, 116, 118
 units of output, 117
 utilization, 114, 118–120
 wait time, 116, 123
 work center, 114–125
 work center load profile report, 121–123

477

478 Index

Capacity planning, 114–115, 126, 418, 445
   lean production, 126, 418
Capacity-related costs, 252–253
Capacity required, 22, 113, 116–117,
   120–123, 145, 249
Capacity requirements planning (CRP), 86,
   115–117, 126
Capital, 1, 114, 192, 216, 254, 347, 382,
   386–387
Capital costs, 251, 256, 276, 316–317, 319,
   343, 347–348, 350, 381, 383–384, 410
Carbon footprint, 181
Carriage, 347–350
   costs of, 347
   types of, 347–350
Carrying costs, 251, 256, 270–273,
   276–277, 279, 363, 445
Cash flow management, 255
Cash-to-cash cycle time, 255
Cause-and-effect diagram, 390, 456
Cellular manufacturing, 410
Center, of distribution, 448–449
Centralized inventory control, 303
Central storage, 321
CEP (cost equalization point), 385–386
Channel master, 172
Chase strategy, 28–29, 33
Check sheets, 455
Circular economy, 180–181
C items, 261, 299, 321, 328
Closed loop system, 26, 216
Cloud computing, 184–185
COGS (cost of goods sold), 200, 254–256
Collaborative planning, forecasting, and
   replenishment (CPFR), 175, 215–216
Commodity, 197, 202, 248, 347, 350, 352,
   376, 382
Common carriers, 349–350
Competition, 2, 16, 170, 196, 205, 215,
   374, 378, 404, 460, 463
Competitive bidding, 196, 201–202
Components, 21, 31, 34, 45, 51–54, 71–75,
   78–80, 82–83, 86–88, 90–92, 137, 149,
   157, 171, 173, 178, 180, 182, 191, 196,
   203, 205, 216, 219, 233, 246–247, 342,
   375–377, 380, 386, 391, 406, 418–419,
   422, 461
Component standardization, 406, 410
Concurrent engineering, 377–378
Configure-to-order, 3–4, 51, 381
Conflict minerals, 206
Conflict of interest, 206
Conflicts, 6–7, 149
Conformance, 441
Consignment inventory, 329
Constant work-in-process (CONWIP), 430
Constraints, 46, 146, 148–149, 349, 421,
   429–430
Consumer's risk, 457
Continuous flow processing, 136, 382
Continuous improvement, 6, 387–388, 405,
   411, 414–415, 419, 425–428, 443–444,
   458, 460
Continuous manufacturing, 136

Continuous process improvement (CPI),
   387–396, 414, 442, 444, 464
   analysis, 393–394
   cause-and-effect diagram, 390
   classes of activity, 391–392
   developing solutions, 394–396
   economic considerations, 388
   environmental factors, 395
   human factor, 388, 395
   learning curve, 396
   maintenance, 396
   motion economy, principles of,
      394–395
   operations process charts, 391–392
   Pareto diagrams, 388–390
   people involvement, 387
   process flow diagrams, 393
   productivity, improving, 387
   root cause, 393–394
   six steps, 387–396
   teams, 387
Contract buying, 203–204
Contract carriers, 350
Control charts, 453–455
Control information, 138
Control limits, 454–455
Control towers, 185
Conveyors, 362
CONWIP (constant work-in-process), 430
Core competencies, 171, 191, 345
Corporate social responsibility, 19, 181–182
Cost equalization point (CEP), 385–386
Cost of goods sold (COGS), 200, 254–256
Costs, 30–31, 35, 51, 58, 80, 170, 172–173,
   191, 195, 198–199, 200–202, 206, 245,
   249–254, 256, 258, 270–277, 342–344,
   347, 350–354, 357–360, 362–365,
   377–378, 383–385, 406–407, 420–421,
   428, 445–447, 457
Costs of failure, 446
Costs of quality, 446–447
Cost tradeoff, 344
Count frequency, 328–329
CPFR (collaborative planning, forecasting
   and replenishment), 175, 215–216
CPI (continuous process improvement),
   387–396, 414, 442, 444, 464
C_{pk} index, 451–453
CR (critical ratio), 156–157
Cranes and hoists, 362
Critical ratio (CR), 156–157
CRM (customer relationship
   management), 175
Cross-docking, 321, 356
Cross-training, 125, 145, 395, 411, 443
CRP (capacity requirements planning),
   86, 115–117, 126,
Cube utilization, 318–320, 362
Customer, 2, 4–5, 122, 169–176, 195–196,
   214, 219, 249, 297, 302, 374, 378–381,
   391, 405, 441–443, 445, 458, 463–464
Customer relationship management (CRM),
   175
Customer service, 7, 48, 51, 58, 249–250,
   297, 316, 323, 343–344

Cycle, 217
Cycle counting, 327–329
Cycle stock, 248
Cycle time, 139, 256, 426

D

Data collection and preparation, 157, 219
Data governance, 327
Data requirements, 137–138
Days of supply, 256–258, 295–296
DBR (drum-buffer-rope), 149, 404, 429
DC (distribution center), 4, 171, 173, 248,
   301–304, 340–341, 343, 346, 352,
   356–359, 362–364
Decentralized inventory control, 303
Decision support systems (DSS), 185
Decline phase, 2, 182, 374
Decoupling inventory, 249
Define-measure-analyze-improve-control
   (DMAIC), 461
Delivery lead time, 2–4, 233, 290
Demand, 71–72, 216–218, 221, 224–227,
   229, 232–233, 250, 288–294, 380, 431
   average demand, 218, 221, 224–227,
      229, 250, 288–294, 431
   characteristics of demand, 216–218
   demand lead time, 232–233
   dependent versus independent
      demand, 218
   deseasonalized demand, 224–227
   patterns, 185, 216–218, 247, 292,
      380, 429, 431
   random variation, 217, 221–222,
      226, 229, 231, 290
   stable versus dynamic, 217–218
   trend, 217, 221–224, 429
   variations in, 289–294, 298, 380
Demand lead time, 232–233
Demand management, 214–216
Demand patterns, 185, 216–218, 247, 292,
   380, 429, 431
Demand risks, 176
Demand time fence, 56–59, 78
Deming circle, 444
Demonstrated capacity, 119–120
Density, 318, 342, 349, 352–353
Dependent demand, 71–72, 91, 218, 288
Deseasonalized data, 226–227
Deseasonalized demand, 224–227
Dispatching, 135, 152, 154–157, 252,
   347, 349
Dispatching rules, 155–157
Dispatch list, 135, 154–155
Dispersion, 230–231, 292–294
Distribution, 4–7, 11, 169, 171–172, 204,
   246, 302–303, 340–365,
   activities, 343–344
   billing and collection costs, 353–354
   common carrier, 349–350
   contract carriers, 350
   distribution center (DC), 4, 171, 173,
      248, 301–304, 340–341, 343,
      346, 352, 356–359, 362–364
   distribution channel, 340–344, 346,
      356–357, 360

Index 479

for hire carrier, 349–350
global distribution, 344–345
inventory, 246, 302–303, 343
line haul costs, 351
material handling, 11, 343, 362–363
multi-warehouse systems, 362–365
packaging, 342, 344, 353, 360–363
physical supply, 11, 340–341, 346, 356
pickup and delivery costs, 352–354
private carriers, 349–350
reverse logistics, 342–343
technology, 365
terminal-handling charges/costs, 352–354
total cost, 344, 353–354, 359, 362–364
transaction channel, 341
transportation, 342–343, 346–356
transportation costs, 344, 347, 350–360, 362–363
warehouses/warehousing, 316–321, 343, 355–360
Distribution center (DC), 4, 171, 173, 248, 301–304, 340–341, 343, 346, 352, 356–359, 362–364
Distribution channel, 340–344, 346, 356–357, 360
Distribution inventory, 246, 302–303, 343
Distribution requirements planning (DRP), 304–305
Distribution warehouse, 316–321, 343, 355–360
DMAIC (define-measure-analyze-improve-control), 461
Drop ship, 173
DRP (distribution requirements planning), 304–305
Drum, 149
Drum-buffer-rope (DBR), 149, 343, 429
DSS (decision support systems), 185
Durability, 206, 441
Dynamic demand, 217–218
Dynamic information, 300

E

Earliest due date (EDD), 155
Earliest operation due date (ODD), 155
Echelon, 171, 173, 184–185
Economic indicators, 220
Economic order quantity (EOQ), 270–280, 288, 420–421
EDD (earliest due date), 155
EDI (electronic data interchange), 170, 204, 345
Efficiency, 118–120
Electronic data interchange (EDI), 170, 204, 345
Employee empowerment, 395, 444
Employee involvement, 415, 442–444, 464
Empowerment, 395, 444
EMS (environmental management system), 460
Engineering drawings, 10, 151, 195, 197
Engineer-to-order (ETO), 3, 52–53, 381

Enterprise resource planning (ERP), 25–27, 126, 138, 170, 192, 296, 299, 331, 405, 428–432, 464–465
Environment, 179, 377, 381, 395
Environmentally responsible business, 179
Environmentally responsible purchasing, 205
Environmental management system (EMS), 460
Environment principles, 182, 460
EOQ (economic order quantity), 270–280, 288, 420–421
E-procurement, 204
ERP (enterprise resource planning), 25–27, 126, 138, 170, 192, 296, 299, 331, 405, 428–432, 464–465
Error management, 59
Ethical procurement, 205–206
Ethical sourcing, 205–206
Ethical standards, 205
ETO (engineer-to-order), 3, 52–53, 381
Exception messages, 92, 94
Expenses, 251, 253–255, 258, 271
Exploding requirements, 80–82, 87–88
Exponential smoothing, 223–224
External failure costs, 446
External setup time, 411
Extrinsic forecasting methods, 219–220

F

Fair price, 2, 200, 442
FAS (final assembly schedule), 52, 423
FCFS (first-come-first-served), 155
Features, 3–4, 51, 374, 406–407, 441–442, 463–464
FIFO (first in, first out), 258
Final assembly schedule (FAS), 52, 423
Financial statements, 253–258, 327
Finished goods, 4, 9, 34, 48, 126, 157, 216, 218, 245–247, 270, 302, 340–341, 343, 346, 375, 418, 430
Finite loading, 141–142
Firm planned orders, 92–93
First-come-first-served (FCFS), 155
First in, first out (FIFO), 258
Fishbone diagrams, 390, 456
Fishyback, 349
Fitness for use, 441
5S, 427
Fixed costs, 200, 347–348, 353, 363, 384–386
Fixed-location storage, 320
Fixed order quantity, 269–270
Fixed-position layout, 383–384
Flexibility, 58, 122, 172, 174–175, 204, 379–380, 410–412, 443
Floating inventory location system, 320
Flow manufacturing, 136, 382–383, 408–410
Flow of material, 1, 4, 7, 10, 91, 149, 191, 246–247, 346, 356, 377, 411–412, 414–417, 425, 464
Flow processing, 382–383
Fluctuation inventory, 247
FOB (free on board), 354–355

Focused factory, 376
Forecast, 21–25, 31, 34, 46, 53, 56, 175, 185, 215–216, 218–219, 225–227, 298, 304, 363, 373–374
Forecast error, 227–232, 290, 293, 419
Forecasting, 51, 145, 214–233, 290, 293, 303, 373–374, 416, 419
  bias, 227–231, 290
  data collection and preparation, 219
  economic indicators, 220
  exponential smoothing, 223–224
  extrinsic forecasting methods, 219–220
  forecast error, 227–232, 290, 293, 419
  intrinsic forecasting methods, 219–224
  mean absolute deviation (MAD), 229–232, 293
  moving average, 221–223
  normal distribution, 230–231
  P/D ratio, 232–233
  principles of, 218–219
  qualitative techniques, 220
  quantitative techniques, 220
  seasonal forecasts, 225–227
  seasonal index, 224–225, 227
  seasonality, 217, 224–227
  smoothing constant, 223
  techniques, 219–227, 373
  tracking, 227–233
  tracking signal, 231–232
Forecast interval, 298
For-hire carrier, 349–350
Form-fit-function, 440
Forward scheduling, 124, 140–141
4PLs (fourth-party logistics) provider, 345
Fourth-party logistics (4PLs) provider, 345
Fraud/dishonesty, 206
Free on board (FOB), 354–355
Frozen zone, 58
Functional layout, 383, 409
Functional specifications, 194–197, 376–377

G

G&A (general and administrative expenses), 254
Gateway operation, 147
GDSS (group decision support system), 185
General and administrative expenses (G&A), 254
General-purpose machinery, 381–383, 410
General warehouse, 355
Global distribution, 344–345
Globalization, 173, 205
Global sourcing, 205, 344
Global Trade Item Number (GTIN), 332
Glocalization, 173
Green manufacturing, 24
Green reverse logistics, 342
Green supply chain, 179
Gross requirements, 82–83, 86–88, 91–92, 116
Group decision support systems (GDSS), 185
Growth phase, 2, 182, 374
GTIN (Global Trade Item Number), 332

H

Hazardous materials, 205–207, 353, 388
Healthcare industry, 145, 174, 202, 204
Hedge inventory, 248
Heijunka, 413–414
Histogram, 291, 292, 447, 455
Hoists, and cranes, 362
HOQ (house of quality), 463–464
Hoshin Planning, 19
House of quality (HOQ), 463–464
Human factor, 388, 395
Human rights principles, 181, 206, 459
100% inspection, 408, 456
Hybrid strategy, 31
Hybrid systems, 320, 430–432

I

Income, 8, 254
Income statement, 20, 190–191, 253–255
Incoterms, 345, 354–355
Indented bill of material, 78, 137
Independent demand, 71, 218, 288–305
Industrial trucks, 360, 362
Infinite loading, 141
Innovation, 6, 173, 183, 446
Input/output control (I/O), 152–154
Input/output report, 152–154
Inspection, 193, 352, 391–392, 408, 456–458
Integrated supply chain, 174–176
Intelligent supply chain, 184
Intermittent manufacturing, 136, 138–139, 145, 152, 247
Intermittent production, 383
Intermodal transport, 349
Internal failure costs, 446
Internal setup time, 411
International Commercial Terminology (Incoterms), 345, 354–355
Internet of Things (IoT), 186, 316, 331, 365
Interorganizational systems, 185
Intrinsic forecasting methods, 219–224
Introduction phase, 2, 182, 373
Inventory, 9–10, 73, 120, 204, 232, 245–261, 269–273, 276–277, 279, 288–299, 301–305, 316–332, 343, 363, 407, 411, 418–419, 445
   ABC, 258–261
   accuracy, 323–330
   actual costs, 258
   anticipation, 247, 249, 279
   average cost, 258
   capacity-related costs, 252–253
   carrying costs, 251, 256, 270–273, 276–277, 279, 363, 445
   costs, 251–253, 270
   cycle stock, 248
   days of supply, 257–258, 295–296
   distribution, 246, 302–305, 343
   first in, first out (FIFO), 258
   flow of material, 246–247
   fluctuation, 247
   functions of, 247–249
   hedge, 248
   housekeeping, 326
   identification, 317, 323–324, 326
   landed cost, 251
   last in, first out (LIFO), 258
   lot-size, 248, 269
   maintenance, repair, and operating (MRO) supplies, 247–249
   management, 10, 245–246, 249–250, 269, 316–332, 418–419
   movement, 248
   objectives of, 249–250, 269
   ordering costs, 250–252
   Pareto's law, 259
   performance measures, 256–258
   physical control and security, 322–323
   physical management of, 316–332
   picking, 317
   pipeline, 248
   profiling, 257
   receipts, 317
   records, 73, 83, 299–300
   record accuracy, 323–329
   safety stock, 73, 120, 232, 247, 249, 257, 261, 288–299, 301, 321, 363
   standard cost, 258
   stockout costs, 252
   storing, 317, 320–321
   supply and demand patterns, 247
   traceability, 331–332
   transactions, 322–324
   transportation, 248, 279
   turns, 256–257, 445
   unit cost, 251
   valuation, 258
   vendor-managed (VMI) 204, 329
Inventory accuracy, 323–330
Inventory costs, 251–253, 270
Inventory management, 10, 245–246, 249–250, 269, 316–332, 418–419
Inventory profiling, 257
Inventory records, 73, 83, 299–300
Inventory turns, 256–257
Inventory valuation, 258
Inventory velocity, 256
I/O (input/output control), 152–154
IoT (Internet of Things), 186, 316, 331, 365
Ishikawa diagram, 390
ISO 9000, 458–459, 462
ISO 9001, 458
ISO 14001, 460
ISO 26000, 459–460
Item inventory management, 245–246
Item master, 73, 137

J

JIT (just in time), 170–171, 404–405, 414, 421
Job costing, 384
Job design, 395
Job enlargement, 395
Job enrichment, 395
Job rotation, 395
Job shop, 136
Just in time (JIT), 170–171, 404–405, 414, 421

K

Kaizen, 425
Kaizen blitz, 425
Kaizen event, 425
Kanban, 299, 421–425, 427–432
Key performance indicator (KPI), 19, 445
Kitting, 322
KPI (key performance indicator), 19, 445

L

Labor principles, 181, 459
Laid down cost, 358
Landed cost, 198–199, 251, 358
Last in, first out (LIFO), 258
LCA (life cycle assessment), 182
LCL (lower control limit), 454
Leading indicator, 220
Lead time, 53, 72–73, 80–81, 91–92, 115–116, 123, 137–140, 143, 152–153, 156, 158, 219, 232–233, 247, 288–291, 294–301, 410, 416–418, 421
Lean accounting, 427–428
Lean capabilities, supplier, 198, 415
Lean environment, 126, 198, 269, 414–428
Lean production, 126, 171, 274, 279, 404–432, 464–465
   adding value, 405–406
   backflush, 418–419
   capacity management, 126, 418
   component standardization, 406, 410
   continuous process improvement, 414, 442, 444, 464
   flow manufacturing, 408–410
   inventory management, 269, 418–419
   just in time, 404–405, 414, 421
   kaizen, 425
   kanban, 421–425, 427–432
   lean accounting, 427–428
   linearity, 412, 414, 418
   line balancing, 412
   machine flexibility, 410
   manufacturing planning and control, 415–428
   master production schedule, 417
   material requirements planning, 417–418, 430–432, 464
   mixed-model scheduling, 413
   objectives, 405
   one-piece flow, 421
   operator flexibility, 411
   order quantities, 279–280, 411, 417–418, 421
   partnering, 414
   poka-yoke (mistake-proof), 408
   post-deduct inventory transaction processing, 418
   process flexibility, 410–411

production planning, 416–417
pull system, 405, 412, 418–421, 429–431
push system, 419
quality at the source, 412
quick changeover, 410–411
supplier partnerships, 414–415, 444
supplier certification, 415
supplier selection, 415
takt time, 426–427
tools, 425–428
total employee involvement, 415
total productive maintenance, 412, 418
total quality management (TQM), 444, 464–465
uniform plant loading, 412
uninterrupted flow, 412–414
value stream mapping, 425–426
visual management, 427–428
warehouse principles, 432
waste, 405–408, 412, 414–415, 418, 425, 427, 432
work cells, 409–411, 418
Learning curve, 396
Least total cost, 344, 364
Less than truckload (LTL), 349, 353–354, 356–359, 363
Level production plan, 32–35
Liabilities, 253–254
Life cycle analysis, 373
Life cycle assessment (LCA), 182
LIFO (last in, first out), 258
Linearity, 412, 414, 418
Line balancing, 412
Line haul cost per hundredweight, 351
Line haul costs, 351
Liquid zone, 58
Little's Law, 394, 419
Load, 113–115, 119–123, 125–126, 139, 144–147
Load leveling, 144
Load profile, 121–122, 141–142, 144
Location audit system, 329
Lost capacity cost, 252
Lot control, 157
Lot-for-lot, 203, 269–270, 417
Lot-size, 137, 248, 269–270, 272, 274–277, 279, 410, 419–421, 425
Lot-size inventory, 248, 269
Lot traceability, 157
Low-cost processing, 377
Lower control limit (LCL), 454
Lower specification limit (LSL), 449–450
Low-level code, 87–88
Low-level coding, 86–88
LSL (lower specification limit), 449
LTL (less than truckload), 349, 353–354, 356–359, 363
Lumpy demand, 279, 303

M

Machine flexibility, 410
MAD (mean absolute deviation), 229–232, 293–294
Maintainability, 441
Maintenance, repair, and operating (MRO) supplies, 192, 247–249
Make-or-buy cost analysis, 191
Make or buy decision, 381
Make-to-order (MTO), 3, 33–34, 51–52, 54, 270, 381
Make-to-order production plan, 33–35
Make-to-stock (MTS), 3–4, 31, 51–52, 54, 141, 381
Make-to-stock production plan, 31–33
Management commitment, 442–443, 464
Manufacturing capability, 198, 415
Manufacturing lead time, 21, 137, 139–140, 142–143, 156, 233, 408, 410
Manufacturing planning and control, 10–11, 17–22, 24, 45, 72, 190, 215, 406, 415–428
Manufacturing resource planning (MRP II), 24–25, 404
Manufacturing strategy, 3–4, 381
Manufacturing systems, 136–137, 406
Market boundary, 358–359
Mass customization, 379–380
Master production schedule (MPS), 17, 21–22, 45, 47–59, 71–73, 114–115, 125, 134, 203, 269, 416–417, 419
available-to-promise (ATP), 54–55
changes to, 58–59
delivery promises, and, 54–56
decisions, 51–53
demand time fence, 56–59
development of, 45, 48–53
differences, resolution of, 51
error management, 59
planning horizon, 17, 21–22, 53, 57–58
preliminary master schedule, 48–49
production plan, relationship to, 46–48, 53–59
projected available balance (PAB), 56–57
rough-cut capacity planning (RCCP), 49–50, 114–115
scheduled receipt, 54–55
time fences, 57–59, 417
Master schedule, 17, 24, 45–49, 51–52, 58, 94, 149, 158, 429–431
Master scheduling, 21, 45–59, 94, 126, 417
Material handling, 11, 316–317, 320, 343, 360, 362–363, 394, 409, 412, 432
Material requirements plan, 11, 17, 21, 45, 71–73, 88–93, 115, 134–135, 269
Material requirements planning (MRP), 21, 45, 71–95, 114–116, 120, 134, 204, 245, 270, 277, 288, 304, 417–418, 430–432
action bucket, 86, 91–92
basic MRP record, 85–86
bills of material, 73–80
bottom-up replanning, 93
capacity requirements planning (CRP), 86, 115
exception messages, 92, 94
exploding requirements, 80–82, 87–88
factors in managing, 92
firm planned orders, 92–93
gross requirements, 82–83, 86–88, 91–92
inputs, 73
inventory records, 73, 83
lead time, 72–73, 80–81, 91–92
lean production, 417–418, 430–432
low-level coding, 86–88
management of, 92–95
multiple bills of material, 75–76, 88–91
net requirements, 82–85, 87–88, 93
objectives of, 72
offsetting, 80–81
open orders, 85, 92, 115–116
planned order receipt, 81–83, 90–91
planned order release, 81–83, 88, 90–95, 116
planned orders, 81–82, 91–92, 115–116
planning factors, 73
planning horizon, 21, 86
priority, 71–72, 86, 92–93
process of, 80–91
releasing orders, 83–85, 92
scheduled receipts, 85, 91–92, 115–116
services, and, 91
software, 72–73, 92
system nervousness, 93–95
time buckets, 85–86, 417–418
transactions, 92
use of, 91–95
Materials management, 7–11, 170, 214, 323, 405
Maturity phase, 2, 182, 374
Maximum inventory level, 301–302
Mean, 229, 292–293, 448–450, 455
Mean absolute deviation (MAD), 229–232, 293–294
Measurement systems, 158, 207, 345
Metrics, 5–6, 183–184, 215, 425–426, 460–462
Minimum order quantities, 279
Min–max system, 269–270
Mixed-model scheduling, 413
Modularization, 375
Motion economy, 394–395
Move card, 422–424
Movement costs, 252, 407
Movement inventory, 248
Move time, 116, 123–124, 139
Moving averages, 221–223
MPS (master production schedule), 17, 21–22, 45, 47–59, 71–73, 114–115, 125, 134, 203, 269, 416–417, 419
MRO (maintenance, repair, and operating) supplies, 192, 247–249
MRP (material requirements planning), 21, 45, 71–95, 114–116, 120, 134, 204, 245, 270, 277, 288, 304, 417–418, 430–432

482   Index

MRP II (manufacturing resource planning), 24–25, 404
MTO (make-to-order), 3, 33–34, 51–52, 54, 270, 381
MTS (make-to-stock), 3–4, 31, 51–52, 54, 141, 381
Multilevel bills of material, 75–78, 137
Multiple bills, 75–76, 88–91
Multiple order quantities, 279
Multiple sourcing, 197
Multi-warehouse systems, 362–365

N
Nesting, 379
Net requirements, 82–85, 87–88, 93, 323
Non-instantaneous receipt model, 274–275
Nonmanufacturing setting, scheduling, 145
Normal curve, 230–231, 292, 295–296
Normal distribution, 230–231, 292, 448

O
ODD (earliest operation due date), 155
Offsetting, 80–81, 418
Offshore, 191, 374
One-piece flow, 421
Open orders, 85, 92, 115–116
Operating environment, 1–4
Operational/process risks, 176
Operation overlapping, 142–143
Operation sequencing, 154–157
Operations management, 1, 373, 379
Operation splitting, 143–144
Operations process charts, 391–392
Operator flexibility, 411
Options, 3, 4, 20, 27, 31, 34, 51, 52, 78–79, 219, 375, 406, 441
Orchestrator, 172
Ordering costs, 250–252, 271–277, 300–301, 420–421
Order picking and assembly, 317–318, 321–322
Order point, 269, 288–291, 295–300, 303, 346
Order preparation, 3, 80, 138–139, 252, 270, 279
Order processing, 215–216, 344–346
and communication, 344
Order qualifiers, 2–3
Order quantities, 73, 135–136, 152, 269–280, 288–289, 297, 299–301, 320, 363, 410–411, 418
    economic order quantity (EOQ), 270–280, 288, 420–421
    fixed order quantity, 269–270
    and lean production, 279–280, 411, 417–418, 421
    lot-for-lot, 203, 269–270, 417
    lot-size, 137, 248, 269–270, 272, 274–277, 279, 410, 419–421, 425
    min–max system, 269–270
    monetary unit lot size, 274
    non-instantaneous receipt model, 274–275
    period order quantity (POQ), 270, 277–279
    quantity discounts, 275–276
    relevant costs, 271–272, 275
    stock keeping unit (SKU), 269, 274
    trial-and-error method, 272–273
    unknown costs, products with, 276–277
Order winners, 2–3
Outsourcing, 191–192, 374
Owners' equity, 253–254

P
PAB (projected available balance), 56–57
PAC (production activity control), 10, 16–17, 21–22, 26, 72, 80, 85, 115, 134–158, 203, 409
Packaging, 4, 180, 205–206, 340, 342, 344, 353, 360–364, 377, 379
Packaging costs, 363–364
Pallet, 316, 318–320, 322, 360–362, 365
Pallet positions, 318–319, 361
Parent–component relationship, 74–75
Parent item, 74
Pareto analysis, 388, 390
Pareto diagrams, 388–390, 455
Pareto's law, 259
Partnering, 170, 414
p chart, 455
PDCA (plan-do-check-act) cycle, 444
P/D ratio, 232–233
Pegging, 79
Pegging report, 79
People involvement, 387
Perceived quality, 442
Performance, 1, 5–6, 27, 72, 135, 148, 152–153, 157–158, 173, 185, 195–196, 205, 207, 215, 408, 414, 425, 428, 440–441, 444–446, 460–461
Performance measure, 5–6, 172, 256–258, 442, 444–446
Performance measurements, 5–6, 158, 183–184, 256–258, 428, 445–446, 464
Performance metrics, 5–6, 184, 215
Performance standards, 6
Periodic inventory, 326–327
Periodic review system, 261, 300–303
Period order quantity (POQ), 270, 277–279
Perishability, 342, 353
Perpetual inventory record, 299–300
Phantom bill of material, 137
Physical distribution, 4, 11, 171, 340–365
    activities, 343–344
    billing and collection costs, 353–354
    common carrier, 349–350
    contract carriers, 350
    distribution centers (DC), 340–341, 343, 346, 352, 356–359, 362–364
    distribution channel, 340–344, 346, 356–357, 360
    for hire carrier, 349–350
    global distribution, 344–345
    interfaces, 346
    line haul costs, 351
    material handling, 343, 362–363
    multi-warehouse systems, 362–365
    packaging, 340, 342, 344, 353, 360–364
    physical supply, 340–341, 346, 356
    pickup and delivery costs, 352–353
    private carriers, 349–350
    reverse logistics, 342–343
    technology, 352–353, 362, 365
    terminal-handling charges/costs, 352–354
    total cost, 342, 344, 353–354, 359, 362–364
    transaction channel, 341
    transportation, 342–344, 346–356, 365
    transportation costs, 344–348, 350–360, 362–364
    warehouses/warehousing, 355–360
Physical supply, 4, 11, 340–341, 346, 356
Picking list, 137
Pickup and delivery costs, 352–354
Piggyback, 349
Pipeline inventory, 248
Pipelines, 348
Plan-do-check-act cycle (PDCA), 444
Planned order receipt, 81–83, 90–91
Planned order release, 81–83, 88, 90–95, 115–116, 304, 417, 419
Planned orders, 81–82, 91–92, 115–116, 121, 192
Planner/buyer, 203
Planning bills, 78–79
Planning factors, 73
Planning horizon, 17, 21–22, 27, 31–32, 34, 53, 57–58, 86, 113, 120, 123, 216, 221, 417
Planning information, 137–138
Planning time fence, 57–59, 445
Point of sale (POS), 175
Point-of-use storage, 320–321
Poka-yoke (Mistake-Proof), 408
POQ (period order quantity), 270, 277–279
POS (point of sale), 175
Post-deduct inventory transaction processing, 418
Postponement, 4, 53, 173, 233, 375, 380
Preliminary master schedule, 48–49
Prevention costs, 446
Preventive maintenance, 248, 412, 415
Price, 2, 30–31, 190–195, 198–202, 205–206, 248, 251, 258, 330, 346, 374, 381, 405, 407, 440, 442
    determining, 192–193, 200–202
    negotiation, 193, 202
Priority, 11, 16–17, 21, 26, 36, 45, 71–72, 86, 92–93, 113–115, 120, 125, 146, 152, 154–155, 215, 261, 418
Private carriers, 349–350
Process batch, 142, 146
Process capability, 449–453, 462
Process capacity index (Cp), 450–451
Process control, 453–456
Process costing, 384
Process design, 373, 377, 379–387, 410

Index

Processes, 16, 27, 30, 58, 90, 148–149, 169, 233, 373–396, 414–415, 442, 444–445, 448, 450, 453, 458–462, 464
Process flexibility, 379, 410–411
Process flow, 376, 391, 411, 425, 432
Process flow diagrams, 393, 455
Process focus, 376
Process industries, 137, 376, 384
Processing equipment, 381–382
Process layout, 383
Process specifications, 10
Procurement, 171, 202, 205–206
    ethical, 205–206
Producer's risk, 457
Product description, 10
Product design, 3–4, 136, 170, 373–382, 406, 410, 429–430, 440–441, 463
Product development, 182, 374–376, 405–406, 440–441
Product focus, 376
Product groups, establishing, 27–28
Production activity control (PAC), 10, 16–17, 21–22, 26, 72, 80, 84, 85, 115, 134–158, 203, 409
    backward scheduling, 140–141
    bottlenecks, 145–148
    constraints, 146, 148–149
    continuous manufacturing, 136
    control, 135, 138, 147, 152–157
    control information, 138
    critical ratio (CR), 156–157
    cumulative variance, 152
    cycle time, 139
    data requirements, 137–138
    dispatching, 135, 152, 154–157
    drum-buffer-rope, 149
    finite loading, 141–142
    flow manufacturing, 136
    forward scheduling, 140–141
    implementation, 134–136, 151–152
    infinite loading, 141
    input/output control (I/O), 152–154
    input/output report, 152–154
    intermittent manufacturing, 136, 138–139, 145, 152
    item master, 137
    load leveling, 144
    manufacturing lead time, 137, 139–140, 142–143, 156
    manufacturing systems, 136–137
    move time, 139
    nonmanufacturing settings, 145
    operation overlapping, 142–143
    operation sequencing, 154–157
    operation splitting, 143–144
    order preparation, 138–139
    planning information, 137–138
    production order, 135–136, 138, 141
    production reporting, 157
    product structure (bill of material), 137
    project manufacturing, 137
    queue time, 138–139, 143
    repetitive manufacturing, 136
    routing, 136–140, 152
    run time, 138–139, 143–144, 147, 151, 156
    scheduling, 134–136, 139–147, 149–150
    setup time, 138–139, 143
    theory of constraints (TOC), 148–150
    throughput, 145–149
    throughput time, 136, 139, 158
    wait time, 139, 143
    work center master, 138
Production card, 422–424
Production control costs, 252, 274
Production lead time, 232–233
Production leveling, 29–32, 250
Production order, 135–136, 138, 141, 418–419, 431
Production plan, 17, 20–22, 25, 27–36, 45–48, 53–54, 113–114, 430
Production planning, 10, 16–36, 46–48, 53–59, 114, 120, 203, 214, 218, 245, 249, 377–378, 416–417
    assemble-to-order (ATO), 34
    backlog, 23, 27, 31, 34
    characteristics, 28
    chase strategy, 28–29, 33
    hybrid strategy, 31
    level production plan, 32–35
    make-to-order (MTO), 33–35
    make-to-stock (MTS), 31–33
    product groups, establishing, 27–28
    production leveling, 29–32
    relationship to master scheduling, 46–48, 53–59
    resource planning, 35–36
    subcontracting, 30–31
Production reporting, 157
Productivity, 6, 19, 119, 316–317, 323, 375–376, 379, 387–388, 394, 405, 419, 426, 432, 446
Product layout, 382
Product life cycle, 2–3, 5, 170, 182, 373–374
Product mix, 16, 117, 136, 356–357, 376, 410, 412–413, 429
Products, 2–4, 16, 24, 51–52, 71–72, 75, 88, 117, 157, 169, 173–174, 180, 196, 202, 205, 214, 224, 233, 247–248, 321, 360, 373–387, 406, 409, 440
    assemble-to-order, 381
    concurrent engineering, 377–378
    configure-to-order, 381
    continuous flow processing, 382
    decline phase, 374
    development, 374–376
    engineer-to-order, 381
    flow processing, 382–383
    focused factory, 376
    growth phase, 374
    intermittent production, 383
    introduction phase, 373
    layout, 382
    low-cost processing, 377
    make-to-order (MTO), 51–52, 381
    make-to-stock (MTS), 51–52, 381
    mass customization, 379–380
    maturity or saturation phase, 374
    modularization, 375
    nesting, 379
    process design, 373, 377, 379–387
    processes, 379
    process focus, 376
    processing equipment, 381–382
    product focus, 376
    project processes, 383–384
    repetitive manufacturing, 382
    simplification, 374–376
    simultaneous engineering, 377–379
    specialization, 375–376
    specification and design, 376–378
    standardization, 375–376
Product specifications, 10, 117, 194–198, 202, 207, 376–379, 406
Product structure, 53, 57, 74–80, 93, 137
Product tracking, 157, 345
Product tree, 71, 74, 79–81, 86–88, 90
Profit, 8–9, 171, 179, 190–191, 200–201, 254, 256, 258, 373–374
Profit leverage, 190–191
Projected available balance (PAB), 56–57
Project manufacturing, 137, 383–384
Protective packaging, 340, 344, 360
Pull system, 303, 405, 412, 418–421, 429–432
    design change, 431–432
    replenishment problems, 431
    with spike control, 430–431
Purchase order, 115, 191–193, 203–204, 252, 274, 323, 352, 420, 425
Purchase order cost, 252
Purchase requisition, 192–193
Purchasing, 7, 17, 21, 30, 72, 83, 85, 91–92, 170–171, 190–207, 415
    approving payment, 193–194
    centralized, 207
    competitive bidding, 196, 201–202
    contract buying, 203–204
    cycle, 192–194
    decentralized, 207
    environmentally responsible, 205
    ethical procurement and sourcing, 205–206
    functional specifications, 194–197
    multiple sourcing, 197
    objectives, 191, 205
    organizational implications and, 206–207
    planner/buyer concept, 203
    price, determining, 192–193, 200–202
    price negotiation, 193, 202
    profit leverage, and, 190–191
    purchase orders, 191–193, 203–204
    purchase requisitions, 192–193
    quotations, 192–193, 201–202
    receiving and accepting goods, 193
    single sourcing, 197
    sole sourcing, 197
    sourcing, 197, 205–206
    specifications, establishing, 194–195, 205

Purchasing (*continued*)
  standard specifications, 196–197
  supplier ranking, 199–200
  suppliers, selecting, 192–193, 197–200, 202–203, 415
  supply chain management, and, 170–171, 206–207
  sustainability principles and guidelines, 205
  trends, 202–205
  value analysis, 195, 206
Purchasing cycle, 192–194
Push system, 419

Q

QFD (quality function deployment), 378, 463–464
QMS (quality management system), 458
QRP (quick-response program), 175
Qualitative forecasting techniques, 220
Quality, 2, 173, 194–195, 342–343, 378, 380, 405, 411–412, 432, 440–442, 445, 455–456
  control tools, 455–456
  management, 411–412
Quality at the source, 412
Quality control tools, 455–456
Quality function deployment (QFD), 378, 463–464
Quality management system (QMS), 458
Quantitative forecasting techniques, 220
Quantity discounts, 201, 248, 250, 275–276
Queue time, 116, 138–139, 143, 410, 445
Quick changeover, 410–411
Quick-response program (QRP), 175
Quotations, 192–193, 201–202

R

Radio frequency identification (RFID), 157, 316, 330–332, 353
Railways, 347
Random-location storage, 320
Random variation, 217, 221–222, 226, 229, 231, 290, 447–448, 453, 455
Range, 221, 293, 449
Rate-based scheduling, 412–413
Rated capacity, 119, 123
Raw materials, 1, 3–4, 7, 9, 16, 21, 51–52, 74–75, 80, 148–149, 157, 169, 171, 175–176, 180, 190, 198, 206, 216, 232, 245–247, 255, 296, 341, 346, 376, 380, 390–391, 407, 418, 430, 447
RCCP (rough-cut capacity planning), 49–50, 114–115, 126
R chart, 453–455
Receiving and accepting goods, 193
Reciprocity, 206
Recovery, 169, 179–180, 342
Recycle, 179–180, 182–183, 205, 207, 342
Reduce, 179–180
Released order, 92, 116, 121
Releasing orders, 83–85, 92

Reliability, 6, 191, 198, 203–204, 415, 441
Remanufacturing, 19, 181, 343
Reorder point (ROP), 288, 419–421, 424
Repetitive manufacturing, 136, 320, 382, 408–409, 412, 418
Request for quote (RFQ), 193
Requisitions, 192–193
Reserve stock, 247, 299, 320, 322
Resource bill, 35, 49
Resource planning, 35–36, 114–115, 126, 216
Retained earnings, 254
Return on investment (ROI), 18, 245, 255–256, 387
Reuse, 19, 24, 169, 179–181, 342
Revenue, 6, 20, 201, 253–254, 427
Reverse logistics, 19, 169, 180, 186, 331, 342–343
Reverse supply chain, 19, 169, 182, 342
RFID (radio frequency identification), 157, 316, 330–332, 353
RFQ (request for quote), 193
Risk acceptance, 178
Risk assessment, 177
Risk avoidance, 177
Risk costs, 251, 276
Risk identification, 177
Risk management, 19, 177–179
Risk mitigation, 177–178, 186
Risk monitoring, 179
Risk pooling, 178
Risk prevention, 178
Risk response, 177–179
Risk sharing, 178
Roadways, 348
ROI (return on investment), 18, 245, 255–256, 387
Root cause, 186, 390, 393–394, 456
ROP (reorder point), 288, 419–421, 424
Rope, 149
Rough-cut capacity planning (RCCP), 49–50, 114–115, 126
Routing, 10, 11, 74, 91, 115–116, 123, 125, 136–140, 152, 247, 409, 445
Run charts, 453–455
Run time, 116, 120, 123, 138–139, 143–144, 147, 151, 156, 250, 382

S

SaaS (software-as-a-service), 184
Safety capacity, 120
Safety factor, 296–297
Safety lead time, 290
Safety stock, 73, 120, 232, 247, 249, 257, 261, 288–299, 301, 321, 363
Sales and operations plan, 17
Sales and operations planning (S&OP), 22–24, 27, 45–46, 53, 94, 216, 416–417, 431
Sample inspection, 456–458
Sampling, 326, 448, 456–457
Sampling plans, 457–458
Saturation phase, 374

Scatter charts, 456
SCEM (supply chain event management), 186
Scheduled receipts, 54–55, 85, 91–92, 115–116
Scheduling, 123–125, 134–136, 139–147, 149–150, 252, 349, 412–414, 417, 445
  bottlenecks, 145–147
  nonmanufacturing setting, 145
  orders, 123–125
SCOR (Supply Chain Operations Reference) model, 184
Seasonal forecasts, 225–226
Seasonal index, 224–225, 227
Seasonality, 217, 223–227
Self-check, 408
Self-directed work teams, 395
Serial numbers, 157
Service, 1–2, 5, 91, 122, 145–146, 169, 198, 290, 295–298, 345, 350, 357, 364–365, 378, 442
  after-sales, 198
  capability, 350, 364–365
  levels, determining, 290, 295–298
  nonmanufacturing setting, 145
  parts, 79
Setup and teardown costs, 250, 252
Setup time, 116, 120, 123, 138–139, 143, 274, 410–411, 419
Shape, 216–217, 292, 448
Shipments, 203, 205, 279, 317, 321, 343, 349–350, 352–357, 362–363
Shop calendar, 116–117
Shop floor control, 10
Shop order, 115–116, 125, 138, 151–152, 203
Shortest processing time (SPT), 155
Simplification, 374–376, 386
Simultaneous engineering, 377–379
Single-level bill of material, 74–77, 137
Single-minute exchange of die (SMED), 411
Single sourcing, 197
Six sigma, 412, 440, 461–462
SKUs (stock keeping units), 258, 269, 317
Slack per operation, 156
Slack time, 156
Slushy zone, 58
SMED (single-minute exchange of die), 411
Smoothing constant, 223
Social responsibility, 19, 181, 459–460
Software-as-a-service (SaaS), 184
Sole sourcing, 197
S&OP (sales and operations planning), 22–24, 27, 45–46, 53, 94, 216, 416–417, 431
Source inspection, 408
Sourcing, 171, 182, 197, 205–206
Specialization, 375–376, 381
Special-purpose machinery, 377, 381–382
Specification, 10, 117, 194–197, 202, 207, 376–378, 406
  and design, 376–378
  establishing, 194–195
  functional, 195–197
  standards, 196–197

Specification doorway, 450, 462
Spread, 230, 292, 448–451, 453–455, 462
SPT (shortest processing time), 155
SRM (supplier relationship management), 175
Stable demand, 217–218, 223, 229, 429
Standard, 5–6, 375
Standard cost, 258
Standard deviation (Sigma), 231, 293–296, 298, 363, 449, 451, 453
Standard hours, 118, 121, 144, 152, 154
Standardization, 201, 233, 375–377, 386, 406, 410–411
Standardized work, 411
Standard specifications, 196–197, 375
Standard time, 10–11, 116, 118
Static information, 299
Statistical control, 448, 455
Stock keeping units (SKUs), 258, 269, 317
Stock location, 316–317, 319–321, 329
Stockout, 247, 249–250, 252, 261, 288–289, 291, 295–298, 421–422
Stockout costs, 252, 288, 297
Storage costs, 251, 320, 323
Strategic business plan, 17, 20–23, 26–27, 53, 216
Strategic plan, 17–20, 22–23, 27, 53, 207, 216, 443
Subcontracting, 30–31, 51, 147, 171, 345
Successive check inspection, 408
Summarized bill of material, 78
Supplier certification, 415
Supplier invoice, 193–194
Supplier location, 198
Supplier partnerships, 170, 197, 414–415, 442, 444, 464
Supplier ranking, 185, 199–200
Supplier relationship management (SRM), 175
Supplier responsiveness, 203–204
Supplier scheduling, 202–203
Supplier selection, 192–193, 197–200, 202–203, 415
    factors in, 198–199
    final selection of, 199–200
    identifying, 199
Supply and demand patterns, 245–247
Supply chain, 4–7, 169, 171–179, 181, 183–186, 195, 204, 215–216, 331, 340–342, 378–379
    agility, 174
    carbon footprint, 181
    collaboration, 175, 195, 204, 378–379
    components of, 169
    conflicts, 6–7
    demand risks, 176
    design, 173–174, 176
    disruptions, 176, 178
    green, 179
    integrated, 174–176
    intelligent, 184
    operational/process risks, 176
    organizations, 171
    performance metrics/measures, 5–6, 183–184
    risk management, 177–179
    strategies, 172–174, 177
    supply risks, 176
    technology and trends, 175, 184–186
Supply chain design, 173–174, 176
Supply chain event management (SCEM), 186
Supply chain management, 169–186, 206–207, 405
    growth of, 170–172
    historical perspective, 170
    life cycle assessment (LCA), 182
    organizational implications, 206–207
    physical distribution, 171, 345
    procurement, 171
    purchasing, 171, 206–207
    sustainability, and, 179–182
    warehousing, 171
Supply Chain Operations Reference (SCOR) model, 184
Supply chain risks, 176–179
Supply chain strategy, 172–174, 177
    agility, 174
    availability, 172
    cost, 173
    customer experience and value, 172
    design, 173–174
    globalization, 173
    glocalization, 173
    innovation, 173
    quality, 173
    traceability, 173
Supply chain technology, 175, 184–186
    artificial intelligence (AI), 185
    big data, 185
    blockchain, 185–186
    cloud computing, 184–185
    control towers, 185
    Internet of Things (IoT), 186
    interorganizational systems, 185
    supply chain event management (SCEM), 186
    virtual try-on, 186
Supply risks, 176
Sustainability, 19, 24, 179–182, 198, 205–206, 377, 459–460
    carbon footprint, 181
    circular economy, 180–181
    conflict minerals, 206
    corporate social responsibility, 19, 181–182
    ethical procurement and sourcing, 205–206
    purchasing and, 205
    triple bottom line (TBL), 179
    waste hierarchy, 179–180
System nervousness, 93–95

T

Takt time, 426–427
Target inventory level, 301–302
TBL (triple bottom line), 179
TCO (total cost of ownership), 199
Teams, 207, 387, 395, 426, 443–444, 462
Technology, 157, 174–175, 184–186, 205, 316, 330–332, 352–353, 362, 365
Terminal-handling charges/costs, 352–354
Terminals, 347–348, 350, 355
Theoretical capacity, 119
Theory of constraints (TOC), 148–150, 404, 428–430
Third-party logistics (3PLs) providers, 4, 343, 345
3PLs (third-party logistics) providers, 4, 343, 345
Threatcasting, 178–179, 185
Throughput, 145–149, 250, 319–320, 322, 356, 362, 394, 430
Throughput time, 126, 136, 139, 158, 248, 256, 394, 418–419
Time buckets, 85–86, 417–418
Time fences, 57–59, 417, 445
TL (truckload), 301, 348–349, 352–354, 356–359, 363, 365
TMS (transportation management system), 349
TOC (theory of constraints), 148–150, 404, 428–430
Tolerance, 197, 324–326, 377, 440–441, 449–451, 455
Total cost, 191, 193, 198–201, 250, 271–272, 275–277, 280, 342, 344, 353–354, 359, 362–364, 385–386, 420–421
Total cost of ownership (TCO), 199
Total employee involvement, 415, 442–444
Total line haul cost, 351
Total productive maintenance (TPM), 412, 418
Total quality management (TQM), 440–465
    assignable variation, 448, 453, 458
    benchmarking, 460
    concepts, 442–446
    continuous process improvement, 442, 444, 464
    controlling quality, cost of, 446–447
    costs of failure, 446
    customer focus, 442–443, 458
    defining quality, 440–442
    employee involvement, 442–444, 464
    empowerment, 444
    ISO 9000, 458–459, 462
    ISO 9001, 458
    ISO 14001, 460
    ISO 26000, 459–460
    lean, ERP and, 464–465
    management commitment, 442–443, 464
    performance measures, 442, 444–446
    process capability, 449–453, 462
    process control, 453–456
    quality control tools, 455–456
    quality function deployment (QFD), 463–464
    random variation, 447–448, 453, 455
    sample inspection, 456–458
    six sigma, 440, 461–462

Total quality management (TQM) (*continued*)
    statistical control, 448, 455
    supplier partnerships, 442, 444, 464
    teams, 443–444, 462
    variation, 440–441, 447–449, 453–455
TPM (total productive maintenance), 412, 418
TQM (total quality management), 440–465
Traceability, 157, 173, 185, 206, 331–332
Tracking signal, 231–232
Training, 125, 250, 252, 323, 326, 387, 396, 442–444, 446, 459, 462
Transaction, 92, 185–186, 207, 299–301, 322–324, 327–331, 418–419, 428
Transaction channel, 341
Transfer batch, 142–143, 146
Transportation, 145, 193, 205, 250, 252, 302, 321, 340, 342–360, 362–365, 376, 407, 432
    agencies, 350
    consolidation, 321, 356
    costs, 250, 252, 302, 344–360, 362–364
    scheduling, 349, 376
    terms, 354–355
Transportation costs, 250, 252, 302, 344–360, 362–364
Transportation inventory, 248, 279
Transportation management systems (TMS), 349
Trend, 217, 221–224, 429
Trial-and-error method, 272–273
Triple bottom line (TBL), 179
Truckload (TL), 301, 348–349, 352–354, 356–359, 363, 365
Two-bin inventory system, 299
Two-card kanban system, 422–424

U

UCL (Upper control limit), 454
Uniform plant loading, 412
Uninterrupted flow, 411–414, 417
Unit cost, 200–201, 251, 259, 274, 374, 384–386
United Nations Global Compact, 19, 181–182
Unitization, 360–362
Unit loads, 360–361, 363
Units of output, 117
Upper control limit (UCL), 454

Upper specification limit (USL), 449–450
USL (upper specification limit), 449–450
Utilization, 114, 118–120, 138, 146–149, 445

V

Value, 1, 4, 9, 172, 191, 194, 206, 215, 251, 253, 255, 258, 341, 353, 405–406
Value analysis, 195, 206
Value stream mapping, 425–426
Variability, patterns of, 448–449
Variable costs, 200–201, 347–348, 353, 384–386
Variable information, 300
Variation, 217, 221–222, 227, 229, 231, 289–294, 298, 447–449, 453
Vehicles, 186, 205, 344, 347–350, 352–353, 361–362
Vendor-managed inventory (VMI), 204, 329
Virtual organization, 185
Virtual try-on, 186
Visual management, 427–428
VMI (vendor-managed inventory), 204, 329
VOC (voice of the customer), 378, 463
Voice of the customer (VOC), 378, 463
Volume, 145, 182, 201, 205, 251, 319, 349, 351, 352, 359, 376, 382–383, 386–387, 410

W

Wait time, 116, 123, 139, 143
Warehouse layout, 319–321
Warehouse management system (WMS), 331
Warehouse/warehousing, 171, 316–321, 340, 343, 355–360, 362–363, 432
    costs, 357–358, 362–363
    effect on transportation costs of, 359
    layout, 319–321
    market boundaries, 358–359
    role of, 356–357
    transportation costs and, 357–358
Warehousing management, 316–332, 355–365
    accessibility, 318–319
    activities, 317
    central storage, 321
    cross-docking, 321, 356
    cube utilization, 318–320, 362
    fixed-location storage, 320
    floating inventory location, 320

    order picking and assembly, 317–318, 321–322
    pallets, 316, 318–320, 322, 360–362, 365
    pallet positions, 318–319, 361
    point-of-use storage, 320–321
    reserve stock, 320, 322
    stock location, 316–317, 319–321, 329
    warehouse management system (WMS), 331
    working stock, 320, 322
Warranty, 342, 378, 442, 446
Waste, 19, 179, 205, 405–408, 412, 414–415, 418, 425, 427, 432, 464
    defects, 407
    inventory, 407
    motion, 407
    overproduction, 407
    processing, 406–407
    transportation, 407
    waiting time, 407
Waste hierarchy, 179–180
Waterways, 348
Ways, 347
Wearable technologies, 316, 331
Where-used list, 78–79
WIP (work in process), 7, 9, 134–136, 139–140, 145, 147, 151–152, 155, 203, 246–247, 255, 316, 382, 409, 411–412, 418
WMS (warehouse management system), 331
Work cells, 409–411, 418, 426, 428
Work center, 21, 49, 114–125, 135–136, 141, 144–145, 152, 154–155, 252, 382–384, 387, 409, 422–424
Work center load profile report, 121–123
Work center master, 138
Working stock, 320, 322
Work in process (WIP), 7, 9, 134–136, 139–140, 145, 147, 151–152, 155, 203, 246–247, 255, 316, 382, 409, 411–412, 418

X

$\bar{X}$ (X) bar, 453–455

Z

Zone counting method, 328
Zone picking, 321–322
Zone random storage, 320